Soil Properties, Conditions and Assessment

Soil Properties, Conditions and Assessment

Editor: Jordan Berg

www.callistoreference.com

Callisto Reference,
118-35 Queens Blvd., Suite 400,
Forest Hills, NY 11375, USA

Visit us on the World Wide Web at:
www.callistoreference.com

© Callisto Reference, 2017

ISBN: 978-1-63239-790-4 (Hardback)

The publisher's policy is to use permanent paper from mills that operate a sustainable forestry policy. Furthermore, the publisher ensures that the text paper and cover boards used have met acceptable environmental accreditation standards.

Trademark Notice: Registered trademark of products or corporate names are used only for explanation and identification without intent to infringe.

Printed in the United States of America.

Cataloging-in-publication Data

Soil properties, conditions and assessment / edited by Jordan Berg.
   p. cm.
Includes bibliographical references and index.
ISBN 978-1-63239-790-4
1. Soils-- Analysis. 2. Soils. 3. Soil science. 4. Soil chemistry. 5. Agriculture. 6. Soil fertility. 7. Soil productivity.
I. Berg, Jordan.
S593 .S65 2017
631.41--dc 23

Table of Contents

Preface...VII

Chapter 1   **Applicability of Different Hydraulic Parameters to Describe Soil Detachment in Eroding Rills**...1
Stefan Wirtz, Manuel Seeger, Andreas Zell, Christian Wagner, Jean-Frank Wagner, Johannes B. Ries

Chapter 2   **The Occurrence, Sources and Spatial Characteristics of Soil Salt and Assessment of Soil Salinization Risk in Yanqi Basin, Northwest China**......................................12
Zhang Zhaoyong, Jilili Abuduwaili, Hamid Yimit

Chapter 3   **Soil Microbial Substrate Properties and Microbial Community Responses under Irrigated Organic and Reduced-Tillage Crop and Forage Production Systems**...................24
Rajan Ghimire, Jay B. Norton, Peter D. Stahl, Urszula Norton

Chapter 4   **Soil Bacterial Community Response to Differences in Agricultural Management along with Seasonal Changes in a Mediterranean Region**...38
Annamaria Bevivino, Patrizia Paganin, Giovanni Bacci, Alessandro Florio, Maite Sampedro Pellicer, Maria Cristiana Papaleo, Alessio Mengoni, Luigi Ledda, Renato Fani, Anna Benedetti, Claudia Dalmastri

Chapter 5   **Strong Discrepancies between local Temperature Mapping and Interpolated Climatic Grids in Tropical Mountainous Agricultural Landscapes**...............................52
Emile Faye, Mario Herrera, Lucio Bellomo, Jean-François Silvain, Olivier Dangles

Chapter 6   **The Culturable Soil Antibiotic Resistome: A Community of Multi-Drug Resistant Bacteria**..63
Fiona Walsh, Brion Duffy

Chapter 7   **Factors Influencing Bank Geomorphology and Erosion of the Haw River, a High Order River in North Carolina, since European Settlement**.....................................74
Janet Macfall, Paul Robinette, David Welch

Chapter 8   **Digital Mapping of Soil Organic Carbon Contents and Stocks in Denmark**...........................86
Kabindra Adhikari, Alfred E. Hartemink, Budiman Minasny, Rania Bou Kheir, Mette B. Greve, Mogens H. Greve

Chapter 9   **Alfalfa (*Medicago sativa* L.)/Maize (*Zea mays* L.) Intercropping Provides a Feasible Way to Improve Yield and Economic Incomes in Farming and Pastoral Areas of Northeast China**...99
Baoru Sun, Yi Peng, Hongyu Yang, Zhijian Li, Yingzhi Gao, Chao Wang, Yuli Yan, Yanmei Liu

Chapter 10   **Determination of Critical Nitrogen Dilution Curve based on Stem Dry Matter in Rice**...................111
Syed Tahir Ata-Ul-Karim, Xia Yao, Xiaojun Liu, Weixing Cao, Yan Zhu

Chapter 11   **Emissions of CH₄ and N₂O under Different Tillage Systems from
Double-Cropped Paddy Fields in Southern China**...123
Hai-Lin Zhang, Xiao-Lin Bai, Jian-Fu Xue, Zhong-Du Chen, Hai-Ming Tang,
Fu Chen

Chapter 12   **Comparison of the Rhizosphere Bacterial Communities of Zigongdongdou
Soybean and a High-Methionine Transgenic Line of this Cultivar**........................134
Jingang Liang, Shi Sun, Jun Ji, Haiying Wu, Fang Meng, Mingrong Zhang,
Xiaobo Zheng, Cunxiang Wu, Zhengguang Zhang

Chapter 13   **How Does Conversion of Natural Tropical Rainforest Ecosystems Affect Soil
Bacterial and Fungal Communities in the Nile River Watershed of Uganda?**.........144
Peter O. Alele, Douglas Sheil, Yann Surget-Groba, Shi Lingling, Charles H. Cannon

Chapter 14   **Can Plants Grow on Mars and the Moon: A Growth Experiment on Mars and
Moon Soil Simulants**...157
G. W. Wieger Wamelink, Joep Y. Frissel, Wilfred H. J. Krijnen, M. Rinie Verwoert,
Paul W. Goedhart

Chapter 15   ***In Situ* Measurement of Some Soil Properties in Paddy Soil using Visible and
Near-Infrared Spectroscopy**..166
Ji Wenjun, Shi Zhou, Huang Jingyi, Li Shuo

Chapter 16   **Soil Type Dependent Rhizosphere Competence and Biocontrol of two Bacterial
Inoculant Strains and their Effects on the Rhizosphere Microbial Community of
Field-Grown Lettuce**...177
Susanne Schreiter, Martin Sandmann, Kornelia Smalla, Rita Grosch

Chapter 17   **Soil Carbon and Nitrogen Fractions and Crop Yields Affected by Residue
Placement and Crop Types**..188
Jun Wang, Upendra M. Sainju

Chapter 18   **Influence of Residue and Nitrogen Fertilizer Additions on Carbon Mineralization
in Soils with Different Texture and Cropping Histories**..199
Xianni Chen, Xudong Wang, Matt Liebman, Michel Cavigelli, Michelle Wander

Chapter 19   **Soil Organic Carbon Loss and Selective Transportation under Field Simulated
Rainfall Events**...210
Xiaodong Nie, Zhongwu Li, Jinquan Huang, Bin Huang, Yan Zhang, Wenming Ma,
Yanbiao Hu, Guangming Zeng

Chapter 20   **Comparison of Soil Quality Index using three Methods**..219
Atanu Mukherjee, Rattan Lal

Chapter 21   **Modeling Spatial Patterns of Soil Respiration in Maize Fields from Vegetation
and Soil Property Factors with the use of Remote Sensing and Geographical
Information System**..234
Ni Huang, Li Wang, Yiqiang Guo, Pengyu Hao, Zheng Niu

Permissions

List of Contributors

Index

Preface

This book provides comprehensive insights into the field of soil. It elucidates the concepts of soil formation, properties, morphology, fertility, etc. It sheds light on the various soil quality assessment techniques. Also included in this text is a detailed explanation of the various concepts and methods required to improve soil conditions. The book presents researches and studies performed by experts across the globe in this area. Coherent flow of topics, student-friendly language and extensive use of examples make this book an invaluable source of information. Readers would gain knowledge that would broaden their perspective about this subject.

The world is advancing at a fast pace like never before. Therefore, the need is to keep up with the latest developments. This book was an idea that came to fruition when the specialists in the area realized the need to coordinate together and document essential themes in the subject. That's when I was requested to be the editor. Editing this book has been an honour as it brings together diverse authors researching on different streams of the field. The book collates essential materials contributed by veterans in the area which can be utilized by students and researchers alike.

Each chapter is a sole-standing publication that reflects each author's interpretation. Thus, the book displays a multi-facetted picture of our current understanding of applications and diverse aspects of the field. I would like to thank the contributors of this book and my family for their endless support.

Editor

Applicability of Different Hydraulic Parameters to Describe Soil Detachment in Eroding Rills

Stefan Wirtz[1]*, **Manuel Seeger**[1,2], **Andreas Zell**[3], **Christian Wagner**[3], **Jean-Frank Wagner**[4], **Johannes B. Ries**[1]

1 Department of Physical Geography, Trier University, Trier, Germany, **2** Department of Land Degradation and Development, Wageningen University, Wageningen, The Netherlands, **3** Department 7.3- Technical Physics, Saarland University, Saarbrücken, Germany, **4** Department of Geology, Trier University, Trier, Germany

Abstract

This study presents the comparison of experimental results with assumptions used in numerical models. The aim of the field experiments is to test the linear relationship between different hydraulic parameters and soil detachment. For example correlations between shear stress, unit length shear force, stream power, unit stream power and effective stream power and the detachment rate does not reveal a single parameter which consistently displays the best correlation. More importantly, the best fit does not only vary from one experiment to another, but even between distinct measurement points. Different processes in rill erosion are responsible for the changing correlations. However, not all these procedures are considered in soil erosion models. Hence, hydraulic parameters alone are not sufficient to predict detachment rates. They predict the fluvial incising in the rill's bottom, but the main sediment sources are not considered sufficiently in its equations. The results of this study show that there is still a lack of understanding of the physical processes underlying soil erosion. Exerted forces, soil stability and its expression, the abstraction of the detachment and transport processes in shallow flowing water remain still subject of unclear description and dependence.

Editor: Vanesa Magar, Plymouth University, United Kingdom

Funding: The research was supported by the "Internationale Graduiertenzentrum" of Trier University and the "Freundeskreis Trierer Universität e.V.". The funders had no role in study design, data collection and analysis, decision to publish, or preparation of the manuscript.

Competing Interests: The authors have declared that no competing interests exist.

* E-mail: wirtz@uni-trier.de

Introduction

Soil erosion models use different composite factors to describe and predict soil detachment and transport capacity. The most frequently used factors are average shear stress [1–4], unit length shear force [5], stream power [4,6–9], unit stream power [10,11] and effective stream power [12,13].

In most cases, a linear equation describes the relation between the hydraulic parameters mentioned above and the detachment rate. By exceeding a certain threshold, erosion by concentrated flow begins and detachment rate increases. This threshold has a positive x-axis intercept, which means that there is no detachment below this point.

Another option is to consider concentrated flow erosion as a nonlinear threshold phenomenon or as a two-part linear threshold phenomenon: below the threshold soil detachment takes place (first linear relationship) but after exceeding the threshold, detachment rate increases much faster (second linear relationship) [14]. But it is unclear if this linear relationship is really suitable.

Knapen et al. [14] calculated the correlation between shear stress, unit length shear force, stream power and Reynolds number and the detachment rate from several WEPP datasets. The best average correlation was determined for stream power with $R^2 = 0.59$. The WEPP-used shear stress is a variable that reaches only low R^2 values for all of the tested data sets. Knapen et al. [14] describes the shear stress as follows (p. 80 f.): "Although the use of flow shear stress as soil detachment predictor can be contested,

critical shear stress (τ_{cr}) and concentrated flow erodibility KC (...) have been selected as the most universal parameters to describe soil erosion resistance to concentrated flow." The correlations between these factors and the soil detachment rate show very varying results. There is not a single parameter that always reveals the best correlation. These considerations lead to two main questions:

1. Are soil erosion, detachment and transport, directly dependent on water flow characteristics?

2. Are these concepts, as implemented in soil erosion models, suitable to describe rill erosion?

These questions have been tackled by many research groups that have been searching for the equation that suits their observations best [1–13,15–42]. However, taking into consideration the numerous and variable results, a deeper insight into the rill erosion processes on hillslopes is essential. To get this insight, different strategies can be applied [43]: (1) Modelling, (2) laboratory experiments (3) field observations and (4) field experiments. Each of these methods shows different advantages and disadvantages.

Due to difficulties to measure certain parameters, models have to be calibrated. During this process, the phenomenon of equifinality can appear: different parameter sets show the same result. Another weakness of rill erosion models is that the model parameters are often adapted from river hydrodynamics equa-

Table 1. Description table of the experiments: Temperature and precipitation with the nearest meteorological station (INM).

Experiment	Meteorological station	Average annual temperature	Annual precipitation	Northing of the rill	Easting of the rill
Freila 1+3	Baza	14.2°C	368 mm	4154368	509860
Freila 2	Baza	14.2°C	368 mm	4154398	509826
Negratin	Baza	14.2°C	368 mm	4156324	505710
Salada	Embalse Valdeinfierno	13.4°C	311 mm	4187266	595761
Belerda	Granada	15.6°C	473 mm	4133440	478070

UTM 30 coordinates of the five tested rills are presented.
Freila 1 and Freila 3 are two experiments in the same rill.

tions. Govers and his colleagues [13,44] showed that these equations are not suitable for rill erosion processes. Therefore, there is often a mismatch between model results and observed or measured "reality" [43]. Additionally, models only project the concepts of the designer, not necessarily the reality.

In laboratory experiments, the initial and boundary conditions are well controlled. Soil parameters are well known and rill forms and slope can be adapted to the specific question. Thus, physical laws can be tested in a well-defined environment. However, Giménez and Govers [5] showed that parameters determined under laboratory conditions are not easily transformable to natural environments. One disadvantage of former laboratory experiments or field observations is the fact that in most cases only total runoff and sediment output are measured while the relative contribution of the individual processes is not considered [45].

Field data currently reflect the reality as close as possible. Nevertheless, observations as well as experiments show certain disadvantages: (1) Measurement techniques may disturb the observed processes, (2) time scale of human observations is shorter than that of the process under study, (3) some processes cannot be measured directly or indirectly and (4) some processes are chaotic and the spatial and temporal variations are difficult to specify [43].

The relationship between soil detachment and hydraulic parameters used in soil erosion models is in most cases deduced from laboratory experiments but the transferability of these results to natural rills is not generally given. Our setup in natural rills enables to measure the input parameters for calculating hydraulic parameters combining the advantages of laboratory experiments with the advantages of testing natural rills.

The main purpose of the field experiments was to quantify in a detailed temporal and spatial resolution the soil erosion dynamics in natural rills under concentrated flow for comparison of the measured sediment dynamics with those calculated by means of the most common detachment and transport equations.

Specifically, this study's objectives are:

1. elucidating the relationship between hydraulic parameters such as shear stress, unit length shear force, unit stream power, stream power, effective stream power and the Reynolds number and soil detachment in natural rills,

2. providing an explanation why physically-based soil erosion models do not capture rill erosion processes and

3. addressing the question whether current modelling approaches are generally suited to describe rill erosion processes.

The overall aim of this study is to have a critical view on concepts for modelling rill erosion based on experiments performed in naturally developed rills.

Materials and Methods

Ethics Statement

No specific permits were required for the described field studies. The mayors of the towns next to the study sites or the owners of the fields were informed about the intended activities and were asked for permission. The test sites Freila, Negratin and Salada are abandoned fields which are sporadically used as pasture for goats or sheep and in Belerda the experiment was accomplished on an almond field. The locations Freila, Negratin end Salada are not privately-owned and permission was granted from the owner of the study site Belerda. None of the study sites are protected in any way and the field studies did not involve endangered or protected species.

Study areas

The four study areas in Andalusia are located at Negratin, Freila, Salada and Belerda. UTM coordinates of the tested rills are given in Table 1.

Negratin and Freila. The areas are located within the Hoya de Baza sedimentary basin and composed of marls, in which calcareous Regosols have developed. The climate is semi-arid and vegetation is dominated by low shrubs and *Stipa tenacissima* grass tussocks. The land cover at the south side of the Negratin-dam is dominated by abandoned cereal fields, which are extensively grazed by sheep and agricultural land comprised mainly of cereal dry-farming and almond grooves [46].

Salada. Located at the SE-margin of the Betic range (SE-Spain), inside the penibetic complex. The area is composed of conglomerates with a clayey to loamy matrix, in which Regosols as well as to fairly developed (Calcic) Cambisols have developed. Vegetation is similar to that found in the Freila and Negratin-area. The climate is semi-arid too, but less accentuated than in the previously mentioned area [46]. Here the land use consists of rain fed agricultural areas (where cereals, olives, and almonds are cultivated), and abandoned or uncultivated areas.

Belerda. This test area is located in the Guadix basin. The parent material consists of tertiary and quaternary conglomerates, sands, silts and clays. The soil texture class following the FAO [47] is a silty clay loam. The land use is separated into cultivated areas, with almond and olive groves, and abandoned agricultural fields [48]. The climate is, though still semi-arid, characterised by higher average annual temperatures and precipitations in comparison with the other test zones.

The climatic parameters of the test fields are summarized in Table 1.

A: Freila 1 + Freila 3

B: Freila 2

C: Negratin

D: Salada

E: Belerda

Figure 1. Photographies of the tested rills. Informations about the rills are presented in Table 2.

Tested rills

The main descriptors of the rills are summarized in Table 2. In this table, grain size class limits are from [49], texture class is determined following [47]. Photographies of the rills are presented in Figure 1.

The tested rills in Freila have developed on a sandy loam with high gravel content. Sand content is 57% with a relatively homogeneous contribution between coarse, medium, fine and very fine sand. The same is true in the silt fraction, the 34% are homogeneously contributed in the complete silt fraction between 63 and 2 μm. The rills show all a dense rock fragment cover and the highest vegetation cover of the four test sides.

In Negratin, the soil material is nearly gravel free, coarse, medium and fine sand also show low amounts, most of the fine material is in the grain size class <20 μm. The rock fragment cover in the rill is higher than the gravel content of the soil material thus it is possible that residual rock fragment accumulation has occurred.

In Salada the grain size distribution is similar to Negratin. The highest account of the fine soil material is in the class <63 μm. The residual rock fragment accumulation is formed even more clearly as in Negratin; the vegetation cover is relatively high compared to the other test sites.

The rill in Salada is the only rill that has developed in a field being used for agriculture. The soil material is composed by a mixture of all particle size classes from gravel to clay. The rock fragment cover is high compared to the other test sites and the vegetation cover comparatively low. This test site shows the highest dry bulk density which can be declared by the actual agricultural use.

Rill experiment (RE)

The rill experiments consist of two runs: first the rill is tested under field conditions (run a); in a second run (run b), approximately 15 minutes later, the same rill is tested under almost saturated soil conditions. A constant discharge of 250 L (or 330 L, respectively) is maintained during 4 minutes (or 3 minutes, respectively), using a motor-driven pump, resulting in a total water inflow of 1000 L. Mobilisation of material at the inflow has been avoided.

The flow velocity within the rill is characterized by the travel time of the waterfront and of two colour tracers (started at 1 and 2 minutes of the experiment), measured for every meter using a chronograph. By means of this procedure, three velocity curves are recorded and changes in flow dynamics can be detected. As colour tracers, food colourings (E 124 (red) and E 13 (blue)) are used for reasons of safety.

The rill's slope is characterized by measuring with a spring bow of 1 m range and a digital spirit level. It must be considered that slope measuring provides only average slopes for 1 meter. A step or a knick-point in the rill is not accounted, but its position and height are recorded.

Four water samples are taken at three different measuring points (MP1–MP3). The first sample is taken as soon as the waterfront has reaching the sampling point, the second 30 seconds later, the third 90 second later, and the fourth 150 seconds later.

The (suspended) sediment concentration SSC is determined by filtration of the samples in laboratory [50].

At each measuring point, rill cross section is measured. With a laser rangefinder, the distance between sensor and rill bottom is measured in 0.002 m steps. This allows an accurate calculation of the rills cross section area and an estimation of the rills volume.

Table 2. Rill parameters: Grain size class limits are from [49], texture class is determined following [47].

		Freila 1	Freila 2	Freila 3	Negratin	Salada	Belerda
	Ø Slope [°]	9.4	7.7	9.4	5.6	25.6	16.9
	Max. Slope [°]	15.2	14.1	15.2	12.9	7.3	12.5
	Tested flow length [m]	16	21	16	30	17	23
	Texture class	SL	SL	SL	SiL	SiCL	L
Gravel	>2000 µm [%]	30	30	30	1	1	13
Sand	2000-630 µm [%]	14	14	14	1	2	10
	630-200 µm [%]	14	14	14	5	2	10
	200-125 µm [%]	13	13	13	6	1	8
	125-63 µm [%]	16	16	16	11	7	17
Silt	63-20 µm [%]	13	13	13	11	17	13
	20-6.3 µm [%]	10	10	10	20	17	13
	6.3-2 µm [%]	11	11	11	24	24	14
Clay	<2 µm [%]	9	9	9	21	29	15
	Starting soilmoisture [% w/w]	3.1	3.5	3.1	3.1	5.8	2.4
	K_t [s^2 $m^{0.5}$ $kg^{-0.5}$]	0.0090	0.0090	0.0090	0.0095	0.0096	0.0093
	Location WEPP dataset	Academy	Academy	Academy	Frederick	Mexico	Caribou
	Maximum width [m]	~0.4	~2.2	~0.4	~0.4	~0.5	~0.3
	Maximum depth [m]	~0.05	~0.7	~0.05	~0.2	~0.25	~0.15
	Vegetation cover [%]	~40	~40	~40	~0	~15	~5
	Rock fragment cover [%]	~80	~80	~80	~5	~20	~50
	Grain density [g cm^{-3}]	2.69	2.69	2.69	2.65	2.66	2.61
	Dry bulk density [g cm^{-3}]	1.44	1.55	1.44	1.57	1.52	1.68
	Org. material [%]	1.29	1.29	1.29	1.75	2.97	1.34
	Critical shear stress [Pa]	1.97	2.07	1.97	2.93	3.20	2.77
	Land use	rangeland	rangeland	rangeland	rangeland	rangeland	cropland

K_t is a transport coefficient, which has been adopted from the WEPP dataset. The WEPP-location is given. Measured values are starting soil moisture, maximum width, maximum depth, grain density, dry bulk density, org. material; parameters estimated in the field are vegetation cover and rock fragment cover; critical shear stress is calculated following WEPP.

Water level is continuously measured by ultrasonic sensors at each measuring point.

Descriptors for soil detachment

Soil detachment can be described by shear stress τ, unit length shear force Γ, stream power ω, unit stream power ω_U and effective stream power ω_{eff}.

$$\tau = \rho * g * R * S \quad [Pa] \qquad (1)$$

$$\Gamma = \rho * g * A * S = \tau * W_P \quad [N\ m^{-1}] \qquad (2)$$

$$\omega = \rho * g * R * S * v = \tau * v \quad [W\ m^{-2}] \qquad (3)$$

$$\omega_U = S * v \quad [m\ s^{-1}] \qquad (4)$$

$$\varpi_{eff} = \frac{(\tau * v)^{1.5}}{d^{\frac{2}{3}}} = \frac{\omega^{1.5}}{d^{\frac{2}{3}}} \quad [W\ m^{-1}] \qquad (5)$$

with ρ = liquid density [kg m^{-3}], g the gravitational acceleration (9.81 m s^{-2}), R the hydraulic radius [m], A the flow cross section area [m^2], S the effective slope (sin(slope angle)), W_P the wetted perimeter [m], v the flow velocity [m s^{-1}] and d the water depth [m]; abbreviations of the units are Pa = Pascal, N = Newton, W = Watt.

Reynolds number describes the balance between the inertial flow forces represented by the product in the numerator and the viscous forces as described by the dynamic viscosity in the denominator. It is a criterion for stability of a flowing medium. When Reynolds number is small, viscous forces dominate the motion and inertial ones can be ignored whereas at high Reynolds numbers inertial forces dominate and it is often possible to ignore viscosity [51]. Reynolds Number Re is calculated as follows:

$$Re = \frac{\rho * v * R}{\eta} \qquad (6)$$

with ρ = liquid density [kg m^{-3}], v = flow velocity [m s^{-1}], R = hydraulic radius [m] and η = dynamic viscosity [Pa s].

Liquid density is calculated using sediment concentration and grain density. The use of water's density is not practicable due to sediment concentrations of more than 400 g L^{-1}. Grain density was measured by a capillary pycnometer following DIN

18124 [52]. Flow velocity for each sample is interpolated between three measured velocities (arrival of the waterfront and arrival of the two colour tracers). Hydraulic radius and wetted cross section area can be calculated by measuring water level and the rill profile.

The viscosity of the sediment suspensions was measured with a shear rate controlled rheometer (Haake MARS from Thermo Fisher Scientific, Karlsruhe, Germany) and a cone-plate geometry with an angle of 2° and a diameter of 60 mm [53]. The shear rate γ is defined as:

$$\gamma = \frac{dv}{dy} \qquad (7)$$

with v = fluid velocity and y = the gap between the cone and base plate. The rheomter controls the shear rate and measures the shear stress τ, from which the viscosity η is calculated via

$$\eta = \frac{\tau}{\gamma} \qquad (8)$$

The sample volume is always 2.0 ml and the cell is tempered to 20°C+/−0.01°C. Data points are taken at shear rates between 150 s^{-1} and 1500 s^{-1}. The viscosity does not depend on the shear rate. This is according to theoretical considerations. For a suspension of monodisperse particles one expects a linear relation [54,55] for volume concentrations up to approximately 10%.

Detachment rate D_R [kg s^{-1} m^{-2}] is calculated from the measured sediment concentrations and different hydraulic parameters:

$$D_R = \frac{SSC * v * A}{L * W_P} \qquad (9)$$

with SSC = sediment concentration [g L^{-1} = kg m^{-3}] and L = flow length [m].

For the calculation of the critical shear stress, the equations from the WEPP model [34] is used. The authors separate between "cropland with sand content >30%" and "rangeland".

$$\tau_{cr}(cropland) = 2.67 + 0.065 * (\%clay) \\ - 0.058 * (\%very\ fine\ sand) \qquad (10)$$

$$\tau_{cr}(rangeland) = 3.23 - 0.056 * (\%sand) - 0.244 * (\%org.\ mat.) \\ + 0.9 * (dry\ bulk\ density) \qquad (11)$$

For quantification of the different processes in the rill, the transport rate T_R [kg s^{-1}] and the transport capacity T_C [kg s^{-1}] are calculated:

$$T_R = SSC * v * A \qquad (12)$$

$$T_C = R * K_t * \tau^{1.5} \qquad (13)$$

Kt [s^2 m$^{0.5}$ kg$^{-0.5}$] is a transport coefficient depending on soil substrate. The Kt value of the WEPP substrate which was most similar to the given test site conditions is used.

Quantification of different erosion processes

Following shear stress based model concepts, the transport rate cannot exceed the transport capacity [56]. Shear stress of the flowing water controls also the detachment. Therefore the transport rate up to the transport capacity is considered here as shear stress dependent uptake. The transport rate exceeding the transport capacity is considered as shear stress independent erosion caused by processes such as bank failure and headcut retreat. The resulting quantities are set into relation and given in percent of total transport rate.

Results

Initial data

The used parameters show a wide range of data. In most cases (12 of 19), the standard deviation is higher than the mean values, the highest standard deviation – mean - percentage reaches the transport capacity (224%), the effective stream power (188.9%), the sediment concentration (168.3%) and detachment and transport rate (both 150%). The lowest percentage is calculated for sample density (0.5%). All initial data are presented in supporting information Tables S1, S2, S3, S4, S5, S6, S7, S8, S9, S10, S11, S12, S13, S14, S15, S16, S17, S18 and the statistical values of the data in Table 3.

Dynamic viscosity

The dynamic viscosity of the liquid shows a clear positive correlation with sediment concentration, i.e. dynamic viscosity increases with sediment concentration (see Figure 2). However, clear deviations from the trend line were observed for samples with low sediment concentrations, which were often rich in transported organic material. The small branchlets with low weight imply a low sediment concentration, but in rheometer measurements, they tilt and a high shear stress is erroneously measured. The trend line equation has been calculated for samples from different test sites, the R^2-value of 0.92 indicates that this equation can be used for further experiments.

Correlations between detachment rate and hydraulic parameters

The R^2 values of the correlations between the detachment rate and different hydraulic parameters show the complete possible range from R^2 = 0 up to R^2 = 0.99 (see Table S19). Trend lines are increasing, decreasing and almost constant and thus it is not possible to find any clear dependency. Notably, only 40 of 252 correlations (about 16%) show an increasing trend line with an R^2 value≥0.7. Table 4 shows that the highest average R^2-value is calculated for the (τ-τ_{cr}) – detachment rate - relationship if all R^2 values are used (0.53), if only the R^2- values with increasing trend line are considered in calculation, the τ – detachment rate relationship shows the highest average R^2 (0.55). Separating the experiments into two groups, Freila 1–3 with low sediment concentrations (LSSC) and Negratin, Salada, Belerda with high sediment concentrations (HSSC), the highest R^2-values of the LSSC-experiments reach τ, Γ and the (τ-τ_{cr}) – detachment rate - relationship (0.65) if all values are used respectively the (τ-τ_{cr}) – detachment rate - relationship (0.39) if only the R^2 values≥0.7 with increasing trend lines are used. In the HSSC-experiments, the Γ reaches the highest value (0.70) if all values are included and ω_{eff} (0.52) if only the R^2 values≥0.7 with increasing trend lines are used, respectively.

Table 3. Descriptive statistics of the initial data.

variable	Maximum	Minimum	Mean	Standard Deviation	Percentage from Mean
SSC [g L^{-1}]	422.30	0.001	52.15	87.78	168.3
D$_R$ [kg s^{-1} m^{-2}]	0.96	0.001	0.10	0.15	150.0
T$_R$[kg s^{-1}]	2.06	0.001	0.16	0.24	150.0
p [g cm^{-3}]	1.26	1.00	1.03	0.005	0.5
Slope [°]	24.50	1.70	9.73	6.90	70.9
T$_C$ [kg s^{-1}]	3.38	0.001	0.25	0.56	224.0
v [m s^{-1}]	2.94	0.04	0.79	0.49	62.0
η [kg s^{-1} m^{-1}]	0.00311	0.00100	0.00126	0.00044	34.9
Water depth [cm]	21.00	0.20	3.99	4.23	106.0
A [cm^2]	877.69	0.80	149.21	195.84	131.3
W$_P$ [cm]	107.58	4.85	38.21	24.16	63.2
R [cm]	9.65	0.10	2.92	2.12	72.6
τ [Pa]	246.70	0.96	52.38	55.18	105.3
Γ [N m^{-1}]	172.58	0.10	23.99	35.10	146.3
ω [W m^{-2}]	365.28	0.31	41.54	55.91	134.6
ω$_U$ [m s^{-1}]	0.88	0.001	0.14	0.17	121.4
ω$_{eff}$ [W m^{-1}]	37864.55	5.81	3807.14	7192.32	188.9
Re [ ]	86918.88	237.00	19053.94	16226.56	85.2
τ - τ$_{cr}$ [Pa]	244.73	−1.46	49.89	55.11	110.5

SSC = sediment concentration, D$_R$ = detachment rate, T$_R$ = transport rate, p = sample density, T$_C$ = transport capacity, v = flow velocity, η = dynamic viscosity, A = flow cross section, W$_P$ = wetted perimeter, R = hydraulic radius, τ = shear stress, Γ = unit length shear force, ω = stream power, ω$_U$ = unit stream power, ω$_{eff}$ = effective stream power, Re = Reynolds-Number, τ$_{cr}$ = critical shear stress.

Quantification of different erosion processes

Figure 3 shows the relationships between the measured transport rates and the predicted transport capacities. From 144 samples, in 82 cases the transport rate exceeds the capacity, corresponding to approximately 57% of all cases. Tables S20 and S21 present the differences between transport rates and transport capacities (S20) and the percentage of transport rate exceeding the capacity (S21) and hence the percentage of processes which are not controlled by the influence of shear stress. The percentage of material which is transported by processes independent of shear stress is on average 41.5% (see Table 5). Remarkably, the distribution is uneven, i.e. in the three Freila-experiments, the mean is 24.3% while in Negratin, Salada and Belerda, the average value is as high as 58.7% (see Table 5). The second group shows clearly higher sediment concentrations, meaning that the processes independent of shear stress provide higher sediment concentrations than the shear stress-based processes. This indicates that the influence of hydraulic parameters is higher for low sediment concentrations, or, in other words that high sediment concentrations are not caused by hydraulic parameters.

Discussion

A comparison with results of other research groups shows that the measured values are in a realistic range. Ghebreiyessus [3] measured shear stress values up to 40 Pa and in the experiments of Nearing et al. [4], Reynolds numbers of up to 100000 and unit stream power values of up to 10 m s^{-1} were reached. Giménez &

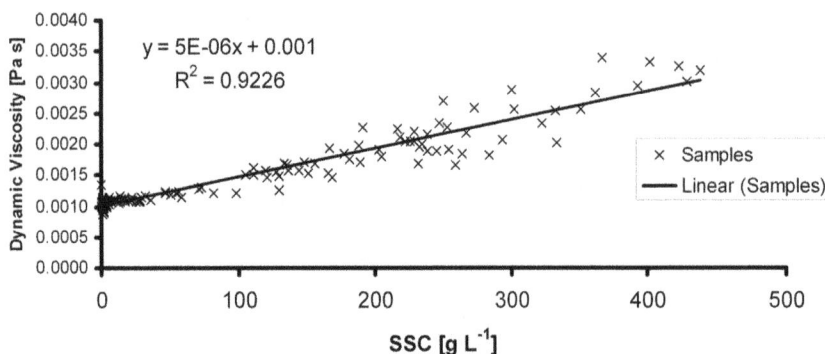

Figure 2. Correlation between sediment concentration of each sample and the measured dynamic viscosity. The linear correlation function and the R^2 value is presented.

Table 4. R^2 - correlation values between different hydraulic parameters and the detachment rate.

	τ	Γ	ω	ω_U	ω_{eff}	Re	τ - τ_{cr}
all values	0.52	0.50	0.37	0.43	0.40	0.39	0.53
only values with increasing trendline	0.55	0.52	0.40	0.46	0.45	0.39	0.53
all Freila experiments	0.65	0.65	0.50	0.53	0.48	0.53	0.65
Negratin, Salada, Belerda all values	0.69	0.70	0.43	0.56	0.56	0.49	0.64
Freila only values with increasing trend line	0.38	0.36	0.25	0.33	0.32	0.26	0.39
Negratin, Salada, Belerda only values with increasing trend lines	0.44	0.41	0.39	0.43	0.52	0.35	0.45

τ = shear stress, Γ = unit length shear force, ω = stream power, ω_U = unit stream power, ω_{eff} = effective stream power, Re = Reynolds number, τ_{cr} = critical shear stress. The complete dataset is presented in table S19.

Govers [5] found unit stream power values of up to 0.4 m s^{-1} and unit length shear force values of up to 6 N m^{-1}. In a study of Zhang et al. [9], shear stress values of up to 30 Pa and unit stream power values of up to 0.5 m s^{-1} were reported. Govers [13] measured shear stress values of up to 100 Pa and effective stream power values of up to 10000 W m^{-1}. While the measurements presented here are in the same order of magnitude compared to the previously published research, there are no clear linear correlations between hydraulic parameters and erosion parameters in the results of the field experiments. Therefore, these outcomes indicate that linear models may generally not be sufficient in order to describe the complex processes in natural rills.

Four possible improvements may help to improve this important concept which has been studied already for over thirty years, (1) including a clear description of the employed parameters, (2) including the turbulence, (3) considering the impact of processes that do not depend on the shear stress and likewise (4) consider the high spatial and temporal variability observed in

natural rills. These potential improvements will be discussed in more detail below.

For instance, the flow shear stress, a hydraulic parameter, and the critical shear stress, a soil parameter (similar to soil strength), must be differentiated. In particular, the flow shear stress must exceed the critical shear stress for erosion to occur. A number of hydraulic parameters, such as the flow velocity or the fluid density, water depth or width and roughness are used for the computation of the flow shear stress. The actual version of the shear stress equation calculates the average shear stress by depth averaging of momentum equation for steady uniform flow per area and time. Some factors used in shear stress calculation have been developed from empirical studies [15–26]. In most cases, the theoretical basis of the equations is however not clear. The formula applied by Chisci et al. [27] is derived from Landau and Lifchitz [57]. Other versions of the Landau-Lifchitz equation can be found in the literature [2–5].

The critical shear stress is the force needed to detach a soil particle. So it corresponds to a soil parameter and therefore, input

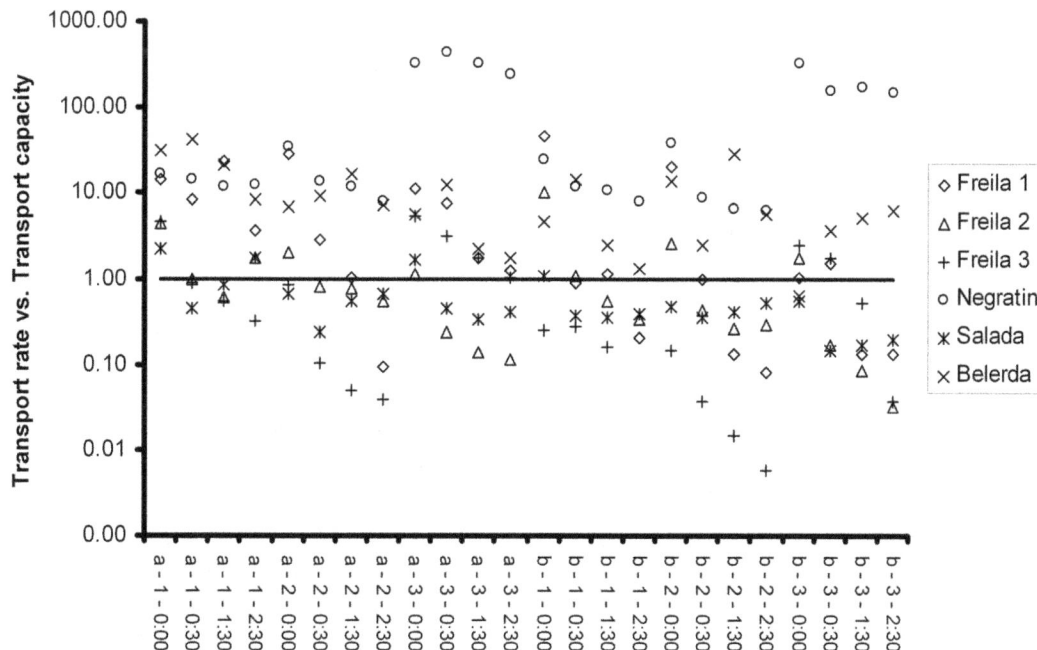

Figure 3. Transport rate vs. Transport capacity for each sample. The different experiments are represented by different symbols. On the x-axis, the following parameters are presented: run a or run b – measuring point 1–3 – sampling time at measuring point. The horizontal line marks the 1:1-relation between transport rate and transport capacity.

Table 5. Percentage of exceedance: Share of transport rate exceeding transport capacity.

Experiment	Value
Freila1	41.4%
Freila 2	16.0%
Freila 3	15.7%
Negratin	94.0%
Salada	6.0%
Belerda	76.1%
Average Freila 1–3	24.3%
Average Negratin, Salada, Belerda	58.7%
Average of all experiments	41.5%

In the first six rows, the exceedance percentage of each experiment is presented, in the next two rows the average percentage of the Freila experiments and of the Negratin-, Salada- and Belerda experiment and in the last row the average of all experiments. The complete dataset is presented in table S20.

for calculations should also depend on soil characteristics. However, this is the case for the WEPP model [34] only, where the critical shear stress is calculated using soil parameters such as texture, organic matter content and dry bulk density. In other cases, both hydraulic and soil parameters are used [34]. The discrepancies in the methods of computation of the shear stress may be due to the conditions under which the equations are deduced, as these equations are based on empirical observations. The empirical nature of the development of the different expressions is clearly highlighted in previous work [30–32]. That means the equations are not deduced from physical laws but from empirical studies.

In many studies [12,35,37–41], neither critical shear stress nor shear stress are used for the calculation of the transport capacity at all. In other studies shear stress is used to calculate transport capacity and detachment capacity [36] or transport rate [42] and critical shear stress to calculate the detachment capacity [36]. In both cases it is clear that shear stress and critical shear stress operate against each other, the important parameter is the difference between these two variables.

A summary of these equations can be found in Reid and Dunne [58], on the EPA-homepage [59] and in Hessel and Jetten [60].

The second reason for the low R^2-values in the correlations between hydraulic parameter and soil detachment can be the lack of turbulence parameters in the equations.

In the study of Knapen et al. [14] the Reynolds number shows very different correlations to the detachment rate, and this holds as well for the results of this study. The reason could be that the turbulence, described by the Reynolds number, does not directly operate on substrate, it influences the acting shear stress, that means the calculated shear stress is much lower than the operating shear stress, a relation which has been confirmed in several studies: Nearing et al. [61] found that turbulence can increase the active shear stress by a factor of several thousands. They measured flow shear stresses ranging from 0.5 to 2 Pa, and tensile strengths ranging from 1 to 2 kPa. Despite the fact that the tensile strengths are 1000 times larger than the flow shear stresses, the authors also measured detachment rates in the order of 300 g m^{-2} s^{-1}. Such large detachment rates were attributed to turbulent burst events. Another study about the influence of turbulence on detachment rates was published by Nearing & Parker [62]. They showed that

under turbulent flow conditions the same shear stress value caused a clearly higher detachment rate. In their flume experiments the difference between detachment rate caused by turbulent and laminar flow increased with increasing shear stress value, i.e., if given hydraulic conditions lead to a high shear stress value, the influence of turbulence on soil erosion is higher than in low shear stress value ranges.

The shear stress equation, as well as the equations describing other hydraulic parameters, assumes that drag forces are dominant for controlling erosion. But rill erosion is the result of the combination of different processes including headcut erosion, sidewall sloughing, tunnelling, micro-piping, slaking piping and sapping [14,45,63–67]. This is the third possible improvement for the problems of the model equations. The percentage of head-cutting in the different studies ranges between "four times higher than the contribution of bed scours" [67] to "60% of total rill erosion" [68]. Stefanovic and Bryan [69] showed that concentrated flow causes sediment production primarily from knickpoints, chutes, meanders and bank failure. Govers [45] distinguished between hydraulic erosion, mass wasting processes on rill sidewalls, gullying and piping. Hydraulic rill erosion mostly occurred during three extreme runoff events. Mass wasting processes caused 37% of total erosion in rills. Gullying, the retreat erosion at knickpoints and headcuts caused about 12% of rill erosion rates. In the experiments presented here, the main mechanisms causing rill erosion were mass wasting and gullying processes, hence the correlations between hydraulic parameters and detachment rate are generally low. However, the hydraulic rill erosion only occurs in extreme runoff events, in most cases, the runoff values are too low to cause hydraulic rill erosion. The percentage of material which is transported independent of shear stress is very high on the water front samples. Here the transport of loose material is probably more important than in the other samples meaning that this process is mainly independent of shear stress. In these cases of transport rate vs. transport capacity <1 the independence of shear stress cannot be excluded, in the other cases the processes controlled by shear stress can occur. Thus, it can be deduced that, in the case of $T_R > T_C$, not only shear stress controlled processes provide the material; at least the difference between T_R and T_C is caused by processes independent of shear stress.

The experiments presented here show that the correlation between hydraulic parameters and detachment rate does neither change from one experiment to another, nor from one run to another, but from one measuring point and run to another. Thus, sediment producing processes have a high spatial and temporal variability. This is the fourth possible improvement for models. It is very difficult to propose a single factor that always describes the soil detachment satisfactory. The high variability of erosion processes, even under controlled experimental conditions, has been highlighted in different studies. Measured variability shows a wide range between 3.4% and 173.2% [70–75]. This is partially the result of non-homogeneous parameters concerning soil characteristics and rainfall. On experimental plots, infiltration rates and soil aggregate stability can be highly variable [76] and rainfall also shows a high spatial and temporal variability [77]. Therefore, the input parameters to the different measurements reflected in the mentioned studies were not really comparable. Nevertheless, the results also make clear that modelling soil erosion has to include uncertainty in model input, as well as in the data used for model calibration and validation.

In field experiments, the spatial and temporal variability of soil conditions cannot be avoided, and is, furthermore, part of the investigations. Thus, additional input parameters as rainfall or flow should be maintained constant in the experiments to generate reproducible data. The high variability in soil erosion processes cannot be represented by a single factor like shear stress.

The results show that there is not a simple linear correlation between a certain hydraulic parameter and soil detachment rate. Depending on model purpose and scale, the factors can be used to predict the magnitude of rill detachment but they are not applicable for the simulation of rill erosion with high-resolution spatial and temporal change in processes.

A newer approach is the use of probability density functions to predict soil detachment [78,79]. Sidorchuk gives two sources of stochasticity in erosion modelling: (1) the necessity of spatial and temporal averaging when determining deterministic equations, which describe concentrated flow erosion and (2) the fact that the main erosion factors, if these can be determined anyway, can only be measured with limited accuracy. This is not the first attempt to model erosion by relating the probability of soil detachment with the excess of erosion driving forces over soil erosion resistance forces, other articles using a stochastic approach to describe soil erosion were published by Nearing [80], Wilson [81] and Sidorchuk [82–87]. Notably, one of the earliest articles about stochastic in erosion processes has been published by Einstein [88]. These stochastic models reduce the number of empirical components. Applying these models to the experiments presented here is beyond the scope of the current study.

Conclusions

The results show that a linear correlation between hydraulic parameter and soil detachment is not sufficient to describe processes in natural rills. The reason for this behaviour is the combination of various processes that can cause different amounts of soil erosion. The shear stress, for instance, only describes one process, while the results clearly show that there is not one fixed parameter that always predicts soil detachment best. Applicability of one certain hydraulic parameter to predict the sediment concentration changes at a certain point in time within a few minutes, because the temporal and spatial distribution of the different erosion processes is highly randomly determined. Therefore, it might be more useful to formulate results in probabilistic terms, an approach which has already been implemented by previous researchers, but is beyond the current work.

Supporting Information

Table S1 Freila 1 erosion data.
(DOC)

Table S2 Freila 1 runoff data.
(DOC)

Table S3 Freila 1 hydraulic data.
(DOC)

Table S4 Freila 2 erosion data.
(DOC)

Table S5 Freila 2 runoff data.

Table S6 Freila 2 hydraulic data.
(DOC)

Table S7 Freila 3 erosion data.
(DOC)

Table S8 Freila 3 runoff data.
(DOC)

Table S9 Freila 3 hydraulic data.
(DOC)

Table S10 Negratin erosion data.
(DOC)

Table S11 Negratin runoff data.
(DOC)

Table S12 Negratin hydraulic data.
(DOC)

Table S13 Salada erosion data.
(DOC)

Table S14 Salada runoff data.
(DOC)

Table S15 Salada hydraulic data.
(DOC)

Table S16 Belerda erosion data.
(DOC)

Table S17 Belerda runoff data.
(DOC)

Table S18 Belerda hydraulic data.
(DOC)

Table S19 R^2 - values between the detachment rate and different hydraulic parameters.
(DOC)

Table S20 Comparison of the transport rate with the transport capacity: Transport rate T_R [kg s^{-1}] - Transport capacity T_C [kg s^{-1}].
(DOC)

Table S21 Comparison of the transport rate with the transport capacity: Percentage of T_R exceeding T_C.
(DOC)

Acknowledgments

We thank all participants of the field trip to Andalusia in September 2009 who supported the performance of the experiments and Olli, Seta and Andreas for revising the whole manuscript.

Author Contributions

Conceived and designed the experiments: SW MS JBR. Performed the experiments: SW MS AZ CW JBR. Analyzed the data: SW MS JFW JBR. Contributed reagents/materials/analysis tools: SW MS AZ CW JFW JBR. Wrote the paper: SW MS AZ CW JFW JBR.

References

1. Lyle WM, Smerdon ET (1965) Relation of compaction and other soil properties to erosion resistance of soils. Transactions of the ASAE 8: 419–422.
2. Torri D, Dfalanga M, Chisci G (1987) Threshold conditions for incipient rilling. Catena Supplement 8: 97–105.
3. Ghebreiyessus YT, Gantzer CJ, Alberts EE, Lentz RW (1994) Soil erosion by concentrated flow: shear stress and bulk density. Transactions of the ASAE 37 (6): 1791–1797.
4. Nearing MA, Norton LD, Bulgakov DA, Larionov GA, West LT, et al. (1997) Hydraulics and Erosion in Eroding Rills. Water Resources Research 33 (4): 865–876.
5. Giménez R, Govers G (2002) Flow Detachment by Concentrated Flow on Smooth and Irregular Beds. Soil Sci Soc Am J 66(5): 1475–1483.
6. Bagnold RA (1977) Bed load transport by natural rivers. Water resources Research 13: 303–312.

7. Hairsine PB, Rose CW (1992) Modeling water erosion due to overland flow using physical principles, 2. Rill flow. Water resources Research 28: 245–250.
8. Elliot WJ, Laflen JM (1993) A Process-based rill erosion model. Transactions of the ASAE 36 (1): 35–72.
9. Zhang G, Liu B, Liu G, He X, Nearing MA (2003): Detachment of undisturbed soil by shallow flow. Soil Science Society of America Journal 67: 713–719.
10. Yang CT (1972) Unit stream power and sediment transport. J Hydraulics Div Am Soc Civil Eng 98: 1805–1825.
11. Moore ID, Burch GJ (1986) Sediment transport capacity of sheet and rill flow: Application to unit stream power theory. Water Resour Res 22: 1350–1360.
12. Bagnold RA (1980): An empirical correlation of bedload transport rates in flumes and natural rivers. Proc. Royal Society Series A372: 453–473.
13. Govers G (1992) Evaluation of transport capacity formulae for overland flow. In: Parsons AJ, Abrahams AD editors. Overland flow: hydraulics and erosion mechanics. London: UCL Press. pp. 243–273.
14. Knapen A, Poesen J, Govers G, Gyssels G, Nachtergaele J (2007): Resistance of soils to concentrated flow erosion: A review. Earth-Science Reviews 80(1–2): 75–109.
15. Foster GR (1982) Modelling the erosion process. In: Johnson HP, Brakensiek DL, editors. Hydraulic Modelling of Small Watersheds. St. Joseph, MI: ASAE monograph 5, Hann CT.
16. Shields A (1936) Anwendung der Ähnlichkeitsmechanik und der Turbulenz-forschung auf die Geschiebebewegung. Berlin: Dissertation, TU Berlin.
17. Miller M, McCave IN, Komar PD (1977) Threshold sediment motion under unidirectional currents. Sedimentology 24: 507–527.
18. Parker G, Klingeman PC, McLean DG (1982) Bedload and size distribution in paved gravel-bed streams. Journal of the Hydraulics Division, American Society of Civil Engineers 108: 544–571.
19. Diplas P (1987) Bedload transport in gravel-bed streams. J. Hydraul. Eng. ASCE 113: 277–292.
20. Parker G (1990) Surface-based bedload transport relation for gravel rivers. J Hydraul Res 28: 417–436.
21. Komar PD (1987 a) Selective grain entrainment by a current from a bed of mixed sizes: A reanalysis. J Sediment Petrol 57: 203–211.
22. Komar PD (1987 b) Selective gravel entrainment and the empirical evaluation of flow competence. Sedimentology 34: 1165–1176.
23. Andrews ED (1983) Entrainment of gravel from naturally sorted riverbed material. Geol Soc Am Bull 94: 1225–1231.
24. Ashworth PJ, Ferguson RI (1989 a) Size-selective entrainment of bed load in gravel bed streams. Water Resources Research 25: 627–634.
25. Ashworth PJ, Ferguson RI (1989 b) Quantifying gravel deposition on river bars using flexible netting. Journal of Sedimentary Research Vol. 59 No. 4: 623–624.
26. Komar PD, Carling PA (1991) Grain sorting in gravel-bed streams and the choice of particle sizes for flow-competence evaluations. Sedimentology 38: 489–502.
27. Chisci G, Sfalanga M, Torri D (1985) An experimental model for evaluating soil erosion on a single-rainstorm basis. In: El-Swaify SA, Moldenhauer WC, Lo A, editors. Soil Erosion and Conservation. Ankeny, Iowa: Soil conservation Society of America. pp. 558–565.
28. Ott WP, van Uchelen JC (1936) Application of similarity principles and turbulence research to bedload movement. Hydrodynamics Laboratory California Institute of Technology Publication No. 167. Soil conservation Service.
29. Graf WH (1971) Hydraulics of sediment transport. New York: McGraw-Hill.
30. Partheniades E, Paaswell RE (1970) Erodibility of channels with cohesive boundary. Journal of the Hydraulic Division, Proceedings of the American Society of civil Engineers: 755–771.
31. Andrews ED (1984) Bed-material entrainment and hydraulic geometry of gravel-bed rivers in Colorado. Geological Society of America Bulletin 95(3): 371–378.
32. Andrews ED, Erman DC (1986) Persistence in the size distribution of superficial bed material during an extreme snowmelt flood. Water Resources Research 22: 191–197.
33. De Ploey J (1990) Threshold conditions for thalweg gullying with special reference to loess areas. Catena Supplement 17: 147–151.
34. Flanagan DC, Livingston SJ (1995) WEPP user summary, USDA-water erosion prediction project. NSERL Report No. 11. National soil erosion research Laboratory.
35. Yalin MS (1963) An expression for bedload transportation. J Hydr Eng Div ASCE 89: 221–250.
36. Foster GR, Flanagan DC, Nearing MA, Lane LJ, Risse LM, et al. (1995) Hillslope erosion component. In: USDA Water Erosion Prediction Project: Hillslope Profile and Watershed Model Documentation, Chapter 11, USDA-ARS NSERL Report 10.
37. Govers G (1991) Rill erosion on arable land in central Belgium: Rates, controls and predictability. Catena 18: 133–155.
38. Abrahams AD, Gao P, Aebly FA (2000) Relation of sediment transport capacity to stone cover and size in rain-impacted interrill flow. Earth Surf Proc Land 25: 497–504.
39. Low HS (1989) Effect of sediment density on bed-load transport. J Hydraul Eng ASAE 115: 124–138.
40. Rickenmann D (1991) Hyperconcentrated flow and sediment transport at steep slopes. J Hydraul Eng ASCE 117: 1419–1439.
41. Yang CT (1973) Incipient motion and sediment transport. J Hydr Eng Div ASCE 99: 1679–1703.
42. Parker G (1979) Hydraulic geometry of active gravel rivers. Journal of the Hydraulics Division, American Society of Civil Engineers 105: 1185–1201.
43. Kleinhans MG, Bierkens MFP, van der Perk M (2010) HESS opinions. On the use of laboratory experimentation: "Hydrologists, bring out shovel and garden hoses and hit the dirt". Hydrology and Earth System Sciences 14: 369–382.
44. Govers G, Giménez R, van Oost K (2007) Exploring the relationship between experiments, modelling and field observations. Earth-Science Reviews 84 (3–4): 87–102.
45. Govers G (1987) Spatial and temporal variability in rill development processes at the Huldenberg experimental site. Catena Supplement 8: 17–34.
46. Seeger M (2007) Uncertainty of factors determining runoff and erosion processes as quantified by rainfall simulations. Catena 71 (1): 56–67.
47. FAO (2006) Guidelines for soil description. Rome: 4. ed. Food and Agriculture Organization of the United Nations.
48. Vandekerckhove L, Poesen J, Oostwoud Wijdenes D, de Figueiredo T (2003) Topographical thresholds for ephemeral gully initiation in intensively cultivated areas of the Mediterranean. Catena 33(3–4): 271–292.
49. Ad-hoc-Arbeitsgruppe Boden (2005) Bodenkundliche Kartieranleitung. Hannover: 5. Auflage, Bundesamt für Geowissenschaften und Rohstoff in Zusammenarbeit mit den staatlichen Geologischen Diensten der Bundesrepublik Deutschland. 438 p.
50. Wirtz S, Seeger M, Ries JB (2010) The rill experiment as a method to approach a quantification of rill erosion process activity. Zeitschrift für Geomorphologie Vol. 54,1: 47–64.
51. Allen JRL (1994) Fundamental properties of fluids and their relation to sediment transport processes. In: Pye K, editor. Sediment transport and depositional processes. Blackwell Scientific Publications. pp. 25–88.
52. DIN 18124 (1997) Baugrund, Untersuchung von Bodenproben - Bestimmung der Korndichte - Kapillarpyknometer, Weithalspyknometer. Deutsches Institut für Normung e.V., Ausgabe : 1997–07, Deutsch.
53. Macosko CW (1994) Rheology: Principles, Measurements and Applications. New York: VCH: Wiley-VCH. New York.
54. Einstein A (1906) Eine neue Bestimmung der Moleküldimensionen. Analen der Physik 19: 289–306.
55. Einstein A (1911) Berichtigung zu meiner Arbeit: "Eine neue Bestimmung der Moleküldimensionen". Analen der Physik 34: 591–593.
56. Scherer U (2008) Prozessbasierte Modellierung der Bodenerosion in einer Lösslandschaft. Karlsruhe: Dissertationsschrift, Fakultät für Bauingenieur-, Geo- und Umweltwissenschaften, Universität Fridericiana zu Karlsruhe (TH). 248 p.
57. Landau L, Lifchitz E (1971) Mécaniques des fluids. Moscow: MIR
58. Reid DM, Dunne T (1996) Rapid evaluation of sediment budgets. Reiskirchen: Catena Verlag.
59. EPA-homepage (2009) Channel Processes: Bedload transport. United States Environmental protection agency. Available: http://water.epa.gov/scitech/datait/tools/warsss/bedload.cfm Accessed 2010 September 14.
60. Hessel R, Jetten V (2007) Suitability of transport equations in modelling soil erosion for a small Loess plateau catchment. Eng Geol 91: 56–71.
61. Nearing MA, Bradford JM, Parker SC (1991) Soil detachment by shallow flow at low slopes. Soil Science Society of America Journal 55 (2): 339–344.
62. Nearing MA, Parker SC (1994) Detachment of soil by flowing water under turbulent and laminar conditions. Soil Science Society of America Journal 58 (6): 1612–1614.
63. Bryan RB, Govers G, Poesen J (1989) The concept of soil erodibility and some problems of assessment and application. Catena 16 (4–5): 393–412.
64. Bryan RB (1990) Knickpoint evolution in rillwash. Catena 17: 111–132.
65. Owoputi LO, Stolte WJ (1995) Soil detachment in the physically based soil erosion process: a review. Transactions of the ASAE 38 (4): 1099–1110.
66. Rapp I. (1998) Effects of soil properties and experimental conditions on the rill erodibilities of selected soils. Pretoria: Ph. D. Thesis, Faculty of Biological and Agricultural Sciences, University of Pretoria, South Africa.
67. Zhu JC, Gantzer CJ, Peyton RL, Alberst EE, Anderson SH (1995) Simulated small-channel bed scour and head cut erosion rates compared. Soil Science Society of America Journal 59 (1): 211–218.
68. Kohl KD (1988) Mechanics of rill headcutting. Ames: Phd. Diss. Iowa State University.
69. Stefanovic JR, Bryan RB (2009) Flow energy and channel adjustments in rills developed in loamy sand and sandy loam soils. Earth Surface Processes and Landforms 34: 133–144.
70. Nearing MA (1998) Why soil erosion models over-predict small soil losses and under-predict large soil losses. Catena 32: 15–22.
71. Ruttimann M, Schaub D, Prasuhn V, Ruegg W (1995) Measurement of runoff and soil erosion on regularly cultivated fields in Switzerland – some critical considerations. Catena 25: 127–139.
72. Wendt RC, Alberts EE, Hjelmfelt AT (1986) Variability of runoff and soil loss from fallow experimental plots. Soil Sci Soc Am J 50: 730–736.
73. Risse LM, Nearing MA, Nicks AD, Laflen JM (1993) Assessment of error in the universal soil loss equation. Soil Sci Soc Am J 57: 825–833.
74. Zhang XC, Nearing MA, Risse LM, McGregor KC (1996) Evaluation of runoff and soil loss predictions using natural runoff plot data. Trans ASAE 39: 855–863.
75. Liu BY, Nearing MA, Baffaut C, Ascough II JC (1996) The WEPP watershed model: III. Comparisons to measured data from small watersheds. Transactions of the ASAE 40: 945–951.

76. Ajayi AE, Horta IDMF (2007) The effect of spatial variability of soil hydraulic properties on surface runoff processes. Anais XIII Simpósio Brasileiro de Sensoriamento Remoto, Florianópolis, Brasil, 21–26 abril 2007, p. 3243–3248.

77. Dunkerley D (2008) Rain event properties in nature and in rainfall simulation experiments: a comparative review with recommendations for increasingly systematic study and reporting. Hydrological Processes 22: 4415–4435.

78. Sidorchuk A (2005 b) Stochastic modelling of erosion and deposition in cohesive soils. Hydrological Processes 19: 1399–1417.

79. Sidorchuk A (2009 b) A third generation erosion model: The combination of probabilistic and deterministic components. Geomorphology 110 (1–2): 2–10.

80. Nearing MA (1991) A probabilistic model of soil detachment by shallow turbulent flow. Transactions of the ASAE 34 (1): 81–85.

81. Wilson BN (1993) Development of a fundamentally-based detachment model. Transactions of the ASAE 36 (4): 1105–1114.

82. Sidorchuk A (2001) Calculation of the rate of erosion in soils and cohesive sediments. Eurasien Soil Science 34 (8): 893–900.

83. Sidorchuk A (2002) Stochastic Modelling of soil erosion and deposition. 12th ISCO Conference, Beijing 2002.

84. Sidorchuk A, Smith A, Nikora V (2004) Probability distribution function approach in stochastic modelling of soil erosion. Sediment transfer trough the fluvial system Vol. 288. IHAS Publ., pp. 345–353.

85. Sidorchuk A (2005 a) Stochastic components in the gully erosion modelling. Catena 63: 299–317.

86. Sidorchuk A, Schmidt J, Cooper G (2008) Variability of shallow overland flow velocity and soil aggregate transport observed with digital videography. Hydrological Processes 22: 4035–4048.

87. Sidorchuk A (2009 a) High-Frequency variability of aggregate transport under water erosion of well-structured soils. Eurasian Soil Science Vol. 42, No. 5: 543–552.

88. Einstein HA (1936) Der Geschiebetrieb als Wahrscheinlichkeitsproblem. Zürich: Diss.-Druckerei A.-G. Gebr. Leemann & Co. 112 p.

The Occurrence, Sources and Spatial Characteristics of Soil Salt and Assessment of Soil Salinization Risk in Yanqi Basin, Northwest China

Zhang Zhaoyong[1,2], Jilili Abuduwaili[1]*, Hamid Yimit[3]

1 State Key Laboratory of Desert and Oasis Ecology, Xinjiang Institute of Ecology and Geography, Chinese Academy of Sciences, Urumqi, China, **2** University of the Chinese Academy of Sciences, Beijing, China, **3** Key Laboratory of Xingjiang Arid Land Lake Environment and Resource, Xinjiang Normal University, Urumqi, China

Abstract

In order to evaluate the soil salinization risk of the oases in arid land of northwest China, we chose a typical oasis-the Yanqi basin as the research area. Then, we collected soil samples from the area and made comprehensive assessment for soil salinization risk in this area. The result showed that: (1) In all soil samples, high variation was found for the amount of Ca^{2+} and K^+, while the other soil salt properties had moderate levels of variation. (2) The land use types and the soil parent material had a significant influence on the amount of salt ions within the soil. (3) Principle component (PC) analysis determined that all the salt ion values, potential of hydrogen (pHs) and ECs fell into four PCs. Among them, PC1 ($C1^-$, Na^+, SO_4^{2-}, EC, and pH) and PC2 (Ca^{2+}, K^+, Mg^{2+} and total amount of salts) are considered to be mainly influenced by artificial sources, while PC3 and PC4 (CO_3^- and HCO_3^{2-}) are mainly influenced by natural sources. (4) From a geo-statistical point of view, it was ascertained that the pH and soil salt ions, such as Ca^{2+}, Mg^{2+} and HCO_3^-, had a strong spatial dependency. Meanwhile, Na^+ and Cl^- had only a weak spatial dependency in the soil. (5) Soil salinization indicators suggested that the entire area had a low risk of soil salinization, where the risk was mainly due to anthropogenic activities and climate variation. This study can be considered an early warning of soil salinization and alkalization in the Yanqi basin. It can also provide a reference for environmental protection policies and rational utilization of land resources in the arid region of Xinjiang, northwest China, as well as for other oases of arid regions in the world.

Editor: Andrew C. Singer, NERC Centre for Ecology & Hydrology, United Kingdom

Funding: This study was supported by the Knowledge Innovation Program of the Chinese Academy of Sciences (KZCX2-EW-308; KZCX2-YW-GJ04). The funders had no role in study design, data collection and analysis, decision to publish, or preparation of the manuscript.

Competing Interests: The authors have declared that no competing interests exist.

* Email: jilil@ms.xjb.ac.cn

Introduction

Soil salinization is a global problem and it is a potential environmental problem in all continents with the exception of unassessed Antarctica. Soil saline levels are found within a wide range, and soil salinization occurs in much of the waterfront, arid and semi-arid zones of more than 100 countries and regions [1–3]. According to statistics done by the United Nations Educational, Scientific and Cultural Organization (UNESCO), and the Food and Agriculture Organization (FAO), salinized soil covers an area of about 9.543×10^6 km^2 on Earth [4–6]. In China alone, the area of salinized soil is about 3.693×10^5 km^2, which accounts for about a third of the total arable land [7–8]. The area of salinized soil in the oasis basin of Xinjiang in northwest China is about 1.05×10^4 km^2, which accounts for 33.4% of the total land in this area and research has found that the salinity of this area is trending upwards [9,10]. Soil salinization restricts agricultural development, especially when sustainable agricultural development and environmental quality improvement strategies are being considered. Studies have found that when the salt ions in the soil attain 8 $g.kg^{-1}$, they can greatly harm and even kill crops in farmland [11,12].

In oases of arid regions of northwest china, the environment is so weak, including a lack of precipitation and high envapotion, that economic activities such as fishing, agriculture, forestry and grassland farming of the oases have been strongly limited, especially for agriculture [13–15]. Therefore, it is necessary to identify the distribution characteristics, sources of the soil properties, such as salt ions, potential of hydrogen (pH) and electrical conductivity (EC), and also the status and causes of salinization of the land of the oasis, in order to provide a scientific basis for protection of the soil that sustains land plants.

Multivariate analyses and other statistical methods have been widely applied in studies to determine the sources of elements found in soil, such as total soil salt content and heavy metals [16–18]. The spatial variation model and spatial distribution are used to make a hazard risk map of soil salt properties in regions of interest. Correlation analysis, principal component (PC) analysis and cluster analysis are classic methods used to identify the natural and man-made sources of salt ions and to simplify data. Additionally, use of the comprehensive index results in a class of data with high correlations that better reflect the associations between the data.

Table 1. Indicators used for risk assessment of soil salinization.

Indicators	Class limits and their ratings score				
	None	Slight	Moderate	Severe	Very severe
*EC (dS.m⁻¹) [49]	<4	4–8	8–16	16–32	>32
**SAR [50]	<8	8–13	13–30	30–70	>70
Total salt content (%) (0–20 cm) [51]	<1	2~3	3–4	4–8	>8

*EC is electrical conductivity; **SAR is sodium adsorption ratio.

The Geostatistical Analyst is based on GIS technology [19,20]. Among these, the ordinary kriging is the most widely used one in the study of soil salt distributions [21,22,29]. In recent years, the Geostatistical Analyst method has been used in the field of hydrology and water resources, including studies of groundwater pollution risk, water potential research and spatial distribution of soil salinization in arid land [23–25,28,32]. Since the 1990s, geostatistical methods have been widely used to study spatial variability characteristics of soil salt properties (salt ions, EC and pH). Sylla et al. [26] studied the spatial variation characteristics of soil salt content of an agricultural ecosystem under different scales in West Africa. Ammari et al. [27] studied the soil salinity changes in the Jordan Valley and the potential threat against the sustainable irrigated agriculture. In China, Bai et al. [30] researched the spatial variation characteristics and composition of soil salt content in Huang Huai Hai plain, northeast China and found that the average influential range of the soil salt content was higher than 200 km, indicating the salt content of the soil is mainly gathered in a large area of the regions. In Xinjiang in northwest China, Lin et al. [33] researched the spatial variation characteristics of soil salt in the Wei Gan He irrigated area and found the agricultural irrigation has resulted in serious soil salinization in this area and these deserve serious attention.

In arid regions, oases in basins are the main places where humans live and life can survive [31]. Therefore, it is important to understand the spatial distribution characteristics of the soil salt properties, including total salt content, salt ions, pH and EC. A quantitative grasp of soil salinity levels could serve as a reference and a basis for maintaining soil quality, which would help to effectively control the human pollution and develop the regional economy in a reasonable and orderly fashion [34,35]. However, previous research has focused on rapidly developing areas, such as coastal plains and large irrigation areas in eastern china and elsewhere of the world with the purpose of assessing land usability, environmental effects and soil salinization risk [36–38]. Since the 1990s, implementation of the "western development policy of China" has led to prodigious economic development in many oases in Xinjiang, and the agriculture in these regions has undergone rapid progress. However, the rational irrigation of the agriculture, lack of precipitation and high envapotion of these regions have negatively influenced soil salt properties, resulting in increased soil salt contents, ECs and pHs, which can result in serious soil salinization [39]. Unfortunately, research on the soil salt property distribution characteristics and soil salinization risk assessment in the oases of arid regions of northwest of China is lacking.

The Yanqi basin is a typical oasis in a basin in the southern Tianshan Mountains, Xinjiang in northwest China. Since the 1990s, both the implementation of the "western development policy of China" and the development policy made by the Xinjiang Province, China, have led to prodigious economic

development in the Yanqi basin, but regional economic development and associated human activity have left the current ecological environment fragile [39,40]. Together with economic development, the blind expansion of farmland and unrestrained surface water irrigation led to a rise in groundwater and an increase in soil salinization in the basin oasis. 64.12% of the area experienced mild soil salinization, 8.25% had moderate salinization, and 27.07% had severe salinization. Research has shown that excessive use of water resources by agriculture has made the soil salinization status severe and decreased agricultural production [41].

After a basic analysis of land use and soil parent materials types in the area, we created land use and soil type geological maps, and, using ArcGIS 10.0 software and combining the grid sampling method with 3S technology, we made sampling points to get soil samples across the whole area. We evaluated the soil salt properties in different land use types in the laboratory, and assessed the soil salinization status and the cause in the Yanqi basin. Then ordinary kriging of Geostatistical Analyst method was used to reveal the spatial distribution characteristics of the soil salt properties in this region. Then by combining these properties with the climate, precipitation, evaporation and temperature, we assessed the soil salinization risk of this area. From this we can provide helpful proposals to prevent the environmental risks that could lead to soil salinization in this area. This research can serve as a helpful reference for environmental protection in this region and for soil salinization prevention in arid regions of northwest China.

Materials and Methods

Study area

The area studied in this work is a desert basin oasis in the arid region of northwest China including four counties in the Yanqi basin: Yanqi County, Hejing County, Bohu County and Heshuo County. This region lies within the geographical coordinates of 85°50′–87°50′E and 41°40′–42°20′N with a length of about 85 km from north to south and width of 130 km from east to west, totaling an area of about 723100 km². The terrain slopes up from the northwest down to the southeast. The northwest is mountainous and the south is low-lying desert that is 1050–2000 m above sea level. The western area has extensive intrusive rock and metamorphic rock from the *Proterozoic era, Neoproterozoic* and *Cenozoic.* Weathering of this rock results in brown earth soil, acidic rocky soil and an acidic soil skeleton in the west. The east is primarily made up of quaternary sediments, which form *Takyic* (Calcisols), *Chemic* (Phaeozems), *Stagnic* (Gleysols), *Irragric* (Anthrosols), *Fragic* (Arenosols), *Eutric* (Gleysols) and *Yemic* (Solonchaks) [42]. The area researched is in a continental desert climate temperate zone with an annual mean temperature of 14.6°C, 186 frost-free days per year, and 50.7–79.9 mm of annual

Table 2. Descriptive statistics of the soil salt properties from Yanqi basin.

Elements	Ranges (g.kg⁻¹)	Contributions(%)	Median (g.kg⁻¹);EC (dS.m⁻¹)	Average (g.kg⁻¹)	Standard deviation (%)	Coefficient of variation (%)	Kurtosis (%)	Skewness (%)
HCO_3^-	0.13-0.98	3.51	0.171	0.19	12.25	35.37	0.14	0.58
CO_3^-	0.18-0.85	4.08	0.252	0.46	10.35	32.08	1.63	8.76
Ca^{2+}	0.59-1.95	6.54	0.681	0.75	9.83	191.67	31.38	42.43
Na^+	0.69-2.43	13.18	0.955	1.19	12.38	23.01	2.45	16.22
Mg^{2+}	0.47-1.89	7.75	0.987	0.58	21.02	20.07	1.10	10.88
K^+	0.49-2.13	12.36	0.855	0.69	23.04	226.25	23.47	51.81
SO_4^{2-}	0.93-1.58	18.95	1.245	1.14	12.56	19.63	1.46	0.76
Cl^-	0.75-2.36	25.63	1.167	0.98	25.7	12.94	2.30	14.79
SAR	3.41-33.241	-	22.417	10.51	121.45	148.56	11.54	15.78
EC	0.7-1.39	-	0.981	0.95	15.09	21.27	1.46	10.99
pH	7.85-8.55	-	8.141	8.15	16.39	30.32	1.17	0.73
Total salt	1.16-14.77	-	8.56	9.73	15.36	126.73	16.84	18.46

precipitation. By calculating the potential evaporation (ET_0) by the method of Hargreaves (1985) [43], we then got the annual average potential evaporation of this area as 2438.9 mm, the $\geq 10°C$ active accumulated temperature 3414.4–3694.1°C and an annual average relative humidity of 72%.

Soil sampling and analyses

In order to perform a basic analysis of the land use and soil type, geological maps were made of the study area using ArcGIS 10.0 software to lay out a grid of soil sampling points on a digital map of the Yanqi basin. All samples were acquired in July 2012 or July 2013 from a collection area. In order to best assess the ecological risk in the Yanqi basin, diverse land use types were encompassed in our study of salt ion distribution. Soil samples were collected at depths of 0–20 cm, where a hard plastic shovel was used to dig a vertical 20×20 cm soil profile. 1 kg uniform samples were collected, and then they were put into a clean cloth, numbered and sealed. The collection position, date, sample vegetation types and surrounding vegetation conditions of each sampling area were recorded. After the soil samples were taken back to the laboratory, they were air dried and impurities, such as plant residues and rocks, were removed. The samples were then pushed through a 20 mesh nylon sieve (0.84 mm) to eliminate the plant residue and stones. We then used agate to grind the soil samples through 100 mesh nylon sieves (0.25 mm) to prevent contamination and then stored the samples in plastic bottles [46].

Total soil salt content, soil salt ions, pH and EC tested are as follows: 50 g of ground sample were removed from the plastic bottles, dissolved in 250 ml of deionized water (CO_2 has been removed) (1:5, soil:water) for 2 hours to fully dissolve salt ions contained in the soil. The samples were then put in a centrifuge tube, vibrated for 3 min with an oscillator and then centrifuged at a speed of 4,500–5,000 r·min⁻¹. To get the supernatant prepared for analysis of total soil salt content, salt ions content, pH and EC, the method described by Lu (2000) was followed [44].

The total salt content of the soil was determined by gravimetry of the evaporation residue. First, the supernatant was absorbed in a porcelain dish, and hydrogen peroxide (H_2O_2) was used to ozidize organic matter. Then, the samples were boiled in a water bath at 105–110°C until it dried, and weighed. The drying quality of the residue is expressed as total salt content of the soil. The pH of the extracted supernatant was tested using a Potentiometric Titrimeter (G20, METTLER, and TOLEDO). Burette drive resolution was 1/20000. Mv/pH electrode measurement range was ±2000 mv. The ECs were tested using a Conductivity Meter (DDSJ-308A, Shanghai, China) with a measurement range of $0–1.999 \times 10^5$ μs/cm and a test error of ±0.5% (FS) ±1.

The supernatant was run through a 0.45 μm drainage cellulose acetate membrane. Then, the cation content (K^+, Na^+, Ca^{2+}, Mg^{2+}) of the solutions was determined using an inductively coupled plasma atomic emission spectrometer (Vista MPX, Varian, USA). The anion content (Cl^-, SO_4^{2-}, CO_3^{2-}, and HCO_3^-) of the solutions was determined using an Ion Chromatograph (ICS-90, Dionex, USA). All tests were conducted using the following protocol: a standard solution was prepared for Na^+, K^+, Ca^{2+}, Mg^{2+}, Cl^-, SO_4^{2-}, CO_3^{2-} and HCO_3^-. The salt ion content was determined by comparing each sample to the standard solution of known concentration. The standard solutions used for the salt ions in this study were national level standard material (Gss series, China). The coefficient of the best fitting curve was determined by the testing equipment based on the standard material and then the amount of the salt ions (Na^+, K^+, Ca^{2+}, Mg^{2+}, Cl^-, SO_4^{2-}, CO_3^{2-}, and HCO_3^-) in the liquid supernatant was tested. After all the samples had been tested for their salt ion

Table 3. Statistical parameters of the soil salt ions found at 0–20 cm depth within the investigated land use and land cover categories of the study area.

LUCC SPM	Parameters	HCO_3^- (g.kg⁻¹)	CO_3^- (g.kg⁻¹)	Na^+ (g.kg⁻¹)	Mg^{2+} (g.kg⁻¹)	K^+ (g.kg⁻¹)	Ca^{2+} (g.kg⁻¹)	SO_4^{2-} (g.kg⁻¹)	Cl^- (g.kg⁻¹)	Total salt (g.kg⁻¹)	EC (dS.m⁻¹)	pH
Farmland (n=51)	Ranges	0.18–0.98	0.27–0.85	0.69–2.43	1.05–1.89	1.31–2.13	1.17–1.95	1.05–1.51	1.47–2.36	8.56–14.77	0.96–1.39	8.05–8.55
	Average	0.32a	0.62a	1.36a	1.15a	1.83a	1.71a	1.21a	1.68a	11.2a	1.02a	8.15a
	SD	22.45	15.23	31.35	23.3	27.57	37.73	24.57	19.19	23.97	18.64	14.77
Forest (n=46)	Ranges	0.13–0.53	0.18–0.78	0.69–0.98	0.47–0.98	0.49–0.97	0.59–1.02	1.09–1.58	0.75–0.97	9.13–13.35	0.7–1.13	7.85–8.45
	Average	0.33a	0.41b	0.75b	0.87b	0.75a	0.77a	1.18b	0.82a	9.98b	0.94b	8.17b
	SD	14.75	17.73	23.53	25.55	23.74	23.73	17.86	28.33	16.79	18.71	14.37
Grassland (n=63)	Ranges	0.21–0.73	0.25–0.82	1.03–1.63	0.54–1.02	0.61–1.31	0.89–1.14	0.93–1.19	0.96–1.61	9.08–11.24	0.76–1.25	7.92–8.42
	Average	0.42a	0.65b	1.32b	0.68b	1.18b	0.98b	1.04b	1.28b	9.45a	0.95b	8.27b
	SD	17.54	23.77	27.52	23.13	11.51	15.23	12.22	22.14	13.75	16.71	21.37
Desert (n=71)	Ranges	0.26–0.83	0.19–0.76	0.98–1.57	0.47–1.44	0.95–1.56	0.99–1.19	1.05–1.24	1.08–2.29	9.31–13.89	0.78–1.34	8.06–8.44
	Average	0.58b	0.32b	1.31b	1.13b	1.21b	1.02b	1.11.6b	1.57b	10.56a	1.08b	8.28a
	SD	17.33	15.65	17.52	25.55	18.85	13.43	17.33	19.25	23.75	12.52	15.31
Urban construction areas (n=40)	Ranges	0.49–0.93	0.33–0.83	0.91–1.54	0.65–1.03	0.51–1.51	0.88–1.21	0.94–1.18	0.79–1.31	6.16–14.13	0.75–1.33	7.94–8.51
	Average	0.53c	0.39c	1.21c	0.91c	0.97c	0.97c	1.06c	0.94c	10.52a	1.11c	8.31c
	SD	15.75	16.51	13.25	15.52	12.35	19.52	23.37	21.54	32.24	22.31	24.87
Sandy shale of weathered material (n=66)	Ranges	0.13–0.86	0.18–0.38	1.19–2.39	0.47–1.75	0.58–2.11	0.59–1.36	0.93–1.53	0.75–2.36	8.69–14.77	0.79–1.26	7.86–8.55
	Average	0.45a	0.42a	1.51a	0.95a	0.91a	0.82a	1.01a	1.21a	10.88a	0.3a	7.93a
	SD	25.55	17.53	13.52	22.31	22.35	23.35	15.75	18.65	11.54	12.25	21.35
Coarse crystalline rock weathered material (n=74)	Ranges	0.18–0.52	0.22–0.79	0.69–2.43	0.51–1.54	0.49–2.13	0.69–1.45	0.97–1.56	0.78–2.12	6.16–12.41	0.7–1.39	7.85–8.32
	Average	0.32b	0.38b	1.72b	0.81b	0.87b	0.95b	1.13b	1.46b	8.72a	1.14b	8.14b
	SD	13.35	11.41	21.25	23.74	11.29	12.52	21.57	22.53	15.54	12.57	24.58
Diluvial material (n=68)	Ranges	0.21–0.93	0.24–0.85	0.81–2.23	0.65–1.89	0.64–1.72	0.71–1.56	0.95–1.58	0.85–2.26	7.89–13.25	0.84–1.28	7.89–8.19
	Average	0.67b	0.51b	1.42b	0.75b	0.98b	1.01b	1.05b	1.62b	9.82a	0.91b	8.01b
	SD	12.15	12.25	23.73	27.35	11.37	12.75	12.73	22.36	21.57	23.37	11.52
Lacustrine deposits (n=63)	Ranges	0.33–0.98	0.23–0.79	0.85–2.35	0.52–1.49	0.71–1.98	0.62–1.95	0.96–1.51	0.96–2.19	8.98–11.51	0.85–1.32	7.97–8.37
	Average	0.74a	0.44b	1.16a	0.91c	1.02c	0.85c	1.06a	1.31b	9.24a	1.24a	8.23a
	SD	13.51	22.35	21.26	23.59	15.34	15.62	21.94	19.31	26.49	23.77	12.35
R^2	LUCC (%)	31.21	9.74	13.49	8.4	71.45	37.85	11.3	10.8	56.71	12.64	12.43
	SPM (%)	58.79	34.59	16.47	28.93	32.7	11.4	17.9	15.9	23.51	17.53	15.2

Different small letters represent a significance of 0.05; LUCC represent land use types; SPM represent soil parent material types.

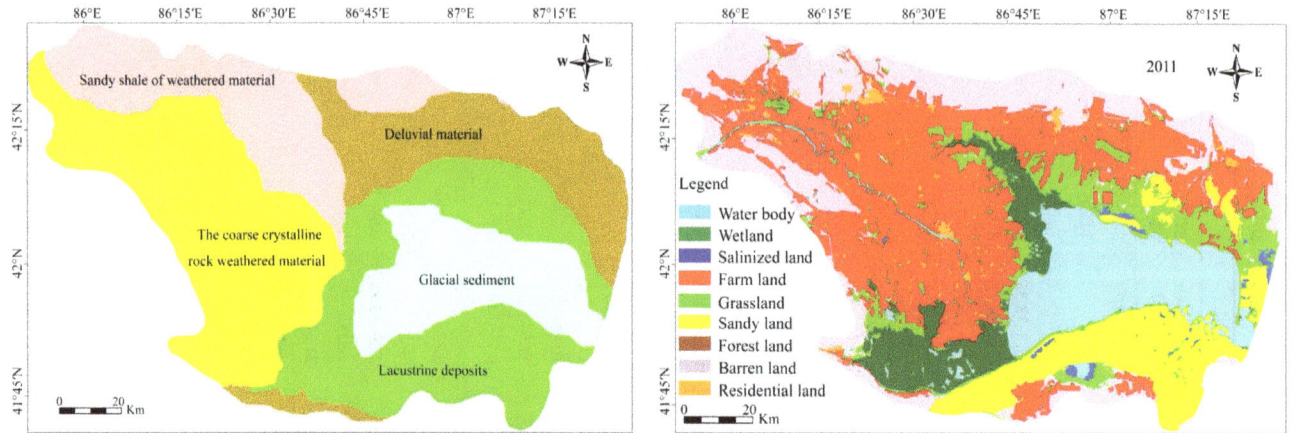

Figure 1. Land use types and parental material pattern in Yanqi basin.

content (Na^+, K^+, Ca^{2+}, Mg^{2+}, Cl^-, SO_4^{2-}, CO_3^{2-}, and HCO_3^-), we chose approximately 20% for retesting and found that 97.3% of the results were repeatable, inspiring confidence in the original data. After all the total salt ion contents (Na^+, K^+, Ca^{2+}, Mg^{2+}, Cl^-, SO_4^{2-}, CO_3^{2-}, and HCO_3^-) of the solution were determined, we recalculated them from unit of μg/ml into mg/g (g/kg) using the method described by Bao (2005) [45]. To prevent contamination during the testing process, all glassware was soaked in 5% HNO_3 for 24 hours, rinsed and then dried.

Statistical analyses

Descriptive and multivariate statistical analysis. Descriptive statistical methods were used to analyze the range, mean, median, standard deviation, coefficient of variation, kurtosis and skewness of the total salt content, each salt ion, SAR, pH and EC of the soil samples. Correlation analysis, PC analysis and cluster analysis of the classic multivariate statistical method were used to process data and identify the soil salinity. Single factor analysis of variance (ANOVA) was used to analyze the differences in the amount of salt ions between different land use types. These analyses were all processed using the software SPSS 19.0.

Ordinary kriging method. Ordinary kriging (OK) is a commonly used linear spatial interpolation method that estimates variables at unsampled locations by using information from neighboring points and assigning weights to these points based on their distance from the point and the spatial variability structure. The OK method can be formulated as

$$Z_{OK}^*(\mathbf{x}_0) = \sum_{i=1}^{n} w_i Z(x_i) \tag{1}$$

where $Z_{OK}^*(\mathbf{x}_0)$ is the OK estimation at an unsampled location (x_0), n is the number of samples in a search neighborhood, and w_i are the weights assigned to the ith observation $Z(x_i)$. Weights are assigned to each sample such that the estimation or kriging variance $E\left[\{Z^*(\mathbf{x}_0) - Z(x_0)\}^2\right]$ is minimized and the estimates are unbiased [47]. Weights are determined after computing a semivariogram that models spatial correlation and covariance structure between data points for each variable using Eq. 1 [48].

$$\gamma(h) = \frac{1}{2N(h)} \sum_{i=1}^{N} [Z(x_i + h) - Z(x_i)]^2 \tag{2}$$

where $\gamma(h)$ is the semivariance between two observation points $Z(x_i)$ and $Z(x_i + h)$ separated by a distance h, and N is number of observation pairs at the distance h.

Soil salinization evaluation criteria used in this research as under below (Table 1).

Results and Discussion

Descriptive statistical analysis of soil salinity

The descriptive statistics concerning the soil properties in Yanqi basin in Table 2 show that the maximum and average values of HCO_3^-, Ca^{2+}, CO_3^-, Na^+, Mg^{2+}, K^+, SO_4^{2-}, Cl^-, SAR, total amount of salts, EC, and pH were 0.98(0.19) g.kg^{-1}, 1.95(0.75) g.kg^{-1}, 0.85 (0.46) g.kg^{-1}, 2.43(1.19) g.kg^{-1}, 1.89(0.58) g.kg^{-1}, 2.13(0.69) g.kg^{-1}, 1.58(1.14) g.kg^{-1}, 2.36(0.98) g.kg^{-1}, 33.241(10.51), 14.77(9.73) g. kg^{-1}, 1.39(0.95) dS.m^{-1}, and 8.55(8.15), respectively. Within the analysis of the soil samples, a large amount of variation occurred, suggesting that the sources and influencing factors of the salt properties in Yanqi basin are complex. This work found that the main salt ions accounted for 84.41% of the total salt content and were K^+, Ca^{2+}, Na^+, Cl^-, Mg^{2+} and SO_4^{2-}. Meanwhile the amount of HCO_3^- and CO_3^- was very low. The pH of the soil of the study areas ranged from 7.85 to 8.55, which had only a small variation. However, there was a dramatic change within the total salt content of the soil, ranging from 1.16 to 14.77 g.kg^{-1}.

The coefficient of variation is the ratio between the standard deviation and average, and it can be used to compare different dimensions of indicators. The coefficients of variation of HCO_3^-, CO_3^-, Na^+, Mg^{2+}, SO_4^{2-}, Cl^-, EC, and pH were 35.37%, 32.08%, 23.01%, 20.07%, 19.63%, 12.94%, 21.27%, and 30.32%, respectively, and were of medium variation (10%<CV<100%). However, the coefficients of variation of SAR, the total amount of salts, Ca^{2+} and K^+ were 148.56%, 126.73%, 191.67% and 226.25%, and, therefore, had high levels of variation (CV> 100%)[52]. In particular, Ca^{2+} and K^+ had higher coefficients of variation as compared to the other elements. From the perspective of skewness, the values of these ten soil properties are ordered as K^+>Ca^{2+}>total amount of salts>Na^+>SAR>Cl^->EC>Mg^{2+}> CO_3^->SO_4^{2-}>pH>HCO_3^-.

The differences in soil salt properties from different land use types and soil parent materials

Land utilization types and soil parent materials are the main examples of human activity and geological background that

Table 4. The correction matrix of soil salt properties in Yanqi basin.

	EC	TSA	TSC	TS	Mg²⁺	Na⁺	K⁺	SO₄²⁻	Cl⁻	CO₃⁻	Ca²⁺	HCO₃⁻	pH
EC	1												
TSA	0.58	1											
TSC	0.40	0.30	1										
TS	0.41	0.52**	0.38**	1									
Mg²⁺	0.10	0.04	0.11	0.06**	1								
Na⁺	0.98**	0.34	0.96**	0.95**	0.01	1							
K⁺	0.65	0.48	0.78*	0.71**	0.51	0.64	1						
SO₄²⁻	0.87**	0.94**	0.42	0.93**	0.001	0.82**	-0.24	1					
Cl⁻	0.66**	0.55*	0.12	0.57**	0.13	0.69**	-0.15	0.24	1				
CO₃⁻	0.48*	0.49	0.27	0.48**	-0.24	0.57	0.12	0.27	0.17	1			
Ca²⁺	0.42	0.55	0.51*	0.54**	0.22	0.25	0.57*	0.23	0.03	-0.14	1		
HCO₃⁻	-0.25	-0.22	-0.20	0.21**	0.03	-0.16	-0.19	-0.42	0.28	0.16	-0.21	1	
pH	0.72**	0.62	0.14	0.82	0.42	0.32*	0.27	0.67**	0.58**	0.10	0.41	0.01	1

EC is electrical conductivity, TSA is the total anionic salt, TSC is the total cationic salt and TS is total salt content.

influence the salt ion content of soil. We analyzed the relationship between soil salt properties, human activity and geological background to further explore the distribution characteristics and sources of the soil salt properties of the Yanqi basin. The salt properties of the soil from each land use type and soil parent material of Yanqi basin are in Table 3. The land use types and soil parent materials of Yanqi basin are shown in Fig. 1. The analysis of these data suggests that the manner in which land is used has a significant influence on the amount of Ca^{2+} and K^+. In farmland, the average content of Ca^{2+} was 1.83 g.kg^{-1}, K^+ was 1.71 g.kg^{-1}, and the total amount of salts was 10.88 g.kg^{-1}. This was significantly higher than in the other land use types including areas of urban construction, forest, grassland, urban construction areas and desert. Meanwhile the maximum average values of Na^+, Mg^{2+}, SO_4^{2-}, Ca^{2+}, the total amount of salts, and K^+ found in farmland were higher than in grassland, desert, and areas of urban construction. The variance test attained a significant level of 0.05, indicating these elements and their distribution are mainly controlled by human activity. The activities that had a significant influence on these soil salt properties include agricultural activities, such as irrigating, fertilizing and farming. The CV calculated indicates that the pH, EC, total amount of salts and all the soil salt ions measured belong to medium variability categories (10–100%). Among the five land use types, the differences between these groups were small and, therefore, the classes of element enrichment were not obvious.

This research also found differences in the amount of organic soil salt ions. For example, HCO_3^- reached a maximum in lacustrine deposits of 0.74 mg.kg^{-1}. The maximum average values of Na^+, Mg^{2+}, K^+, SO_4^{2-}, total amount of salts, and Cl^- were found in coarse crystalline rock weathered material, sandy shale of weathered material and lacustrine deposits, while the maximum average values of CO_3^-, Ca^{2+}, and EC were found to be significantly higher in diluvial material than in sandy shale or coarse crystalline rock from weathered material or lacustrine deposits. For the five land use types, the class of element enrichment was not obvious as the differences between these groups were small.

R^2 represents the ratio of the sum of squares in groups and the total error of the sum of squares. It reflects the contribution of different factors on the soil salt properties [53]. For this study, the land use types explained the variances in Ca^{2+} (37.85%), K^+ (71.45%) and total amount of salts (56.71%), which were higher than that of the soil parent material. This demonstrates that the way land was used played a major role in the accumulation of Ca^{2+}, K^+ and total amount of salts in Yanqi basin. This analysis also found that the variances of HCO_3^-, CO_3^-, Mg^{2+}, Na^+, SO_4^{2-}, and Cl^- in the soil parent material are higher than those of the land use factors (Table 4), indicating that the soil parent material played a major role in the accumulation of these elements. However, there was little difference in the variances of EC and pH, indicating that these soil salt properties were mainly influenced by land use and soil parent material. Overall, the R^2 analysis fits well with the results of the multivariate statistical analysis.

Multivariable statistics

Correlation analysis. Table 4 shows the Pearson correlation coefficients between the soil salinity variables. There is a significant correlation of 0.96 (P<0.01) between the total amount of salt cations and Na^+, as well as the total amount of salt cations and Ca^{2+} at 0.51 (P<0.05) in the soil in the Yanqi basin, but there is no significant correlation with other salt cations. Furthermore, we found a significant correlation between the amount of salt

Table 5. Factors matrix of soil salt properties from Yanqi basin.

Soil salt properties	Principal components			
	PC 1	PC 2	PC 3	PC 4
K^+	−0.18	0.65	0.37	0.56
Mg^{2+}	0.37	0.80	−0.07	0.14
Ca^{2+}	0.63	0.74	0.50	0.45
CO_3^{2-}	−0.05	−0.02	0.59	−0.10
SO_4^{2-}	0.61	0.54	0.42	0.17
pH	0.49	0.15	0.06	−0.79
Cl^-	0.91	−0.04	0.29	0.13
Na^+	0.68	−0.08	0.56	0.31
EC	0.46	−0.01	−0.73	0.13
HCO_3^-	−0.09	−0.09	0.29	0.37
Total salt	0.21	0.78	0.14	0.62
Percentage of variance (%)	33.75	28.54	18.25	15.18
Percentage of cumulative variance (%)	33.75	62.29	80.54	95.72

anions and SO_4^{2-} of 0.94 (P<0.01), and the correlation coefficients between the total amount of salt anions and Cl^- or CO_3^{2-} are 0.55 and 0.49 (P<0.05), respectively. This indicates that SO_4^{2-} was the primary salt anion, Cl^- was the secondary and CO_3^{2-} was the tertiary. Together, the correlation coefficients between the total amount of salts and Cl^-, and the total amount of salts and SO_4^{2-} are 0.57 and 0.92, respectively (P<0.01), indicating that the main types of salt in the soil were sulfate and chloride. This study also found that there are close correlation coefficients between the EC and Na^+, the EC and Cl^-, and the EC and SO_4^{2-} content, where the coefficients are 0.98, 0.66, and 0.87, respectively (P<0.01). Additionally, the correlation coefficient between the soil EC and CO_3^{2-} is 0.48, which is also significant (P<0.05). Overall, a significant correlation was found between pH and EC, and total amount of salts (TS) and salt anions (Na^+, Ca^{2+}, Ca^{2+}, Ca^{2+}, Cl^-, SO_4^{2-}, HCO_3^-, and CO_3^{2-}) (P<0.01). Further analysis shows that the correlation coefficient between the pH and SO_4^{2-} is 0.67 and between pH and Cl^- is 0.58 (P<0.01), indicating the pH of the soil is primarily influenced by the SO_4^{2-} and Cl^- content.

Principal component analysis and clustering analysis. All of the elements studied were found to fall into four PCs (Table 5) with a cumulative variance of 95.72%. This is a reflection of their sources and main influences, particularly for the ten indicators of Cl^-, Na^+, EC, pH, K^+, Mg^{2+}, Ca^{2+}, HCO_3^-, CO_3^{2-}, total amount of salts, and SO_4^{2-}. Among these, the variance contribution rate of the first PC (Cl^-, Na^+, SO_4^{2-}, EC and pH) was 33.75% and the second PC (Ca^{2+}, K^+, Mg^{2+}, and total amount of salts) was 28.54%. The primary contribution was by agricultural development, in particular from herbicide application and acid salt fertilizer [54,55]. The variance contribution rates of the third (CO_3^-) and fourth PCs (HCO_3^{2-}) were at 18.25% and at 15.18%, respectively. Upon combining the sampling sites within their respective land use types, it was found that the samples that had a high content of salts were often taken from the desert, grassland and forest, and appear to have originated from natural sources. This analysis also showed there were larger loads of Ca^{2+} in the first PC, SO_4^{2-} in the second PC and K^+, total salt content in the fourth PC, indicating these elements were influenced by both artificial and natural sources. We used clustering analysis to see if the results are consistent with the results from the PC analysis

(Fig. 2). All the elements were classified into four categories: the first category consisted of Cl^-, Na^+, SO_4^{2-}, pH, and EC; the second of Mg^{2+}, Ca^{2+}, K^+, and the total amount of salts; the third of HCO_3; and the fourth of CO_3^{2-}.

Spatial distribution of soil salt ions in Yanqi basin. The spatial dependence of the soil salt properties was determined by semivariance analysis in order to quantitate the spatial variability. The parameters of the semivariogram included model type, nugget, sill and effective range. The nugget value (C_0) represents the random variation derived from measurement inaccuracy or variations in properties that cannot be detected in the sample range [54]. The sill value is the upper limit of the fitted semivariogram model [47]. The ratio of nugget to sill was a criterion to classify the spatial dependence of soil properties and it reflects the influence of regional factors (nature) and the role of the non-regional factors (human factors). The range of the semivariogram (A_0) represents the average distance through which the variable semivariance reaches its peak value. A small effective

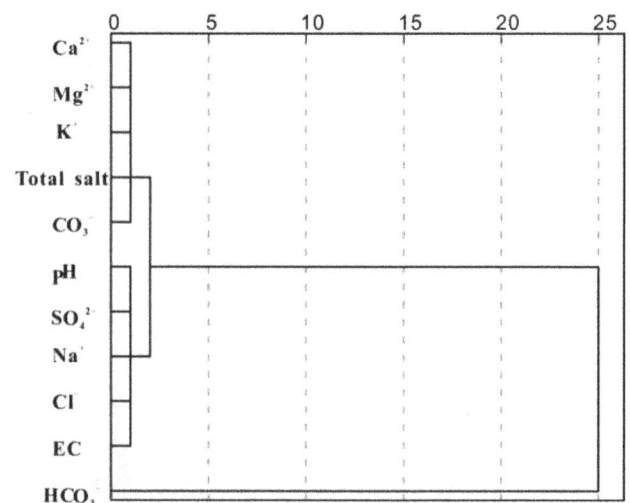

Figure 2. Clustering tree of soil salt properties of Yanqi basin.

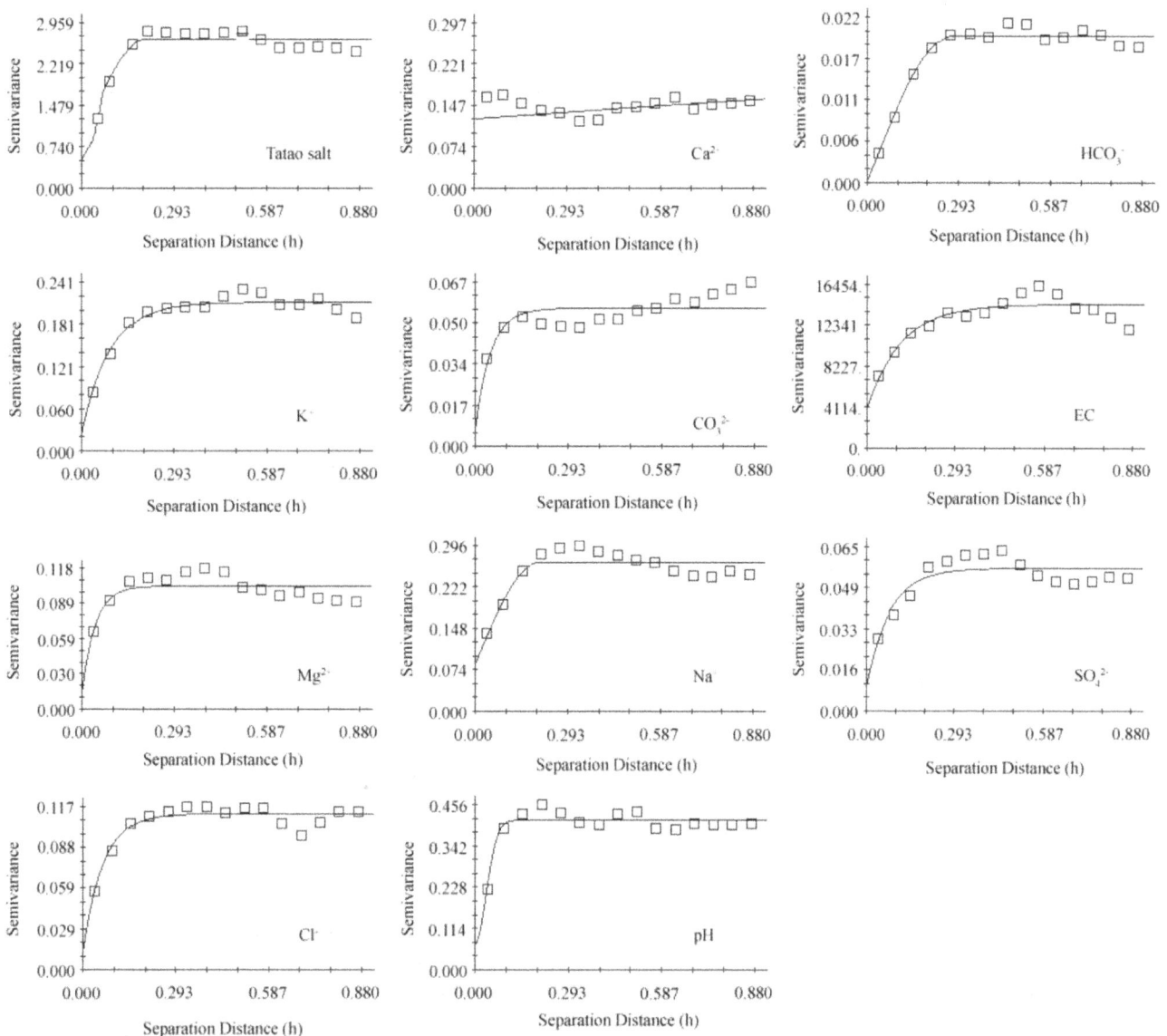

Figure 3. The semivariance function diagram of the soil salt properties of Yanqi basin.

range implies a distribution pattern composed of small patches. The cross-validation value is the coefficient of determination (R^2) of the correlation between the measured values and the cross-validation values, which were predicted based on the semivariogram and neighbor values [53]. Previous studies have shown that the semivariogram often differs considerably from its regional counterpart [33,53]. Fig. 3 and Table 6 show the tested variables over the study area that were modeled using spherical, gaussian and exponential semivariograms with the lower nugget effect based on the coefficients of determination (R^2) and residual sum of square (RSS). In this respect, a low ratio (less than 25% as was found for pH, Ca^{2+}, Mg^{2+}, and HCO_3^-) means that a large part of the variance is introduced spatially. This implies a strong spatial dependency of the variable, most likely due to intrinsic factors, including soil formation factors, such as soil parent materials, topography, and/or climate. A high ratio (more than 75% was found with Na^+ and Cl^-) often indicates a weak spatial dependency in the present sampling that was most likely due to extrinsic factors, including contamination, irrigation, and soil management

practices, where the fertilization and the soil chemical properties were continuously shifting. In this research the Nug/Sill ratios of CO_3^{2-}, and total salt content are 67.27 and 67.53% in the range between 25% and 75%, indicating the total salt content of the soil in Yanqi basin was influenced by the above factors. Additionally, all the other variables have a moderate spatial dependency for both intrinsic and extrinsic factors. The effective range calculated for variograms of different soil salt properties was 50–280 m, indicating that the sample distance was adequate for the characterization of the spatial variability of these properties.

The main application of geostatistics in soil science has been estimating and mapping the chemical properties in the soil of unsampled areas. Maps for each of the soil properties can be obtained using an ordinary kriging interpolation based on the best-fit semivariogram model. The skewness, defined as more than +1 or less than -1, indicates that some soil properties were not normally distributed. For these properties, it is difficult to estimate the semivariogram and doing so would result in a high value of kriging standard deviations [53]. Lognormal kriging with non-

Table 6. The spatial variation parameters of the soil salt properties of Yanqi basin.

Soil salt properties	Model	Nugget (Co)	Sill (Co+C)	Nug/Sill ratios $C_0/(C_0+C)$ (%)	Range A_0 (m)	R^2	RSS
Na^+	Spherical	0.084	0.102	82.35	200	0.921	4.053E-03
K^+	Exponential	0.068	0.212	32.08	90	0.861	2.900E-03
Ca^{2+}	Exponential	0.034	0.217	15.67	101	0.954	1.131E-02
Mg^{2+}	Spherical	0.023	0.103	22.33	100	0.733	1.288E-03
Cl^-	Gaussian	0.087	0.111	78.38	100	0.891	4.104E-04
CO_3^{2-}	Exponential	0.037	0.055	67.27	50	0.914	4.014E-04
HCO_3^-	Spherical	0.004	0.021	19.05	280	0.927	2.446E-05
SO_4^{2-}	Exponential	0.012	0.041	29.27	80	0.823	1.826E-04
pH	Gaussian	0.071	0.415	17.11	50	0.872	5.215E-03
EC	Exponential	0.082	0.106	77.36	120	0.791	2.081E-05
Total salt	Spherical	0.497	0.836	59.45	180	0.833	1.47E-02

linear transformations is an alternative method for dealing with a data set with outliers or a non-normal distribution [31,33]. Since the concentration of the variables with the non-normal distribution had a lognormal distribution, their concentrations were log-transformed, resulting in more regular variograms. The kriging interpolations were performed on the log-concentrations and the estimated values were back-transformed by an exponential function. As seen in the results, total amounts of salt, pH, EC, and the concentrations of Na^+, K^+, Mg^{2+}, CO_3^{2-}, and Cl^- were relatively higher when from farmland and grassland. The spatial patterns of these variables had a significant geographical distribution with their primary occurrence in the western, north and central areas, which were higher than in other areas. This further proves the effect of the intrinsic factors of topography, soil forming factors and soil type, and extrinsic factors of soil management practices, such as fertilization, use of organic fertilizer and land management in farms, on the spatial distribution of soil chemical properties [33,53]. The soil salinity tended to increase from the margin to the center across the study area, where agricultural wells are denser. It is clear that substantial soil salinization has taken place in these areas due to the effects of land management, farming and climate conditions, such as rainfall and high evaporation. This shows that more attention should be paid to these areas to prevent future problems. Soil pH in the study area ranged from 7.95 to 8.55 in most parts of the Yanqi basin. This pH range falls into the middle level of values that meets the fundamental conditions for plant growth and fertility. The ECs of the soil in all sections were within the acceptable value of <4 dS/m as based on the soil quality standard given by Bao et al. [45]. Areas presently not affected by salinity, but near saline areas, are potential areas for the development of salinity in the future, especially if they are also low-lying. In this respect, it is important to take necessary precautions and implement proper land use plans and cultivation practices.

Assessment of soil salinization risk in Yanqi basin. Soil salinization occurs mainly in arid and semi-arid regions. It may arise due to climate, but is more likely to occur when irrigation practices alter the natural salt balance. Irrigation promotes soil salinization by raising the water table of the underlying aquifer, thus carrying salts upwards. Salts, unlike water, remain in the soil as evaporation and plant transpiration take place, thereby amplifying the salinity. The accumulation of salts in the surface and near-surface zones of soil is a major issue of environmental degeneration and is one of the main causes of low crop yields, loss of land and decreased production. The risk of soil salinization in the Yanqi basin is presented in Table 7. We observed that the whole area has a low salinization risk, as previously mentioned, mainly due to anthropogenic activities and climatic variation [39,40]. The overall salinization risks for farmland, grassland, forest, urban construction areas and desert were none, none, none, low, and moderate, respectively. The mean values of the salinization risk indicators, except the SAR in farmland and desert, are lower than the corresponding maximum limitations that are predicted for the risk (Table 7). This indicates that farmland and desert in the study area have a low salinization risk, but they have mean values of 15.2 and 32.7 for SAR, which showed a moderate grade of salinization (Table 1, Table 7). This indicates a potential risk for environment declination. This research showed the grassland and forest have no risk of salinization, while the areas of urban construction have a moderate risk of salinization. In terms of spatial distribution, the soil salinization risk and the soil salt ions present at higher concentrations in the study area were similar. Environment declination due to soil salinity within the study area almost all

Table 7. Classification of soil salinization risk taken at 0–20 cm depth within the investigated land use and land cover categories in Yanqi basin.

LUCC categories	EC	Total salt content	SAR	Levels of risk
Farmland	1.981	0.552	15.2	Low
Grassland	0.901	0.715	7.5	None
Forest	0.857	0.382	6.8	None
Urban construction areas	1.035	2.761	8.2	Low
Desert	1.024	3.657	32.7	Moderate

took place on the edge of a lake, pond or river. These areas are heavily affected by climate warming, which in turn results in an increase in the amount of evapotranspiration that exceeds 2438.9 mm/a and in the average annual rainfall <100 mm (the drought and waterlog) (Fig. 4). Apart from natural factors, the main driving factors that jointly determined how local dwellers changed the landscape pattern were land use policies, economic systems and population growth. Human activity increases salinization through excessive application of irrigation water without adequate drainage. In addition, the cultivation of grassland is another major cause of inland salinization [40,41]. On one side of the study area, plenty of grass landscape suffered damage, and, in some regions, the land salinization and desertification problem were serious enough to destroy the harmony between the material cycle and energy flow of the ecosystems [42]. On the other side of the study area, the habitat for wildlife was deteriorating, thus seriously threatening the biological diversity. Therefore, establishing and modifying policy, adjusting the irrigation system, improving drainage, dredging the surface water system, promoting circulation between surface water and underground water, and setting up wetland resource monitoring systems are necessary in order to restore damaged wetland and grassland [49,50,55].

Conclusions

(1) From all the soil samples taken in this study, it can be gathered that there are numerous salt properties that vary largely between different samples. This analysis shows that the main salt ions in the soil were K^+, Ca^{2+}, Na^+, Cl^-, Mg^{2+}, and SO_4^{2-}, which accounted for 84.41% of the total salt content of the soil samples. Conversely, the concentrations of HCO_3^- and CO_3^- were very low. Except for the high variation found for the amounts of Ca^{2+} and K^+ (191.67% and 226.25%, respectively), the other soil salt properties of Yanqi basin had moderate levels of variation (10%< CV<100%).

(2) From the analysis shown that the average values of Na^+, Mg^{2+}, SO_4^{2-}, Ca^{2+}, total amounts of salt and K^+ being higher in farmland than grassland, desert, and urban construction areas. This work also determined that the maximum average values of Na^+, Mg^{2+}, K^+, SO_4^{2-}, total amount of salts, and Cl^- were found in coarse crystalline rock weathered material, sandy shale of weathered material and lacustrine deposits. Within the five land use types examined, they were not obvious which classes of elements were enriched and the differences between these groups were small.

(3) PC analysis determined that PC1 (Cl^-, Na^+, SO_4^{2-}, EC, and pH) and PC2 (Ca^{2+}, K^+, Mg^{2+}, and total amount of salts) originated from artificial sources, while PC3 and PC4 (CO_3^- and HCO_3^{2-}) originated from natural sources. Together, this research

shows that Ca^{2+}, K^+, SO_4^2 and the total amount of salts were influenced by both artificial and natural sources. Clustering analysis is consistent with the results from the PC analysis.

(4) From the geo-statistical point of view, it can be speculated that pH and soil salt ions, such as Ca^{2+}, Mg^{2+} and HCO_3^-, had a strong spatial dependency. Meanwhile, Na^+ and Cl^- had only a weak spatial dependency, which was probably due to extrinsic factors, such as contamination, irrigation, and current soil management practices. We evaluated the EC, SAR and total salt content standard to reveal the risk of soil surface salinization. Soil salinization indicators suggest that the entire area had a low risk of salinization as mentioned previously, and this risk was mainly due to anthropogenic activities and climate variation. It is recommended that management of salinized land be preceded by an assessment of local factors and processes that may affect land composition.

Although the overall soil environment was healthy in the Yanqi basin, human activity, such as excessive groundwater pumping, have negatively impacted conditions by inducing soil salinization in the oasis. This matter deserves increased attention. This study can be considered an early warning of soil salinization and alkalization in the Yanqi basin. It can also provide a reference for environmental protection policies and for rational utilization of land resources in the arid region of Xinjiang, northwest China, as well as for other oases of arid regions in the world.

Acknowledgments

Many thanks to the members from the Institute of Geographical Science and Tourism of Xinjiang Normal University, Urumqi, China, and the

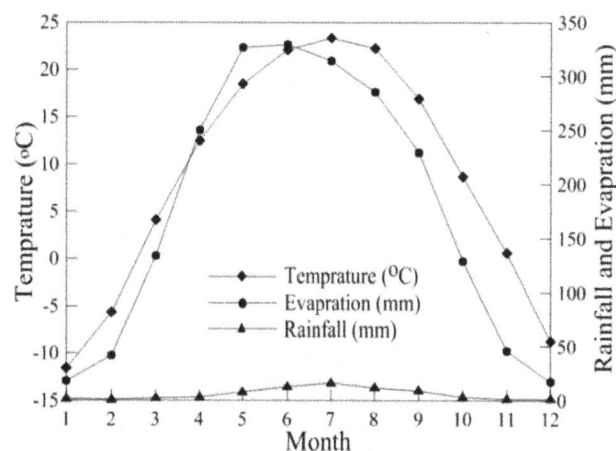

Figure 4. The monthly evaporation, precipitation and temperature of Yanqi basin in 2011.

members form College of Resources and Environment Sciences of Xinjiang University, Urumqi, China for the data collection and assistance in processing.

References

1. Yang YJ, Yang SJ, Liu GM, Yang XY (2005) Space-Time Variability and Prognosis of Soil Salinization. Pedoshpet 15(6):797–804.

2. Amezketa E (2006) An integrated methodology for assessing soil salinization, a pre-condition for land desertification. Journal of Arid Environments 67(4):594–606.

3. Masoud AA, Koike K (2006) Arid land salinization detected by remotely-sensed landcover changes: A case study in the Siwa region, NW Egypt. Journal of Arid Environments 66(1):151–167.

4. Yang JS (2008) Development and prospect of the research on salt-affected soils in China. Acta Pedologica Sinica 45(5):837-845. doi:10.3321/j.issn:0564-3929.2008.05.010

5. Stirzaker RJ, Cook FJ, Knight JH (1999) Where to plant trees on cropping land for control of dryland salinity: some approximate solutions. Agricultural Water Management 39(2):115–133.

6. Ghassemi F, Jakeman AJ, Nix HA (1995) Salinisation of land and water resources: human causes, extent, management and case studies. CAB international.

7. Chai S X, YangBZ, Wang XY, Wei L, Wang P, et al. (2008) Analysis of salinization of saline soil in west coast area of Bohai gulf. Rock and Soil Mechanics 29(5):1217–1221. (In Chinese).

8. Wang XL, Zhang FR, Wang YP, Feng T, Lian XJ, et al. (2013) Effect of irrigation and drainage engineering control on improvement of soil salinity in Tianjin. Transactions of the Chinese Society of Agricultural Engineering (20):82–88. (In Chinese).

9. Ren JG, Heng XL, Xi JM, Li JL (2005) Salinization characteristics of the soil in Yeerqiang river valley, Xinjiang. Soils 37 (6): 635–639. (In Chinese).

10. Chen XB, Yang JS, Liu CQ, Hu SJ (2007) Soil Salinization Under Integrated Agriculture and Its Countermeasures in Xinjiang. Soils 39(3):347–353. (In Chinese).

11. Wang YG, Xiao DN, Li Y (2008) Spatial and Temporal Dynamics of Oasis Soil Salinization in Upper and Middle Reaches of Sangonghe River, Northwest China. Journal of Desert Research 28(3):478–484. (In Chinese).

12. Jiang L, Li PC, Hu AY, Yi X (2009) Analysis and evaluation of soil salinization in oasis of arid region. Arid Land Geography 32(2):234–239. (In Chinese).

13. Sawut M, Eziz M, Tiyip T (2013) The effects of land-use change on ecosystem service value of desert oasis: a case study in Ugan-Kuqa River Delta Oasis, China. Canadian Journal of Soil Science 93(1): 99–108.

14. Ling H, Xu H, Fu J, Fan Z, Xu X (2013) Suitable oasis scale in a typical continental river basin in an arid region of China: A case study of the Manas River Basin. Quaternary International 286:116–125.

15. Eziz M, Yimit H, Mamat Z, Li JT (2013) Driving forces of farmland dynamics and its ecological effects in Keriya Oasis in recent 60 years. Agricultural Research in the Arid Areas 3:033. (In Chinese).

16. Triki I, Trabelsi N, Zairi M, Dhia HB (2014) Multivariate statistical and geostatistical techniques for assessing groundwater salinization in Sfax, a coastal region of eastern Tunisia. Desalination and Water Treatment 52(10-12):1980–1989.

17. Fu S, Wei CY (2013) Multivariate and spatial analysis of heavy metal sources and variations in a large old antimony mine, China. Journal of Soils and Sediments 13(1): 106–116.

18. Gil PM, Saavedra J, Schaffer B, Navarro R, Fuentealba C, et al. (2014) Quantifying effects of irrigation and soil water content on electrical potentials in grapevines (Vitis vinifera) using multivariate statistical methods.Scientia Horticulturae173:71–78.

19. Hu W, Shao MA, Wan L, Si BC (2014) Spatial variability of soil electrical conductivity in a small watershed on the Loess Plateau of China. Geoderma 230: 212–220.

20. Oliver MA, Webster R (2014) A tutorial guide to geostatistics: Computing and modelling variograms and kriging.Catena 113:56–69.

21. Emadi M, Baghernejad M (2014) Comparison of spatial interpolation techniques for mapping soil pH and salinity in agricultural coastal areas, northern Iran. Archives of Agronomy and Soil Science 60(9):1315–1327.

22. Elbasiouny H, Abowaly M, Abu-Alkheir A, Gad A (2014) Spatial variation of soil carbon and nitrogen pools by using ordinary Kriging method in an area of north Nile Delta, Egypt. Catena 113:70–78.

23. Wu Y, Wang Y, Xie X (2014) Spatial occurrence and geochemistry of soil salinity in Datong basin, northern China. Journal of Soils and Sediments 1–11.

24. Li SJ, Sun YN, Wang HB, Chen ZW (2013) Impact of Soil Nutrient Contents and Spatial Variability in Different Sampling Schemes under the Conservation Tillage.Advanced Materials Research 718:316–320.

25. Bilgili AV (2013) Spatial assessment of soil salinity in the Harran Plain using multiple kriging techniques. Environmental monitoring and assessment 185(1):777–795.

26. Sylla M, Stein A, Van Breemen N, Fresco LO (1995) Spatial variability of soil salinity at different scales in the mangrove rice agro-ecosystem in West Africa. Agriculture, ecosystems & environment 54(1), 1–15.

27. Ammari TG, Tahhan R, Abubaker S, Al-Zu'Bi Y, Tahboub A, et al. (2013) Soil salinity changes in the Jordan Valley potentially threaten sustainable irrigated agriculture.Pedosphere 23(3):376–384.

28. Walter C, McBratney AB, Douaoui A, Minasny B (2001) Spatial prediction of topsoil salinity in the Chelif Valley, Algeria, using local ordinary kriging with local variograms versus whole-area variogram. Soil Research 39(2), 259–272.

29. Jordán MM, Navarro-Pedreno J, García-Sánchez E, Mateu J, Juan P (2004) Spatial dynamics of soil salinity under arid and semi-arid conditions: geological and environmental implications. Environmental Geology 45(4): 448–456.

30. Bai YL, Li BG, Hu KL (1999) Spatial variability of soil salt and its composing ions in salt-affected soil in Huang-Huai-Hai plain. Soil and fertilizer (3): 22–26. (In Chinese).

31. Hu Kl, Li BG, Chen DL (2001) Spatial variability of soil water and salt in field and estimating soil salt using CoKriging. Advances in water science 12(4): 460–466. (In Chinese).

32. Xu Y, Chen YX, Shi HB (2004) Scale effect of spatial variability of soil water-salt. Transactions of the Chinese Society of Agricultural Engineering 20(2): 1–5. (In Chinese).

33. Lin J, Anwar M, Dilbar S (2007) Investigation of the spatial variability of soil salts in saline soil in Xinjiang. Research of Soil and Water Conservation 14(6): 189–192.

34. Aragüés R, Medina ET, Clavería I, Martínez-Cob A, Faci J (2014) Regulated deficit irrigation, soil salinization and soil sodification in a table grape vineyard drip-irrigated with moderately saline waters. Agricultural Water Management 134, 84–93.

35. Bouksila F, Bahri A, Berndtsson R, Persson M, Rozema J, et al. (2013) Assessment of soil salinization risks under irrigation with brackish water in semiarid Tunisia. Environmental and Experimental Botany 92, 176–185.

36. Liang SY, Wu TN, Wu YS, Chou YC, Lee CH (2013) Assessment of Aquifer Salinization Beneath an Offshore Industrial Park Based on Solute Transport Calculation. Advanced Materials Research 779, 1285–1288.

37. Chen L, Feng Q (2013) Geostatistical analysis of temporal and spatial variations in groundwater levels and quality in the Minqin oasis, Northwest China.Environmental Earth Sciences 70(3):1367–1378.

38. Wang LC, Wu RW, Gao J (2014) Spatial coupling relationship between settlement and land and water resources–based on irrigation scale–A case study of Zhangye Oasis.Advanced Engineering and Technology 225.

39. Mamat Z, Yimit H, Eziz M, Ablimit A (2013) Analysis of the Ecology-Economy Coordination Degree in Yanqi Basin, Xinjiang, China. Asian Journal of Chemistry 25(16): 9034–9040.

40. Mamat Z, Yimit H, Eziz A (2014) Oasis land-use change and its effects on the eco-environment in Yanqi Basin, Xinjiang, China. Environment Monitoring and Assessment 186(1): 335–348.

41. Wu JS, Zhang YQ, Liu ZH, Peng J, He JF (2010) Land salnization monitorng with remote sensing on Yanqi County, Xin jiang. Arid Land Geography 32(7):251–257. (In Chinese).

42. IUSS Working Group, WRB (2006). World reference base for soil resources. World Soil Resources Report,103.

43. Hargreaves G H, Allen R G (2003) History and evaluation of hargreaves evapotranspiration equation. Journal of Irrigation and Drainage Engineering 129(1): 53–63.

44. Lu RK (2000) Soil and agricultural chemistry analysis. Beijing/China Agricultural.

45. Bao SD (2005) Soil Agricultural Chemistry Analysis. Beijing/China Agriculture Press. pp: 17–200.

46. Carter M R (1993) Soil sampling and methods of analysis. CRC Press.

47. Lark RM (2001) Geostatistics for environmental scientists.European Journal of Soil Science 52(3), 526–526.

48. Cressie N (1988) Spatial prediction and ordinary kriging. Mathematical Geology 20(4): 405–421.

49. Fan LQ, Yang JG, Xu X, Sun ZJ (2012) Salinity characteristics and correlation analysis of saline soil in irrigation area of Ningxia. Soil and Fertilizer Sciences in China 6: 003. (In Chinese).

50. Metternicht G, Zinck JA (1997) Spatial discrimination of salt-and sodium-affected soil surfaces. International Journal Remote Sensing 18(12): 2571–2586.

51. Agriculture Department of Xinjiang (1996) Soil Survey Office in Xinjiang. Xinjiang Soil. Beijing/Science press 458–464.

52. Wilding LP (1984) Spatial variability: Its documentation, accommodation and implication to soil surveys//MNielson D R, Bouma J. Soil Spatial Variability. Purdoc, Wageningen: 166–193.

Author Contributions

Conceived and designed the experiments: ZZ JA. Performed the experiments: ZZ JA HY. Analyzed the data: ZZ JA. Contributed reagents/materials/analysis tools: ZZ JA HY. Contributed to the writing of the manuscript: ZZ.

53. Fan XM, Liu GH, Liu HG (2014) Evaluating the spatial distribution of soil salinity in the Yellow river delta based on Kriging and Cokriging Methods. Resources Science 36(2):0321–0327. (In Chinese).
54. Wang SX, Dong XG, Liu YF (2009) Spatio-Temporal variation of subsurface hydrology and groundwater and salt evolution of the oasis area of Yanqi Basin in 50 Years recently. Geological Science and Technology Information 28(5):101–108. (In Chinese).
55. Li XG, Lai N, Chen SJ, Mamattursun E (2014) Spatial variability of soil salt based on geostatistics and GIS in the oasis of the lower reaches of Kaidu River: a case study on Yanqi County. Gco-graphy and Gco-information Science 30(1):105–109. (In Chinese).

Soil Microbial Substrate Properties and Microbial Community Responses under Irrigated Organic and Reduced-Tillage Crop and Forage Production Systems

Rajan Ghimire[1], Jay B. Norton[1]*, Peter D. Stahl[1], Urszula Norton[2]

1 Department of Ecosystem Science and Management, University of Wyoming, Laramie, Wyoming, United States of America, **2** Department of Plant Sciences, University of Wyoming, Laramie, Wyoming, United States of America

Abstract

Changes in soil microbiotic properties such as microbial biomass and community structure in response to alternative management systems are driven by microbial substrate quality and substrate utilization. We evaluated irrigated crop and forage production in two separate four-year experiments for differences in microbial substrate quality, microbial biomass and community structure, and microbial substrate utilization under conventional, organic, and reduced-tillage management systems. The six different management systems were imposed on fields previously under long-term, intensively tilled maize production. Soils under crop and forage production responded to conversion from monocropping to crop rotation, as well as to the three different management systems, but in different ways. Under crop production, four years of organic management resulted in the highest soil organic C (SOC) and microbial biomass concentrations, while under forage production, reduced-tillage management most effectively increased SOC and microbial biomass. There were significant increases in relative abundance of bacteria, fungi, and protozoa, with two- to 36-fold increases in biomarker phospholipid fatty acids (PLFAs). Under crop production, dissolved organic C (DOC) content was higher under organic management than under reduced-tillage and conventional management. Perennial legume crops and organic soil amendments in the organic crop rotation system apparently favored greater soil microbial substrate availability, as well as more microbial biomass compared with other management systems that had fewer legume crops in rotation and synthetic fertilizer applications. Among the forage production management systems with equivalent crop rotations, reduced-tillage management had higher microbial substrate availability and greater microbial biomass than other management systems. Combined crop rotation, tillage management, soil amendments, and legume crops in rotations considerably influenced soil microbiotic properties. More research will expand our understanding of combined effects of these alternatives on feedbacks between soil microbiotic properties and SOC accrual.

Editor: Jose Luis Balcazar, Catalan Institute for Water Research (ICRA), Spain

Funding: Agricultural Prosperity for Small and Medium-Sized Farms Competitive Grant no. 2009-55618-05097 from the USDA National Institute of Food and Agriculture. The funders had no role in study design, data collection and analysis, decision to publish, or preparation of the manuscript.

Competing Interests: The authors have declared that no competing interests exist.

* Email: jnorton4@uwyo.edu

Introduction

Many changes in soil properties after conversion from one agricultural management system to another result from changes in soil microbiotic properties, defined here as the quality of microbial substrate and its effects on soil microbial communities [1,2]. It is well known that management practices such as reduced-tillage, cover crops, and crop diversification increase soil microbial activity in general, and microbial biomass and diversity in particular [2]. Similarly, practices used in certified-organic food and feed production, including amendments and legume crops in rotations, support increased microbial biomass [1,4], arbuscular mycorrhizal fungi (AMF) [5], and soil fauna [4]. It is not as clear how beneficial these management practices are under marginally productive conditions of cold, semiarid agroecosystems. In the study reported here, we evaluated whole-system effects on soil microbiotic properties after conversion from irrigated maize monoculture to conventional, reduced-tillage, and organic crop rotation systems in the central High Plains region of North America. Each management system combines a different suite of practices, including cultivation methods, crop rotations, and soil amendments.

Soil microbiotic properties are influenced by soil amendments, crop rotations, and tillage practices by different mechanisms. Organic amendments contribute diverse microbial substrates as heterogeneous organic materials in different states of decomposition [1,4], while crop rotations diversify the supply of plant residues, including fine roots, root exudates, sloughed off tissues, and rhizodeposited materials, which drive diversification of soil microbial communities [2,6,7]. Intensive tillage drives pulses of microbial activity that mineralize labile soil organic matter (SOM) and shift microbiotic properties toward C-limited conditions that favor bacteria and reduce SOM concentrations, while reduced-tillage conserves labile substrates and creates a more consistent soil environment for microbial activity [2,3,8,9]. In reduced-tillage systems, plant- and root-derived residues provide nucleation sites for fungal and bacterial growth, which further colonize soil particles to form aggregates and increase aggregate-protected, labile SOM and efficiency of substrate utilization (less C respired per unit of microbial biomass) [10,11]. Perennial legume and non-legume forage crops in rotations further reduce soil disturbance compared with annual crops, and stimulate SOM accrual and microbial activity through increases in root biomass and residues [11–13]. Combinations of organic amendments and perennial legumes in rotations, which are common practices in certified organic crop and forage production, support more efficient soil N utilization than conventional, synthetic fertilizer-based management [4,14] and can shift microbiotic properties toward N-limited conditions that favor fungi and accrue SOM.

Such management systems may be especially important in the central High Plains agroecosystem, where the semiarid environment, with inherently low SOM, cold winters, hot, dry summers, and irrigation-driven wetting-drying cycles, exacerbate mineralizing microbiotic conditions that drive losses of SOM [15–18]. Improved understanding of how reduced-tillage and organic crop and forage production systems affect soil microbial substrate quality, microbial biomass, community structure, substrate utilization, and soil organic C (SOC) sequestration in this cold and dry agroecosystem will help to design more sustainable systems during a time of uncertainty due to the changing climate, increasing operation costs, and changing markets [24–26].

The aim of this study was to evaluate SOC, DOC, C:N ratios of microbial substrates, soil microbial biomass and community structure, and substrate utilization after transition from monocropped corn to crop rotations under conventional, organic, and reduced-tillage crop and forage production. The experiments were set up on inherently low fertility, irrigated soils in the dry and cold central High Plains agroecosystem. We hypothesized that crop rotations developed in the previously monocropped field would increase microbial biomass and microbial community diversity by increasing the quantity and changing the quality of microbial substrates. In addition, organic and reduced-tillage management systems would favor greater increases in soil microbial biomass and more diverse microbial communities with higher substrate utilization efficiency compared with conventional management.

Materials and Methods

Experimental Site

The four-year study was established in 2009 at the University of Wyoming Sustainable Agriculture Research and Extension Center (SAREC) near Lingle, Wyoming (42°7'15.03"N; 104°23'13.46"W). The study area has cool temperatures and a short growing season with an average frost-free period of about 125 days and 60-year average maximum and minimum temperature of 17.8°C and 0.06°C, respectively, and precipitation of 332 mm [19]. In addition, maximum and minimum air temperature and precipitation were monitored at the SAREC weather station within 1 km of the research plots during the study period. Monthly average maximum and minimum air temperature and monthly total precipitation throughout the study period are presented in Figure S1. Soil at the study site is mapped as Mitchell loam (loamy, mixed, active, mesic Ustic Torriorthent) with low SOM content (<1%), and slightly alkaline soil pH [20]. Soil texture of the study site was loamy with sand, silt and clay content of 41.0 (13.5)%, 41.4 (10.5)% and 17.6 (4.0)%, respectively (standard deviation in parentheses; n = 24).

Experimental Design and Treatments

The study was designed as two independent randomized complete block experiments (row-crop production and forage production) laid out on a 15-ha half-circle under an irrigation pivot (305-m radius) that was divided into four wedge-shaped blocks (replications) (Figure S2). Each block was further separated into six plots consisting of three 0.405-ha crop production plots (outer three circles) and three 0.81-ha forage production plots (inner three circles). The three management-system treatments (conventional, certified organic, and reduced-tillage) were then randomly assigned to the crop and forage production plots. Before establishment of the experiment the entire area was under conventionally managed corn for at least six years.

All treatments were managed under four-year rotations starting in 2009. Table 1 shows the specific rotations, which were determined by a project advisory committee consisting of local producers and the SAREC management team. Under the conventional system inputs are applied as needed to maximize production, namely commercial synthetic fertilizer based on soil-test recommendations to supply nutrients, and chemical pesticides to control weeds, insects, and diseases. Specific management details are provided in Table S1. Conventional plots were moldboard ploughed, disked, and harrowed, which typically incorporates crop residues into soils leaving <15% of the soil surface covered by residues. The reduced-tillage system used conservation tillage that does not invert surface soil and leaves > 15% residue cover on the soil surface. In the organic system, tillage was done as in conventional plots, and pest control and nutrient management were based on practices allowed by the USDA National Organic Program standards (http://www.ams.usda.gov/AMSv1.0/nop). Conventional and reduced-tillage systems had chemical weed and pest control.

For soil fertility management, conventional and reduced-tillage plots received chemical fertilizer based on soil-test recommendations (Table S1). Organic management received composted cattle manure (dry matter 78% and C:N:P:S = 24.6:0.88:0.22:0.25%) in both crop and forage system in 2010. Because of the limited availability of composted cattle manure in 2011 and 2012, the organic crop system received raw manure (dry matter 29.2% and C:N:P:S = 21.3:1.42:0.35:0.40%) and the organic forage system received composted manure.

In the crop production experiment, management systems had different crops in rotation (Table 1). In forage system plots, a legume-grasses mixture was planted at 22 kg ha^{-1} in all plots in 2009, and included 50% alfalfa (*Medicago sativa* L.), 30% orchard grass (*Dactylis glomerata* L.), 10% meadow brome (*Bromus riparius* Rehmann), and 10% oat (*Avena sativa* L.) by weight. The forage production system plots were winter grazed for three months during 2011–2012 at stocking density of 1.6 fall-weaned calves ha^{-1}.

Soil Sampling

Soil samples were collected during spring, early summer, late summer, and fall seasons of the first (Year 1; 2009) and the fourth year (Year 4; 2012) from each of the 24 plots. During each sampling event, soil cores (3.2-cm diameter) were collected from 0–15 cm at 16 sampling points along a 50-m transect set in each plot, composited, thoroughly homogenized, subsampled (~500 g), and placed on ice for transport to the laboratory. The 0–15 cm depth was considered to be sufficient because the focus was on near-surface microbial properties. Sampling transects were mapped using GPS (Trimble GeoXT, Sunnyvale, CA) to locate transects for subsequent sampling. In the laboratory, soil samples were stored at −20°C for PLFA analysis and at 4°C for DOC, TDN, and potential soil respiration. Phospholipid fatty acid contents in soil were analyzed within two weeks of soil sample collection. Soil bulk density was measured in a separate set of 2.1×15 cm cores collected from 8 sampling points along the 50-m

Table 1. Crop rotations and management practices under different conventional (CV), organic (OR), and reduced-tillage (RT) management systems for crop and forage production (see Table S2 for detailed dates and management activities).

System		year	Crop in rotation	Management practices
Crop	CV	2009	Pinto bean	Tillage with moldboard plow and disk (5–7 passes each year), use of chemical fertilizers based on soil test recommendation for each crop, pesticides application as needed, and no livestock grazing.
		2010	Corn	
		2011	Sugar beet	
		2012	Corn	
	OR	2009	Alfalfa	Tillage with moldboard plow and disk (5–7 passes each year) and use of USDA-NOP certified practices for soil fertility (organic manure application) and pest management (e.g., cultivation), and no livestock grazing.
		2010	Alfalfa	
		2011	Corn	
		2012	Pinto bean	
	RT	2009	Pinto bean	Reduced-tillage (1–2 tillage passes each year that leave >15% crop residue on surface), use of chemical fertilizers based on soil test recommendation, pesticides application as needed, and no livestock grazing.
		2010	Corn	
		2011	Sugar beet	
		2012	Corn	
Forage	CV	2009	Alfalfa/grasses	Conventional tillage (5–7 passes in year 1 and 4), use of chemical fertilizers based on soil test recommendation, pesticides application as needed, and grazing with fall weaned calves during winter 2011/12.
		2010	Alfalfa/grasses	
		2011	Alfalfa/grasses	
		2012	Corn	
	OR	2009	Alfalfa/grasses	Conventional tillage and use of USDA certified practices for soil fertility (compost application) and pest management (no pesticides), and grazing with fall weaned calves during winter 2011/12.
		2010	Alfalfa/grasses	
		2011	Alfalfa/grasses	
		2012	Corn	
	RT	2009	Alfalfa/grasses	Reduced-tillage in the first year and no-tillage after, use of chemical fertilizers based on soil test recommendation, pesticides application as needed, and grazing with fall weaned calves during winter 2011/12.
		2010	Alfalfa/grasses	
		2011	Alfalfa/grasses	
		2012	Corn	

transects. Soil samples from the first and the last sampling dates were analyzed for other soil properties described below.

Laboratory Analysis

Total soil C and N were analyzed by dry combustion (EA1100 Soil C/N analyzer, Carlo Erba Instruments, Milan, Italy), inorganic C by modified pressure-calcimeter [21], and soil moisture by the gravimetric method [22]. Soil organic C was determined by subtracting inorganic C from total soil C. Soil pH was measured in a 1:1 soil:water mixture using an electrode [23]. Soil texture was determined by the hydrometer method [24]. Microbial substrate quality was determined as the ratio of DOC to total dissolved N (TDN) present in soils expressed as the C:N ratio of microbial substrate. For this, 10 g of field-moist soil was extracted with 50 ml of 0.5 M K_2SO_4 and amounts of DOC and TDN were determined by 720°C combustion catalytic oxidation/chemiluminescence with a Schimadzu TOC Analyzer (TOC-VCPH with TNM-1, Schimadzu Scientific Instruments, Inc.) coupled with TOC-Control V Ver.2 analysis software. Dissolved inorganic C was removed by automatic acidification and sparging

within the instrument. Potential soil respiration was determined as the amount of CO_2-C mineralized during a two-week incubation period [25]. Soil bulk density was determined by the core method [26] and water filled soil pore space was calculated from bulk density and gravimetric moisture content [27].

Microbial biomass and community structure was analyzed by the Blight and Dyre [28] method of fatty acid methyl ester (FAME) analysis as modified by Frostegård et al. [29] and Buyer et al. [30]. Fatty acids were directly extracted from soil samples using a 1:2:0.8 chloroform:methanol:phosphate buffer mixture (0.15 M, pH 4.0), and PLFAs were separated from neutral and glycolipid fatty acids in a solid-phase extraction column (Agilent Technologies Inc.). The PLFAs were methylated using a mild methanoic KOH, and the FAMEs were analyzed using an Agilent 6890 gas chromatograph with autosampler, split-splitless injector (7683B series), and flame ionization detector (Agilent Technologies Inc.). The system was controlled with Agilent Chemstation and MIDI Sherlock software, and the fatty acid peaks were identified using the MIDI peak identification software (MIDI, Inc., Newark, DE, USA). All solvents and chemicals used were of analytical grade, and all glassware used was rinsed 10 times with deionized water, and sterilized overnight in 450°C in a Blue M lab heat box type muffle furnace (Blue M Electric, Richardson, TX). The PLFA signatures of 16 different fatty acids, which were quantified in almost all the field plots, were used to study soil microbial community structure and these fatty acids were grouped into gram positive, gram negative and other bacteria, AMF and other fungi, and protozoa (Table S3). In addition, the Shannon diversity index [31] was calculated as an index of soil microbial diversity as influenced by management systems in crop and forage production. The ratio of potential soil respiration to total PLFA microbial biomass was also calculated as an index of microbial substrate utilization.

Statistical Analysis

Crop and forage production experiments were each analyzed as separate randomized complete block designs (RCBD) with three management-system treatments (conventional, organic, reduced-tillage) and four replicates. The analysis of soil properties that were measured at the beginning and end of the study, such as SOC, STN, pH and EC, were analyzed as split plot in time analysis of variance set in an RCBD for each system (p = 0.05). This analysis considered year as a repeated observation and replication as a random term in the model. Soil properties measured four times each year, such as soil microbial PLFA contents, DOC, C:N ratio of microbial substrate, potential soil respiration, water filled pore space, and soil bulk density, were analyzed as a split plot in time analysis of variance that considered season and year as repeated observation terms in the model. Statistical computations for both designs were facilitated by the mixed model (Proc Mixed) procedure of the Statistical Analysis System (SAS, ver. 9.3, SAS Institute, Cary, NC). Means were separated using the PDIFF test in the LSMEANS procedure (p = 0.05) unless otherwise stated. There were no significant season×management system interactions for either system in the three way split plot in time analysis of variance, therefore, results are reported as average of all four seasons within a year. In addition, PLFA data for individual microbial groups were normalized to the total microbial PLFAs and the data (mole percent of total PLFAs) were reanalyzed through a multivariate method (principal component analysis) to compare shifts in microbial community structure. Relationships between soil microbial substrate properties, microbial biomass and community structure, and substrate utilization were analyzed using Pearson correlation. Principle component and Pearson correlation analyses were performed using a Minitab V.16.0 (Minitab Inc., State College, PA, USA) and the first two principal components are graphed to summarize the results.

Results

Monthly average maximum and minimum temperatures during growing seasons (May to September) of 2009–2012 varied from year to year (Figure S1). The average minimum temperature was lowest in December 2009 and February 2011. Average precipitation was the lowest in 2012, followed by 2009 and 2010, compared to that in 2011. The amount of irrigation water depended on crop demand and the amount of precipitation received, and more water was applied to meet the crop water requirement in 2012 than in 2009–2011. All plots were irrigated to 60% of field capacity. Water filled porosity was consistent across management systems, seasons, and study years in both production systems (data not presented).

Soil pH was consistent across management systems and study years (range 7.3–7.8) under both crop and forage production, as was SOC concentration (Table 2). Soil organic C concentrations

were, however, significantly influenced by a management system× year interaction. Soils under reduced-tillage (p = 0.034) and organic (p = 0.004) crop production had significantly more SOC than soils under conventional crop production. In addition, soils under organic crop production in the fourth year had significantly more SOC than in the first year (p = 0.02). Soils under reduced-tillage forage production had significantly more SOC than those under conventional forage production (p<0.01). Soil total N concentrations were not significantly influenced by management systems, years, or management system×year interactions under either crop or forage production. Soil bulk density was not significantly influenced by management system, season, or year, but was significantly influenced by a management system×year interaction under crop production. Specifically, soil bulk density was significantly higher under reduced-tillage than organic crop production in the fourth year (p = 0.007).

Soil microbial biomass concentrations were significantly influenced by a management system×year interaction, but not by season. Specifically, in the fourth year under crop production there was significantly more soil microbial biomass under organic than conventional and (p<0.001) and reduced-tillage management (p = 0.01) (Fig. 1a). In the fourth year under forage production there was significantly more microbial biomass in soils under reduced-tillage than conventional (p = 0.002) and organic management (p = 0.047) (Fig. 1b). Under crop production, the greatest increase in total microbial biomass over the four-year period was observed in soils under organic management (353%) followed by conventional (262%) and reduced-tillage (202%) (based on year-one values). Under forage production, the increase in total microbial biomass concentrations were statistically similar at 396, 378 and 361% higher in the fourth year than in the first year in soils under conventional, organic, and reduced-tillage management systems, respectively.

We also observed increases in soil bacterial PLFAs, fungal biomarker PLFAs, DOC, and TDN across all treatments (only DOC data presented in Table 2), but the increases differed in magnitude. Dissolved organic C per unit SOC was 0.28–0.56% in the first year and 1.49–2.04% in the fourth year, with highest amount of DOC per unit SOC under conventional management. These changes corresponded with significantly higher fungal to bacterial ratios (F:B ratios) (Figure 2) and C:N ratios of microbial substrates (Figure 3) in the fourth year than in the first year. In addition, C:N ratios of microbial substrates were greater in soils under organic forage production than under conventional and reduced-tillage forage production. Similarly, microbial substrate utilization (potential soil respiration per unit PLFA) was consistent across management systems (Figure 4), but was 85–90% lower under crop production and 61–77% lower under forage production in the fourth year than in the first year.

Principal component analysis of microbial community structure revealed that the first two principle components explained 64.9% and 25.4% of the total sample variance under the crop production, and 74.3% and 21.0% of total sample variance under the forage production (Figure 5). The soil samples collected in the first year clustered on the left side of the figures 5.a1 and b1, and those collected in the fourth year clustered on the right side, corresponding to the increase in microbial substrate quality and decrease in potential soil respiration. There was greater variance in microbial community data collected in the first year than in the fourth year under both crop and forage production. In addition, loading scores for management systems separated more clearly along PC2 in the fourth year than in the first year. Among microbial groups, protozoa, other bacteria, and other fungi had positive loadings, while AMF and gram-negative bacteria had

Table 2. Soil properties as influenced by conventional (CV), organic (OR), and reduced-tillage (RT) management systems for crop and forage production one and four years after transition from continuous conventional corn.

System		SOC‡			STN			DOC			Db		
		Y1	Y4	Δ%	Y1	Y4	Δ%	Y1	Y4	Δ%	Y1	Y4	Δ%
		g kg⁻¹ soil			g kg⁻¹ soil			mg kg⁻¹ soil			g cm⁻³		
Crop	CV	6.70aA(1.31)	5.83bA(0.98)	−13.0	0.65(0.08)	0.73(0.12)	+12.3	19.5aB(2.08)	119cA(4.46)	+510	1.46aA(0.02)	1.43abA(0.02)	−2.05
	OR	6.52aB(0.16)	8.95aA(0.37)	+37.3	0.79(0.05)	0.90(0.08)	+13.3	36.4aB(5.45)	154aA(4.43)	+323	1.48aA(0.02)	1.40bA(0.02)	−5.41
	RT	7.31aA(0.86)	8.00aA(0.52)	+9.44	0.72(0.04)	0.80(0.05)	+11.8	20.6aB(3.78)	137bA(8.49)	+565	1.42aA(0.02)	1.48aA(0.03)	+4.23
Forage	CV	8.96aA(1.36)	7.23bA(0.46)	−19.3	0.77(0.09)	0.81(0.08)	+5.19	38.6aB(3.85)	139aA(13.3)	+360	1.47aA(0.03)	1.44aA(0.02)	−2.04
	OR	9.26aA(0.76)	8.50abA(0.35)	−8.31	0.77(0.02)	0.89(0.07)	+15.5	26.5aB(5.73)	147aA(7.36)	+455	1.47aA(0.02)	1.42aA(0.03)	−3.40
	RT	8.13aA(0.70)	10.2aA(1.04)	+25.5	0.86(0.14)	1.04(0.09)	+20.9	27.7aB(3.31)	152aA(2.67)	+449	1.39bA(0.02)	1.42aA(0.02)	+2.16

†Number in parenthesis indicates standard error.
‡SOC = soil organic carbon, STN = soil total nitrogen, DOC = dissolved organic carbon, Db = soil bulk density (g cm⁻³). Different lowercase letters within a column indicate significant difference among management systems within a year and different uppercase letters indicate significant difference among years within a management system in crop as well as forage production (p = 0.05). No letter within a column or row indicates no significant management system×year difference.

negative loadings along the PC1 axis (Figure 5.a2 and b2). Along the PC2 axis, gram-positive bacteria under crop production, and both gram-positive bacteria and AMF under forage production, had positive loadings. Across management systems and crop and forage production, gram-positive bacteria, gram-negative bacteria, and AMF together constituted of 93% of total soil microbial biomass in the first year and 76–78% in the fourth year (Table 3). After four years under alternative management systems, biomarker PLFAs for these three microbial groups had increased 2–4 fold, while biomarker PLFAs for other bacteria, fungi, and protozoa had increased up to 26, 9, and 36 fold, respectively, corresponding with significant shifts in microbial community structure. Changes in microbial community structure over the four-year study period are also indicated by Shannon's diversity index in Table 3, which was significantly higher in the fourth year than in the first year across both crop and forage production and all three management systems.

Increases in microbial biomass and F:B ratios, along with other changes in microbial community structure, along PC1 were strongly positively correlated with substrate availability (DOC concentrations) and quality (C:N ratio of microbial substrate) (Table 4) under both crop and forage production. Microbial substrate utilization decreased significantly with increasing substrate availability, increasing C:N ratios of microbial substrates, and increasing microbial biomass. Similarly, microbial community changes along PC2 were not related with substrate properties and substrate utilization.

Discussion

Our results support our hypotheses and indicate that conversion from continuous corn to crop rotations positively impacted soil microbiotic properties across all three management systems, with higher substrate availability, substrate C:N ratios, and soil microbial biomass contents, but lower substrate utilization in the fourth year than in the first year following transition (Tables 2 and 4; Figures 1a, 3a, and 4a). Both reduced-tillage and organic management systems added to these effects.

Under organic crop production, combined effects of perennial legumes in the rotations, which eliminated tillage for two of the four years, with additions of manure and compost, apparently offset losses of microbial substrates due to heavy tillage during the annual crop phases and supported the highest year-four microbial biomass concentrations (Table 2). Inclusion of legumes in rotations and organic amendments typically favor microbial growth, SOC and N accumulation, and diversification of microbial substrates [6,7,33]. Under reduced-tillage crop production, a more consistent soil environment apparently facilitated higher soil microbial biomass concentrations and diversity, as well as higher fungal productivity than under conventional management (Table 3; Figure 1a). Similar effects of reduced disturbance have been noted [2,9,34] in which fungal hyphae improve soil aggregation, which protects labile SOM components and regulates microbial substrate utilization [3,12,13,33].

Under forage production, reduced-tillage management had the highest year-four microbial biomass concentrations of the three systems, probably due to lack of plowing with conversion from perennial forage to corn. Organic management, with applications of composted manure, created significantly higher C:N ratios of microbial substrates in the fourth year than under conventional and reduced-tillage with chemical fertilizer application (Figure 3b). The fact that this difference in substrate quality did not occur under the crop production experiment (Figure 3a) suggests that it resulted from a combined effect of the alfalfa-grasses mixture and

a.

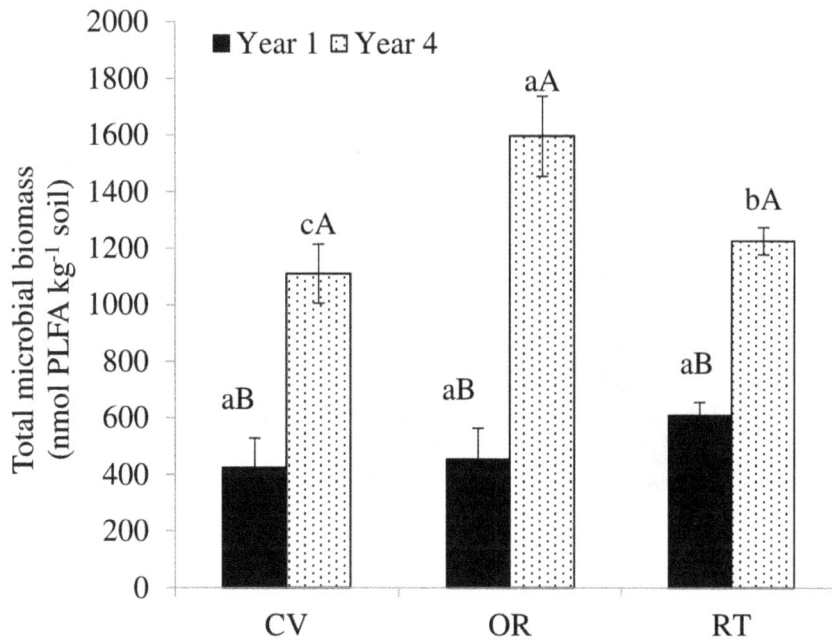

b.

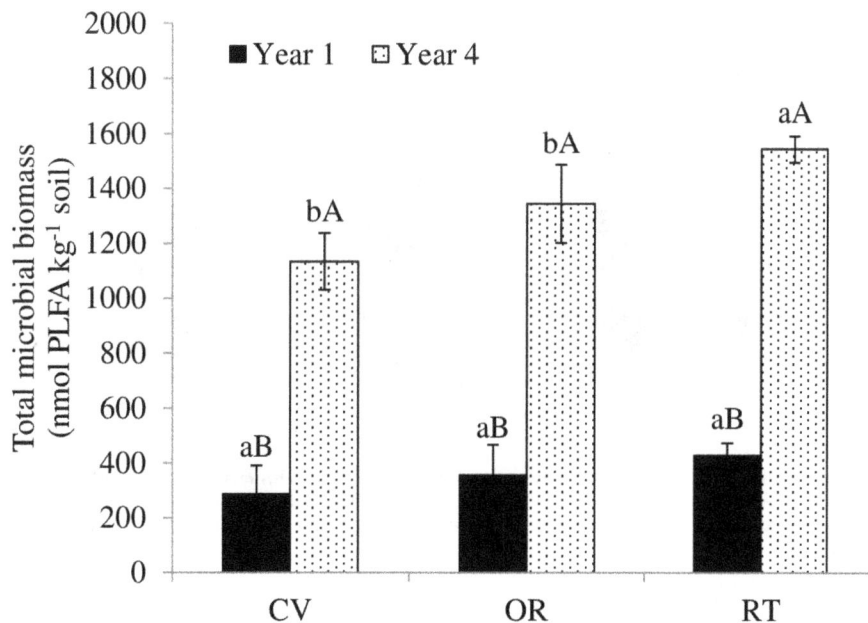

Figure 1. Total soil microbial biomass as influenced by conventional (CV), organic (OR), and reduced-tillage (RT) management systems for crop (a) and forage (b) production in the first and fourth years. Different lowercase letters indicate significant differences among management systems within a year and different uppercase letters indicate significant difference among years within a management system (p = 0.05).

compost applications. In the conventional and reduced-input forage systems, N from chemical fertilizer and alfalfa may have contributed to lower substrate C:N ratios than under the organic forage system.

Changes in soil microbiotic properties we observed in response to alternative management systems are consistent with results of previous studies, but greater in magnitude. Previous studies have reported two- to three-fold increases in microbial biomass with

a.

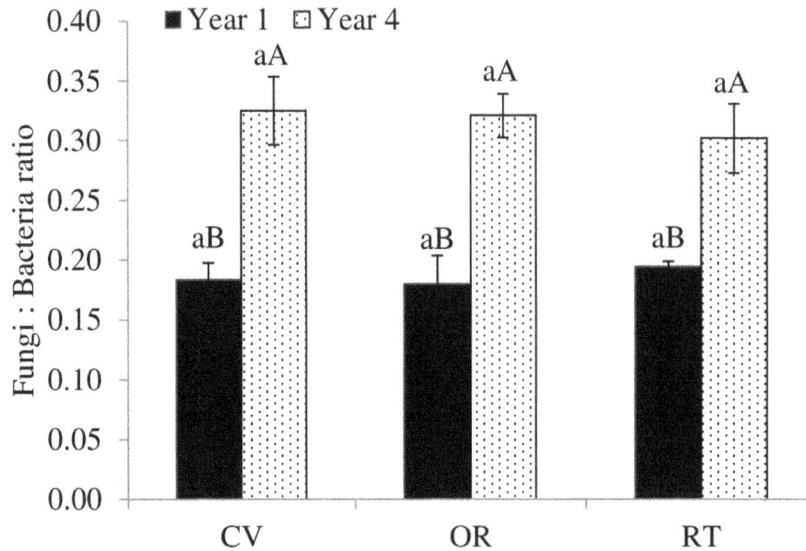

b.

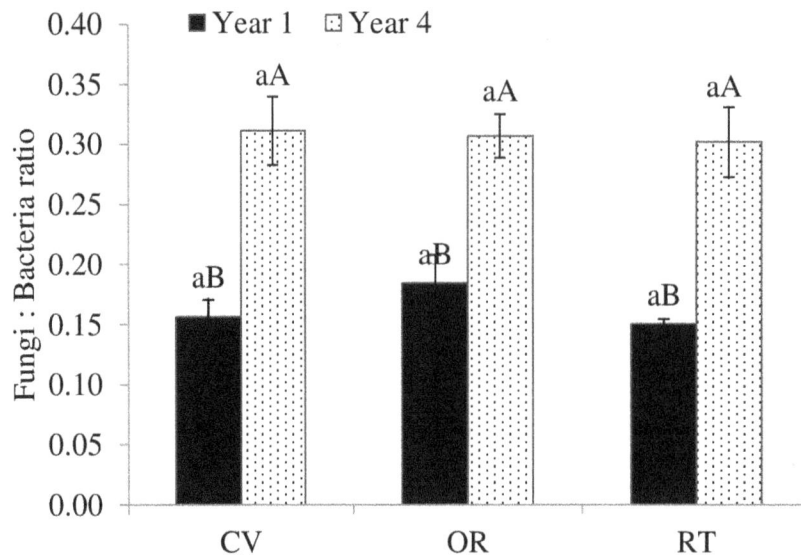

Figure 2. Fungal-to-Bacterial ratio as influenced by conventional (CV), organic (OR), and reduced-tillage (RT) management systems for crop (a) and forage (b) production in the first and fourth years. Different lowercase letters indicate significant differences among management systems within a year and different uppercase letters indicate significant difference among years within a management system (p = 0.05).

diversified rotations or reduced disturbance in place for several years [2,8,32,35]. The greater magnitudes we observed probably resulted from combined effects of transition from continuous corn to rotations with legumes, manure applications, and reduced-tillage on depleted, inherently-low-fertility soils. While variable precipitation during the seasons of the study may have influenced comparisons between years 1 and 4, air temperatures (Figure S1) and soil water filled pore space were consistent among management systems across years under both crop and forage production. Therefore, we believe that the changes in management, rather

than annual climatic variability, drove the observed changes in soil microbiotic properties.

The changes in quantity and quality of microbial substrates during the study period drove notable shifts in microbial community structure, including greater increases in fungal relative to bacterial PLFAs (Table 3; Figure 2), in gram-positive relative to gram-negative bacterial PLFAs, and in saprophytic fungi and protozoa relative to other groups. Higher DOC and C:N ratios of microbial substrates at the end of the study drove greater increases in saprophytic fungi, which rely on carbonaceous substrates, than AMF, which are often associated with more mineral-rich, low C:N

a.

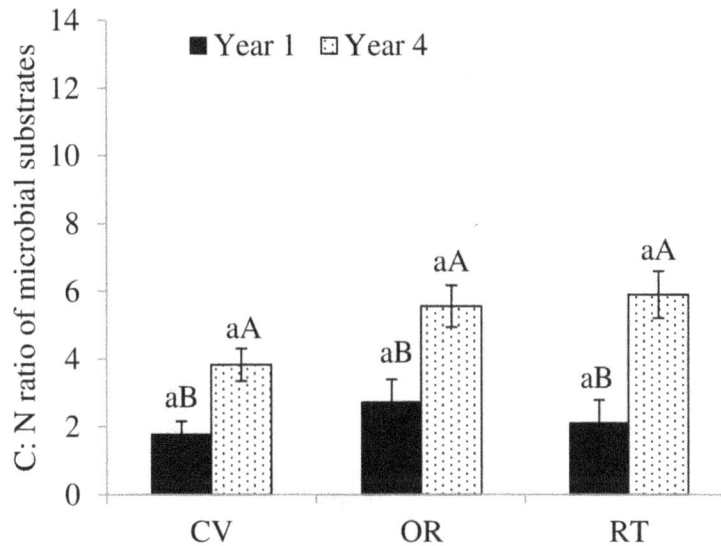

b.

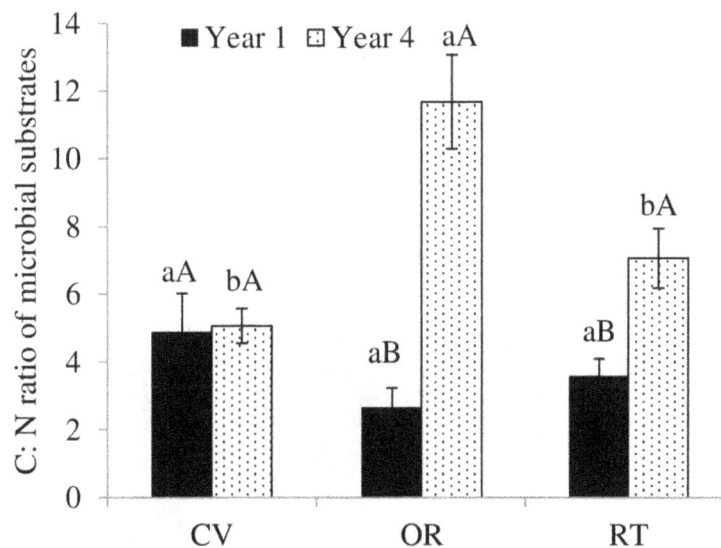

Figure 3. Carbon-to-nitrogen ratio of microbial substrate as influenced by conventional (CV), organic (OR), and reduced-tillage (RT) management systems for crop (a) and forage (b) production in the first and fourth years. Different lowercase letters indicate significant differences among management systems within a year and different uppercase letters indicate significant difference among years within a management system (p = 0.05).

substrates [36–38]. The observed increases in the non-mycorrhizal fungi strongly correlated with increased substrate availability, indicating changes in C:N ratios favored more fungal growth (saprotrophic fungi) than bacterial growth.

Increases in gram-positive relative to gram-negative bacteria are also often associated with increases in diversity of C sources in soils and decreases in mechanical soil disturbance, [36–39]. Gram-positive bacteria are associated with low substrate availability (high C:N) environments, while gram-negative bacteria dominate soils with more easily decomposable substrates [38]. Decreases in relative abundance of gram-negative bacteria may be beneficial

because many plant pathogenic microorganisms such as *Pseudomonas* and *Xanthomonas* species are gram negative [40,41]. Large increases in protozoa parallel the increases soil bacteria, which are their food source [42].

Greater amounts of higher C:N-ratio microbial substrates, more diverse communities of microorganisms with higher F:B ratios, and reduced potential soil respiration in the fourth year across all three management systems in general, and under reduced-tillage forage and organic crop systems in particular, suggests that minimum soil disturbance, application of organic amendments, and more legume crops in rotation increase soil microbial biomass,

a.

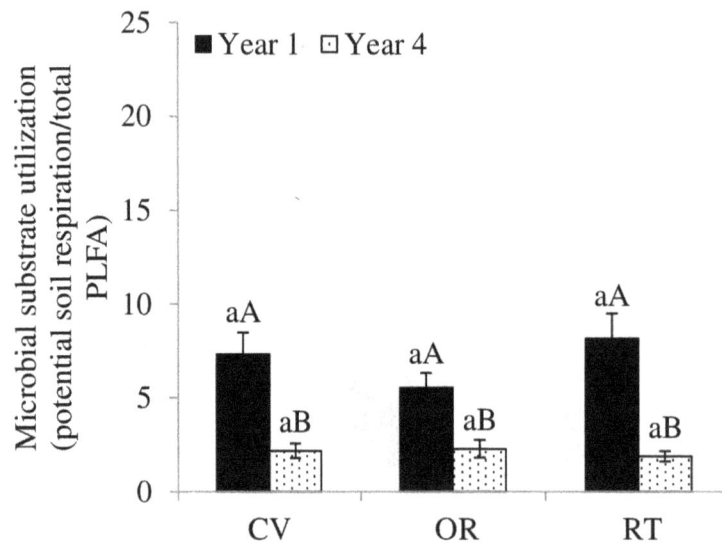

b.

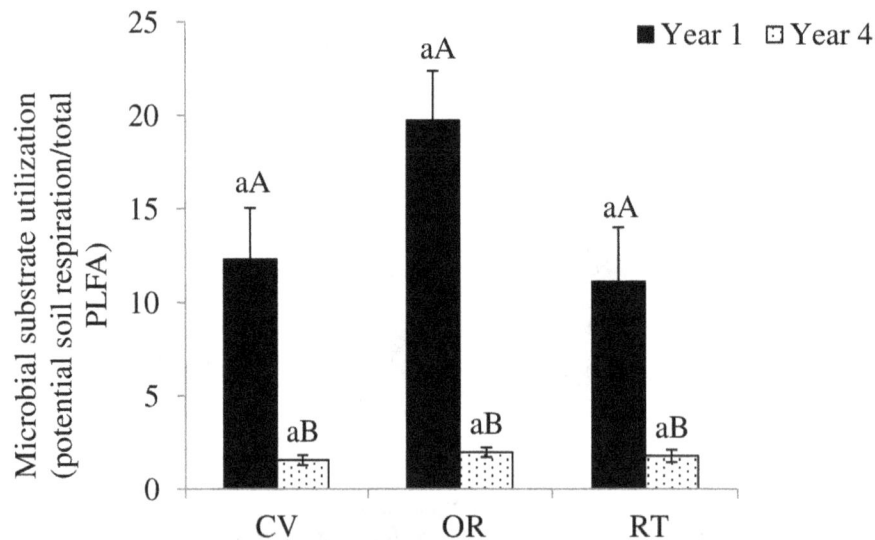

Figure 4. Microbial substrate utilization as influenced by conventional (CV), organic (OR), and reduced-tillage (RT) management systems for crop (a) and forage (b) production in the first and fourth years. Different lowercase letters indicate significant differences among management systems within a year and different uppercase letters indicate significant difference among years within a management system (p = 0.05).

alter community structure, and thereby influence SOC accrual (Figure 6). Although mechanisms of SOC regulation by specific groups of microorganisms are not well defined, given similar site characteristics, higher SOC sequestration potential is typically observed in soils with higher F:B ratios [43,44]. Higher SOC in fungal-dominated systems is mainly attributed to higher biomass C production per unit of C metabolized by fungus than by soil bacteria [44]. The changes in microbiotic properties we observed

indicate that substrate C was mainly transformed into microbial biomass or less labile SOM components, which may be reflected in year-four SOC contents.

Under crop production, both our organic system, with two years of alfalfa, and reduced-tillage system involved considerably less soil disturbance than our conventional system, and both had more year-4 SOC than the conventional system (Table 1, S1). Under forage production, with the same rotation across the three

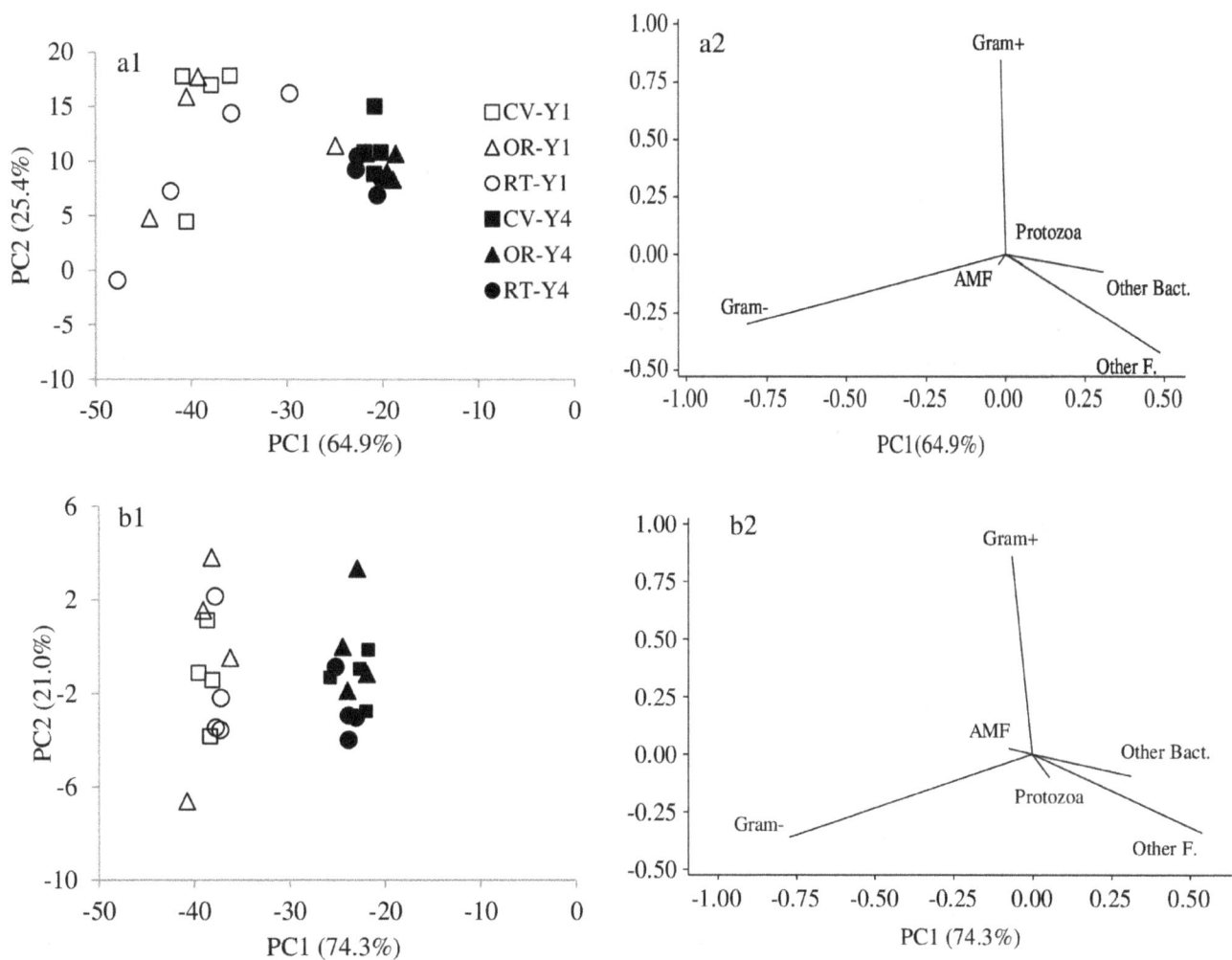

Figure 5. Score plots of the first two principle components (1) and loading of different microbial groups (2) as influenced by conventional (CV), organic (OR), and reduced-tillage (RT) management systems for crop (a) and forage (b) production. Gram+ = gram-positive bacteria, Gram− = gram-negative bacteria, AMF = arbuscular mycorrhizal Fungus, Other Bact. = other bacteria and Other F. = other fungus.

management systems, reduced tillage had significantly more year-4 SOC than the other two system, indicating that organic amendments combined with intensive tillage did not increase SOC. Taken together, these results suggest that reduced-tillage combined with legumes in rotation had the largest impacts on SOC accrual. Repeated tillage to plow down the grasses and alfalfa and establish corn in year 4 might have caused significant loss of SOC accrued during three years under forages in both organic and conventional systems.

Overall, increases in microbial substrate availability and microbial biomass over the 4-year study represent a small fraction of SOC reservoirs, even in this low-SOM environment. Therefore, longer-term evaluation of the effects of tillage, crop rotations, soil amendments, and legume integration in crop and forage production may further our understanding of the influence of microbiotic properties on SOC sequestration. Results of this cropping systems study bundle effects of reduced tillage, crop rotation, legumes in rotation, and soil fertility options into three management systems for crop and forage production. While overall effects are crucial to understanding how management alternatives affect system sustainability, evaluating individual

components will complement these results and contribute to design of best management practices for irrigated agriculture in cold, semiarid agroecosystems like the central High Plains.

Conclusions

In this study, management systems that included reduced tillage, perennial legumes, and organic amendments improved soil microbiotic properties that support SOC accrual. The greatest changes occurred with transition from continuous corn to crop rotation. The changes were enhanced under organic management in cash-crop production and reduced-tillage management under forage production. Under the different rotations of our crop production systems, more legume crops in rotation had greater influence on soil microbiotic properties than fewer legume crops. Under the same rotations of our forage production systems, reduced-tillage management had the greatest influences on soil microbiotic properties. These effects were driven by interactions between soil microbial community structure and microbial substrate quantity and quality that resulted in increases in fungal biomass that support SOC accrual. The results indicate that

Table 3. Soil microbial communities as influenced by conventional (CV), organic (OR), and reduced-tillage (RT) management systems for crop and forage production one and four years after transition from continuous conventional corn.

System		Gram positive bacteria[†]			Gram negative bacteria			Other Bacteria			AMF			Other Fungi			Protozoa			Shannon's diversity index		
		nmol kg⁻¹ soil			nmol kg⁻¹ soil			nmol kg⁻¹ soil			nmol kg⁻¹ soil			nmol kg⁻¹ soil			nmol kg⁻¹ soil					
		Y1	Y4	Δ%	Y1	Y4	Δ%	Y1	Y4	Δ%	Y1	Y4	Δ%	Y1	Y4	Δ%	Y1	Y4	Δ%	Y1	Y4	Δ%
Crop	CV[‡]	200aB (53.3)	414bA (43.7)	+107	160bB (33.8)	391cA (60.4)	+144	12.2aB (5.64)	60.3bA (11.7)	+394	28.7aB (6.21)	76.5bA (7.24)	+167	20.0bB (5.01)	152bA (15.8)	+665	3.30 (1.81)	14.4bA (2.70)	+336	3.50aB (0.25)	5.53aA (0.23)	+58.2
	OR	213aB (51.6)	555aA (48.8)	+161	166bB (34.4)	559aA (84.4)	+237	12.0aB (4.26)	105aA (18.9)	+775	28.8aB (7.00)	118aA (13.9)	+310	29.1aB (6.70)	230aA (16.1)	+201	3.02 (1.63)	28.7aA (2.17)	+850	3.55aB (0.29)	5.89aA (0.20)	+66.0
	RT	270aB (60.7)	467bA (54.1)	+72.9	236aB (45.9)	479bA (78.2)	+106	22.4aB (6.97)	70.9bA (13.5)	+217	44.3aB (8.88)	89.6bA (10.5)	+116	31.7aB (7.17)	173bA (23.0)	+147	4.38 (2.36)	21.4abA (3.28)	+754	3.80aB (0.27)	5.71aA (0.27)	+50.2
Forage	CV	124aB (21.5)	391bA (23.7)	+215	117aB (17.5)	408bA (55.1)	+249	2.30aB (1.61)	51.5aA (10.6)	+2139	24.9aB (3.50)	86.7bA (7.99)	+248	18.2aB (3.59)	176bA (15.8)	+867	0.00	21.1 (2.31)	-	3.85aB (0.22)	5.69aA (0.20)	+47.9
	OR	157aB (22.3)	474abA (55.3)	+197	142aB (18.8)	490bA (93.0)	+245	3.69aB (2.26)	70.7aA (11.5)	+1575	29.5aB (3.64)	111aA (22.7)	+276	23.0aB (3.77)	177aB (16.1)	+670	0.00	22.8 (3.87)		4.00aB (0.16)	5.64aA (0.22)	+41.0
	RT	180aB (24.2)	526aA (41.6)	+192	177aB (19.5)	566aA (72.1)	+219	3.11aB (2.19)	80.4aA (19.4)	+2485	36.9aB (4.22)	116aA (12.3)	+214	29.1aB (3.30)	218aB (23.0)	+649	1.13 (1.13)	37.1 (2.94)	+3597	4.19aB (0.04)	5.71aA (0.22)	+36.1

†Number in parenthesis indicates standard error.

‡AMF = Arbuscular Mycorrhizal Fungi. Different lowercase letters within a column indicate significant difference among management systems within a year and different uppercase letters indicate significant difference among years within a management system in crop as well as forage production (p = 0.05). No letter within a column or row indicates no significant management system × year difference.

Table 4. Significant correlation coefficients (r) between soil microbial substrate properties, microbial community and substrate utilization in crop and forage production systems.

	DOC[†‡]	C:N MAS	Microbial biomass	F:B ratio	PC1[§]
Crop system					
Substrate C:N	0.69(<0.001)	-			
Microbial biomass	0.83(<0.001)	0.86(<0.001)	-		
F:B ratio	0.83(<0.001)	0.81(<0.001)	0.91(<0.001)	-	
PC1	0.78(<0.001)	0.63(0.001)	0.83(<0.001)	0.91(<0.001)	-
Soil Resp.	−0.64(0.001)	−0.47(0.02)	−0.70(<0.001)	−0.74(<0.001)	−0.84(<0.001)
Forage system					
Substrate C:N	0.65(0.001)	-			
Microbial biomass	0.85(<0.001)	0.56(0.006)	-		
F:B ratio	0.91(<0.001)	0.64(0.001)	0.83(<0.001)	-	
PC1	0.93(<0.001)	0.67(0.001)	0.91(<0.001)	0.97(<0.001)	-
Soil Resp.	−0.79(<0.001)	−0.40(<0.001)	−0.87(<0.001)	−0.73(<0.001)	−0.84(<0.001)

[†]Number in parenthesis indicates Pearson correlation p values.
[‡]DOC = dissolved organic carbon, F:B ratio = fungi to bacteria ratio, PC1 = First principal component and Soil Resp. = potential soil respiration.
[§]PC1 explains shift in soil microbial community structure from the first to the fourth year.

reducing disturbance, including legumes, and applying organic amendments positively impact soil processes in ways that enhance sustainable productivity of inherently low-fertility soils in a cold, semiarid environment, even over a relatively short time period. Further research may confirm the combined effects of crop rotations and alternative management systems on SOC accrual

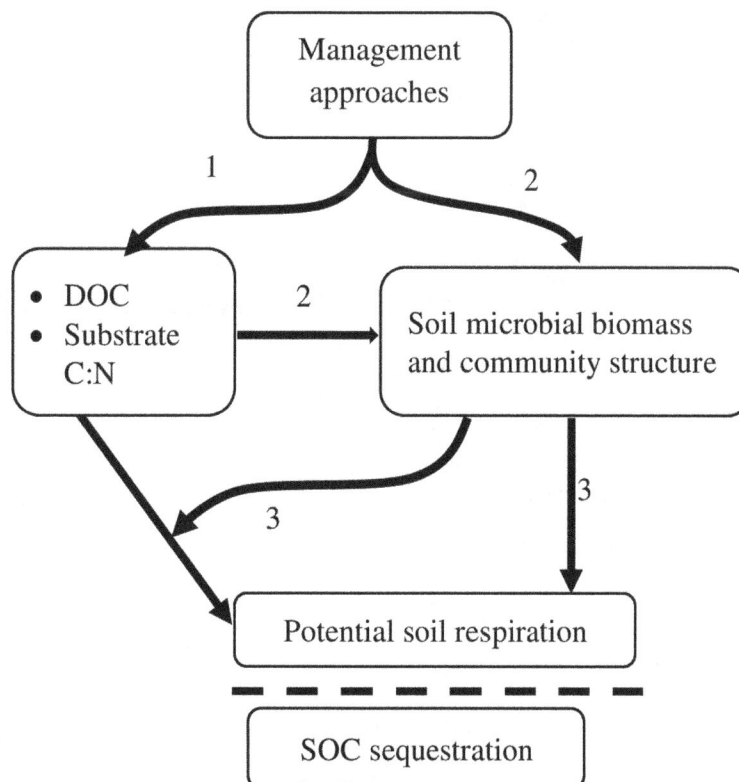

Figure 6. Conceptual framework illustrating influence of management system on microbial substrate properties, microbial communities, and SOC sequestration. Alternative management systems influence (1) microbial substrate availability and quality, (2) soil microbial biomass and community structure, and (3) soil respiration and SOC sequestration. DOC = dissolved organic carbon (microbial substrate).

and ecosystem services as influenced by soil microbial biomass and fungal productivity.

Supporting Information

Figure S1 Monthly average maximum and minimum temperature, and total precipitation at the SAREC weather station, Lingle, Wyoming (Apr. 2009–Sept. 2012).
(TIFF)

Figure S2 Experimental design of crop and forage production plots under alternative management systems (C, conventional; O, organic; R, reduced tillage). Tiers A, B, and C are 1-acre cash-crop plots; tiers D, E, and F are 2-acre forage plots.
(TIFF)

Table S1 Crop rotations under conventional (CV), organic (OR) and reduced-tillage (RT) management systems in crop and forage production.
(DOCX)

Table S2 Timing of management practices by under conventional (CV), organic (OR), and reduce-tillage (RT) management systems on crop and forage production (2009–2012).
(DOCX)

Table S3 Biomarker phospholipid fatty acids used for identifying taxonomic microbial groups.
(DOCX)

Acknowledgments

We thank Jenna Meeks and Caley Kristian Gasch for field and laboratory assistance, Professor David Legg for assistance with statistical analyses, and Prakriti Bista and Professor Elise Pendall for their helpful comments to improve earlier drafts of this manuscript.

Author Contributions

Conceived and designed the experiments: JBN RG UN PDS. Performed the experiments: RG JBN. Analyzed the data: RG JBN. Contributed reagents/materials/analysis tools: JBN PDS UN. Wrote the paper: RG JBN. Reviewed the manuscript: UN PDS.

References

1. Berthrong ST, Buckley DH, Drinkwater LE (2013) Agricultural management and labile carbon additions affect soil microbial community structure and interact with carbon and nitrogen cycling. Microbial Ecol 66: 158–170.
2. Acosta-Martinez V, Mikha MM, Vigil MF (2007) Microbial communities and enzyme activities in soils under alternative crop rotations compared to wheat-fallow for the Central Great Plains. Appl Soil Ecol 37: 41–52.
3. Stahl PD, Parkin TB, Christensen M (1999) Fungal presence in paired cultivated and undisturbed soils in central Iowa. Biol Fertil Soils 29: 92–97.
4. Drinkwater LE, Letourneau DK, Workneh F, Vanbruggen AHC, Shennan C (1995) Fundamental differences between conventional and organic tomato agroecosystems in California. Ecol Appl 5: 1098–1112.
5. Oehl F, Sieverding E, Ineichen K, Mader P, Boller T, et al. (2003) Impact of land use intensity on the species diversity of arbuscular mycorrhizal fungi in agroecosystems of Central Europe. Appl Environ Microbiol 69: 2816–2824.
6. Kandeler E, Tscherko D, Spiegel H (1999) Long-term monitoring of microbial biomass, N mineralisation and enzyme activities of a Chernozem under different tillage management. Biol Fertil Soils 28: 343–351.
7. Shi Y, Lalande R, Ziadi N, Sheng M, Hu Z (2012) An assessment of the soil microbial status after 17 years of tillage and mineral P fertilization management. Appl Soil Ecol 62: 14–23.
8. Acosta-Martínez V, Dowd SE, Bell CW, Lascano R, Booker JD, et al. (2010) Microbial community structure as affected by dryland cropping systems and tillage in a semiarid sandy soil. Diversity 2: 910–931.
9. Halvorson AD, Wienhold BJ, Black AL (2002) Tillage, nitrogen, and cropping system effects on soil carbon sequestration. Soil Sci Soc Am J 66: 906–912.
10. Blanco-Canqui H, Lal R (2007) Regional assessment of soil compaction and structural properties under no-tillage farming. Soil Sci Soc Am J 71: 1770–1778.
11. Ghimire R, Norton J, Pendall E (2014) Alfalfa-grass biomass, soil organic carbon, and total nitrogen under different management systems in an irrigated agroecosystem. Plant Soil: 374: 173–184.
12. Six J, Feller C, Denef K, Ogle SM, Sa MJC, et al. (2002) Soil organic matter, biota and aggregation in temperate and tropical soils: Effects of no-tillage. Agronomie 22: 755–775.
13. Liebig M, Carpenter-Boggs L, Johnson JMF, Wright S, Barbour N (2006) Cropping system effects on soil biological characteristics in the Great Plains. Renew Agric Food Syst 21: 36–48.
14. Jenkinson DS, Parry LC (1989) The nitrogen-cycle in the broadbalk wheat experiment - a model for the turnover of nitrogen through the soil microbial biomass. Soil Biol Biochem 21: 535–541.
15. Ghimire R, Norton JB, Norton U, Ritten JP, Stahl PD, et al. (2013) Long-term farming systems research in the central High Plains. Renew Agric Food Syst 28: 183–193.
16. Krall JM, Delaney RH, Taylor DT (1991) Survey of nonirrigated crop production practices and attitudes of Wyoming producers. J Agron Educ 20: 120–122.
17. Norton JB, Mukhwana EJ, Norton U (2012) Loss and recovery of soil organic carbon and nitrogen in a semiarid agroecosystem. Soil Sci Soc Am J 76: 505–514.
18. Frey SD, Elliott ET, Paustian K (1999) Bacterial and fungal abundance and biomass in conventional and no-tillage agroecosystems along two climatic gradients. Soil Biol Biochem 31: 573–585.
19. Western Regional Climate Center (2013) Historical climate information. Reno, NV: Desert Research Institute. http://www.wrcc.dri.edu/. Accessed: 06-15, 2013.
20. Soil Survey Staff (2013) Web Soil Survey. Natural Resources Conservation Service, United States Department of Agriculture. http://websoilsurvey.sc.egov.usda.gov/ Accessed: 06-15, 2013.
21. Sherrod LA, Dunn G, Peterson GA, Kolberg RL (2002) Inorganic carbon analysis by modified pressure-calcimeter method. Soil Sci Soc Am J 66: 299–305.
22. Gardner WH (1986) Water content. In: Klute A, editor. Methods of Soil Analysis Part 1: Physical and Mineralogical Methods. 2nd ed. Madison, WI: Agronomy Monograph 9. ASA, SSSA pp. 493–541.
23. Thomas GW (1996) Soil pH and soil acidity. In: Sparks DL, editor. Methods of Soil Analysis, part 3: Chemical Methods. Madison, WI: Agronomy Monograph 9. ASA, SSSA pp. 475–490.
24. Gee GW, Bauder JW (1986) Particle-size analysis. In: Klute A, editor. Methods of Soil Analysis Part 1: Physical and Mineralogical Methods. 2nd ed. Madison, WI: Agronomy Monograph 9. ASA, SSSA pp. 383–411.
25. Nie M, Pendall E, Bell C, Gasch CK, Raut S, et al. (2012) Positive climate feedbacks of soil microbial communities in a semi-arid grassland. Ecol Letters 16: 234–241.
26. Blake GR, Hartge KH (1986) Bulk density. In: Klute A, editor. Methods of Soil Analysis, part 1: Physical and Mineralogical Methods. 2nd ed Madison, WI. ASA, SSSA 363–375.
27. Linn DM, Doran JW (1984) Effect of water-filled pore-space on carbon-dioxide and nitrous-oxide production in tilled and nontilled soils. Soil Sci Soc Am J 48: 1267–1272.
28. Blight EG, Dyre WJ (1959) A rapid method of total lipid extraction and purification. Can J Biochem Physiol 37: 911–917.
29. Frostergard A, Tunlid A, Baath E (1991) Microbial biomass measured as total lipid phosphate in soils of different organic content. J Microbiol Methods 14: 151–163.
30. Buyer JS, Roberts DP, Russek-Cohen E (2002) Soil and plant effects on microbial community structure. Can J Microbiol 48: 955–964.
31. Shannon CE, Weaver W (1949) The Mathematical Theory of Communication. Urbana, IL: Univ. of Illinois Press. 132 p.
32. Minoshima H, Jackson LE, Cavagnaro TR, Sanchez Moreno S, Ferris H, et al. (2007) Soil food webs and carbon dynamics in response to conservation tillage in California. Soil Sci Soc Am J 71: 952–963.
33. Delate K, Cambardella CA (2004) Agroecosystem performance during transition to certified organic grain production. Agron J 96: 1288–1298.
34. Ngosong C, Jarosch M, Raupp J, Neumann E, Ruess L (2010) The impact of farming practice on soil microorganisms and arbuscular mycorrhizal fungi: Crop type versus long-term mineral and organic fertilization. Appl Soil Ecol 46: 134–142.
35. Reganold JP, Andrews PK, Reeve JR, Carpenter-Boggs L, Schadt CW, et al. (2010) Fruit and soil quality of organic and conventional strawberry agroecosystems. Plos One 5: e12346.
36. Fliessbach A, Oberholzer HR, Gunst L, Maeder P (2007) Soil organic matter and biological soil quality indicators after 21 years of organic and conventional farming. Agric Ecosyst Environ 118: 273–284.
37. Carpenter-Boggs L, Stahl PD, Lindstrom MJ, Schumacher TE, Barbour NW (2003) Soil microbial properties under permanent grass, conventional tillage, and no-till management in South Dakota. Soil Till Res 71: 15–23.
38. Fierer N, Schimel JP, Holden PA (2003) Variations in microbial community composition through two soil depth profiles. Soil Biol Biochem 167–176.
39. Schaad NW, Jones JB, Chun W, editors (2001) Laboratory guide for identification of plant pathogenic bacteria: APS press, Minnesota. 398 p.

40. De Vos P, Goor M, Gills M, De Ley J (1985) Ribosomal ribonucleic acid Cistron similarities of phytopathogenic Pseudomonas species. Int J Syst Evol Microbiol 35: 169–184.

41. Stout JD (1980) The role of Protozoa in nutrient cycling and energy flow. Adv Microbial Ecol 4: 1–50.

42. Jastrow JD, Amonette JE, Bailey VL (2007) Mechanisms controlling soil carbon turnover and their potential application for enhancing carbon sequestration. Clim Change 80: 5–23.

43. Bailey VL, Smith JL, Bolton Jr H (2002) Fungal:bacterial ratios in soils investigated for enhanced C sequestration. Soil Biol Biochem 34: 997–1007.

44. Strickland MS, Rousk J (2010) Considering fungal: bacterial dominance in soils - Methods, controls, and ecosystem implications. Soil Biol and Biochem 42: 1385–1395.

Soil Bacterial Community Response to Differences in Agricultural Management along with Seasonal Changes in a Mediterranean Region

Annamaria Bevivino[1]*, Patrizia Paganin[1], Giovanni Bacci[2,3], Alessandro Florio[2],

Maite Sampedro Pellicer[1], Maria Cristiana Papaleo[3], Alessio Mengoni[3], Luigi Ledda[4], Renato Fani[3],

Anna Benedetti[2], Claudia Dalmastri[1]

1 ENEA (Italian National Agency for New Technologies, Energy and Sustainable Economic Development) Casaccia Research Center, Technical Unit for Sustainable Development and Innovation of Agro-Industrial System, Rome, Italy, 2 Consiglio per la Ricerca e la Sperimentazione in Agricoltura - Research Centre for the Soil-Plant System, Rome, Italy, 3 Laboratory of Microbial and Molecular Evolution, Department of Biology, University of Florence, Florence, Italy, 4 Dipartimento di Agraria, University of Sassari, Sassari, Italy

Abstract

Land-use change is considered likely to be one of main drivers of biodiversity changes in grassland ecosystems. To gain insight into the impact of land use on the underlying soil bacterial communities, we aimed at determining the effects of agricultural management, along with seasonal variations, on soil bacterial community in a Mediterranean ecosystem where different land-use and plant cover types led to the creation of a soil and vegetation gradient. A set of soils subjected to different anthropogenic impact in a typical Mediterranean landscape, dominated by *Quercus suber* L., was examined in spring and autumn: a natural cork-oak forest, a pasture, a managed meadow, and two vineyards (ploughed and grass covered). Land uses affected the chemical and structural composition of the most stabilised fractions of soil organic matter and reduced soil C stocks and labile organic matter at both sampling season. A significant effect of land uses on bacterial community structure as well as an interaction effect between land uses and season was revealed by the EP index. Cluster analysis of culture-dependent DGGE patterns showed a different seasonal distribution of soil bacterial populations with subgroups associated to different land uses, in agreement with culture-independent T-RFLP results. Soils subjected to low human inputs (cork-oak forest and pasture) showed a more stable bacterial community than those with high human input (vineyards and managed meadow). Phylogenetic analysis revealed the predominance of *Proteobacteria*, *Actinobacteria*, *Bacteroidetes*, and *Firmicutes* phyla with differences in class composition across the site, suggesting that the microbial composition changes in response to land uses. Taken altogether, our data suggest that soil bacterial communities were seasonally distinct and exhibited compositional shifts that tracked with changes in land use and soil management. These findings may contribute to future searches for bacterial bio-indicators of soil health and sustainable productivity.

Editor: Jack Anthony Gilbert, Argonne National Laboratory, United States of America

Funding: This research was funded by MIUR (Integrated Special Fund for Research - FISR) in the frame of the Italian National Project SOILSINK "Climate change and agro-forestry systems, impacts on soil carbon sink and microbial diversity", and partially supported by MIUR (Research Department of Italian Government) in the framework of the Agreement Program ENEA-CNR (Articolo 2, comma 44, Legge 23.12.2009 n. 191 - Legge Finanziaria 2010). The funders had no role in study design, data collection and analysis, decision to publish, or preparation of the manuscript.

Competing Interests: The authors have declared that no competing interests exist.

* Email: annamaria.bevivino@enea.it

Introduction

Soil microorganisms play an important role as regulators of major biogeochemical cycles and can significantly affect the ecosystem functioning [1], being involved in organic matter dynamics, nutrient cycling and decomposition processes [2]. The anthropogenic activities affect the diversity of natural habitats modifying the number of species occurring in the environment at the landscape scale. Soil management strongly influences soil biodiversity in agricultural ecosystems. Different practices can alter the below-ground ecosystem, often leading to depletion of soil carbon and loss of biodiversity, and thus affecting the structure of

the resident microbial communities [3]. Therefore, characterizing genetic and functional diversity of soil bacterial communities in response to agricultural practices and/or climate is fundamental to better understand and manage the ecosystem processes.

The Mediterranean area is one of the most important biodiversity hotspots in the world and is increasingly threatened by intensive land use [4]. The high environmental diversity that characterizes the Mediterranean region is related to the integration of natural ecosystems and traditional human activities such as the agroforestry practices [5]. The collapse of the traditional agro-silvo-pastoral system that occurred during the past century has led to major changes in the extension of woodlands dominated by typical Mediterranean species, i.e. cork oak (*Quercus suber*) and/ or holm oak (*Quercus ilex*) woodlands [6], [7]. Research on the influence of management practices on the biodiversity of these agro-silvo-pastoral systems is increasing but it has focused mostly on plants [8] and vertebrates [5], [9]. In the frame of the Italian Project SOILSINK (Climatic changes and agricultural and forest systems: impact on C reservoirs and on soil microbial diversity), a hilly basin in Gallura (Berchidda site, Sardinia, Italy) was selected as a reference Mediterranean site for studying the influence of land-use changes on diversity, function and seasonal variations of soil microbial communities [10], [11]. The site is within an area of about 1,450 ha and is characterised by extensive agro-silvo-pastoral systems, typical of north-eastern Sardinia (Italy) and similar areas of the Mediterranean basin [12]. The chosen site represents a sustainable balance between human activities and natural resources that have created a landscape of high heterogeneity and cultural value, whose importance has been recognized at the European level [13], [14]. Indeed, it is considered climatically (Mediterranean zone) and pedologically homogeneous with vegetation patterns similar to those called *dehesas* or *montados* of the south-western Iberian Peninsula [15], [16], [17]. In the past, this area was covered by cork-oak forests, which gradually were subjected to increasing under-storey grazing and usage for the extraction of cork. Today, there are different land-use and plant cover types that lead to a soil and vegetation gradient with an ecological progression: from a cork-oak forest undergoing minimum disturbance to managed vineyards with an intensive agricultural practice (grass covered and ploughed), passing through areas with temporary grassland, and pasture.

The different land uses altered soil potential, making possible to discriminate the role of human management on soil functioning. When forests are converted to grasslands, and grasslands turned into agricultural lands, a sharp switch from one type of soil microbial community to another one occurs. Since the ability of an ecosystem to withstand serious disturbances may partly depend on its microbial component(s), characterizing bacterial community composition and/or structure might help to better understand and manipulate ecosystem processes. The aim of the present study was to investigate the effects of soil characteristics and different agricultural managements on soil bacterial community in two seasons. Sampling was carried out in spring and autumn 2007,

when the plant cover-growing season usually starts and ends. A combination of culture-based and molecular techniques along with statistical analysis of data obtained was applied to interrogate the diversity, function, and ecology of soil bacterial communities. The results obtained in the present study, along with the other ones obtained within the SOILSINK Project [11], [18], [19], [20], provide useful data on the impact of soil type, cover vegetation, and human activities on the distribution of the bacterial genetic resources in soil communities for this Mediterranean region. Our results confirm that the environments with low inputs (cork-oak forest and pasture) show a more stable soil microbial community than those subjected to increasing human input (vineyards and managed meadow) and suggest that soil bacterial communities are seasonally distinct with compositional shifts that track with changes in land use and soil management.

Materials and Methods

Ethics Statement

We carried out the study on the hilly basin in Gallura (Olbia-Tempio municipalities, Sardinia, Italy). Five soils uses were identified in private farms within Berchidda site (40°49′ 15″N, 9°17′ 32″ E) and were obtained. The soil sampling was carried out in the frame of a national research project (SOILSINK Project) and soils used in this study were collected under consent of the landowners. The responsible of the study site was Prof. P.P. Roggero (University of Sassari, Italy). We confirm that our study did not harm the environment and did not involve endangered or protected species. Specific geographic coordinates (referred to World Geodetic System, 1984) of our study area are reported in Table S1.

Sampling site

The study area (Olbia-Tempio) is representative of the climate, vegetation type and management of some of the most common agro-forestry systems in the Mediterranean basin [16].

The Berchidda site is made up of hydromorphic and granitic soil with a loamy sand texture. The altitude ranges from 275 m to 300 msl. This area is referred to as a meso-thermo Mediterranean, subhumid phytoclimatic belt with a mean annual rainfall of 862 mm and the mean annual temperature of 13.8°C [10]. In the

Figure 1. Long-term effects of different land-use with increasing level of intensification in spring and autumn. Both pasture and managed meadow included spotted cork oak trees, which are key components of the Dehesatype landscape typical of this area of Sardinia. The cork-oak formation, pasture, and managed meadow have been converted to the current use and maintained unchanged for more than 30 years, whereas the non-tilled cover cropped vineyard and the tilled one were planted in 1985 and 1994, respectively. From the left to right: cork-oak forest (CO), hayland pasture rotation (PA), managed meadow (MM), grass covered vineyard (CV), tilled vineyard (TV).

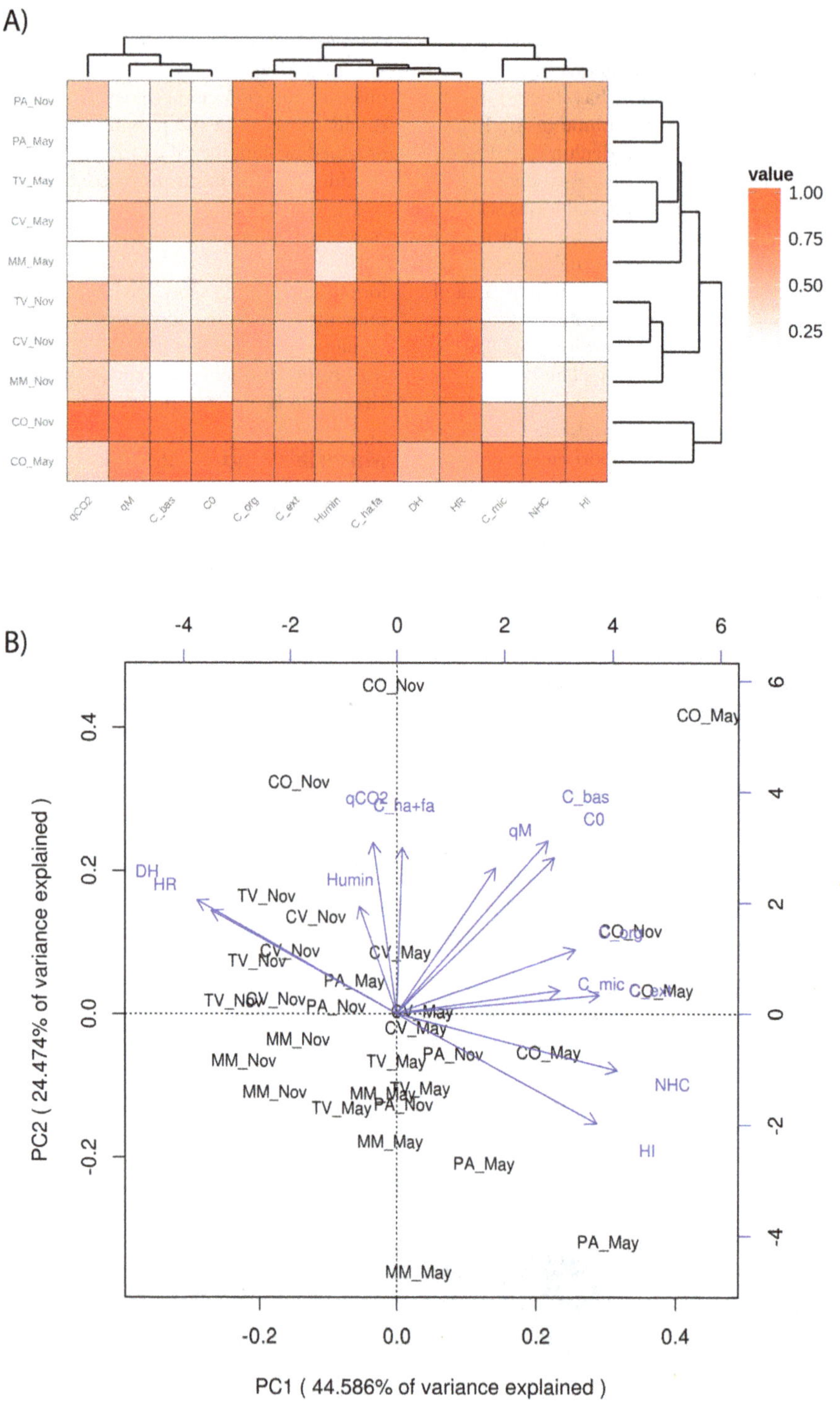

Figure 2. Effect of land-use and season on soil physical-chemical and biological parameters. A) Heat map with hierarchal clustering of physical-chemical and biological parameters across the five Sardinia soils with different land uses at the two different sampling time points (May and November). The heat map was constructed using a maximum-minimum normalization of the data in order to represent each value in a range between 0 and 1. Higher values are represented by darker colors whereas lower ones are represented by lighter colors. CO = cork-oak forest; PA = hayland-pasture rotation; MM = managed meadow; TV = tilled vineyard; CV = grass covered vineyard. B) PCA ordination of data (axes 1 and 2) generated from physical-chemical and biological properties of the different types of land use in May and November.

past, the Berchidda area was covered by cork-oak forests (dominated by *Quercus Suber* L.), which were subjected to intense usage for the extraction of cork. Today, there are five main different land-use units close together: cork-oak forest (CO),

hayland-pasture rotation (PA), managed meadow (MM), and tilled (TV) and grass covered vineyard (CV). The cork-oak formation, pasture, and managed meadow have been converted to the current use and maintained unchanged for more than 30 years,

A)

B)

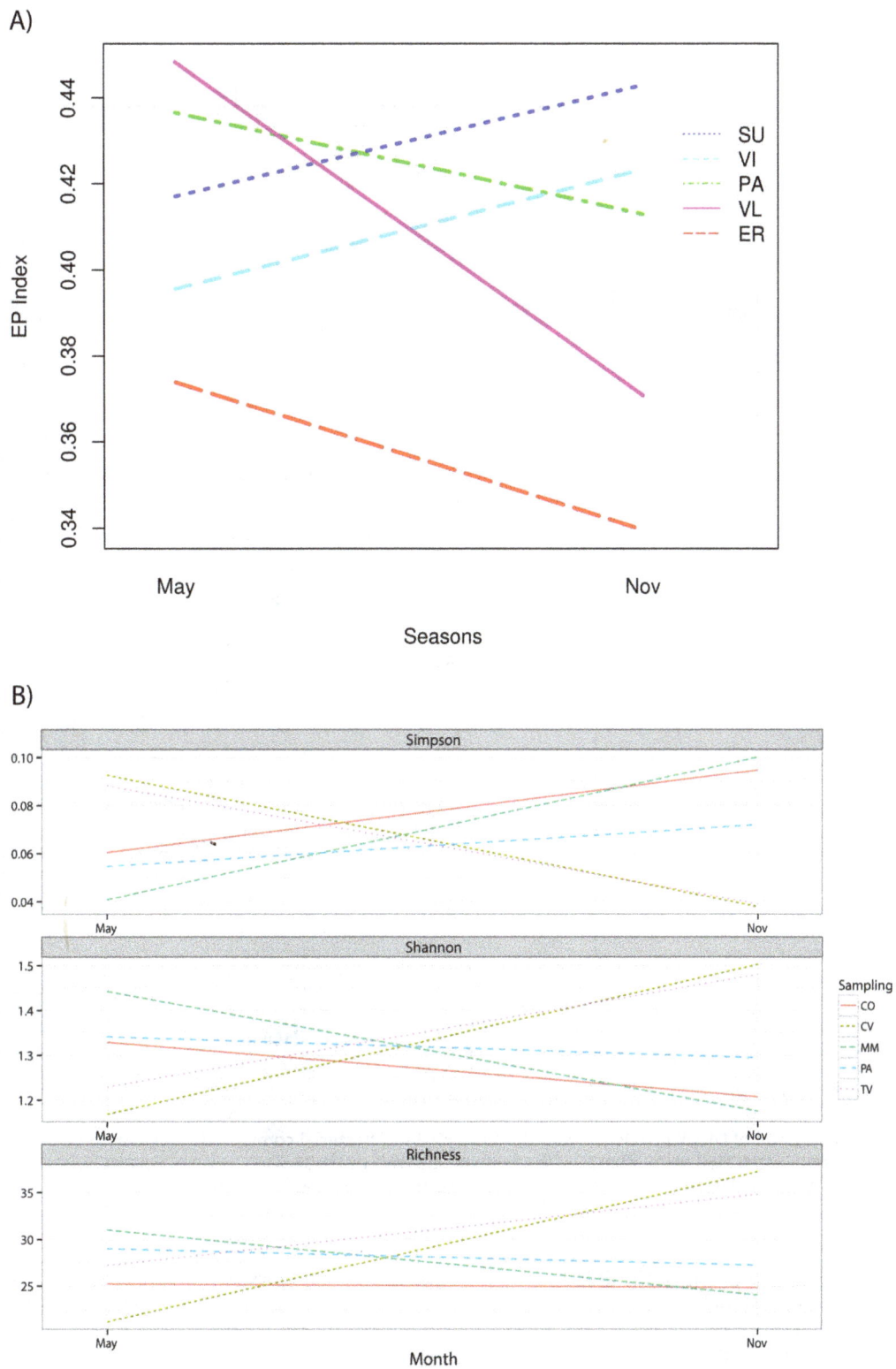

Figure 3. Effect of land-use and season on Eco-Physiological (EP) index of culturable bacteria (A) and diversity indices from CD-DGGE profiles (B).

whereas the non-tilled cover cropped vineyard and the tilled one were planted in 1985 and 1994, respectively. The five soils are located inside an area of 161.5 km^2. Detailed characteristic and management of the five soils have been previously described [12], [20], [21], [22], [23].

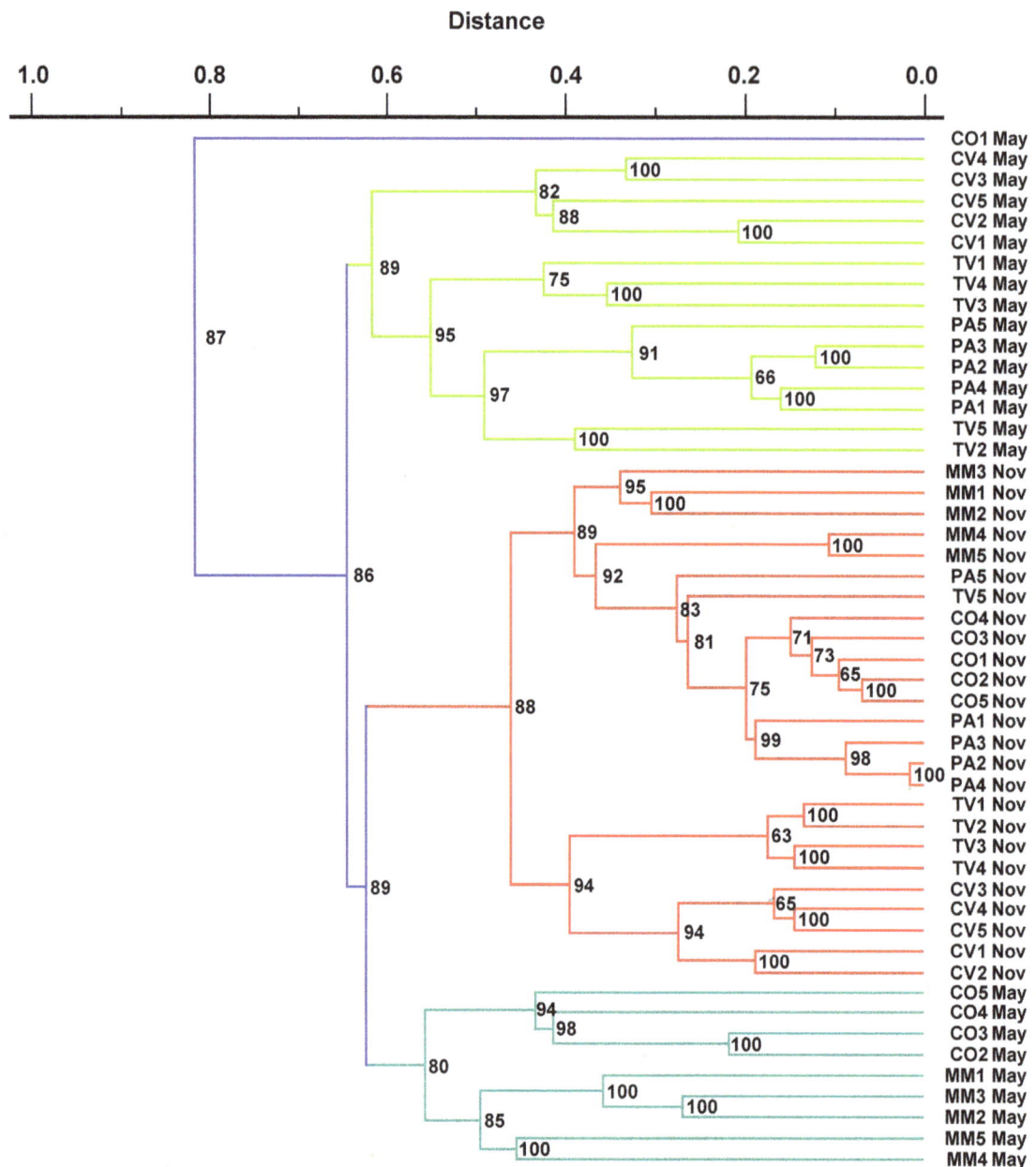

Figure 4. UPGMA dendrogram of DGGE profiles of amplified 16S rDNA of bacterial communities recovered in CO, PA, MM, CV and TV, in spring and autumn, generated using Phoretix ID advanced analysis package. Designation of samples is the same as in Fig. 1. The scale bar represents dissimilarity among samples. Consistency of each cluster was measured by the Cophenetic correlation coefficient shown at each node.

Pedological characterization of the study area

The pedogenic substrate of the study area consists of medium-grained granite, affected by localized presence of veins of quartz and porphyry. The morphology of the Berchidda area varies from flat to undulating. The processes of soil erosion by water channeled are evident only in the short-term forage crops made on soil with more than 15% of slope. All profiles have a horizons sequence of the type A-Bw-C or A-Bw-BC-C or more rarely, A-Bw-C-R (Table S1). The power of the profiles, limited to horizons A and Bw, varies from a minimum of 38 cm to a maximum of 100 cm. In three soil profiles in TV, the sequence between the horizon A and Bw and the substrate is gradually altered by the presence of a horizon BC, characterised by coarse texture, the

power of which varies from about 30 cm to 90 cm. Direct contact with the unaltered rock, R horizon, was observed only in one profile in CV. The prevailing textural classes are sandy and sandy loam. The content of organic substance in the horizons A is never very high, with average value around to about 3%. The maximum values of 11.8% and 8.6% were observed on the A horizons in two soil profiles in CO. The exchange complex, in agreement with the reduced content in clays, is never high. The degree of base saturation is predominantly less than 60% (Dystric conditions of the USDA Soil Taxonomy). The profiles with the exchange complex with a degree of base saturation below 60% in all horizons between 25 and 75 cm (TV, CV, MM, and PA) were

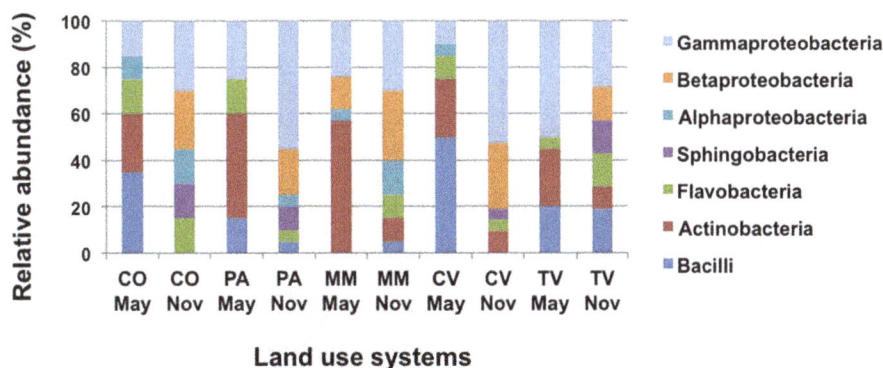

Figure 5. Relative abundances of major taxonomic groups across land use systems in spring and autumn. Detailed data of each class are listed in Additional file 1: Table S6. CO = cork-oak forest; PA = hayland-pasture rotation; MM = managed meadow; TV = tilled vineyard; CV = grass covered vineyard. Values presented are the mean percent.

classified as Typic Dystroxerepts [24]. The other one (CO) was classified as Lithic Xerorthents.

Soil sampling

Five soil replicates were collected from bulk soil of the five different managements (CO, PA, MM, CV, and TV) (Fig. 1). Soils were collected in May and November 2007. After removal of litter layer, soil core samples (50 to 100 g; diameter, 5 cm) were taken from each of the five locations, using a 5-on-dice sampling pattern with ca. 70 m distance between each sampling point. Sampling was performed at 20 cm depth, where most microbial activity is known to occur [25], [26]. In the vineyard soils, samples from along the rows and between the rows were pooled together to form a field replicate. The other soil samples (CO, PA, MM) were collected out of trees influence. At each season and for each soil type, five randomly field replicates were collected for a total of 25 soil samples (5 replicates × 5 land uses), each one being a composite sample of five soil cores. A total of 50 composite samples were taken for the two seasons. Soil samples were immediately sieved (<2 mm) to remove fine roots and large organic debris, air dried, and transported to the labs for microbiological analysis. The moisture content was adjusted to 60% of their water holding capacity (WHC) and soil samples were then left to equilibrate at room temperature in the dark for one day prior to analyses, in order to restore, within limits [27], the microbial activity of air-dried soils to that of soils in the field.

Chemical and biochemical analyses of soil samples

The chemical and biochemical analyses were performed on three replicates for each land use for a total of 15 soil samples (3 replicates × 5 land uses), each one being a composite sample of five soil cores. Total organic carbon (C_{org}) was estimated after oxidation with $K_2Cr_2O_7$ and subsequent titration of unreduced $Cr_2O_7^{2-}$ with $Fe(NH_4)_2(SO_4)_2$, as reported by Springer and Klee [28]. Soil Organic Matter (SOM) was determined by C_{org} multiplied by 1.724 van Belem coefficient. The C_{org} fractionation was set up as reported by Ciavatta and co-workers [29]. In particular, solid samples were extracted at 65°C for 24 h using 0.1 mol l^{-1} NaOH plus 0.1 mol l^{-1} $Na_4P_2O_7$ solution (1:50, solid:liquid ratio). The samples were then centrifuged at 5000×g and the supernatants were filtered through a 0.20-μm Millipore filter (Millipore, Billerica, MA) (total extractable C, C_{ext}). The humic-like acid (HA) fraction was separated from the fulvic-like acid (FA) and the non-humified carbon fractions (NHC) by

precipitation after acidification of the alkaline solution (supernatant) to pH<2. Chromatography on a column of polyvinylpyrrolidone (PVP, Aldrich, Germany) was used to separate the NHC from the FA. The FA was then combined with the HA to obtain total humified fraction (HA+FA). Total extractable C (C_{ext} %), humic and fluvic acid C (C_{HA+FA} %) were determined by the dichromate oxidation method. The non-humified carbon (C_{NH}) was determined as the difference between C_{ext} and C_{HA+FA}. Humification indexes HI, DH, and HR were determined according to previous works [29], [30].

Microbial biomass C (C_{mic}) was determined by the fumigation-extraction method of Vance and co-workers [31] with some slight modifications. The measurements were performed on air-dried soils, pre-conditioned by a 10-d incubation in open glass jars, at −33 kPa water tension, and 30°C. The incubation was employed for restoring, within limits [27], the microbial activity of air-dried soils to that of soils in the field. Four replicates of each soil sample were used. Average values are given in mg C kg^{-1} of soil. For measuring microbial respiration, 20 g (oven-dry basis) of moist sample were placed in 1 L stoppered glass jars. The CO_2 evolved was trapped, after 1, 2, 4, 7, 10, 14, 21, 28 days of incubation, in 2 ml 1 M NaOH and determined by titration of the excess NaOH with 0.1 M HCl [32].

Non-linear least square regression analysis was used to calculate parameters affecting C mineralization from daily CO_2 evolution data (Stat Win 4.0 for Windows). The best fit was obtained with the exponential model of CO_2-C accumulation according to the negative exponential decay model:

$$C_m = C_0(1 - e^{-kt}),$$

where C_m is the cumulative value of mineralized C during t days, k is the rate constant, and C_o is the potentially mineralizable C [33]. The CO_2 emitted in 28 days of incubation was used as cumulative respiration (C_{cum}). The CO_2 evolved during the 28th day of incubation was used as the basal respiration value (C_{bas}). Microbial indices were calculated as follows [34], [35], [36]:

$$qCO_2 = [(\mu g\ CO_2 - C_{bas} \times \mu g^{-1} C_{mic})h^{-1}]10^3,$$
$$qM_{cum} = (\mu g\ CO_2 - C_0 \times \mu g^{-1} C_{org}),$$
$$qM_{bas} = (\mu g\ CO_2 - C_{bas} \times \mu g^{-1} C_{org})$$

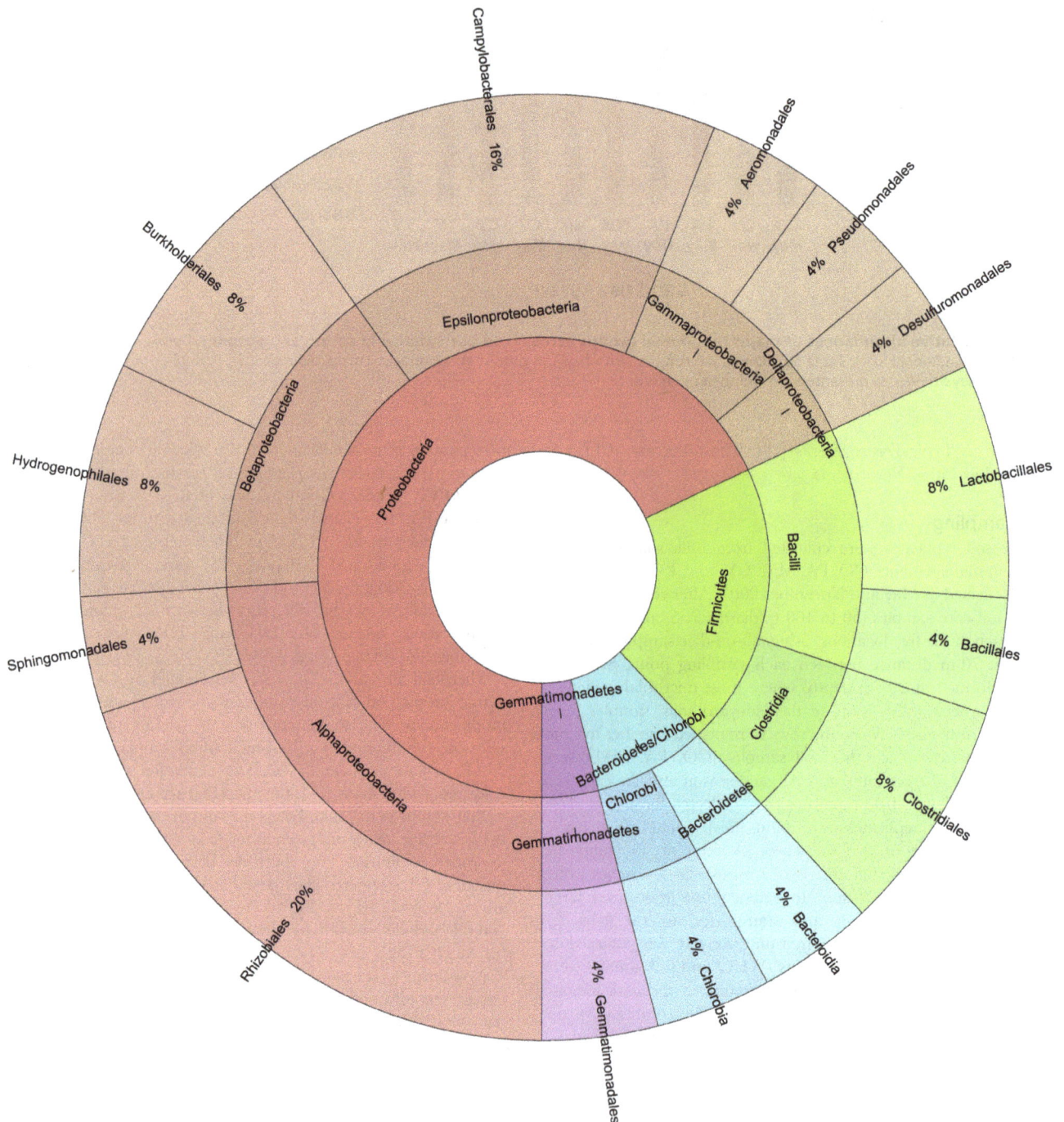

Figure 6. Plot of the taxonomic composition of total bacterial community as inferred from taxonomic interpretation of T-RFLP profiles.

Recovery of cultured bacterial cells from soil samples

Soil bacterial community analysis was performed on five soil replicates for each land use, as described above. Bacterial cell extraction was performed according to the recommendations of Smalla et al. [37] with minor modifications. Briefly, 1 g of soil was placed in a sterile 15 ml plastic tube containing 10 ml of phosphate buffered saline (PBS, pH 7.0). This mixture was homogenized for 30″ at low speed by using the Ultra-Turrax Thyristor Regle 50 (Janke & Kunkel IKA-Labortechnik). After homogenization, suspension was placed into a Erlenmeyer flask

(100 ml) containing 10 g of glass beads (0,2 mm) and shaken for 1 h at 180 r.p.m. and 28°C to disperse bacteria. The flasks and glass beads were autoclaved for 20 min at 121°C before use. The soil suspension was transferred into a sterile 15 ml plastic tube and serially diluted with sterile saline solution (9 g l^{-1}NaCl) from 10^{-1} up to 10^{-7}. The, 100 μl aliquots of serially diluted soil suspensions were plated in triplicate on 0.1 Triptic Soy Broth (TSB; Difco) amended with 15 g l^{-1} agar (0.1 TSA) and 100 g μl^{-1} of cycloheximide (Sigma) to inhibit fungal growth. Plates were incubated at 28°C for 6 days.

Growth strategy and total bacterial populations

To determine the changes in the structure of culturable fraction of soil bacteria, the r/K-strategy concept proposed by De Leij and co-workers [38] was used. Bacterial colonies appearing within 48 h were designated as r-strategists, and the remaining as K-strategists. Colonies were enumerated at 1, 2 and 6 days of growth on 0.1 TSA; in this way, three counts (or classes) were generated per sample. Plates containing between 30 and 300 colonies were then selected for enumeration. Total bacterial counts obtained were expressed as colony forming units (CFU) per gram of soil. Distribution of bacteria in each class as a percentage of the total counts gave insight into the distribution of r- and K-strategists in each sample.

To evaluate the changes in the biodiversity of bacterial populations in soils, the eco-physiological (EP) index [38] was used. The EP index of each soils tested was calculated using the equation:

$$H' = -\sum (P_i)(log_e P_i),$$

where P_i represents the CFU on each day (1, 2 and 6 days of incubation) as a proportion of the total CFU in that sample after 6 days incubation, i.e., the proportion of colonies appearing on counting day i ($i = 1, 2, 6$) with $EP_{min} = 0$. Higher values of EP index imply a more even distribution of proportions of bacteria developing on different days (i.e., different classes of bacteria).

Terminal-Restriction Fragment Length Polymorphism (T-RFLP)

DNA was extracted from soil samples by using the FastDNA SPIN Kit for Soil (QBiogene). Terminal-Restriction Fragment Length Polymorphism (T-RFLP) was performed on 16SrRNA genes amplified from extracted DNA with primer pairs P0 and P6 as previously reported [39], [40]. Purified amplification products were digested separately with restriction enzymes *Rsa*I and *Msp*I and digestions were resolved by capillary electrophoresis on an ABI310 Genetic Analyzer (Applied Biosystems, Foster City, CA, USA) using LIZ 500 (Applied Biosystems) as size standard. T-RFLP analysis was performed as previously reported [41]. Diversity indices were calculated with PAST software [42] as previously reported [19], taking into account peak intensities of T-RFLP fragments. Taxonomic interpretation of T-RFLP profiles was performed by querying the Ribosomal Database Project Database by using MiCA3 web tool (http://mica.ibest.uidaho.edu/), as previously described [40].

Culture-dependent DGGE (CD DGGE) analysis

Culture-dependent DGGE (CD DGGE) fingerprinting of 16S rRNA gene was used to characterize mixed bacterial communities recovered on agar plates.

Collection of cultured bacterial communities for DGGE analysis. Cultured bacterial communities were collected following procedure proposed by Duineveld and co-workers [43] with minor modifications. Briefly, after one week of incubation at 28°C, colonies were removed from plates containing between 100 and 1000 colonies by adding 3.0 ml sterile physiological solution (0.9% NaCl) on each plate and scraping off all grown colonies with a sterile Drigalski spatula. The cell suspensions thus obtained were aliquoted into 1.5 ml Eppendorf tubes and centrifuged at 8,000 r.p.m. for 10 minutes. The pellets were stored at −80°C for subsequent DNA extraction and PCR-DGGE analysis.

DNA extraction and PCR amplification of 16S rRNA genes. Genomic DNA was extracted with sodium dodecyl sulfate-proteinase K lysis buffer, followed by a treatment with cetyl-trimethylammonium bromide (CTAB) as described in *Current Protocols for Molecular Biology* [44]. Briefly, the pellet was resuspended in 567 µl of TE [10 mM Tris-HCl - 1 mM EDTA (pH 8.0)] buffer, and glass beads 0.3 mm in diameter (250 mg) were added, followed by bead beating for 20 s. Then, 30 µl of 10% sodium dodecyl sulfate and 3 µl proteinase K 20 mg/ml (Sigma) were added and samples were incubated at 37°C for 1 h. Glass beads were removed by centrifugation 2 min at 2,800 rpm and samples were then incubated at 65°C for 10′ with100 µl of 5 M NaCl prepared with sterile water and 80 µl of CTAB/NaCl (10% CTAB in 0.7 M NaCl). Following incubation, extracts were purified by using phenol/phenol-chloroform/isoamyl alcohol (49.5:49.5:1) extraction and DNA was recovered by isopropanol precipitation at 4°C o/n. Pelleted DNA was washed twice with cold 70% ethanol, allowed to air dry, and re-suspended in 50 µl of sterile water. Quantity and purity of DNA were checked by NanoDrop (NanoDrop Technologies, USA) and gel electrophoresis. The DNA samples were stored at −20°C until required for use.

The 16S rRNA gene was amplified using 20 ng of lysate suspension and the universal bacterial primers P0 and P6 [45]. Dilutions 1:100 (2 µl) of the1450 bp PCR products were then used as template for the second PCR amplification with the forward primer 63F (5′- AGGCCTAACACATGCAAGTC -3′), with a GC clamp (5′-CGCCCGCCGCGCGCGGCGGGCGGGGCG-GGGGCACGGGGGG -3′) incorporated at the 5′ end, and the reverse primer 518R (5′-ATTACCGCGGCTGCTGG-3′), to produce 495 bp fragments suitable for DGGE analysis [46]. Both PCR reactions were performed in Qiagen Taq buffer (10X) containing 1.5 mM MgCl₂, with 150 ng of each primer, 250 µM (each) deoxynucleoside triphosphates, and 0.5 U of *Taq* DNA polymerase (Qiagen, Hilden, Germany) in a 25 µl reaction volume. Cycle parameters for PCR with the primer pairs P0–P6 and 63F-GC and 518R were previously described by Di Cello and co-workers [45] and El Fantroussi and co-workers [46], respectively.

Denaturing Gradient Gel Electrophoresis. 16S rRNA gene amplicons were separated by double gradient denaturing gradient gel electrophoresis (DG-DGGE) as described by Cremonesi et al. [47], in a DCode universal mutation detection system (Bio-Rad, CA, USA). Separation of purified PCR products (700 ng) was achieved in6%–12% polyacrylamide (acrylamide: N,N-methylenebisacrylamide, 37.5:1) gels containing an increasing linear gradient of denaturants ranging from 30% to 60% (100% denaturant corresponds to 7 M urea and40% deionized formamide). Each gel also included marker lanes represented by DGGE profiles containing a large number of discrete bands spanning the entire gradient, suitable for within- and between-gel alignment. Electrophoresis were carried out for 16 h at 75 mV in 1X TAE buffer at 60°C, stained with 50 µg/ml ethidium bromide for 30 min, destained in water and photographed with the UVIpro Platinum Gel Documentation System (GAS7500/7510; Eppendorf, Cambridge,UK).

Cluster analysis and diversity indices

Quantity One software package (Bio-Rad) and Phoretix 1D PRO software (Phoretix International, Newcastle upon Tyne, United Kingdom) were used for CD-DGGE profile analysis. The cluster analysis and dendrogram generation were carried out by using the Phoretix 1D Pro software according to the manufacturer's instructions (Phoretix International, Newcastle upon Tyne,

United Kingdom). Bands of CD-DGGE patterns were aligned and normalized using reference lanes. Background noise was subtracted by rolling ball algorithm with a radius of 50 pixels; the automatic band detection was performed with a minimum slope of 200 and a noise reduction of 10, and peaks smaller than 2% of the maximum peak were discarded. Bands were manually corrected and matched to create an absent/present binary matrix. The similarity between the band patterns was calculated using the Dice coefficient and the clustering analysis was performed with the unweighted pair group method with arithmetic averages (UP-GMA) to generate a dendrogram by using mathematic averages algorithm programs integral to the Phoretix 1D Pro software. Coefficient of cophenetic correlation was used to measure the consistency of clusters.

DGGE banding data were used to estimate three diversity indices by treating each band as an individual operational taxonomic unit (OTU). The number of DGGE bands present in each sample was used to measure the Richness index (R). The Shannon-Weaver index of general diversity (H') [48] and the Simpson index of dominance (D) [49] were calculated from the number of bands present and the relative intensities of each band (P_i) in each lane. Relative signal intensities of detected bands, in each gel track, were determined by using the Quantity One software package (Bio-Rad) and calculated from the peak area of the densitometric curves.

The Shannon-Weaver diversity (H') was calculated using the following equation:

$$H' = -\sum (P_i)(log_e P_i),$$

where P_i (the proportion of abundances of the ith band) is measured as:

$$P_i = n_i/N,$$

where n_i is the peak height of a band, and N is the sum of all peak heights in the densitometry profiles.

The Simpson index (D) was calculated with the formula:

$$D = \sum (P_i x P_i);$$

it measures the strength of dominance because it weights towards the abundance of the OTUs and varies inversely with species diversity [50].

Bacterial isolation and identification

A total of 100 bacterial colonies were randomly picked up for each soil sample from 0.1 TSA plates (containing approximately 50 to 500 colonies), previously used for the determination of the CFU counts and EPI-index, and repeatedly streaked onto 0.1 TSA fresh plates to obtain pure cultures. Isolated colonies were then grown overnight (o/n) in TSB medium at 28°C and 200 r.p.m., and stored at −80°C in 30% glycerol until further analysis. From all five soil samples, 500 colonies were isolated in each season, for a total of 1000 bacterial colonies. A total of 203 colonies with different morphologies (about 20 colonies per each sample) were taken up to investigate their taxonomic affiliation.

Genomic DNA and PCR amplification of the 16S rRNA gene were performed as described above. Sequencing reactions were prepared from PCR products using an Applied Biosystem Big Dye Terminator sequencing kit version 3.1, according to the manufacturer's instructions and analysed using a 3730 DNA Analyzer

Applied Biosystem apparatus. The sequences were compared with those in the GenBank databases by using the BLAST program and Seqmatch tool of the RDP (http://www.ncbi.nlm.nih.gov/BLAST/and http://rdp.cme.msu.edu/, respectively) and aligned with the closest relatives with the Clustal W function of the BioEdit package [51]. Bacterial identification by 16S rRNA gene sequences assignment was performed using the RDP Classification Algorithm (http://rdp.cme.msu.edu/classifier/classifier.jsp).

Statistical analysis

Bacterial population data (CFU/g of soil) were log transformed and subsequently analysed by one-way ANOVA (STATISTICA, Release 3.0b, Copyright StatSoft Inc., CA, USA). Percentage data of EP index value were *logit*-transformed, as follows:

$$Logit\,(p) = log[p/(1-p)]$$

for the proportion p, and compared using one-way ANOVA (STATISTICA, Release3.0b, Copyright StatSoft Inc., CA, USA).

Analyses of variance (ANOVA) on biodiversity indexes (Shannon-Weaver, Richness and Simpson), principal component analysis (PCA) and clustering analysis on biochemical data were performed using R packages "stats" and "vegan" (http://cran.r-project.org/and http://cran.r-project.org/web/packages/vegan/index.html). All data clustering were performed using the "UPGMA" algorithm implemented in the "hclust" function of the R "stats" package. Distances among samples were calculated using "Bray-Curtis" distance implemented in "vegan" package as:

$$i = sample\,``i"; \ j = sample\,``j", \ BC_{ij} = 2C_{ij}/(S_i + S_j),$$

where C_{ij} = sum of the smaller value for species in common between samples i and j, and S_i and S_j = total number of species in samples i and j, respectively [52]. Variation of biodiversity indexes of cultured bacteria was inspected using ANOVA analysis. Biodiversity indexes were first divided into groups depending on sampling season and different managements of soils and then the analysis was performed. Clustering analysis (UPGMA) on biochemical parameters was performed. Each parameter was first divided in groups, in the same way of previous ANOVA analysis, and averaged. Then, each result obtained was normalized using the maximum-minimum normalization technique in order to make the data comparable. PCA analysis using each biochemical data was performed.

Results and Discussion

Effect of land use on soil chemical and biochemical properties

Soil organic matter (SOM) represents a dynamic system influenced by several factors, including climate, clay content, mineralogy and soil management, which all affect the processes of organic matter transformation and evolution in soil [53], [54]. Both soil fertility and stability are related to the organic matter content of soil. Many functions of SOM are due to its more stabilised fractions, humified materials and balance between the labile and the stabilised fractions [55]. Changes in SOM content are related to changes in microbial biomass turnover, because they reflect the balance between rates of microbial organic matter accumulation and rates of organic matter degradation. The extent of organic matter's organization not only impacts the amount of carbon mineralized but also the type of carbon that is consumed by the microorganisms.

In this study two categories of soil quality indicators were used: organic matter quality indicators and microbial biomass activity indicators (Tables S2 and S3). Our results (Fig. S1) revealed that both land use and sampling season affect the chemical and structural composition of the most stabilised fractions of SOM. A higher content of C_{org} occurred in CO soil in May, and C_{ext} and C_{HA+FA} showed the same pattern, whereas the average values of these parameters were slightly higher in PA soil in November. These results were reflected in the humification parameters, where the humification rate (HR%) can provide quantitative information about the humic substances content normalised with respect to total SOM, while the degree of humification (DH%) provides the amount of the humified carbon in the extracted organic fraction and the humification index (HI) can be considered as an index of soil humification activity as well as of availability of non humified labile fractions [29]. Overall, land use change reduced soil C stocks and labile organic matter at both sampling times except for PA in November. Pasture has a great potential soil organic C stock and, in the long term, grass management systems have nearly equivalent potential to store soil organic C as forest [56]. Potentially mineralizable C (C_0), which indicates the amount of C in the labile fraction of soil organic matter, decreased over sampling time in all soils, and C_{mic} similarly declined. Cultivated soils are characterised by low microbial activity, mainly due to the disappearance of easily decomposable organic compounds through tillage and soil disturbance. As previously found for chemical properties, the metabolic activity responds to the different land uses; in both May and November samplings, qCO_2 was higher in CO soil when compared to the others, resulting in increases stress. Hence, unfavourable conditions result in a decrease in the size of the microbial biomass and the efficiency of C substrates degradation, conducting to an increase in respiration rate per unit of microbial biomass [57].

As beneficial and negative effects of soil management practices are strongly linked to microbial activities and regulate soil quality and functioning [58] the relationship existing between soil management practices and variation in chemical and biochemical parameters was evaluated, by performing a clustering analysis (UPGMA) as reported in Tables S2 and S3. The dendrogram obtained was linked to a heat-map representing all the biochemical parameters variation (Fig. 2A). Data analysis revealed that chemical and biochemical parameters clustered in two different groups each of which corresponding to one of the two parameters analyzed (chemical and biochemical). In both sampling times, TV and CV clustered together, suggesting that seasonal change rather than management regime was a major driving force contributing to vineyard soil fertility. Furthermore, according to the UPGMA clustering, CO formed a separate cluster from that of the other soil uses (Fig. 2A). Interestingly, the chemical and biochemical parameters clustered together in both CO and PA, regardless of the season, in contrast to MM, CV and TV, where they clustered separately in relation to the sampling season. This finding suggests that soils subjected to low human inputs (pasture and cork-oak forest) showed a more stable chemical and biochemical soil composition than those with high human input (managed meadow and vineyard).

The different positions of the variables in the plane of the first two principal components, as revealed by PCA analysis (Fig. 2B), indicated that chemical and biochemical parameters were differentially affected by the various land use types. The first and the second principal components (PC1 and PC2) accounted for 44.59% and 24.47% of the total variance in the data, respectively (Fig. 2B). Both chemical and biochemical parameters were positively affected by PC1 except for HR and DH, whereas PC2

was able to discriminate between the two variable sets, with the most of chemical and biochemical parameters being positively and negatively affected, respectively. The above analyses revealed an important effect of land uses on both chemical and biochemical; on the contrary, only biochemical parameters were affected by seasons.

Influence of land use on soil bacterial community

Community structure analysis by EP-index and r-K strategy. Environmental conditions select organisms that either grow rapidly in uncrowded, nutrient-rich conditions (r-strategists), or can efficiently exploit resources in crowded conditions (K-strategists). In this work, we applied the method developed by De Leij et al. [38], who used the r/K-strategy concept for the characterization of soil bacterial communities. First, we compared the microbial community structures found in the different land uses during each of the two seasons. Since sampling was performed at 20 cm depth, where most microbial activity is known to occur [26], the main changes are expected through conversion from one soil management system to another one [59]. Microbial community structure found in the different land uses during each of the two seasons was investigated by means of EP index that is a measure of both richness (i.e. total number of species in the community) and evenness (i.e. how evenly individuals in the community are distributed over the different species) of groups of microorganisms with similar developmental characteristics.

The cultured bacteria belonging to fast-growing organisms, especially the r-strategists detected on day 2, dominated in all land uses and in both seasons (Table S4). The MM soil exhibited a lower EP index and a lower percentage of r-strategists in both seasons, possibly due to amensalism from a dominant bacterial group in the community. Variation of EP index of cultured bacteria was inspected using an ANOVA analysis of variance (Fig. 3A). Interestingly, statistical analysis revealed a significant effect of land uses on bacterial community structure ($P<0.001$) as well as an interaction effect between land uses and season change ($P<0.001$) on EP index. Variation of EP index due to seasonal changes was significant in the soil with higher human impact (TV) when compared with the other soils. Total bacterial concentrations varied significantly in respect of season ($P<0.001$) being higher in spring than in autumn (Table S2) in all but CO soils, with significant differences between CO and PA in spring, and between CO and MM, and MM and PA in autumn.

Overall results suggest that land uses affected cultured bacterial communities. Soils with low human impact (cork-oak forest) have a more stable bacterial density and show a less variation of bacterial community structure across seasons than soils subjected to high human impact such as tilled vineyard.

Bacterial community profiling by CD-DGGE and T-RFLP analyses. Culture dependent DGGE (CD-DGGE) fingerprinting of the 16S rRNA was used to characterize mixed bacterial communities recovered on agar plates. CD-DGGE represents a useful technique to follow the dynamics of distinct culturable fractions of the soil bacterial community in relation to physical, chemical and biological changes in the soil environment [60]. Since culture-dependent and culture-independent methods likely profile distinct fractions of the soil bacterial community with unique ecological roles [61], culturable bacteria may provide an ecologically relevant complement to culture-independent community characterization [62].

By pooling the bacterial cells growing on individual agar plates, we obtained a culture-dependent bacterial community. The analysis of DGGE profiles revealed clear banding patterns for each land-use and plant-cover types of sufficient complexity to investigate differences in soil microbial communities and identified stable communities with highly reproducible profiles (Fig. S2). A different community composition among land-use types was found as evidenced by the presence of different dominant signals; this finding indicated a compositional shift among soils examined in spring and autumn and subjected to different anthropogenic impact. Bacterial diversity was investigated through richness (R), Shannon-Weaver (H'), and Simpson (D) indices. Analysis of variance (ANOVA) confirmed the differences in the distribution of bacterial species due to land uses and sampling seasons ($P<0.001$). Both R and H' indices decreased from May to November in MM samples and increased in TV and CV samples, and the complementary opposite trend was observed for D index (Fig. 3B). Otherwise, cultured bacterial community in CO and PA soils did not vary significantly over seasons. Most likely, these results reflect the impact of both land-use and vegetation type and coverage on soil microbial communities. In particular, the low shift of biodiversity indices observed in CO and PA seems to be correlated to a higher stability of bacterial populations in natural habitats with low human impact, whereas populations inhabiting more anthropogenic areas tend to be more variable. Land-use type and, in particular, differences in vegetation dynamics may have a large role in modulating the temporal variability in soil bacterial communities. As observed by Lauber and co-workers [3], soils from the different land-use types did not exhibit identical temporal dynamics even though all the soils were located in close proximity and exposed to the same climatic conditions. Diversity indices obtained from T-RFLP analysis performed on total bacterial DNA (Fig. S3) partially confirmed the trends of diversity shown by CD-DGGE, though observed differences were not statistically significant (Table S5).

The unweighted pair-group method using arithmetic averages (UPGMA) of bacterial community profiling by CD-DGGE revealed a high diversity of the bacterial communities in each land-use and plant-cover soils. The similarity between the DGGE patterns of the soil bacterial communities revealed three distinct clusters (Fig. 4). Samples collected in autumn grouped into a separate cluster with about 52% similarity, whereas samples collected in spring grouped into two separate clusters, each of which composed by samples sharing about 40% similarity. The highest similarity values were shown by clusters based on soil land uses, suggesting a low bacterial variation within each soil type. This finding is consistent with previous reports showing that land-use type was the most important factor in determining the composition of soil microbial community [3].

The clustering results based on CD-DGGE were in agreement with culture-independent T-RFLP analysis that confirmed a different seasonal distribution of soil bacterial populations with subgroups associated to different land uses. Most of samples retrieved from the same soil (CO, PA, and MM) clustered together, whereas samples retrieved from TV and CV were often intermixed in the UPGMA dendrogram (Fig. S4). As culture-dependent and culture-independent profiles can separately resolve unique, diverse, and equally complex fractions of the soil bacterial community [60], our combined results from both CD-DGGE and T-RFLP methods, applied on culturable and total fractions, respectively, revealed an interaction effect between land uses and season change in affecting soil bacterial communities.

Taxa responses to land use determined by phylogenetic affiliation of soil bacterial isolates. Taxonomic affiliation was investigated on a total of 203 bacterial isolates. In detail, 20–21 colonies recovered from each soil sample at each season and showing different morphologies (for a total of 101 bacterial isolates in spring and 102 in autumn) were subjected to DNA extraction

and PCR amplification of 16S rRNA gene. An amplicon of the expected size (about 1500 bp) was obtained from each isolate, and its nucleotide sequence was determined and submitted to GenBank (Table S6). Although it is generally accepted that not all bacteria, including types of soil bacteria, are culturable, the isolation of bacteria by agar plate cultivation and subsequent phylogenetic analysis permit to isolate and identify previously uncultured representatives or even new members of certain bacterial species for further analysis of their metabolic function. Therefore, even if the metagenome sequencing is becoming the most powerful tool to investigate microbial communities, the ability to isolate indigenous strains actually remains the unique way to further characterize and select soil bacteria showing interesting properties. Additionally, standard cultivation techniques have been shown to be able to capture members of the soil rare biosphere which could not be detected by metagenome sequencing [63].

The 203 sequences obtained were compared with those present in the GenBank databases by using the BLAST [64] program. Results showed that most of 16S rRNA gene sequences matched NCBI database sequences at 99–100% of similarity at the genus level, with *Arthrobacter, Bacillus, Stenotrophomonas, Pseudomonas* and *Burkholderia* as the most representative genera. So, identification at the genus level was achieved in 103 isolates and at the species level in 72 isolates, while 12 isolates were only affiliated to taxa level higher than genus and 16 remained unidentified (Table S6). Further comparison with GenBank databases by using the Seqmatch tool of the RDP indicated that 16S rRNA gene sequences were affiliated to four phyla: *Proteobacteria* (classes α, β and γ), *Bacteroidetes* (classes *Flavobacteria* and *Sphingobacteria*), *Actinobacteria* and *Firmicutes*. Even though the RDP analysis does not permit to affiliate a bacterial isolate to a given species, it allowed to assess the genus of almost all isolates, including those not identified through the BLAST search. In fact, a boostrap value of at least 80% is satisfactory for RDP requirement. Almost all our sequences gave rise to 100% boostrap, 12 ranged from 89% to 99%, and only two of them only resulted in low values (24 and 54% respectively, referred to two spring isolates).

Data obtained from both approaches allowed to assess that Berchidda soil is colonized by bacteria included in the classes of γ-*Proteobacteria* (with a prevalence of *Pseudomonas* and *Stenotrophomonas* genera), *Actinobacteria* (in particular, the genus *Arthrobacter*), β-*Proteobacteria* (with a prevalence of *Burkholderia* genus), *Bacilli* (with the genus *Bacillus* as the dominant one), Flavobacteria and Sphingobacteria. The relative abundances of the different classes identified by RDP across the different samples related to soil uses and season are represented in Figure 5. A putative taxonomic description of total bacterial community, derived from the interpretation of T-RFLP data (Fig. S3) is reported in Figure 6, in which the largest fraction of the community is represented by members of *Proteobacteria*, with α-*Proteobacteria* as the most abundant class.

Differences in class composition across the site were observed suggesting that the microbial composition changes in response to land uses. In fact, all the seven classes (α, β and γ *Proteobacteria*, *Sphingobacteria Flavobacteria, Actinobacteria* and *Bacilli*) were present in CV, PA, and CO soils, while all classes but α-*Proteobacteria* in TV and all classes but *Sphingobacteria* in MM were found. The observed large diffusion of *Proteobacteria* in all soil-uses is in agreement with previously reported data concerning soil bacterial communities in different land-use systems [65]. Within *Proteobacteria*, the majority of the isolates fell into the gamma subgroup, which showed higher relative abundance compared to that of any other taxa at the class level while a few were alfa proteobacteria, especially in vineyards and pasture. In all soil-uses, were also found *Actinobacteria*, in agreement with previous work [65], and β-*Proteobacteria*, like *Burkholderia* sp., previously detected by Pastorelli and co-workers [19] who analysed the denitrifying bacterial communities present in the same Berchidda soil samples. Among Bacteroidetes, it has to be noted the relative abundance of the genera *Chryseobacterium* sp. (in all soil uses) and *Flavobacterium* sp. (in all soil uses, but MM), which include isolates already observed in Korea soils [66], and bacteria with plant growth promoting properties recovered in Iran soil [67], respectively. As already pointed out by Fierer and co-workers [68], the β-*Proteobacteria* and *Bacteroidetes* follow copiotrophic lifestyles and their relative abundance were highest in soils with high C availability. In general, copiotrophic bacteria should have higher growth rates and traditional culturing methods are likely to select for microorganisms that can grow rapidly in high resource environments.

Strong differences in class composition were observed in each soil when sampled in the two seasons: for instance, *Bacilli* and *Actinobacteria* dominate in spring (particularly, *Bacilli* in CV and CO, while *Actinobacteria* in MM and PA), whilst β-*Proteobacteria* tend to dominate in autumn. Six out of the seven identified classes (β and γ-*Proteobacteria*, *Flavobacteria*, *Actinobacteria* and *Bacilli*) were recovered in both seasons whilst *Sphingobacteria* were recovered only in autumn. In all soils, *Actinobacteria* were prevalent in spring, β *Proteobacteria* predominated in autumn, while β and γ-*Proteobacteria* in both spring and autumn. When genera composition was used to infer the diversity indices of cultured bacteria (Table S7) an increase of diversity (estimated as Shannon H and Evenness) from May to November was present for all soils, but CO, where Shannon H was slightly higher in May than in November. In particular, MM soil showed the highest increases for most indices, especially for alpha diversity (*i.e.* the species richness and evenness within a sample) that has often been correlated with ecosystem stability and functionality [69]. It must be noted that cultivated bacterial populations did not fluctuate with seasonal changes in soils with low human input (cork-oak forest). The bacterial communities of the cork-oak forest soil were similar in richness and composition, furthering the point that in a community with moderate disturbance, new individuals and groups could be introduced in a manner that promotes competition and diversity of the community, thus establishing a more stable community [65].

These data suggest that shifts from forest to managed meadow and vineyard result in changes of bacterial communities composition. Previous studies have also shown that shifts from forest to grassland soil [70], from cultivated system to pasture [71] and from grazed pine forest to cultivated crop and grazed pasture [65] resulted in significant changes in bacterial community composition. Our results suggest that the use of culture-dependent 16S rRNA gene sequencing along with traditional analysis of soil physiochemical properties may provide insight into the ecological relevance of soil bacterial taxa.

Conclusion

Overall, data obtained in this work revealed an important effect of land uses on both chemical and biochemical soil parameters. Soil bacterial communities were seasonally distinct and exhibited compositional shifts that tracked with changes in land use and soil management. This study, combining the pedological and biochemical data with microbiological and molecular analysis, furnishes a good methodological approach to describe the influence of different soil managements on soil microbial

community structure. In fact, the results demonstrate that, in the same pedological conditions, long-term soil management influence the community structure; i.e., soils subjected to low human inputs (cork-oak forest and pasture) showed a more stable chemical and biochemical soil composition as well as bacterial community than those with high human input (vineyards and managed meadow). Further research is required to determine whether the observed shifts in bacterial community composition produce parallel changes in the functional attributes of these communities across soil types under different long-term management regimes. The use of culture-independent approaches, like metabarcoding and metagenome sequencing, will make it possible to identify the specific drivers of land-use dynamics exhibited by soil bacterial communities and to give a complete picture of the bacterial communities in a typical Mediterranean agro-silvo-pastoral system.

Supporting Information

Figure S1 **Box-plot analysis showing the frequency distribution of physical-chemical and biological properties of the five Sardinia soils.**
(TIFF)

Figure S2 **Examples of CD-DGGE profiles of the soil bacterial communities associated to the different land uses in May and November.** A) From the left to right: hayland pasture rotation (PA), tilled vineyard (TV), grass covered vineyard (CV) in May; B) managed meadow (MM), cork-oak forest (CO), hayland pasture rotation (PA) in May; C) grass covered vineyard (CV) in May, grass covered vineyard (CV) in November, tilled vineyard (TV) in May; D) cork-oak forest (CO), hayland pasture rotation (PA), managed meadow (MM) in November.
(TIFF)

Figure S3 **Examples of T-RFLP profiles obtained after digestion with *Msp*I (A) and *Rsa*I (B) restriction enzymes of amplified of 16S rRNA gene sequences.**
(PDF)

Figure S4 **Cluster analysis of T-RFLP patterns generated by *Msp*I and *Rsa*I digestion of 16S rRNA gene sequences.** The UPGMA cluster analysis based on Jaccard similarity matrix was calculated for each set of samples using the "hclust" function of the R "stats" package. The scale bar represents the percent of dissimilarity.
(TIF)

Table S1 **Pedological profiles and classification of the soils investigated.**
(DOCX)

Table S2 **Determination of total organic carbon soil (C$_{org}$), extractable carbon (C$_{ext}$), humified carbon (C$_{HA+FA}$), non humified carbon (C$_{NH}$) and humification parameters of the five Sardinian soils.**
(DOCX)

Table S3 **Biochemical parameters measured in the five Sardinian soils.**
(DOCX)

Table S4 **r/k bacterial strategists, total culturable bacteria and EPI index measured in soils under different long-term management practices.**
(DOCX)

Table S5 **Diversity indices of total bacterial communities as inferred from T-RFLP profiles.**
(DOCX)

Table S6 Phylogenetic affiliations of 203 randomly selected soil bacterial isolates based on comparative analysis of their 16S rRNA gene sequences.
(DOCX)

Table S7 The ratio of diversity indices related to the abundance of the different genera detected in cultured isolates between November and May.
(DOCX)

Acknowledgments

We are grateful to P.P. Roggero (University of Sassari) and his team (G. Seddaiu, G. Urracci and L. Doro) for their contribution to collect the Sardinian soils and for providing data on experimental area. The experimental site was chosen on the basis of vegetation and soil surveys made in collaboration with S. Madrau, S. Bagella, R. Filigheddu, M.C. Caria, and E. Farris (University of Sassari). We thank R. Pastorelli for her help in soil sampling, S. Tabacchioni and L. Chiarini for bacterial isolation, C. Cantale and M. Sperandei for figure preparation, M.T. Rubino for chemical-biochemical analysis, and S. Cesarini and L. Pirone for their valuable suggestions. The research was carried out in the context of the FISR SOILSINK research project coordinated by R. Francaviglia (CRA-RPS, Rome) (?http://soilsink.entecra.it).

Author Contributions

Conceived and designed the experiments: A. Bevivino RF CD A. Benedetti. Performed the experiments: A. Bevivino PP MSP MCP CD LL. Analyzed the data: A. Bevivino PP GB AF AM CD LL. Contributed reagents/materials/analysis tools: A. Bevivino PP GB AF MSP AM LL MCP RF CD A. Benedetti. Contributed to the writing of the manuscript: A. Bevivino RF GB AF CD LL A. Benedetti.

References

1. Tiedje JM, Asuming-Brempong S, Nusslein K, Marsh TL, Flynn SJ (1999) Opening the black box of soil microbial diversity. Appl Soil Ecol 13: 109–122. doi:10.1016/S0929-1393(99)00026-8.
2. Nannipieri P, Ascher J, Ceccherini MT, Landi L, Pietramellara G, et al. (2003) Microbial diversity and soil functions. Eur J Soil Sci 54: 655–670. doi:10.1046/j.1365-2389.2003.00556.x.
3. Lauber CL, Ramirez KS, Aanerud Z, Lennon J, Fierer N (2013) Temporal variability in soil microbial communities across land-use types. ISME J 7: 1641–1650. doi:10.1038/ismej.2013.50.
4. Myers N, Mittermeier RA, Fonseca GAB, Fonseca GAB, Kent J (2000) Biodiversity hotspots for conservation priorities. Nature 403: 853–858. doi:10.1038/35002501.
5. Puddu G, Falcucci A, Maiorano L (2011) Forest changes over a century in Sardinia: implications for conservation in a Mediterranean hotspot. Agrofor Syst 85: 319–330. doi:10.1007/s10457-011-9443-y.
6. Médail F, Quézel P (1999) Biodiversity Hotspots in the Mediterranean Basin: Setting Global Conservation Priorities. Conserv Biol 13: 1510–1513. doi:10.1046/j.1523-1739.1999.98467.x.
7. Blondel J, Aronson J (1999) Biology and wildlife of the Mediterranean region. Oxford University Press, Oxford.
8. Salis L, Marrosu M, Bagella S, Sitzia M RP (2010) Grassland management, forage production and plant biodiversity in a Mediterranean grazing system. In: Porqueddu C, Ríos S, editors. The contributions of grasslands to the conservation of Mediterranean biodiversity. Zaragoza: CIHEAM/CIBIO/FAO/SEEP, Vol. 185. 181–185.
9. Gonçalves P, Alcobia S, Simões L, Santos-Reis M (2012) Effects of management options on mammal richness in a Mediterranean agro-silvo-pastoral system. Agrofor Syst 85: 383–395. doi:10.1007/s10457-011-9439-7.
10. Bacchetta G, Bagella S, Biondi E, Farris E, Filigheddu R, et al. (2004) A contribution to the knowledge of the order Quercetalia ilicis Br. -Bl. ex Molinier 1934 of Sardinia. Fitosociologia 41: 29–51.
11. Orgiazzi A, Lumini E, Nilsson RH, Girlanda M, Vizzini A, et al. (2012) Unravelling soil fungal communities from different Mediterranean land-use backgrounds. PLoS One 7: e34847. doi:10.1371/journal.pone.0034847.
12. Francaviglia R, Benedetti A, Doro L, Madrau S, Ledda L (2014) Influence of land use on soil quality and stratification ratios under agro-silvo-pastoral Mediterranean management systems. Agric Ecosyst Environ 183: 86–92.
13. Council of the European Communities (1992) Council Directive 92/43/EEC of 21 May 1992 on the conservation of natural habitats and of wild fauna and flora. Off J Eur Communities 35: 7–50.
14. Council of the European Communities (2001) Commission Regulation (EC) No 1808/2001 of 30 August 2001 laying down detailed rules concerning the implementation of Council Regulation (EC) No 338/97 on the protection of species of wild fauna and flora by regulating trade therein. Off J Eur Communities L250: 1–43.
15. Aru A, Baldaccini P, Delogu G, Dessena M, Madrau S, et al. (1990) Carta dei Suoli della Sardegna, scala 1/250.000. Assessorato alla programmazione e all'assestamento del territorio, Centro Regionale Programmazione, Dip. Sc. della Terra, Università di Cagliari, Italy.
16. Bacchetta G, Bagella S, Biondi E, Farris E, Filigheddu R, et al. (2009) Forest vegetation and serial vegetation of Sardinia (with map at the scale 1:350,000). Fitosociologia 46: 3–82.
17. Bagella S, Caria MC (2011) Vegetation series: a tool for the assessment of grassland ecosystem services in Mediterranean large-scale grazing systems. Fitosociologia 48: 47–54.
18. Lumini E, Orgiazzi A, Borriello R, Bonfante P, Bianciotto V (2010) Disclosing arbuscular mycorrhizal fungal biodiversity in soil through a land-use gradient using a pyrosequencing approach. Environ Microbiol 12: 2165–2179. doi:10.1111/j.1462-2920.2009.02099.x.
19. Pastorelli R, Landi S, Trabelsi D, Piccolo R, Mengoni A, et al. (2011) Effects of soil management on structure and activity of denitrifying bacterial communities. Appl Soil Ecol 49: 46–58. doi:10.1016/j.apsoil.2011.07.002.
20. Lagomarsino A, Benedetti A, Marinari S, Pompili L, Moscatelli MC, et al. (2011) Soil organic C variability and microbial functions in a Mediterranean agro-forest ecosystem. Biol Fertil Soils 47: 283–291. doi:10.1007/s00374-010-0530-4.
21. Francaviglia R, Coleman K, Whitmore AP, Doro L, Urracci G, et al. (2012) Changes in soil organic carbon and climate change - Application of the RothC model in agro-silvo-pastoral Mediterranean systems. Agric Syst 112: 48–54.
22. Seddaiu G, Porcu G, Ledda L, Roggero PP, Agnelli A, et al. (2013) Soil organic matter content and composition as influenced by soil management in a semi-arid Mediterranean agro-silvo-pastoral system. Agric Ecosyst Environ 167: 1–11. doi:http://dx.doi.org/10.1016/j.agee.2013.01.002.
23. Lai R, Lagomarsino A, Ledda L, Roggero PP (2014) Variation in soil C and microbial functions across tree canopy projection and open grassland microenvironments. Turkish J Agric For 38: 62–69. doi:10.3906/tar-1303-82.
24. Soil Survey Staff (2006) Keys to soil taxonomy. In: United States Department of Agriculture NRCS, editor. Natural Resoiurces Conservation Service.
25. Doran JW, Elliott ET, Paustian K (1998) Soil microbial activity, nitrogen cycling, and long-term changes in organic carbon pools as related to fallow tillage management. Soil Tillage Res 49: 3–18.
26. O'Brien HE, Parrent JL, Jackson JA, Moncalvo J-M, Vilgalys R (2005) Fungal community analysis by large-scale sequencing of environmental samples. Appl Environ Microbiol 71: 5544–5550.
27. Stotzky G, Goos RD, Timonin MI (1962) Microbial changes occurring in soil as a result of storage. Plant Soil 16: 1–18. doi:10.1007/BF01378154.
28. Springer U KJ (n.d.) Prüfung der Leistungsfähigkeit von einigen wichtigeren Verfahren zur Bestimmung des Kohlenstoffs mittels Chromschwefelsäure sowie Vorschlag einer neuen Schnellmethode. J Plant Nutr Soil Sci 64: 1–26 (in German).
29. Ciavatta C, Govi M, Vittori Antisari L, Sequi P (1990) Characterization of humified compounds by extraction and fractionation on solid polyvinylpyrrolidone. J Chromatogr 509: 141–146.
30. Sequi P, De Nobili M, Leita L, Cercignani G (1986) A new index of humification. Agrochimica 30: 175–179.
31. Vance ED, Brookes PC, Jenkinson DS (1987) An extraction method for measuring soil microbial biomass C. Soil Biol Biochem 19: 703–707. doi:10.1016/0038-0717(87)90052-6.
32. Badalucco L, Grego S, Dell'Orco S, Nannipieri P (1992) Effect of liming on some chemical, biochemical, and microbiological properties of acid soils under spruce (Picea abies L.). Biol Fertil Soils 14: 76–83. doi:10.1007/BF00336254.
33. Riffaldi R, Saviozzi A, Levi-Minzi R (1996) Carbon mineralization kinetics as influenced by soil properties. Biol Fertil Soils 22: 293–298. doi:10.1007/BF00334572.
34. Dilly O, Munch J-C (1998) Ratios between estimates of microbial biomass content and microbial activity in soils. Biol Fertil Soils 27: 374–379. doi:10.1007/s003740050446.
35. Anderson T-H, Domsch KH (1989) Ratios of microbial biomass carbon to total organic carbon in arable soils. Soil Biol Biochem 21: 471–479. doi:10.1016/0038-0717(89)90117-X.
36. Pinzari F, Trinchera A, Benedetti A, Sequi P (1999) Use of biochemical indices in the mediterranean environment: Comparison among soils under different forest vegetation. J Microbiol Methods 36: 21–28.
37. Smalla K, Wieland G, Buchner A, Zock A, Parzy J, et al. (2001) Bulk and Rhizosphere Soil Bacterial Communities Studied by Denaturing Gradient Gel Electrophoresis: Plant-Dependent Enrichment and Seasonal Shifts Revealed. 67: 4742–4751. doi:10.1128/AEM.67.10.4742.

38. De Leij FAAM, Whipps JM, Lynch JM (1994) The use of colony development for the characterization of bacterial communities in soil and on roots. Microb Ecol 27: 81–97.

39. Grifoni A, Bazzicalupo M, Di Serio C, Fancelli S, Fani R (1995) Identification of Azospirillum strains by restriction fragment length polymorphism of the 16S rDNA and of the histidine operon. FEMS Microbiol Lett 127: 85–91. doi:10.1016/0378-1097(95)00042-4.

40. Trabelsi D, Mengoni A, Elarbi Aouani M, Mhamdi R, Bazzicalupo M (2009) Genetic diversity and salt tolerance of bacterial communities from two Tunisian soils. Ann Microbiol 59: 25–32. doi:10.1007/BF03175594.

41. Mengoni A, Tatti E, Decorosi F, Viti C, Bazzicalupo M, et al. (2005) Comparison of 16S rRNA and 16S rDNA T-RFLP approaches to study bacterial communities in soil microcosms treated with chromate as perturbing agent. Microb Ecol 50: 375–384. doi:10.1007/s00248-004-0222-4.

42. Hammer Ø, Harper DAT, Ryan PD (2001) Past: Paleontological Statistics Software Package for Education and Data Analysi. Palaeontol Electron 4: 1–9. doi:10.1016/j.bcp.2008.05.025.

43. Duineveld B, Kowalchuk G, Keijzer A, van Elsas J, van Veen J (2001) Analysis of bacterial communities in the rhizosphere of chrysanthemum via denaturing gradient gel electrophoresis of PCR-amplified 16S rRNA as well as DNA fragments coding for 16S rRNA. Appl Environ Microbiol 67: 172–178.

44. Ausubel FM, Brent R, Kingston RE, Moore DD, Seidman JG, et al. (1987) Unit 2.4 Preparation of genomic DNA from bacteria. Curr Protoc Mol Biol: 2.4.1–2.4.2.

45. Di Cello F, Pepi M, Baldi F, Fani R (1997) Molecular characterization of an n-alkane-degrading bacterial community and identification of a new species, Acinetobacter venetianus. Res Microbiol 148: 237–249.

46. El Fantroussi S, Verschuere L, Verstraete W, Top EM (1999) Effect of phenylurea herbicides on soil microbial communities estimated by analysis of 16S rRNA gene fingerprints and community-level physiological profiles. Appl Environ Microbiol 65: 982–988.

47. Cremonesi L, Firpo S, Ferrari M, Righetti PG, Gelfi C (1997) Double-gradient DGGE for optimized detection of DNA point mutations. Biotechniques 22: 326–330.

48. Shannon C, Weaver W (1963) The mathematical theory of communication. University of Illinois Press, Urbana. doi:10.1145/584091.584093.

49. Simpson E (1949) Measurement of diversity. Nature 163: 688–688. doi:10.1038/163688a0.

50. Whittaker RH (1972) Evolution and measurement of species diversity. Taxon 21: 213. doi:10.2307/1218190.

51. Hall TA (1999) BioEdit: a user-friendly biological sequence alignment editor and analysis program for Windows 95/98/NT. Nucleic Acids Symp 41: 95–98. doi:citeulike-article-id:691774.

52. Bray J, Curtis J (1957) An ordination of the upland forest communities of southern Wisconsin. Ecol Monogr 27: 325–349.

53. Haider K (1992) Problems related to the humification processes in soils of temperate climates. In: Stotzky G, Bollag J-M, editors. Soil Biochemistry. Marcel Dekker, New York. 55–94.

54. Oades JM (1995) An overview of processes affecting the cycling of organic carbon in soils. In: Zepp R, Sonntag C, editors. Role of non living organic matter in the earth's carbon cycle. John Wiley, New York. 293–303.

55. Stevenson FJ, Cole MA (1999) Cycles of soil: carbon, nitrogen, phosphorus, sulfur, micronutrients. 2nd ed. John Wiley, New York.

56. Franzluebbers AJ, Stuedemann JA, Schomberg HH, Wilkinson SR (2000) Soil organic C and N pools under long-term pasture management in the Southern

Piedmont USA. Soil Biol Biochem 32: 469–478. doi:10.1016/S0038-0717(99)00176-5.

57. Anderson T, Domsch KH (1993) The metabolic quotient for CO2 (qCO2) as a specific activity parameter to assess the effects of environmental conditions, such as pH, on the microbial biomass of forest soils. Soil Biol Biochem 25: 393–395.

58. Benedetti A, Brookes P, Lynch J (2005) Conclusive remarks. Part I: Approaches to defining, monitoring, evaluating and managing soil quality. In: Bloem J, Hopkins D, Benedetti A, editors. Soil Quality. CABI; Wallingford. 63–70.

59. Conant RT, Paustian K, Elliott ET (2001) Grassland management and conversion into grassland: effects on soil carbon. Ecol Appl 11: 343–355. doi:10.1890/1051-0761(2001)011[0343:GMACIG]2.0.CO;2.

60. Edenborn SL, Sexstone AJ (2007) DGGE fingerprinting of culturable soil bacterial communities complements culture-independent analyses. Soil Biol Biochem 39: 1570–1579.

61. Garland JL (1999) Potential and limitations of BIOLOG for microbial community analysis. In: Bell C, Brylinsky M, Johnson-Green P, editors. Microbial Biosystems: new frontiers. Proceedings of the 8th international symposium on microbial ecology. Atlantic Canada Society for Microbial Ecology. Halifax, Canada. 1–7.

62. Ellis RJ, Morgan P, Weightman AJ, Fry JC (2003) Cultivation-dependent and -independent approaches for determining bacterial diversity in heavy-metal-contaminated soil. Appl Environ Microbiol 69: 3223–3230. doi:10.1128/AEM.69.6.3223.

63. Shade A, Hogan CS, Klimowicz AK, Linske M, McManus PS, et al. (2012) Culturing captures members of the soil rare biosphere. Environ Microbiol 14: 2247–2252. doi:10.1111/j.1462-2920.2012.02817.x.

64. Altschul SF, Madden TL, Schäffer AA, Zhang J, Zhang Z, et al. (1997) Gapped BLAST and PSI-BLAST: a new generation of protein database search programs. Nucleic Acids Res 25: 3389–3402.

65. Shange RS, Ankumah RO, Ibekwe AM, Zabawa R, Dowd SE (2012) Distinct soil bacterial communities revealed under a diversely managed agroecosystem. PLoS One 7: e40338. doi:10.1371/journal.pone.0040338.

66. Weon H-Y, Kim B-Y, Yoo S-H, Kwon S-W, Stackebrandt E, et al. (2008) Chryseobacterium soli sp. nov. and Chryseobacterium jejuense sp. nov., isolated from soil samples from Jeju, Korea. Int J Syst Evol Microbiol 58: 470–473. doi:10.1099/ijs.0.65295-0.

67. Soltani A, Khavazi K, Asadi-Rahmani H, Omidvari M, Abaszadeh Dahaji P (2010) Plant growth promoting characteristics in some Flavobacterium spp. isolated from soils of Iran. J Agri Sci 2: 106–115. Available: http://www.ccsenet.org/journal/index.php/jas/article/view/8404/6221.

68. Fierer N, Bradford MA, Jackson RB (2007) Toward an ecological classification of soil bacteria. Ecology 88: 1354–1364. doi:10.1890/05-1839.

69. Girvan MS, Campbell CD, Killham K, Prosser JI, Glover LA (2005) Bacterial diversity promotes community stability and functional resilience after perturbation. Environ Microbiol 7: 301–313. doi:10.1111/j.1462-2920.2005.00695.x.

70. Nacke H, Thürmer A, Wollherr A, Will C, Hodac L, et al. (2011) Pyrosequencing-based assessment of bacterial community structure along different management types in German forest and grassland soils. PLoS One 6: e17000. doi:10.1371/journal.pone.0017000.

71. Acosta-Martínez V, Dowd S, Sun Y, Allen V (2008) Tag-encoded pyrosequencing analysis of bacterial diversity in a single soil type as affected by management and land use. Soil Biol Biochem 40: 2762–2770. doi:10.1016/j.soilbio.2008.07.022.

Strong Discrepancies between Local Temperature Mapping and Interpolated Climatic Grids in Tropical Mountainous Agricultural Landscapes

Emile Faye[1,2,3]*, Mario Herrera[3], Lucio Bellomo[4], Jean-François Silvain[1], Olivier Dangles[1,3,5]*

1 Institut de Recherche pour le Développement (IRD), UR 072, Laboratoire Evolution, Génomes et Spéciation, UPR 9034, Centre National de la Recherche Scientifique (CNRS), Gif sur Yvette, France et Université Paris-Sud 11, Orsay, France, 2 UPMC Univ Paris06, Sorbonne Universités, Paris, France, 3 Facultad de Ciencias Exactas y Naturales, Pontificia Universidad Católica del Ecuador, Quito, Ecuador, 4 Mediterranean Institute of Oceanography (MIO) CNRS/INSU, IRD, UM 110, Université de Toulon, La Garde, France, 5 Instituto de Ecología, Universidad Mayor San Andrés, Cotacota, La Paz, Bolivia

Abstract

Bridging the gap between the predictions of coarse-scale climate models and the fine-scale climatic reality of species is a key issue of climate change biology research. While it is now well known that most organisms do not experience the climatic conditions recorded at weather stations, there is little information on the discrepancies between microclimates and global interpolated temperatures used in species distribution models, and their consequences for organisms' performance. To address this issue, we examined the fine-scale spatiotemporal heterogeneity in air, crop canopy and soil temperatures of agricultural landscapes in the Ecuadorian Andes and compared them to predictions of global interpolated climatic grids. Temperature time-series were measured in air, canopy and soil for 108 localities at three altitudes and analysed using Fourier transform. Discrepancies between local temperatures vs. global interpolated grids and their implications for pest performance were then mapped and analysed using GIS statistical toolbox. Our results showed that global interpolated predictions over-estimate by 77.5±10% and under-estimate by 82.1±12% local minimum and maximum air temperatures recorded in the studied grid. Additional modifications of local air temperatures were due to the thermal buffering of plant canopies (from −2.7°K during daytime to 1.3°K during night-time) and soils (from −4.9°K during daytime to 6.7°K during night-time) with a significant effect of crop phenology on the buffer effect. This discrepancies between interpolated and local temperatures strongly affected predictions of the performance of an ectothermic crop pest as interpolated temperatures predicted pest growth rates 2.3–4.3 times lower than those predicted by local temperatures. This study provides quantitative information on the limitation of coarse-scale climate data to capture the reality of the climatic environment experienced by living organisms. In highly heterogeneous region such as tropical mountains, caution should therefore be taken when using global models to infer local-scale biological processes.

Editor: Michael Sears, Clemson University, United States of America

Funding: This work was partly conducted within the project "Adaptive management in insect pest control in thermally heterogeneous agricultural landscapes" (ANR-12-JSV7-0013-01) funded by the Agence Nationale pour la Recherche (ANR, http://www.agence-nationale-recherche.fr/). A financial support of the McKnight Foundation (http://www.mcknight.org/) to EF during the fieldwork of this study is greatly acknowledged. The funders had no role in study design, data collection and analysis, decision to publish, or preparation of the manuscript.

Competing Interests: The authors have declared that no competing interests exist.

* Email: ehfaye@gmail.com (EF); olivier.dangles@ird.fr (OD)

Introduction

Bridging the gap between the predictions of coarse-scale climate models and the fine-scale climatic reality of species is increasingly recognized as a key issue of climate change biology research [1,2,3,4]. Despite decades of study on microclimates [5,6,7,8] and evidence for habitat-related and topographical variations in local temperatures and their relevance for species ecology [2,9,10,11,12,13], most attempts to understand and model species distributions still do not integrate spatially-explicit fine-scale climatic data (e.g. [14,15,16]). Many work use global model of temperature interpolation to examine species vulnerability to climate change and, doing so, ignore the critical issue of habitat complexity in climate buffering [4,5,17]. Indeed, climate surfaces used in species distribution models (SDMs) are rarely generated or interpolated to a resolution finer than 1 km^2 (e.g. WorldClim database), a resolution that is still very coarse relative to the home ranges or body size of most species [13,18]. For instance, [8] showed that climate grid lengths used in SDMs are, on average, ~10,000-fold larger than studied animals, and ~1,000-fold larger than studied plants. Their meta-analysis showed that the WorldClim was the most widely used climatic dataset in global SDMs. As this commonly used coarse scale climatic data in SDMs overlook the spatiotemporal thermal heterogeneity experienced by organisms, there is an urgent need for a more sophisticated use of these datasets for making inferences about biological processes that are driven by hour to hour operative temperatures of organisms.

An important yet poorly studied issue in climate change biology is to quantify to what extent climatic conditions differ between widely used 1 km^2 interpolated grid cells of global climatic database and real-world landscapes of similar areas. While it is now well-known that most organisms, especially tiny ectotherms such as insects and other arthropods, do not experience the climatic conditions recorded at weather stations [9,12,18], there is little quantitative information on the spatial and temporal heterogeneity at the landscape scale of local climatic conditions (i.e. conditions at biologically relevant scales, e.g., from cm to km for insects) and their consequences for organisms' performance. A better quantification of the climatic conditions of ecologically-relevant habitats over relatively large landscape scales (e.g., 1 km^2) is therefore a necessary first step to better incorporate dynamical microclimate into global distribution models.

Here, we investigate the sources of variance between global interpolated and local temperatures by examining 1) how well WorldClim predicts local air temperatures in our study region (the tropical Andes), 2) to what extent temperatures in crop canopies and soils differ from local air temperatures, and 3) how relevant is to use WorldClim to infer the potential performance of an insect crop pest. Addressing these questions is not an easy task as the mosaic of climatic habitats relevant for small ectothermic species at a 1-km^2 scale in real-world landscapes may be outstandingly complex. In this study, we focused on highland agricultural landscapes of the tropical Andes as most prior similar data came from low elevation and temperate agroecosystems. In such systems, most crop pests experience, over their entire life cycle, climatic conditions in three well-defined environmental layers (air, air inside-canopy and soil) and these conditions are remarkably stable over the year [19]. In this context, we firstly decided to map over replicated 1-km^2 climatic grid cells the ecologically relevant local temperatures for ectothermic crop pests in agricultural landscapes, and to compare these maps to interpolated temperature grid cells of the widely used WorldClim database. We used Fourier analysis applied to local temperature time-series as a tool to fit daily variations of temperature and to feature microclimate discrepancies in space and in time (both in terms of amplitude and phase). We then explored the implication of our thermal landscape mapping for pest performance by comparing temperature frequencies in our grid cells with the temperature-dependent growth curve of the potato tuber moth (*Phthorimaea operculella*) a major crop pest species in the region and worldwide.

Materials and Methods

1. Study area

The Ecuadorian Andes are characterized by a low seasonality, with mean temperatures varying more within days (up to 30°K variation) than within months and years (less than 0.6°K and 0.2°K variations, respectively, see [19]). This region exhibits a marked altitudinal gradient in temperatures (between 2000 and 4000 m) with mean monthly air temperature roughly decreasing by 0.6°K every 100 m of elevation [20]. Agricultural landscapes dominate the altitudinal belt between 2600 and 3800 m, and are typically composed by small field crops (mainly potato *Solanum tuberosum* L., broad bean *Vicia faba* L., corn *Zea mays* L., alfalfa *Medicago sativa* L., and pasture), natural grasslands (páramos) and a few forest patches [21]. Under the climatic conditions of the region, crops can be planted and harvested all year round, thereby creating a landscape mosaic of a wide variety of crops at different phenological stages.

Our study area was located 115 km south from the equatorial line (01°01′36″S, 78°32′16″W) in the Cotopaxi province of Ecuador. It spread out on a 20-km^2 elevation transect (2.35×8.5 km), ranging from 2,600 to 3,800 m a.s.l. The gradient had a Southwest exposure and an average slope of 9.5° (±5.2) (based on a 30 m resolution digital elevation model). To investigate the elevation effect on local vs. global interpolated temperature variations, we divided our study area into three 400 m altitudinal belts which correspond to natural floors in the hillside (2,600–3,000 m, 3,000–3,400 m, and 3,400–3,800 m) with a mean monthly temperature of 13.2±0.4°C, 10.8±0.6°C, and 9.3±0.4°C, respectively. Beyond temperature, these belts also differed in terms of landscape composition (Appendix S1 in Supporting Information), with lower elevations dominated by small fields (0.3±0.1 Ha) of potato, corn, broad bean, and pasture while the higher band had larger fields (0.7±0.3 Ha) of mainly potato and pasture. Working in these agricultural landscapes no requires specific permissions expect the kind agreement of the field owner. The presented study did not involve endangered or protected species.

2. Temperature data collection

In each of the three-altitudinal belts, we measured temperature regimes in six habitats (five crops and natural grasslands) where insect pests can be found. In each habitat, we defined three layers: air, air inside-canopy (referred as "air canopy" in the text) and soil. These layers are all used by most insect pests over their life cycle: air layer by adults, air canopy layer by adults and leaf-eating larvae and pupae, soil layer by tuber feeding larvae and pupae. In each layer of each habitat, temperature was recorded with a 1 min time step using data loggers (Hobo U23-001-Pro-V2 internal temperature loggers, Onset Computer Corporation, Bourne, USA) with an accuracy of ±0.21°K over the 0–50°C range and a resolution of 0.02°K at 25°C. According to [4], 1) air loggers were fixed on a wooden stake at 1 m high to overstep most crop canopies and sheltered by a 20 cm^2 white plastic roof to minimize solar radiation heating; the roof was itself placed 5 cm above the logger to avoid warming by greenhouse effect, 2) air canopy loggers were placed 0.3 m high inside vegetation 5 cm bellow large leaves to minimize the effect of direct solar radiation and 3) soil loggers were buried 0.1 m into the ground where roots and tubers grow (see Appendix S2 for photographs). In each field, only one logger per layer measured the temperatures. Those triplets of loggers were located at the centre of the field to avoid edge effect (see Appendix S3 for an analysis of the spatial variability of temperatures within a field and [22]). As vegetation land cover influences microclimate beneath and around plants, see [5,6], we repeated these 54 measurements (3 elevations ×6 habitats ×3 layers) for three classes of leaf area index (LAI) [23] defined as follows: 0 (bare soil), 0.01–0.5 for and >0.5 of LAI. Minimum LAI was fixed to 0.01 to avoid confusion with bare soil and allowed enough leaf area to place the loggers underneath. At each measurement site, LAI values were visually estimated (twice) measuring the ratio of leaf area within a 1-m^2 quadrant sub-divided into 0.1 m^2 cells delimited by strings. This indirect method did not account for leaves that lie on each other however it relates to shaded areas that influence inside-canopy and soil microclimates [23].

Each of the 162 measurement combinations (3 altitudinal belts ×6 habitats ×3 layers × 3 LAI classes) was replicated 1–3 times depending on availability of habitats at a given elevation and phenology stage. In total 324 independent temperature time series were acquired over 15 days between September and December 2011 (data available in Appendices S9, S10 and S11). Importantly, under the climatic conditions of the study area, 15-days time series characteristics did not differ from those obtained over one year (see Appendix S4 for details). At each measurement site, we

Temperature time-series

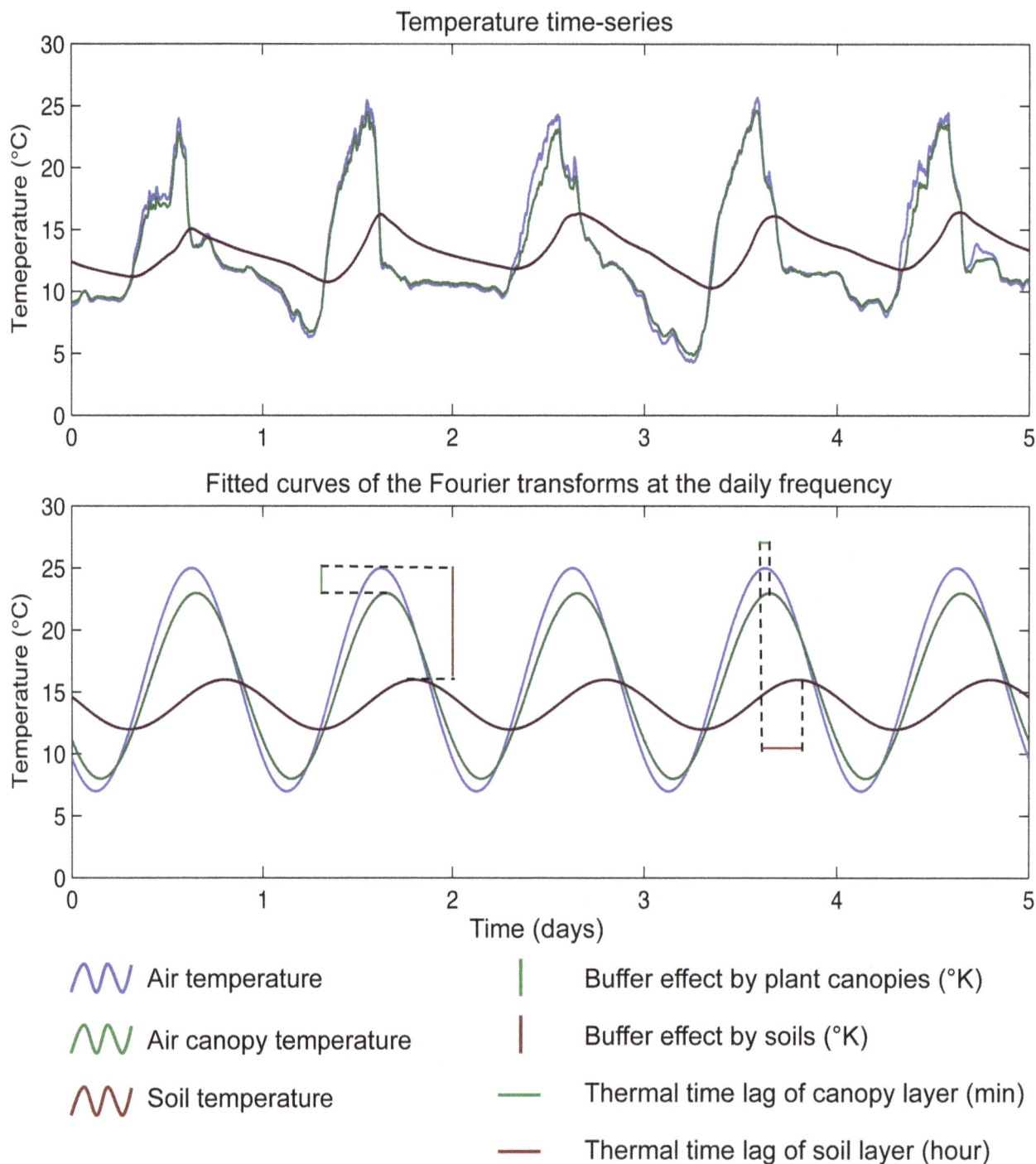

Fitted curves of the Fourier transforms at the daily frequency

Figure 1. Fit of temperature time series with discrete Fourier transforms at the daily frequency K_d. Air temperatures are in blue, crop canopy temperatures are in green and soil temperatures are in brown.

recorded the UTM-WGS84 geographic coordinates with a handheld GPS Garmin Oregon 550 (Garmin, Olathe, USA).

3. Global solar radiations

Infrared and visible radiations (expressed in Watt/m^2) were monitored in each altitudinal belts using a LI-1400 LI-COR datalogger equipped with a LI-200 pyranometer sensor (LI-COR, Lincoln, USA) placed perpendicular to gravity. Between 9:00 AM and 4:00 PM, mean global solar radiations ranged from 500 to

1000 watts/m^2, with temporal variability mainly induced by short-term changes in cloud cover.

4. Data analyses

4.1. Times series analyses using Fourier transforms. Air and air canopy temperature time series showed extreme events during a few minutes that were certainly due to strong radiations experienced at the study sites − these affected loggers recording despite their plastic roofs. Therefore, we found

A. For minimum temperatures

B. For maximum temperatures

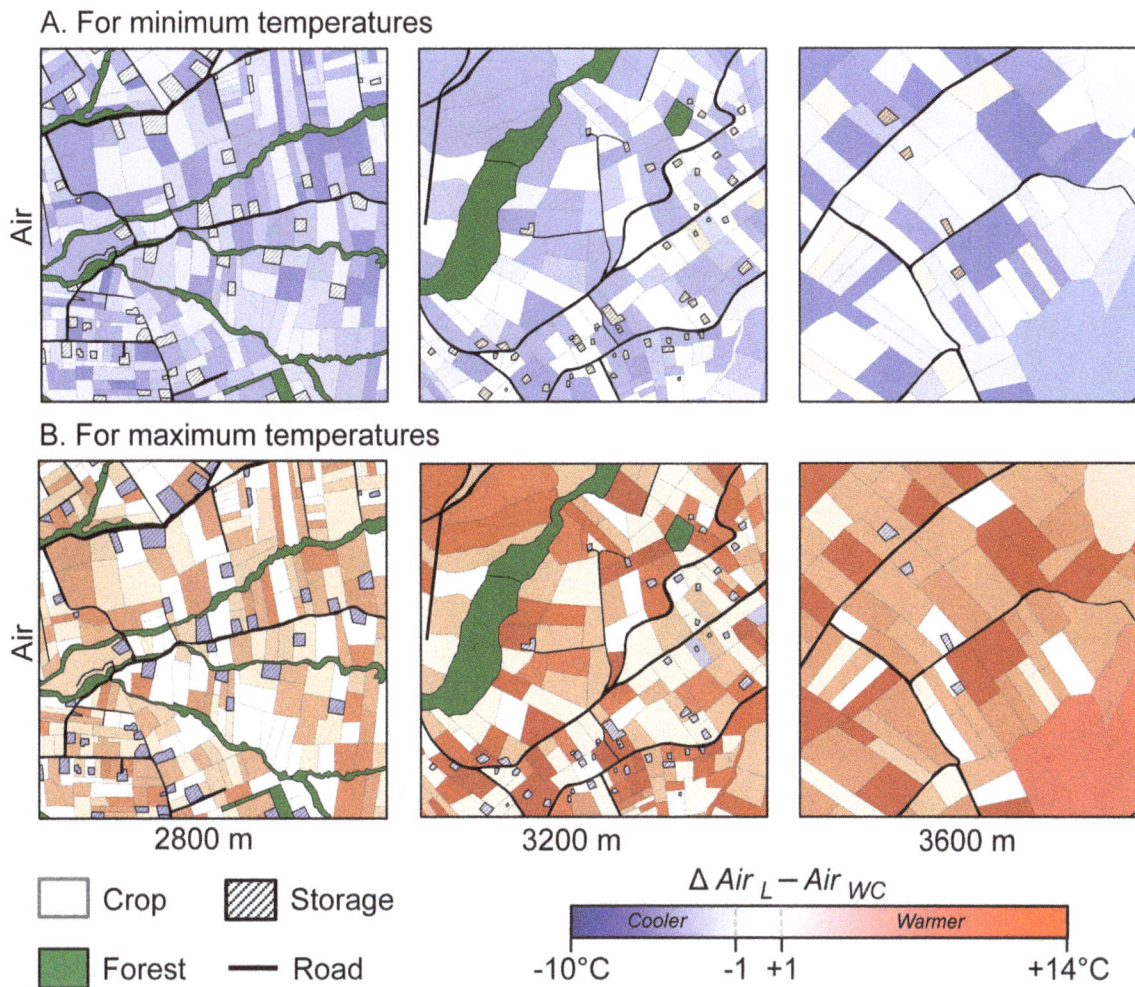

Figure 2. Maps showing the differences between local air temperatures and the WorldClim interpolated minimum (A) and maximum (B) ($\Delta\ Air_L - Air_{WC}$). Blue colours indicate $\Delta\ Air_L - Air_{WC} < 0$, i.e. area where local air temperatures are cooler than those gave by WorldClim. Red colours indicate $\Delta\ Air_L - Air_{WC} > 0$, i.e. area where air local temperatures are warmer than the ones gave by the WorldClim. White colours $\Delta\ Air_L - Air_{WC} = 0$ indicate areas where air WorldClim temperatures equate air local temperatures ($\pm 1°$C). The extent and position of each square is equal to the spatial resolution of the WorldClim database: 30-arc sec that is the equivalent of 0.86 km^2 for the study area. Temperatures in storages were obtained from [26].

relevant to fit our time series data with a discrete Fourier transform (DFT) at the daily frequency k_d (Fig. 1) as this allowed averaging daily minimum and maximum temperatures while limiting the effect of short extremes (mainly for maximum). Moreover fitting temperature time series with the DFT allowed us to circumvent (or partially resolve) the issue of comparing time series with different temporal resolution: a sinusoid built from a daily time step time series will be accurate enough to compare with another sinusoid built from a one minute time step time series (our operative temperatures vs. global climatic models).

DFT analyses allowed us estimating two important descriptors of the time series at the daily frequency k_d: the amplitude A_d and the phase ϕ_d of the DFT (see Appendix S5 for details). The thermal amplitude allowed us to measure the thermal buffer effect in Kelvin between air and canopy layers and air and soil layers (Fig. 1 and Appendix S5). The phase allowed us to measure the thermal time lag expressed in minute in inside-canopy and soil layers with respect to the air layer (Fig. 1 and Appendix S5). Thermal time lag therefore quantifies the time delay in time series to reach their maximum between air vs. canopy and air vs. soil

layers. This is an important climatic parameter to test whether microclimate conditions below canopy (canopy and soil layers) would track air conditions with some time lag depending on habitat characteristics.

We also ran DFT analyses on a four-year monitoring (2008–2012) of air temperatures (recorded at one meter high with half an hour time step with the same shelter process described above) to measure the seasonality. Analyses were performed for the three-altitudinal belts of the study area (2800, 3200, 3600 m) by reading the amplitude at the seasonal frequencies (91, 182 and 364 days, see Appendix S6). On average the Fourier transform amplitudes at 91, 182 and 364 days were 0.14 (+/−0.01), 0.44 (+/−0.04), 0.97 (+/−0.03)°K indicating that the seasonality was negligible in the study area [24].

All Fourier analyses were performed in MATLAB R2011a (Mathworks, Natick, USA). The effects of habitat, elevation, LAI classes and the interaction "elevation × LAI classes" on daytime and nigh-time DFT amplitudes and on DFT thermal time lag were assessed using a two-way ANOVA with Bonferroni corrections. When habitat was found significant, we ran post-hoc

A. For minimum temperatures

B. For maximum temperatures

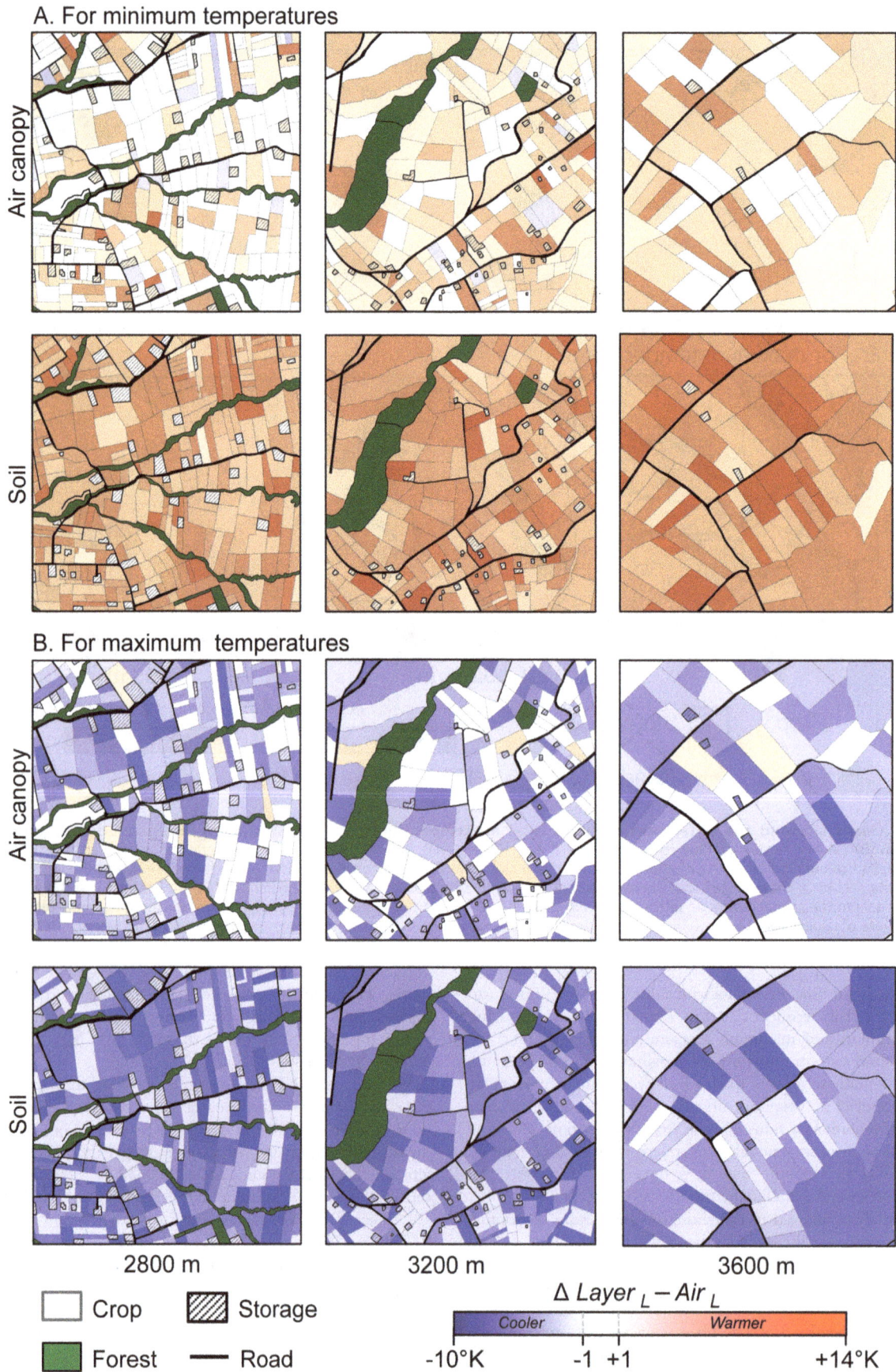

Figure 3. Maps showing the differences between local air canopy and soil temperatures with the air local for minimum (A) and maximum (B) (*Δ Layer L − Air L*)**.** Colour code is given in Figure 2.

multiple comparisons using a Tukey HSD test to identify differences among habitats. All statistical analyses were performed in R version 3.0.0 (R Development Core Team 2012).

4.2. Thermal landscape analyses. To compare local temperatures with global interpolated climate data employed in species distribution models, we considered one of the most widely used and readily available climate database, WorldClim [25]. The WorldClim database is a set of global climate layers (interpolated averages of monthly minimum, maximum and mean 1.5 m high air temperatures from weather stations spread out worldwide) with a spatial resolution of 30 arc seconds. Close to equator, this resolution is equivalent to squares of 0.86 km. In each altitudinal belt, we selected one WorldClim grid cell with homogenous slope (between 5.4° and 7.9°), micro-topography and exposition (south-west). Based on a digitized municipal cadastre (from the town council of Salcedo, Cotopaxi province) and a 5-m resolution digital orthophoto (Ecuadorian Military Geographical Institute, www.igm.gob.ec/site/index.php), we built the digital landscape of each grid cell in ArcGIS 10.01 (ESRI, Redlands, USA). In addition to the six studied habitats, crop storage infrastructures were also included into the digital maps as they significantly modify air temperature patterns, offering optimal conditions for crop pest development [26]. Outside air vs. inside air storage-temperature relationships for different elevations were derived from measurements made by [26] within the same area with

similar temperature data design (see Fig. 1 in Appendix A2 of their paper). Roads and woodlots were also indicated on the maps even if they were not included in the temperature comparison analysis, as they do not constitute relevant habitats for crop pests.

In order to simulate landscape thermal heterogeneity, crop habitats were attributed with one crop type (potato, broad bean, corn, alfalfa or pasture) and one LAI classes (0, 0.01–0.5, >0.5) based on a survey of 85 sites in the region, in which we quantified landscape composition (% of each crop and LAI classes) in 100-m radius sampling circles (see Appendix S7). For each habitat, we assigned the corresponding air, air canopy and soil temperature values at each elevation. Finally, since we were particularly interested in minimal and maximal values, as they are the most biologically relevant for ectothermic crop pests [4], we focused on minimum and maximum temperatures obtained from the DFT analyses and the WorldClim database.

Afterwards, we decomposed the variance of temperatures between global interpolated grids and local temperatures measured in agricultural landscapes by mapping the differences in minimum and maximum temperatures between the air local temperatures (Air $_L$) and the WorldClim interpolated temperatures (Air $_{WC}$) for the three studied grid cells. Then, to illustrate the part of the variance due to microclimate effects, we mapped the differences in minimum and maximum temperatures between measured local air canopies, soil temperatures (Layer $_L$) and the air local temperatures (Air $_L$) for the three studied grid cells.

4.3. Pest performance in thermal landscape. As a final step of our analysis, we explored the implication of our thermal landscape mapping for pest performance by comparing temperature frequencies in our grid cells with the temperature-dependent growth curve of a major crop pest species in the region: *Phthorimaea operculella* (Lepidoptera: Gelechiidae). This pest is considered one of the most important potato pests worldwide, but also attacks a wide variety of other crops such as tomato (*Solanum lycopersicum* L.), eggplant (*Solanum melongena* L.) or tobacco (*Nicotiana tabacum* L.) (see [27] for a review). *P. operculella* feeds on different part of the plant (leaves, stems, and tubers) and also tubers in storage structures [26,28]. In agricultural landscapes, *P. operculella* is abundant in virtually all types of habitats (even far from its host plant) because 1) this pest is able to fly over large distances (100–250 m) to infest suitable host plants [29] and 2) a significant quantity of tubers are left in the field after harvest, and are rapidly colonized by the moth before the following crop is planted. It is therefore common to observe infested potato plants in corn or broad bean fields. These left-over potatoes are well know by farmers and agronomists as significant obstacle to the control of these pests [28].

The temperature-dependent growth rate curve of *P. operculella* larvae (in day-1) over a 0–40°C range was obtained using published temperature-response data of laboratory experiments performed in the Andean region (see [30] for details). PTM development rate data were then modeled with the [31] equation as modified by [32]:

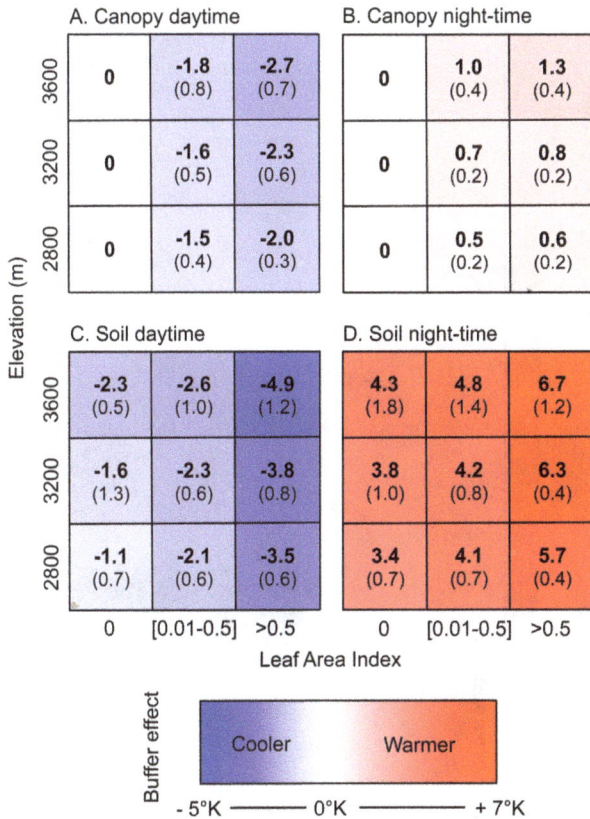

Figure 4. Mean thermal buffering from Fourier transforms at the daily frequency for canopy (A, B) and soil temperatures (C, D) as a function of elevation and leaf area index. (A, C) show the daytime temperature excursion with respect to air, whereas (B, D) are the equivalent results for night-time temperatures. The 95% interval of confidence is given between brackets. Blue colours show colder temperatures than air. Red colours show warmer temperatures than air.

$$D(T) = \frac{\frac{dT}{298.16}\exp\left[\frac{e}{R}\left(\frac{1}{298.16}-\frac{1}{T}\right)\right]}{1+\exp\left[\frac{f}{R}\left(\frac{1}{g}-\frac{1}{T}\right)\right]+\exp\left[\frac{h}{R}\left(\frac{1}{i}-\frac{1}{T}\right)\right]} \quad (1)$$

where T is temperature in Kelvin (°C+273.15), $R = 1.987$, and d, e, f, g, h, and i estimated parameters. This model has been widely used to describe the kinetics of insect development based on several assumptions about the underlying developmental control

Table 1. Results of the two-way ANOVA with a Bonferroni correction on the effects of habitat, elevation, LAI and elevation × LAI terms on daytime and nigh-time DFT amplitudes and thermal time lag on inside-canopy and soil temperature time series.

Effect	Canopy				Soil			
	Df	Mean sq	F value	P value	Df	Mean sq	F value	P value
Daytime amplitude								
Habitat	5	6.282	3.370	**0.007***	5	5.745	2.466	0.036
Elevation	2	12.491	6.701	**0.002***	2	5.722	2.456	0.089
LAI	1	40.171	21.551	**<0.001***	1	136.78	58.705	**<0.001***
Elevation × LAI	2	2.513	1.348	0.263	2	0.292	0.125	0.882
Residuals	132	1.864			127	2.330		
Night-time amplitude								
Habitat	5	0.936	3.895	**0.002***	5	1.390	0.839	0.525
Elevation	2	4.539	18.896	**<0.001***	2	2.143	1.293	0.278
LAI	1	1.083	4.509	0.035	1	157.52	95.041	**<0.001***
Elevation × LAI	2	1.754	7.302	0.010	2	0.097	0.059	0.943
Residuals	132	0.240			127	1.657		
Thermal Time Lag								
Habitat	5	0.001	1.297	0.269	5	0.009	3.881	**0.003***
Elevation	2	0.001	5.777	**0.004***	2	0.024	10.139	**<0.001***
LAI	1	0.001	29.322	**<0.001***	1	0.022	9.165	**0.003***
Elevation × LAI	2	0.001	2.374	0.097	2	0.005	2.334	0.101
Residuals	132	0.001			127	0.002		

Bold* indicates significant results (P<0.05).

A. Canopy

B. Soil

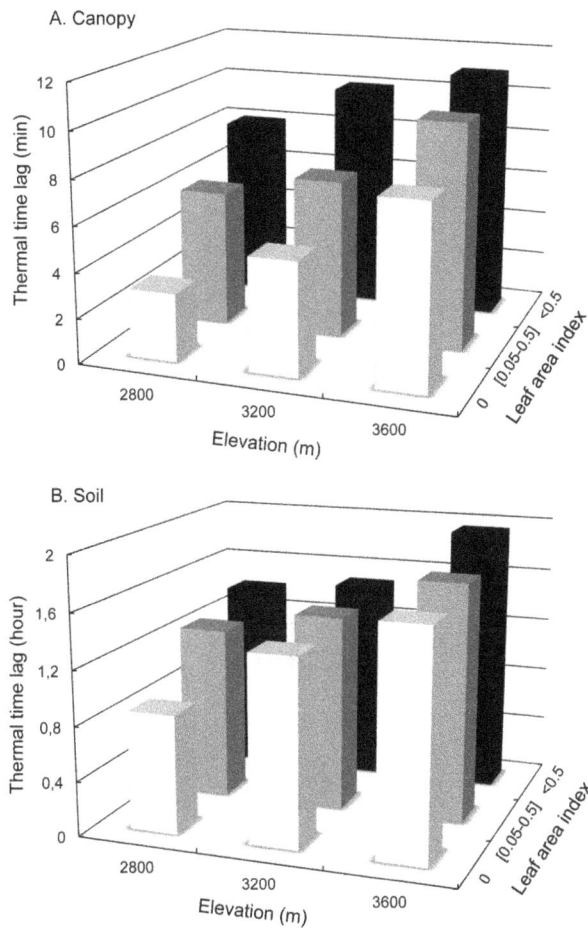

Figure 5. Thermal time lag from Fourier transforms at the daily frequency for canopy (A) and soil temperatures (B) as a function of elevation and leaf area index. The z-axis (log+1 transformed) is expressed in minutes (A) and in hours (B).

enzymes. For instance, it has been used to describe poikilotherms' temperature-dependent development [33].

We then compared the growth rate performance curve of *P. operculella* for local temperature distribution (canopy and soil layer temperatures) and for global interpolated ones (e.g., Fig. 3 in [3]). Distributions of canopy and soil minimum, maximum and mean temperatures were extracted from the three digitized landscapes using the geostatistical analyst extension of ArcGIS. Canopy and soil temperature frequencies were expressed as the percent of total grid cell area. The growth performance model of *P. operculella* given by Eqn. 1 was implemented with WorldClim minimum and maximum temperatures and the local minimum, maximum and mean temperature distribution. This allowed estimating insect growth rate within the range of WorldClim and measured field data.

Results

1. Local vs. global air temperature discrepancies in thermal landscapes

Differences in average minimum and maximum temperatures between local air temperatures and the global coarse grain interpolated air temperatures from the WorldClim (Δ Air $_L$ – Air $_{WC}$) were mapped for the three studied grid cells (Fig. 2). While

minimum local air temperatures were cooler than those predicted by WorldClim in $77.5 \pm 10\%$ of the studied areas (blue areas, average min Δ Air $_L$ – Air $_{WC} = -2.9°$K) maximum local air temperatures were warmer than extrapolated temperatures in $82.1 \pm 12\%$ of the studied areas (red areas, average max Δ Air $_L$ – Air $_{WC} = +5.6°$K). This pattern was not influenced by elevation. Notably, for all elevations, local mean air temperatures were quite well predicted by the WorldClim ($+/-1°$K) as in average $55.3 \pm 3.4\%$ of the studied areas felt in the range of Air $_L$ – Air $_{WC} \leq 1°$K (Appendix S8).

2. Temperature discrepancies due to microclimate in agricultural landscapes

Differences in average minimum and maximum temperatures between local canopy and soil temperatures and local air temperatures (Δ Layer $_L$ – Air $_L$) were mapped for the three studied grid cells (Fig. 3). Overall, canopy and soil areas were always cooler than maximum air temperature and were always warmer than air minimum temperatures resulting in a general buffer effect of minimum and maximum air temperatures by canopy and soil layers. The buffer effect on air temperatures was significantly stronger for soil than for canopy layer (see Fig. 4, Student's t-test, $t = -27.10$ and $t = 4.52$, $P<0.001$ for night-time and daytime, respectively). Interestingly, the buffer effect on air temperatures by soil was higher during night-time than daytime (Fig. 4D) while the opposite pattern was found in crop canopy (Fig. 4A).

Elevation had a significant effect on air temperature buffering in the canopy layer but not in the soil layer (Table 1). Contrastingly, LAI had a highly significant thermal buffering effect in both soils (night and daytime) and canopies (daytime, see Table 1). Buffer effect on air temperatures by bare soil (e.g. without plant cover, LAI = 0) ranged from $-1.1°$K to $-2.3°$K for daytime and from $3.4°$K to $4.3°$K for night-time. Crop type had no significant effect on buffering patterns except for potato in which higher buffer effects were recorded (Post-Hoc HSD test, $P<0.05$).

Overall, thermal time lag was much shorter in canopies (7.5 ± 2.6 min) than in soils (1.5 ± 0.3 hours, Fig. 5). LAI classes had a significant positive effect on thermal time lag for both canopy and soil layers (Table 1). On average, thermal time lag increased by 2 min. in canopies and 30 min. in soils between two LAI classes. Similarly, elevation had a significant positive effect on thermal time lag for both canopy and soil layers (Table 1) with an average increase of 2 ± 0.3 min. in canopies and of 60 ± 31 min. in soil between two altitudinal belts (Fig. 5).

3. Thermal performance curve using local vs. interpolated temperatures

To assess the implication of local vs. global interpolated temperature discrepancies for crop pest performances, we plotted the frequency distribution of the minimum (blue bars), maximum (red bars) and mean local (stripped bars) temperatures and those given by WorldClim (from minimum to maximum temperature, shaded region in the background) with the temperature-dependent growth rate curve of the potato moth *P. operculella* (Fig. 6). As a general pattern, global interpolated temperature ranges predicted lower growth rates of *P. operculella* than those predicted by local temperatures at all elevations, in both inside-canopy and soil layers (where the pest lives most of their time). While mean temperature distribution generally fell within the WorldClim min-max range, extreme temperatures (and especially maximum ones) largely exceeded this range.

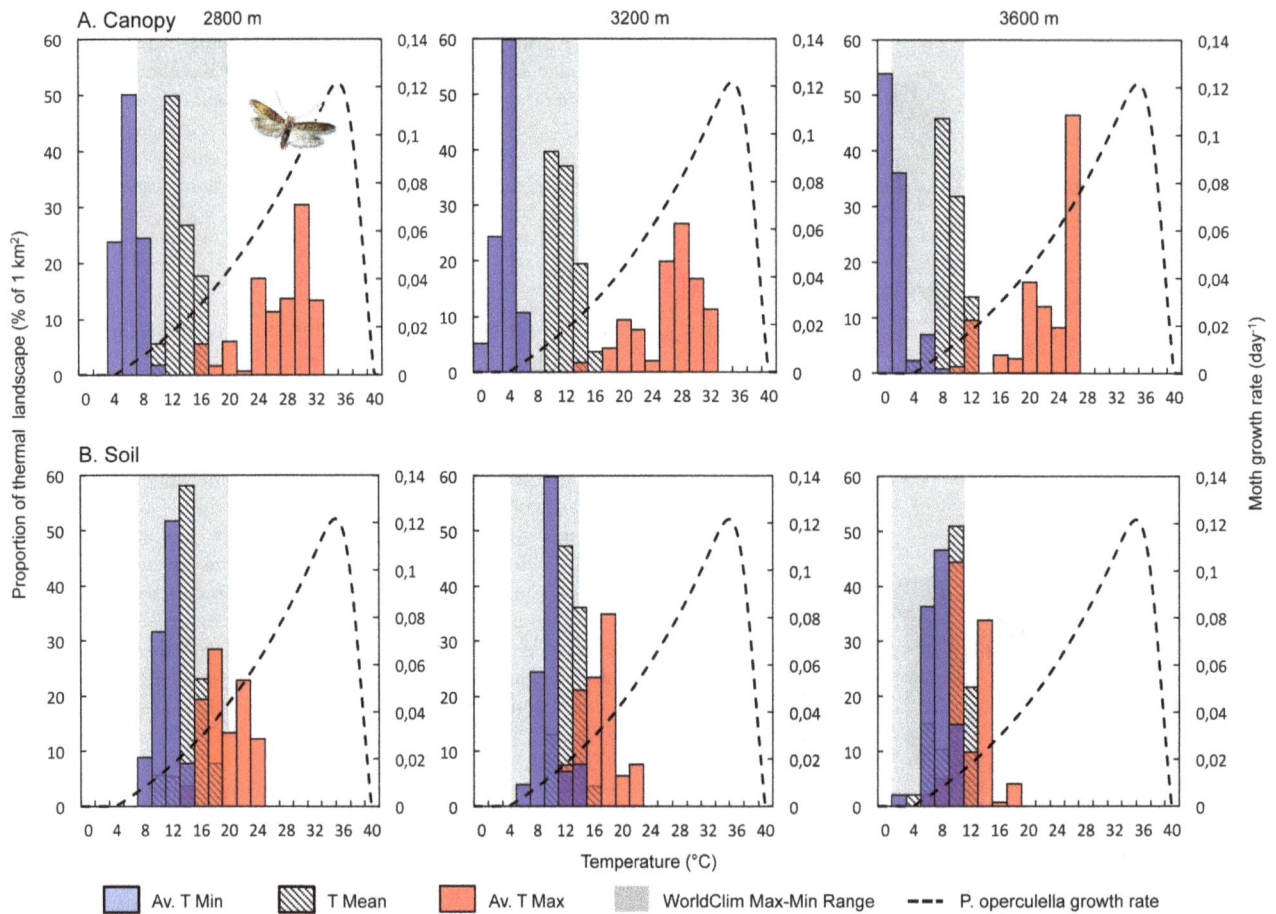

Figure 6. Superimposed plot of the temperature-dependent growth rate curve of the potato moth Phthorimaea operculella (dashed line) and the frequency distribution (% of area) of average minimum (blue), maximum (red) and mean (striped) temperatures for canopy and soil layers at the three studied elevations. Grey (shaded) bands in the background represent the WorldClim minimum and maximum temperature range.

The WorldClim estimations predicted *P. operculella* growth rates ranging between 0.007 and 0.045 day^{-1} at 2800 m, and between 0 and 0.018 day^{-1} at 3600 m, the maximum rates being slightly lower than those predicted by soil temperatures (0.068 day^{-1} at 2800 m and 0.037 day^{-1} at 3600 m). These differences were exacerbated in canopy layers where estimated maximum growth rates were 2.6–4.3 times higher than those predicted by WorldClim (0.118 day^{-1} at 2800 m and 0.079 day^{-1} at 3600 m). Discrepancies between WorldClim and local temperature-based growth rate estimations were not significantly affected by elevation (One-way ANOVA, $F = 7.79$, $P = 0.219$ and $F = 1.67$, $P = 0.419$ for canopies and soils, respectively).

Discussion

Accurate predictions of the responses of organisms to climate change using SDMs require knowledge of microclimates at spatial and temporal scales relevant for studied organisms [13,34,35]. To our knowledge, our study is the first to quantify the thermal heterogeneity among a set of agricultural habitats at fine spatial and temporal scales and to compare those thermal microhabitats to the most widely used global climatic dataset in SDMs. By documenting the mosaic of thermal habitats found in tropical agricultural landscapes, our study confirms previous evidence that microclimates strongly differ from nearby macroclimates due to

the variability of air motion and solar radiation patterns created by complex topographies with heterogeneous elevation, slope angle, exposure or roughness [1,7,18,36]. Our study therefore supports the view that results from the long tradition of agrometeorological studies on microclimates (e.g. [6,17,22]) have to be revived in the new context of microhabitat modelling for predicting the response of organisms to climate change.

1. LAI-based and elevation-based climate heterogeneity

In contrast to many previous studies (see [7] for a review), our objective was not to examine the well-documented effect of topography on local temperatures but rather to examine the less-known effects of habitat types and vegetation land cover on thermal landscape features. We found significant thermal time lag and buffer effects on air temperatures by plant and soil layers below crop canopies during night-time and daytime. The top of canopies reflects and absorbs part of the solar radiation during the day, allowing less energy to reach the layers (plants and soils) below canopies. During the night, infrared heat released from both the ground and plants is partly held back by the canopy above [5]. As a consequence plants and soils limit night-time cooling and daytime warming [6], leading to a significant buffer effect of minimum and maximum temperatures [1,4,17]. That is also why we found a buffer effect on air temperatures by soil higher during night-time than daytime and the opposite pattern for crop canopies.

Our results indicate a strong effect of elevation on thermal buffering and thermal time lag by canopy and soil layers. This could result from the combination of a negative relationship between elevation and air temperature and a positive relationship between elevation and solar radiation exposure, part of which is absorbed by plants and soils [6]. As a result, the difference between air temperature and canopy and soil temperature increased with elevation. Interestingly, the modifications of local temperatures by habitats and LAI were of the same magnitude (from −2.70 to 4.82°C in average) than that generated by topography-related factors [7,36], supporting the need to better consider habitat effects on microclimates.

2. Fine scale variations in temperature vs. climatic units

Our findings show that the complex agricultural mosaic resulting from habitat types and LAI classes at the landscape scale was a major modifier of the thermal patterns in the studied tropical highlands. More importantly, our findings revealed that, at best, 55% of landscape habitats had real mean air temperatures that were well estimated by WorldClim predictions while in average less than 20% of these areas had minimum and maximum air temperatures well estimated. Additional thermal discrepancies between large and fine-scale temperatures resulted from hetero-geneity in crop types and phenologies. This strongly supports the view that the common use of the WorldClim database arrayed into 1-km^2 grids may not adequately capture the reality of the climatic environment experienced by living organisms, in particular tiny ectothermic species [2,3,13,18]. It is important to note that to obtain the highest level of thermal heterogeneity we chose a complex mountainous agricultural study area that provided boundary conditions for climate modelling. Indeed, these moun-tainous areas provide strong climatic gradients and extreme habitat fragmentation which combined with un-seasonal agrosys-tem make up a mosaic of thermal patches that expanded the difficulties to faithfully assess climatic parameters for modelling [25]. In view of the urgent need of fine scale climate data with large extent [2,8,35] more research is necessary to develop accurate up- or down-scaling methods, in mountainous locations where thermal heterogeneity is large, and may be needed to properly describe the ecologically significant microclimates [7,37].

3. Microclimates and species distribution models

From tiny insects to mega-herbivores, it is well recognized that species ecology is strongly influenced by micro-climatic features of the landscape [2,10,11,12,13] yet quantitative information on how thermal landscape heterogeneity may affect species performance is scarce. Short-scale differences in temperatures may provide opportunities for individual organisms, even with limited dispersal capabilities, to escape unfavourable microclimates or to maximize physiological performances by selecting preferred microclimates [38,39]. Our analysis showed that predictions on *P. operculella* growth rates strongly differed between Wordclim-based and locally-measured temperatures, suggesting that global species distribution models using global coarse-scale climatic datasets without further microclimate modelling could be strongly limited to accurately predict species occurrence and performance, in particular that of ectotherms living in habitats such as mountain slopes. Such a spatial heterogeneity in thermal patches, where climatic conditions are strongly modified, provides a mosaic of favourable, sub-optimal or lethal thermal habitats that directly influences the performance of natural populations of ectotherms.

Coarse-extent modeling of microclimate is currently one of the major obstacles to predicting how organism will react to their experienced environments and forecast their distribution under climate change [8]. To date, two main types of models have been shown to provide relatively accurate, continent-wide calculations of microclimate: statistical model and mechanistic model [13]. The first one is statistical as the variables are not deterministically but stochastically related. These models perform statistical correlation of species occurrences with climatic data and have proven to be powerful interpolative tools for defining and projecting climatic envelopes [40,41]. A disadvantage of these statistical models is that they can only be applied to the conditions under which they are fitted. On the other hand, mechanistic models of the climatic responses of organisms [13,34] use fundamental knowledge of the interactions between process variables to define the model structure. Therefore they do not require much data for model development and validation. One of them is the Microclim model recently developed by [35,42] for all terrestrial landmasses − except Antarctica− which quantify key microclimatic parameters at macro-scales, with a relatively fine spatial (15 km^2) and temporal resolution (hours). The microclimatic parameters such as wind velocity, humidity, and solar radiation allow building energy and mass budgets of organisms, and therefore serve as key inputs for biophysical models of species distributions.

It is important to highlight that a better spatiotemporal resolution in temperature patterns should go in pair with the development of more accurate temperature-based population dynamics models to integrate it [2,13,34,43]. Existing predictions of models based on insect response measured in constant temperatures may yield different and less realistic results than those from predictions of models that include the effect of real temperature fluctuation on insect biology [33]. For example, to date, we still do not know the impact of a few hours of warm temperature for the performance of ectotherm species at longer time scales [33]. In this context, fine-scale spatiotemporal temperature mapping has revealed a key step for any studies aiming at understanding, predicting and managing the responses of species distributions to climate change.

Supporting Information

Appendix S1 Habitat and field size distribution in the three studied altitudinal belts.
(PDF)

Appendix S2 Photos of the temperature recording experiment.
(PDF)

Appendix S3 Spatial variability of temperatures within a field.
(PDF)

Appendix S4 Comparison of time series analysis out-puts using 15 days vs. 1-year temperature data.
(PDF)

Appendix S5 Fourier analysis description.
(PDF)

Appendix S6 Seanonality measured on four year air temperature time series with Discrete Fourier Trans-form.
(PDF)

Appendix S7 Crop habitat composition survey used in the study area.
(PDF)

Appendix S8 Local and global air mean temperature discrepancies mapping.
(PDF)

Appendix S9 Microclimate temperature time-series used in this work #1.
(ZIP)

Appendix S10 Microclimate temperature time-series used in this work #2.
(ZIP)

Appendix S11 Microclimate temperature time-series used in this work #3.
(ZIP)

Acknowledgments

We are grateful to the city council of Salcedo (Cotopaxi, Ecuador) for providing the digital shape files of the study area cadastre. We also thank all farmers who collaborated with us during fieldwork. And finally, we gratefully acknowledge Dr. Pincebourde S. and Dr. Duyck F. for their constructive scientific comments and suggestions.

Author Contributions

Conceived and designed the experiments: EF OD. Performed the experiments: EF MH. Analyzed the data: EF LB OD. Contributed reagents/materials/analysis tools: EF LB OD. Contributed to the writing of the manuscript: EF LB JFS OD.

References

1. Scherrer D, Korner C (2011) Topographically controlled thermal-habitat differentiation buffers alpine plant diversity against climate warming. J Biogeogr 38: 406–416.
2. Bennie J, Hodgson JA, Lawson CR, Holloway CTR, Roy DB, et al. (2013) Range expansion through fragmented landscapes under a variable climate. Ecol Lett 117: 285–229.
3. Logan ML, Huynh RK, Precious RA, Calsbeek RG (2013) The impact of climate change measured at relevant spatial scales: new hope for tropical lizards. Global Change Biol 19: 3093–3102.
4. Scheffers BR, Edwards DP, Diesmos A, Williams SE, Evans TA (2013) Microhabitats reduce animal's exposure to climate extremes. Global Change Biol 20: 495–503.
5. Geiger R (1965) The climate near the ground. Cambridge, USA.
6. Jones HG (1992) Plants and microclimate: a quantitative approach to environmental plant physiology. Cambridge University Press, Cambridge.
7. Dobrowski SZ (2011) A climatic basis for microrefugia: the influence of terrain on climate. Global Change Biol 17: 1022–1035.
8. Potter KA, Woods HA, Pincebourde S (2013) Microclimatic challenges in global change biology. Global Change Biol 19: 2932–2939.
9. Cloudsley-Thompson JL (1962) Microclimates and the distribution of terrestrial arthropods. Annu Rev Entomol 7: 199–222.
10. Tracy CR (1977) Minimum size of mammalian homeotherms: role of the thermal environment. Science 198: 1034–1035.
11. Willmer PG (1982) Microclimate and the environmental physiology of insects. Adv Insect Physiol 16: 1–57.
12. Unwin DM, Corbet SA (1991) Insects, plants and microclimate. Richmond Publishing Company Ltd.
13. Kearney M, Porter W (2009) Mechanistic niche modelling: combining physiological and spatial data to predict species' ranges. Ecol Lett 12: 334–350.
14. Beaumont LJ, Pitman AJ, Poulsen M, Hughes L (2007) Where will species go? Incorporating new advances in climate modelling into projections of species distributions. Global Change Biol 13: 1368–1385.
15. Deutsch CA, Tewksbury JJ, Huey RB, Sheldon KS, Ghalambor CK, et al. (2008) Impacts of climate warming on terrestrial ectotherms across latitude. P Natl Acad Sci USA 105: 6668–6672.
16. Warren RJ, Chick L (2013) Upward ant distribution shift corresponds with minimum, not maximum, temperature tolerance. Global Change Biol 19: 2082–2088.
17. Suggitt AJ, Gillingham PK, Hill JK, Huntley B, Kunin WE, et al. (2011) Habitat microclimates drive fine-scale variation in extreme temperatures. Oikos 120: 1–8.
18. Sears MW, Raskin E, Angilletta MJ (2011) The world is not flat: defining relevant thermal landscapes in the context of climate change. Integr Comp Biol 51: 666–675.
19. Dangles O, Carpio C, Barragan AR, Zeddam JL, Silvain JF (2008) Temperature as a key driver of ecological sorting among invasive pest species in the tropical Andes. Ecol Appl 18: 1795–1809.
20. McCain CM (2007) Could temperature and water availability drive elevational species richness patterns? A global case study for bats. Global Ecol Biogeogr 16: 1–13.
21. Dangles O, Carpio FC, Villares M, Yumisaca F, Liger B, et al. (2010) Community-based participatory research helps farmers and scientists to manage invasive pests in the Ecuadorian andes. Ambio 39: 325–335.
22. Baldocchi DD, Verma SB, Rosenberg NJ (1983) Microclimate in the soybean canopy. Agr Meteorol 28: 321–337.
23. Wilhelm WW, Ruwe K, Schlemmer MR (2000) Comparisons of three-leaf area index meters in a corn canopy. Crop Sci 40: 1179–1183.
24. Fitzpatrick EA (1964) Seasonal distribution of rainfall in Australia analysed by Fourier methods. Archiv Meteorologie, Geophysik und Bioklimatologie 13: 270–286.
25. Hijmans RJ, Cameron SE, Parra JL, Jones PG, Jarvis A (2005) Very high resolution interpolated climate surfaces for global land areas. Int J Climatol 25: 1965–1978.
26. Crespo-Perez V, Rebaudo F, Silvain JF, Dangles O (2011) Modeling invasive species spread in complex landscapes: the case of potato moth in Ecuador. Landscape Ecol 26: 1447–1461.
27. Rondon S (2010) The potato tuberworm: a literature review of its biology, ecology, and control. Am J Potato Res 87: 149–166.
28. Hanafi A (1999) Integrated pest management of potato tuber moth in field and storage. Potato Res 42: 373–380.
29. Cameron PJ, Walker GP, Penny GM, Wigley PJ (2002) Movement of potato tuberworm within and between crops, and some comparisons with diamondback moth. Environ Entomo 31: 65–75.
30. Crespo-Pérez V, Dangles O, Régnière J, Chuine I (2013) Modeling temperature-dependent survival with small datasets: insights from tropical mountain agricultural pests. Bul Entomol Res 103(03): 336–343.
31. Sharpe PJH, DeMichele DW (1977) Reaction-kinetics of poikilotherm development. J Theor Biol 64: 649–670.
32. Schoolfield RM, Sharpe PJH, Magnuson CE (1981) Non-linear regression of biological temperature-dependent rate models based on absolute reaction-rate theory. J Theor Biol 88: 719–731.
33. Gilbert E, Powell JA, Logan JA, Bentz BJ (2004) Comparison of three models predicting developmental milestones given environmental and individual variation. B Math Biol 66(6): 1821–1850.
34. Buckley LB, Urban MC, Angilletta MJ, Crozier LG, Rissler IJ, et al. (2010) Can mechanism inform species' distribution models? Ecol Lett 13: 1041–1054.
35. Kearney M, Shamakhy A, Tingley R, Karoly DJ, Hoffmann AA, et al. (2013) Microclimate modelling at macro scales: a test of a general microclimate model integrated with gridded continental-scale soil and weather data. Global Change Biol Doi : 10.1111/2041–210X.12148.
36. Scherrer D, Korner C (2010) Infra-red thermometry of alpine landscapes challenges climatic warming projections. Global Change Biol 16: 2602–2613.
37. Fridley JD (2009) Downscaling climate over complex terrain: high finescale (< 1000 m) spatial variation of near-ground temperatures in a montane forested landscape. J Appl Meteorol Clim 48: 1033–1049.
38. Kinahan AA, Pimm SL, van Aarde RJ (2007) Ambient temperature as a determinant of landscape use in the savanna elephant, Loxodonta africana. J Therm Biol 32: 47–58.
39. Dillon ME, Liu R, Wang G, Huey RB (2012) Disentangling thermal preference and the thermal dependence of movement in ectotherms. J Therm Biol 37: 631–639.
40. Guisan A, Thuiller W (2005) Predicting species distribution: offering more than simple habitat models. Ecol Lett 8: 993–1009.
41. Elith J, Leathwick JR (2009) Species distribution models: ecological explanation and prediction across space and time. Annu Rev Ecol Evol S 40: 677–697.
42. Kearney MR, Isaac AP, Porter WP (2014) microclim: Global estimates of hourly microclimate based on long-term monthly climate averages. Scientific Data 1.
43. Bakken GS, Angilletta MJ (2014) How to avoid errors when quantifying thermal environments. Func Ecol 8: 96–107.
44. Bloomfield P (2004) Fourier analysis of time series: an introduction. Wiley-Interscience, New York, USA.

6

The Culturable Soil Antibiotic Resistome: A Community of Multi-Drug Resistant Bacteria

Fiona Walsh*, Brion Duffy

Bacteriology Research Laboratory, Federal Department of Economic Affairs, Education and Research EAER, Research Station Agroscope Changins-Wädenswil ACW, Wädenswil, Switzerland

Abstract

Understanding the soil bacterial resistome is essential to understanding the evolution and development of antibiotic resistance, and its spread between species and biomes. We have identified and characterized multi-drug resistance (MDR) mechanisms in the culturable soil antibiotic resistome and linked the resistance profiles to bacterial species. We isolated 412 antibiotic resistant bacteria from agricultural, urban and pristine soils. All isolates were multi-drug resistant, of which greater than 80% were resistant to 16–23 antibiotics, comprising almost all classes of antibiotic. The mobile resistance genes investigated, (ESBL, bla_{NDM-1}, and plasmid mediated quinolone resistance (PMQR) resistance genes) were not responsible for the respective resistance phenotypes nor were they present in the extracted soil DNA. Efflux was demonstrated to play an important role in MDR and many resistance phenotypes. Clinically relevant *Burkholderia* species are intrinsically resistant to ciprofloxacin but the soil *Burkholderia* species were not intrinsically resistant to ciprofloxacin. Using a phenotypic enzyme assay we identified the antibiotic specific inactivation of trimethoprim in 21 bacteria from different soils. The results of this study identified the importance of the efflux mechanism in the soil resistome and variations between the intrinsic resistance profiles of clinical and soil bacteria of the same family.

Editor: A. Mark Ibekwe, U. S. Salinity Lab, United States of America

Funding: This project was funded by the Swiss Federal Office for Agriculture, the Swiss Federal Office for the Environment and the Swiss Expert Committee for Biosafety (SECB). Work was conducted within the European research network COST TD0803 Detecting evolutionary hotspots of antibiotic resistances in Europe (DARE). The funders had no role in study design, data collection and analysis, decision to publish, or preparation of the manuscript.

Competing Interests: The authors have declared that no competing interests exist.

* E-mail: fiona.walsh@agroscope.admin.ch

Introduction

Antibiotic resistance has developed over time from resistance to single classes of antibiotics to multi-drug resistance and extreme drug resistance. Until recently, antibiotics and antibiotic resistance were discussed in terms of treatments of infections and the prevention of successful treatment, respectively. The mechanisms of action of antibiotics and antibiotic resistance mechanisms have been studied almost exclusively in pathogenic bacteria. It is only in recent years that antibiotic resistance research has focused on the environment from which the antibiotics were initially extracted: soil microorganisms and the soil ecosystem. With an every decreasing supply of novel antibiotics and increasing resistance research has started to focus on investigating the natural antibiotic resistome and understanding the ecology and evolution of antibiotic resistance in the non-clinical environment in order to identify reservoirs of both known and novel antibiotic resistance mechanisms.

Despite the belief that the soil antibiotic resistome bacteria play an increasingly important role in the evolution, development and spread of antibiotic resistance in humans and animals, there is little known about the natural bacterial resistome in soil. There have been many calls for more information about the natural resistome and these have also highlighted the importance of understanding the soil resistome in the preservation of antibiotics for the treatment of infections [1–4]. However, to date there have been few studies which have investigated the culturable soil resistome

and these have been limited to the antibiotic producing bacteria *Streptomyces*, and an isolated cave microbiome [5,6].

Culture based antibiotic susceptibility testing is the gold standard of antibiotic resistance testing in hospitals throughout the world. It is a relatively cheap and easy technique with little need for sophisticated or expensive equipment. Therefore, using susceptibility testing would enable the comparison of non-clinical data with clinical data. Antibiotic susceptibility testing also enables the phenotypic detection of as yet uncharacterized resistance mechanisms and complex resistance mechanisms such as efflux, which are frequently mediated by several genes. However, antibiotic resistance and antibiotic breakpoints have been defined within the context of their medical functions. Breakpoints define the thresholds of response of bacteria to an antibiotic [7]. Clinical breakpoints define bacteria as susceptible, intermediate or resistant to an antibiotic and are calculated using several factors, including clinical results from studies, antibiotic dosing and pharmacokinetic (PK) and pharmacodynamics (PD) measurements [8]. Breakpoints are used as a guide for the clinician to decide how to treat the patient, with antibiotic resistance meaning treatment failure. Clinical resistance is a complex concept in which the type of infecting bacterium, its location in the body, the distribution of the antibiotic in the body and its concentration at the site of infection, and the immune status of the patient all interact. The difficulty arises when we apply these definitions of resistance to soil bacteria or non-pathogenic bacteria, for which no breakpoints exist. Antibiotic resistant bacteria detected in soil to date have been defined as bacteria capable of growth at 20 µg/ml, as no breakpoints exist for the non-pathogenic bacteria found in soil [5,6].

The bacteria that can be grown in the laboratory are only a small fraction of the total diversity that exists in nature. Approximately only 1% of bacteria on Earth can be readily cultivated *in vitro* [9,10]. Therefore, non-culture based tools such as PCR and metagenomics are required to capture the non-culturable section of the non-clinical antibiotic resistome. However, these tools are limited to screening for known resistance genes and are not sufficient to characterize the intrinsic resistance or efflux resistance mechanisms, that are mediated by several genes.

Mechanistic commonalities between soil bacteria and clinical pathogens were first identified in the 1970s, including identical molecular aminoglycoside resistance mechanisms in *Streptomyces* and clinical pathogens [11]. However, the resistance mechanisms are not limited to antibiotic producing soil bacteria. The soil may be a reservoir of resistance genes, which are already present in human pathogens or which may emerge to increase the current arsenal of antibiotic resistance mechanisms in pathogens. Most antibiotics used in human medicine have been isolated from soil microorganisms. Therefore, soil is thought of as a potential reservoir of antibiotic resistance genes. The presence of antibiotics in soil is believed to have promoted the development of highly specific antibiotic resistance mechanisms in antibiotic producing and non-producing bacteria [4]. This belief is based on studies, which have identified resistance genes such as bla_{CTX-M}, $qnrA$ and bla_{NDM} as originating in the environmental bacteria *Kluyvera* sp., *Shewanella algae* and *Erythrobacter litoralis*, respectively [12–14]. These genes are clinically relevant resistance genes and are currently causing difficulties in the treatment of bacterial infections. The origins of other plasmid mediated resistance genes are still unknown. The commonly believed theory of the role of the soil resistome is based on the belief that antibiotic production and resistance co-exist in soil bacteria, as demonstrated by studies of antibiotic biosynthetic pathways and genome analysis [5,15]. The theory relies upon the idea that without the resistance gene the antibiotic producing bacteria would self-destruct, on production of the antibiotic. However, Davies and Davies identified that this theory remains to be proven [16]. In order to understand the importance of soil as a potential reservoir of antibiotic resistance mechanisms we need to investigate the soil bacteria.

The soil is a reservoir of antibiotic resistance genes, but not all resistance mechanisms are necessarily a threat to the continued use of antibiotics in all pathogens [17]. Intrinsic resistance is a characteristic of almost all isolates of the bacterial species and occurs when the antimicrobial activity of the drug is clinically insufficient or antimicrobial resistance is innate, rendering it clinically ineffective [18]. The most intrinsically-resistant bacteria have a non-clinical origin (e.g. soil), which are less likely to have antibiotic selective pressures equivalent to hospitals. This suggests that the main physiological role of the elements involved in the intrinsic resistance phenotype is not conferring resistance to antibiotics [19,20].

We aimed to identify the levels of culturable resistant bacteria, both antibiotic producers and non-antibiotic producers, and identify the roles of the different mechanisms of resistance such as efflux, novel enzymatic resistance mechanisms and selected plasmid mediated resistance genes in conferring multi-drug resistance within the soil bacterial community. We linked the bacterial phylogeny to the antibiotic resistance profiles in order to compare the intrinsic resistance profiles of soil and clinical bacteria from the same bacterial order.

Results

Four hundred and twelve antibiotic resistant bacterial isolates were cultured from ten soils from agricultural, urban and pristine environments (Table 1). All isolates were multi-drug resistant, of which greater than 80% were resistant to 16–23 antibiotics, comprising almost all classes of antibiotic (Figures 1, 2, S1, S2). The 23 antibiotics tested are used in agriculture, human and veterinary medicine and covered all known antibiotic classes and mechanisms of action. They included natural antibiotics such as penicillin and streptomycin, semi-synthetic antibiotics such as cefotaxime and cephalexin, synthetic antimicrobials such as ciprofloxacin and sulfamethizole, and antibiotics used as the last line of defense, vancomycin and colistin, to treat multi-drug resistant infections such as methicillin resistant *Staphylococcus aureus* (MRSA) and *Acinetobacter*, respectively. Multi-drug resistance is defined as resistance to three or more different classes of antibiotics.

Beta-lactam resistance ranged from 84% to 100% and was greater than 90% on average for the ten soils (Figure 1), indicating a natural high level of resistance to the β-lactams, both the natural penicillins and the semi-synthetic cephalosporin, in all soils and little variation between the soils. In contrast, an average of less than 20% of the soil populations were resistant to the fluoroquinolones; levofloxacin and ciprofloxacin, while individual soil resistance levels ranged from 0–48% (Figures 1, S3). However, resistance to the non-fluorine quinolone, nalidixic acid, was on average 84% and ranged from 65% to 97% (Figures 1, S3). Greater than 70% of the bacterial populations in all soils were resistant to the tetrahydrofolate synthesis inhibitor antibiotics; sulfisoxazole, sulfamethoxazole and trimethoprim and to rifampicin, vancomycin, chloramphenicol, D-cycloserine, novobiocin, and erythromycin. Resistance to the aminoglycosides; amikacin, streptomycin, kanamycin, gentamicin and sisomicin and to colistin and tetracycline were variable, dependent on the soil. Soils S2 (urban), S9 (agriculture) and S10 (urban) had higher levels of resistance to one or more aminoglycosides than the other soils and soils S4 (agriculture), S6 (pristine) and S7 (urban) (Figure S4) had lower resistance levels to tetracycline than the other soils (Figure S5). Colistin resistance varied from 34% in soil S6 (pristine) to 100% in soil S8 (pristine) (Figure S6). The variation in resistance levels were not due to the anthropogenic use of the soil.

Multi-drug resistance conferred by efflux was identified in 83% of the soil isolates and 410 of 412 isolates were capable of conferring resistance to at least one class of antibiotics by efflux. Over 76% of the fluoroquinolone (levofloxacin and ciprofloxacin) resistance was mediated by efflux (Figure 3). In comparison, efflux mediated resistance to nalidixic acid, the non-fluorine quinolone antibiotic, was identified in 36% of the resistant population. Efflux accounted for 80% and 72% of the resistance mechanisms to tetracycline and rifampicin, respectively and 83% of colistin resistance. Efflux was identified as the main mechanism of resistance to the aminoglycosides amikacin, gentamicin, sisomicin and kanamycin in greater than 56% of the total soil populations but only in 20% of streptomycin resistant isolates. Efflux mediated resistance to the β-lactams penicillin, dicloxacillin and cephalexin and to vancomycin was identified in very few isolates (≤5%). However, efflux mediated resistance to the cephalosporin cefotaxime was conferred in 40% of the resistant isolates. A low percentage of efflux mediated resistance was detected for the tetrahydrofolate reductase pathway inhibitors sulfamethoxazole (5%), sulfisoxazole (2%) and trimethoprim (17%). The antibiotics D-cycloserine, novobiocin and erythromycin also were also infrequently the substrates of efflux pumps. Whether the

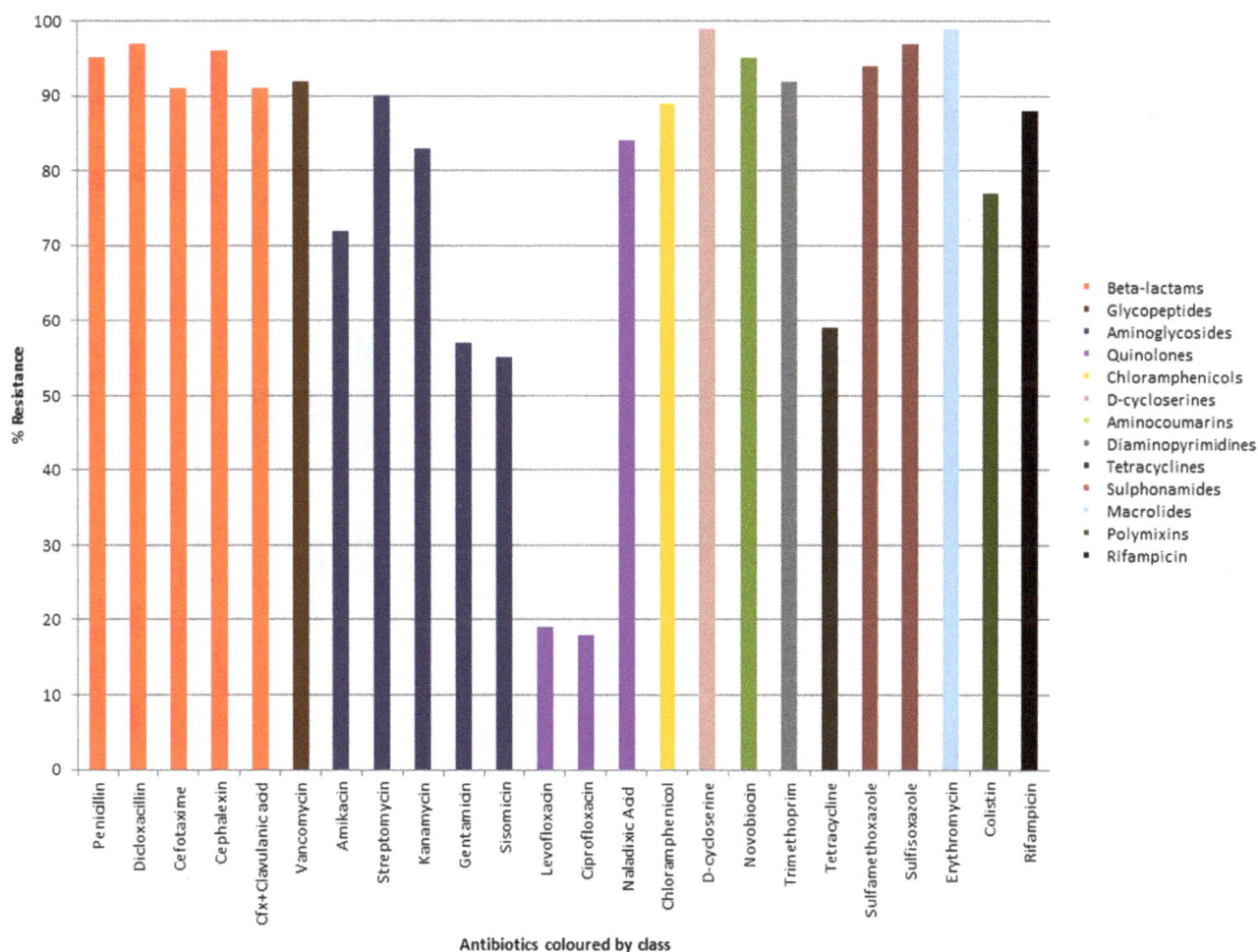

Figure 1. The antibiotic resistance profiles of all the study bacteria from ten soils. Isolates were individually screened for resistance to each of 23 antibiotics.

antibiotics were natural, semi-synthetic or synthetic did not play a role in the likelihood of them being a substrate for efflux.

The phylogenetic profiling identified four phyla: Proteobacteria (81.7%), Firmicutes (6.3%), Actinobacteria (1.5%) and Bacteroidetes (1%) (Figure 4). The remaining 9.5% were unclassified bacteria. These phyla were repesented by ten orders including Pseudomonadales, Aeromonadales, Burkholderiales, Xanthomonadales, Bacillales, Enterobacteriales and Vibrionales. Over 50% of the isolates belonged to the order Pseduomonadales.

It is important to factor in the role of intrinsic resistance in the levels of resistance, especially to the β-lactams. The pathogenic bacteria *Pseudomonas aeruginosa*, *Stenotrophomonas maltophilia*, *Burkholderia cepacia*, *Acinetobacter* species, *Achromobacter xylosoxidans*, *Serratia marcescens* and *Aeromonas* species are intrinsically resistant to penicillin and the β-lactam antibiotics [18]. We defined intrinsic resistance as resistance of the majority of the population to the antibiotic as no definition exists for environmental bacteria. The Bacillales, Pseudomonadales, Burkholderiales, Xanthomonadales (*Stenotrophomonas* species), Enterobacteriales (78% *Serratia* species) and Aeromonadales constituted 87% of the soil bacterial population within this study. The classes of antibiotics used to treat *Pseudomonas* species infections include the fluoroquinolones, amikacin, gentamicin and colistin as an antibiotic of last resort. Intrinsic resistance to these antibiotics has not yet been identified

in clinical *Pseudomonas* species and levels of resistance remains relatively low [22]. The dominant bacterial species investigated were intrinsically resistant to the β-lactams, which are most frequently constitutively or inducibly expressed β-lactamases. The results of the penicillin enzyme assay and the efflux tests confirmed that the resistance mechanisms were enzyme based and not efflux mediated, which concur with the intrinsic mechanisms identified in clinical bacteria from these orders.

In order to identify the impact of intrinsic resistance and efflux in the resistance profiles of the soil bacteria we separated the resistance levels according to most frequently identified bacterial orders (Table 2). Resistance to vancomycin, rifampicin, trimethoprim and erythromycin was mediated by intrinsic resistance or efflux in all six orders consistent with the intrinsic resistance of the clinically relevant species of these orders. Clinical isolates of *Burkholderia cepacia* and *Stenotrophomonas* are intrinsically resistant to amikacin and gentamicin, and *Serratia marcescens* to amikacin. The soil bacteria belonging to the orders Burkholderiales, Xanthomonadales (*Stenotrophomonas* species) and Enterobacteriales (almost all *Serratia* species) were also intrinsically resistant to the same aminoglycosides as their clinically relevant counterparts. In the soil bacterial populations the Bacillales, Pseudomonadales and the Aeromonadales were also intrinsically resistant to amikacin and gentamicin in contrast to clinical findings for the bacteria and the

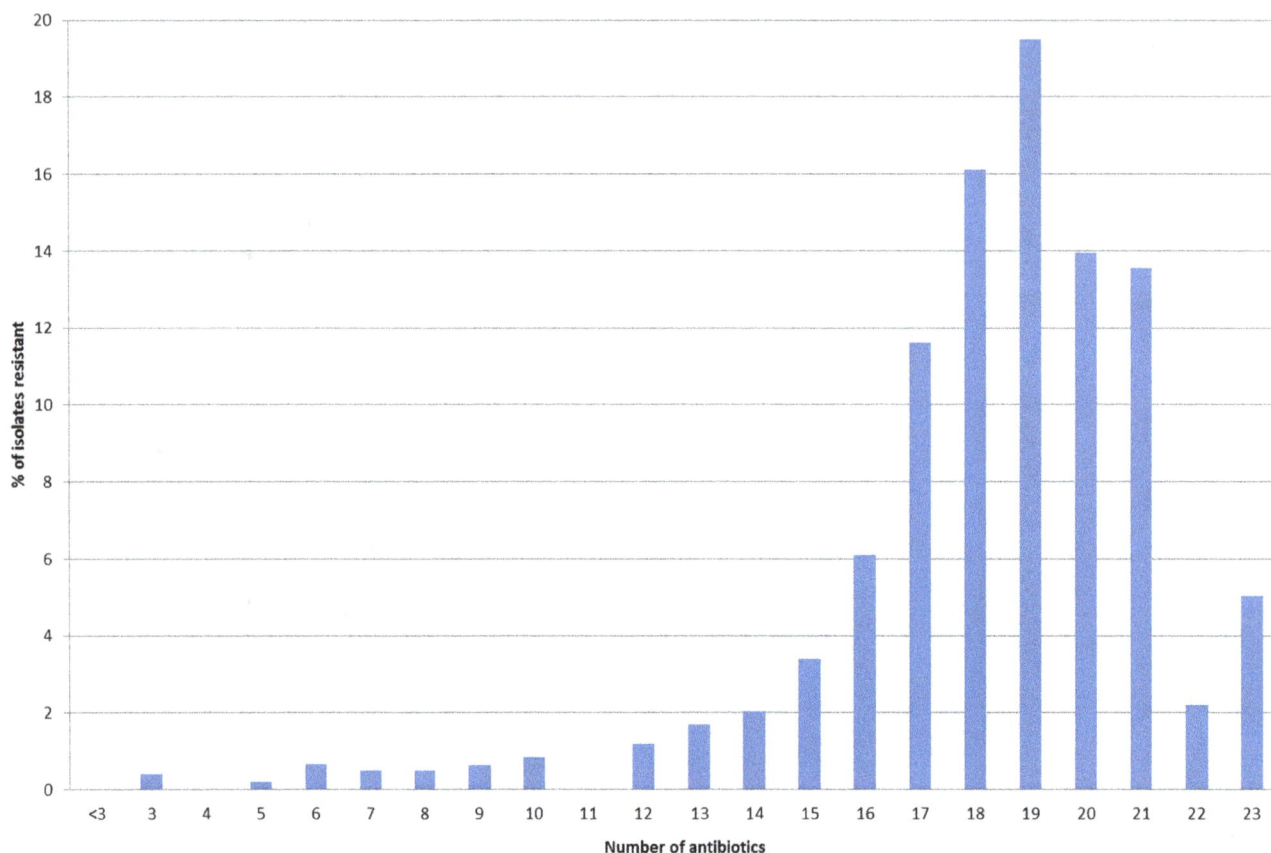

Figure 2. Resistance spectrum of the bacterial soil isolates. Antibiotic resistance was measured as growth on 20 µg/ml of antibiotic.

Enterobacteriales were additionally intrinsically resistant to gentamicin. Resistance to gentamicin was mainly mediated by efflux.

Innate resistance to ciprofloxacin has been described only in the *Burkholderia cepacia* complex species of clinical bacteria [18]. However, in this study only ten of 34 Burkholderiales isolates were resistant to ciprofloxacin, nine of which were positive for efflux. In contrast to their clinically relevant species, the Xanthomonadales and Bacillales were both intrinsically resistant to ciprofloxacin. The remaining species had low levels of resistance

to ciprofloxacin, mainly due to the non-specific efflux resistance mechanism.

Clinical *Pseudomonas aeruginosa* are intrinsically resistant to the tetracycline class [18]. However, the Burkholderiales, Xanthomonadales and Enterobacteriales, in addition to the Pseudomonadales, from the soils were also intrinsically resistant to tetracycline by efflux. Intrinsic resistance to colistin in clinical isolates has been described in gram-positive bacteria, such as Bacillales and the gram-negative *Burkholderia* and *Serratia marcescens* species. Our results identified that in contrast to the clinical findings, these

Table 1. Geographical and descriptive characteristics of the analyzed soil samples.

Sample ID	Description	Environmental matter	Elevation (m)	Latitude	Longitude	pH
S1	Wädenswil Apple Orchard	Agriculture	407	47.2333	8.6667	7.0
S2	Einsiedeln	Urban	880	47.1167	8.75	7.1
S3	Hospital garden	Urban	667	46.7167	9.4333	7.3
S4	Lindau Apple Orchard	Agriculture	485	47.4833	8.2	4.1
S5	Güttingen Apple Orchard	Agriculture	503	47.6	9.2833	5.3
S6	Ruetli meadow	Pristine	835	46.9667	8.6	7.1
S7	Area beside Lake Zürich	Urban	408	47.3667	8.55	7.0
S8	Matterhorn mountain trail	Pristine	1936	46.0167	7.75	7.2
S9	Farmland treated with pig manure	Agriculture	592	47.1833	8.3167	5.7
S10	Mountain near Zürich	Urban	700	47.3496	8.492	7.0

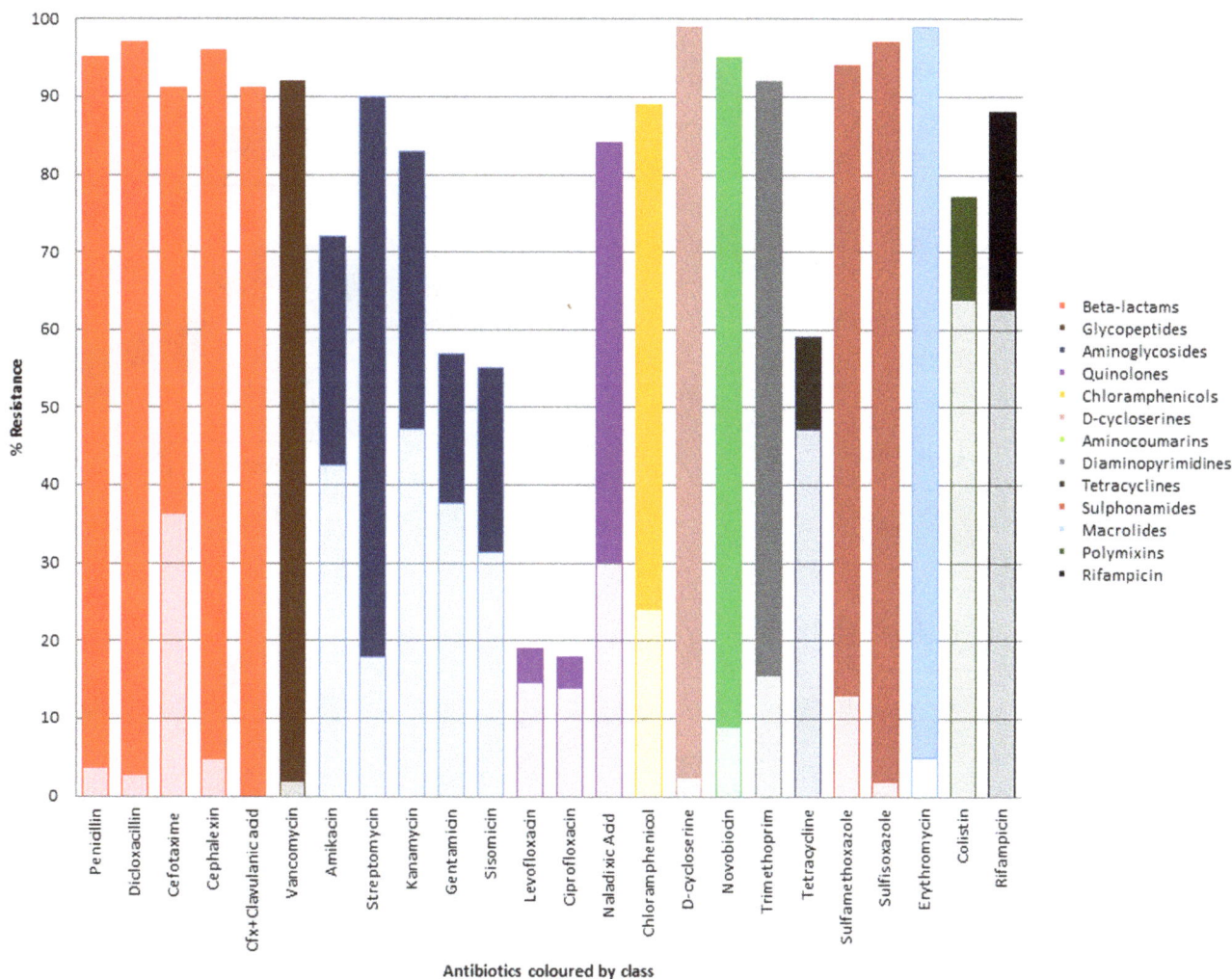

Figure 3. The antibiotic resistance profiles and percentage mediated by efflux of all the study bacteria. The antibiotics are color coded according to class. The total bar is the percentage resistance of all isolates to each antibiotic. The lightly shaded bar is the percentage of resistance mediated by efflux for each antibiotic.

species in the soil populations were not intrinsically resistant to colistin and the Xanthomonadales and Aeromonadales bacteria were intrinsically resistant to colistin and was mainly mediated by efflux in all species, except the Xanthomonadales.

The dissemination of extended spectrum β-lactamases (ESBL) and plasmid mediated quinolone resistance (PMQR) multi-drug resistant plasmids is causing increasing difficulty in the treatment and management of both community and hospital acquired infections. As these antibiotic resistance genes are thought to have their origins in environmental bacteria it was important to investigate the role of these genes in the soil resistome [11–14]. Extended spectrum β-lactamases, unlike AmpC β-lactamases, are inhibited *in vitro* by clavulanate. All β-lactam resistant isolates were tested for the presence of ESBL by comparison of growth on cefotaxime with growth on cefotaxime and clavulanate. Growth on cefotaxime but inhibition on cefotaxime and clavulanate suggested the presence of ESBL. Thirty isolates were ESBL positive, 50% of which were isolated from the urban soil S3. None of the plasmid mediated ESBL genes nor β-lactamase genes were detected in any of the isolates or the extracted soil DNA (Table 3). No inhibitor resistant β-lactamases were detected. The extracted DNA from the ten soil samples were screened for the presence of

$bla_{\text{NDM-1}}$, all samples were negative. Extended spectrum β-lactamase production was identified in a wide variety of orders: Enterobacteriales (6/18 bacterial isolates), Pseudomondales (6/221), Aeromonadales (5/31), Burkholderiales (5/34), unclassified (5/39), Actinomycetales (1/6), Bacilliales (1/26) and Xanthomonadales (1/28).

One hundred and four isolates resistant to the fluoroquinolones, levofloxacin and ciprofloxacin, and the total DNA extracted from the soils were screened for the presence of PMQR genes (Table 3). All fluoroquinolone resistant isolates and soil DNA extracted from the ten soils were negative for the *qnrA*, *qnrB*, *qnrS* and *qep* genes by PCR. All isolates were negative by sequencing for the *aac(6′)Ib-cr* gene. The DNA extracted from five soils contained novel *aac(6′)Ib* genes, with between two and seven amino acid mutations in comparison to the closest previously described protein or amino acid sequence in GenBank. These novel gene sequences have been deposited in GenBank (accession numbers KC916938, KC916939, KC916940). The DNA extracted from soils S1 (agriculture), S7 (urban) and S10 (urban) contained the same novel *aac(6′)Ib* gene (KC916938). Soils S4 (KC916939, agriculture) and S9 (KC916940, agriculture) each contained another novel *aac(6′)Ib* resistance gene containing two and three amino

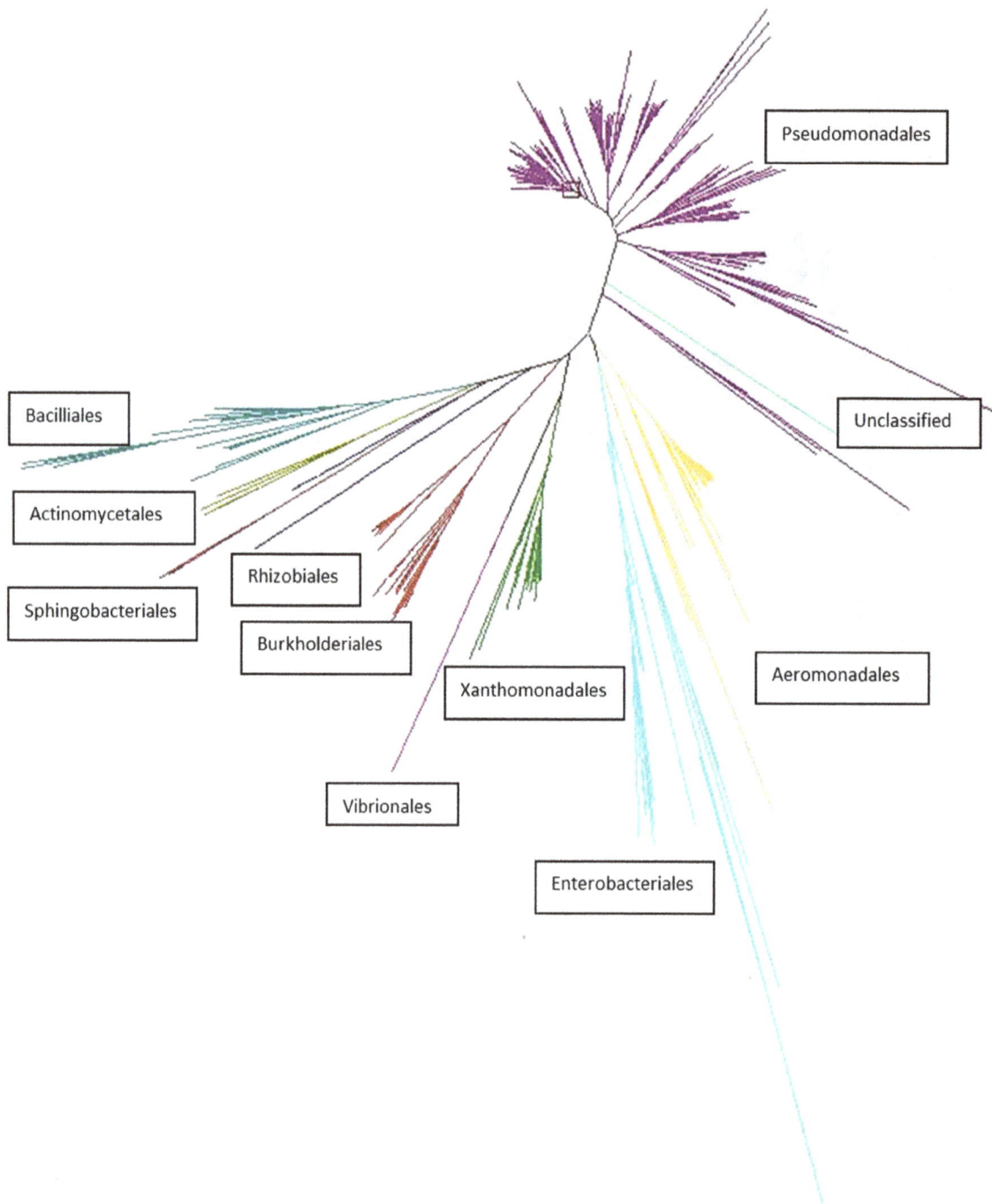

Figure 4. The phylogenetic distribution and relatedness of the soil bacteria. The 16S rRNA genes were sequenced from all bacteria and the phylogenetic tree was created using these sequences and the ARB program package. Branches of the tree are color coded by bacterial orders, with each line at the branch end representing an individual bacterial isolate. The orders with the largest number of lines indicate the largest number of bacterial members within this study. The relatedness of the orders is defined by the distance from one branch to the next. The unclassified bacteria are therefore, closest to the Pseudomonadales.

acid mutations, respectively, in comparison to their closest described amino acid sequence in GenBank.

Antibiotic inactivation assays were performed in order to identify novel mechanisms of antibiotic resistance in the soil resistomes (Figure 5). Inactivation of penicillin and trimethoprim were identified in 100% and 20% of the investigated bacteria, respectively. The trimethoprim inactivating bacteria were inves-

tigated for inactivation of all other antibiotics assayed and did not result in inactivation of any except penicillin. Thus, this inactivation mechanism is a trimethoprim specific effect and not a non-specific effect, such as indole production [21]. The bacterial species containing the inactivation mechanism are indole negative. This is the first description of an inactivating resistance mechanism against trimethoprim. The principle of the assay was that

Table 2. The variations in the intrinsomes of soil and clinical bacteria.

Bacterial order	Penicillin (Soil/Clinical bacteria)	Amikacin (Soil/Clinical bacteria)	Gentamicin (Soil/Clinical bacteria)	Ciprofloxacin (Soil/Clinical bacteria)	Tetracycline (Soil/Clinical bacteria)	Colistin (Soil/Clinical bacteria)	Vancomycin (Soil/Clinical bacteria)	Rifampicin (Soil/Clinical bacteria)	Chloramphenicol (Soil/Clinical bacteria)	Trimethoprim (Soil/Clinical bacteria)
Bacillales	I/N[a]	I/N	I/N	I/N	N/N	N/I	I/I	I/N	I/N	I/N
Pseudomonadales	I/I	I/N	I/N	N/N	I/N	N/N	I/I	I/I	I/I	I/I
Burkholderiales	I/I	I/I	I/I	N/I	I/N	N/I	I/I	I/I	I/I	I/I
Xanthomonadales	I/I	I/I	I/I	I/N	I/N	I/N	I/I	I/I	I/N	I/I
Enterobacteriales (78% Serratia sp.)	I/I	I/I	I/N	N/N	I/N	N/I	I/I	I/I	N/N	I/N
Aeromonadales	I/I	I/N	I/N	N/N	N/N	I/N	I/N	I/I	N/N	I/N

The patterns of intrinsic resistance for each bacterial order are compared to those of their clinical counterparts. Those in bold highlight the variations in intrinsic resistance between soil bacteria of this study and their clinical counterparts.

[a] I = intrinsically resistant N = not intrinsically resistant.

antibiotic inactivating enzymes produced by the test bacteria would diffuse into the agar, in a similar fashion to antibiotics in the disk diffusion assay (Figure 5). None of the remaining antibiotic classes were inactivated using this assay, indicating that enzymatic inactivation of these classes of antibiotics within these bacteria was unlikely.

Discussion

In order to analyze the soil antibiotic resistome, we investigated soils closest to human activities i.e. farming and urban, and compared these to a pristine soil population. While culture techniques will always bias the population of bacteria under study they have the advantage of being able to detect phenotypic resistances, including efflux mediated resistance and novel resistance mechanisms, and the resistance mechanism can be investigated in further detail. Culture techniques are also the gold standard for clinical testing of antibiotic resistance levels and certain resistance mechanism such as efflux or ESBL.

The resistance levels to almost all classes of antibiotics were extremely high. We identified the multi-drug resistant nature of soil bacteria, with greater than 80% of all isolates resistant to 16–23 antibiotics. Antibiotic resistance of bacterial pathogens consists of acquired resistance mechanisms and intrinsic resistance mechanisms. In order to elucidate the resistance mechanisms responsible for these high levels of resistance we investigated the bacteria and soil DNA for different resistance mechanisms, including efflux, production of ESBL and the presence of PMQR genes and antibiotic inactivation.

Intrinsic resistance in pathogenic bacteria has been well characterized in order to ensure that the most appropriate antibiotic therapy is administered [18]. In order to identify the role of the intrinsic resistance in the soil bacteria, the resistance phenotypes were linked to bacterial order. These were compared to the intrinsic resistances associated with the clinically relevant species of the same order (e.g., Pseudomonadales with *Pseudomonas aeruginosa*). We expected that the intrinsic resistance patterns of the clinically relevant species would correspond to the soil bacteria, in particular for bacterial pathogens that originated from soil, such as *Stenotrophomonas maltophilia*, *Burkholderia* species and *Pseudomonas* species. Resistance to the β-lactam and macrolide classes of antibiotics is mediated by intrinsic resistance in these clinically relevant species and soil bacteria. However, the pattern of intrinsic resistance of the soil bacteria of this study to the aminoglycosides, ciprofloxacin (for Burkholderiales only), tetracyclines, and colistin were not those associated with the corresponding pathogens. In some cases the soil bacteria were not intrinsically resistant (e.g., Burkholderiales to ciprofloxacin), and in others they were intrinsically resistant to antibiotics used to treat pathogenic bacteria (e.g., Pseudomonadales were intrinsically resistant to amikacin). *Burkholderia* species and *Stenotrophomonas maltophilia* are intrinsically resistant to all aminoglycosides, attributed to poor permeability and putative efflux [18]. These results suggest that while species such as *Pseudomonas* and *Stenotrophomonas* have their origins in soil, their closely related soil species do not necessarily contain the same intrinsic resistance mechanisms.

Multi-drug resistance was most frequently mediated by the efflux resistance mechanism and was identified in non-antibiotic producing bacteria. Resistance to colistin and tetracycline in particular was frequently mediated by efflux. Resistance to colistin in clinically relevant *Pseudomonas* species is most frequently associated with alterations in the outer membrane lipopolysaccharide (LPS) [25]. Efflux mediated colistin resistance has to date only been identified in *Yersinia* species but has not been well

Table 3. Genotypic screening of mobile antibiotic resistance genes in the bacterial isolates and extracted soil DNA.

Antibiotic resistance phenotype	Study samples investigated	Resistance genes	Detected (n[b])
ESBL[a]	Total soil DNA (n = 10) Bacteria isolates (n = 30)	bla_{TEM}	0
ESBL[a]	Total soil DNA (n = 10) Bacteria isolates (n = 30)	bla_{SHV}	0
ESBL[a]	Total soil DNA (n = 10) Bacteria isolates (n = 30)	bla_{OXA},	0
ESBL[a]	Total soil DNA (n = 10) Bacteria isolates (n = 30)	bla_{CTX-M} Groups 1, 2, 8, 9, 25	0
ESBL[a]	Total soil DNA (n = 10) Bacteria isolates (n = 30)	bla_{VEB}	0
ESBL[a]	Total soil DNA (n = 10) Bacteria isolates (n = 30)	bla_{PER}	0
ESBL[a]	Total soil DNA (n = 10) Bacteria isolates (n = 30)	bla_{GES}	0
Quinolone	Total soil DNA (n = 10) Bacteria isolates (n = 346)	aac(6')Ib-cr	0
Quinolone	Total soil DNA (n = 10) Bacteria isolates (n = 346)	qnrA	0
Quinolone	Total soil DNA (n = 10) Bacteria isolates (n = 346)	qnrB	0
Quinolone	Total soil DNA (n = 10) Bacteria isolates (n = 346)	qnrS	0
Quinolone	Total soil DNA (n = 10) Bacteria isolates (n = 346)	qep	0
Carbapenem	Total soil DNA (n = 10)	bla_{NDM}	0

[a]ESBL = extended spectrum β-lactamase.
[b]n = number.

characterized [26]. Tetracycline resistance in Burkholderiales, Xanthomonadales, Enterobacteriales and Aeromonadales was most frequently mediated by efflux and in all case in greater than 50% of the resistant populations. Although efflux mediated resistance to tetracycline has been described in clinically relevant species it is infrequently identified in the populations. Intrinsic resistance (over 50% of the population resistant) to tetracycline in clinically relevant species of these orders has not been described. The identification of efflux mediated colistin and intrinsic tetracycline resistance in the soil bacteria identifies potential novel mechanisms of resistance, which could evolve in or be transferred to clinically relevant species over time.

Bacteria have the ability to survive in antibiotics due to their evolution in the presence of natural antibiotics over millennia but in our study this also holds true for the synthetic antibiotics, which have not been present in soils. These soil bacteria have developed or utilize universal mechanisms of resistance to overcome the inhibtion by natural antibiotics or toxins, which also work efficiently against many classes of antibiotics [27]. Efflux pumps are evolutionary ancient elements, highly relevant for the physiology and ecological behaviour of all living things [2]. However, efflux resistance is not confined to a chromosomal resistance mechanism but may be transferred on mobile elements, e.g. qep or recently a resistance-nodulation-cell division/multidrug resistance efflux mechanism [28,29].

None of these ESBL or quinolone mediated plasmid mediated resistance genes were present in either the bacteria or the DNA. These results suggest that these resistance genes are either location or sample type specific or in extremely low numbers in the environment. The lack of bla_{NDM} β-lactamases in the soil DNA is similar to findings from a study conducted in another European site (Cardiff wastewater treatment plant), which also contained no bla_{NDM} genes but are in contrast to environmental sites in New Delhi, India, where bla_{NDM} was detected in two of 50 drinking-water samples and 51 of 171 seepage samples [23]. The first bla_{NDM-1} infections in Switzerland were isolated from three patients in Geneva university hospitals between 2009 and 2010 [24]. Thus, it is unlikely that Swiss soils were the sources of these clinical bla_{NDM} positive bacteria.

This study identified the first inactivating resistance mechanism against trimethoprim in 21 different bacterial isolates from five different soils, comprising farmland and urban environments. Although previous studies, restricted to cave bacteria and Actinomycetes, have investigated soil bacteria for trimethoprim inactivating enzymes they did not result in their identification [5,6]. Trimethoprim is a synthetic antibiotic, which is not used in agriculture and thus is restricted to clinical use and it is unlikely that the soil bacteria were exposed to trimethoprim. The presence of an inactivation mechanism against a synthetic antibiotic suggests that it may have evolved for an alternative function. The trimethoprim inactivation mechanism was identified as trimethoprim specific, as these bacteria did not inactivate any other class of antibiotics. Penicillin inactivating enzymes were also identified in all investigated resistant isolates. Gram-negative bacteria frequently contain chromosomally mediated β-lactamases and have most likely developed these mechanisms through millennia of co-evolution with penicillin producing soil dwelling fungi. High levels of β-lactam inactivation were identified in penicillin resistant bacteria isolated from an isolated cave [6]. Thus, the high frequency of penicillin inactivating enzymes is due to intrinsic resistance of these bacteria.

There have been many calls for more information about the natural resistome and these have also highlighted the importance of understanding the soil resistome in the preservation of antibiotics for the treatment of infections [3,4,27]. Our study provided a comprehensive description of the soil resistome in relation to multi-drug resistance together with phylogenetic analysis of culturable soil bacteria under a range of anthropogenic influences. The novel inactivation mechanism detected in this study suggest that the soil bacteria could be a reservoir of resistance mechanisms to synthetic antibiotics as well as natural antibiotics.

While soil bacteria are naturally resistant to many different classes of antibiotics, natural, snythetic and semi-synthetic, the resistance mechanisms are most frequently the non-specific efflux or intrinsic resistance mechanisms. The efflux mechanisms could be present on mobile elements or may be transferred to mobile DNA and cause increased difficulties in the treatment of bacterial pathogens. We have identified that the patterns of intrinsic resistance of the soil bacteria differ from the clinical isolates of the same phyla. Clinical isolates of Burkholderiales are intrinsically resistant to ciprofloxacin and colistin but soil isolates of the order Burkholderiales were not intrinsically resistant to ciprofloxacin or colistin. Similarly, clinical isolates of *Serratia marcescens* and *Bacillus* species are intrinscially resistant to colistin but were not intrinscially resistant to colistin in the soil population. The results of this study identified that the intrinsic resisome of soil bacteria is in contrast to that of their closely related clinically relevant species. These results suggest that the evolution of intrinsic resistance in soil and clinical bacteria has not developed along the same lines, although there may be similarities in certain resistance genes detected in both environments.

The conclusions of this study are that there was a high level of multi-drug resistance in soil bacteria to a wide variety of antibiotics and was not dependent on the soil use. This MDR was most frequently conferred by efflux, which if present on or transferred to mobile elements could cause increased difficulties in the treatment of human bacterial pathogens. These results can be used to enhance the understanding of the emergence and dissemination of novel antibiotic resistance from the natural reservoir to the clinical setting, which may aid the development of inhibitors of resistance mechanisms and resistant bacteria.

Methods

Antibiotics

The following antibiotics and antibiotic resistance mechanism inhibitors were used in this study: Penicillin, dicloxacillin, amikacin, cephalexin, kanamycin (Carl Roth GMBH and CO. KG), gentamicin, sisomicin, streptomycin, vancomycin (Carl Roth GMBH and CO. KG), levofloxacin, ciprofloxacin, sulfamethizole, nalidixic acid, chloramphenicol, d-cycloserine, cefotaxime, novobiocin, trimethoprim, sulfisoxazole, tetracycline, erythromycin, colistin, rifampicin, cefotaxime, clavulanic acid, phenylboronic acid and carbonyl cyanide 3-chlorophenylhydrazone (CCCP). All antibiotics and chemicals, except kanamycin and vancomycin, were obtained from Sigma-Aldrich Chemie Gmbh, Buchs SG, Switzerland. All antibiotic solutions were prepared according to manufacturer's instructions.

Description of Soils

The soils were collected from four agricultural sites, four urban sites and two pristine environments as described in Table 1. No specific permissions were required for these locations, as they were part of the research centers field trials or were not protected land. The field studies did not involve endangered or protected species.

Culturing of Antibiotic Resistant Soil Bacteria

The antibiotic resistant bacteria were isolated as previously described [5]. Soil samples (1.9 g) were suspended in 15 ml LB

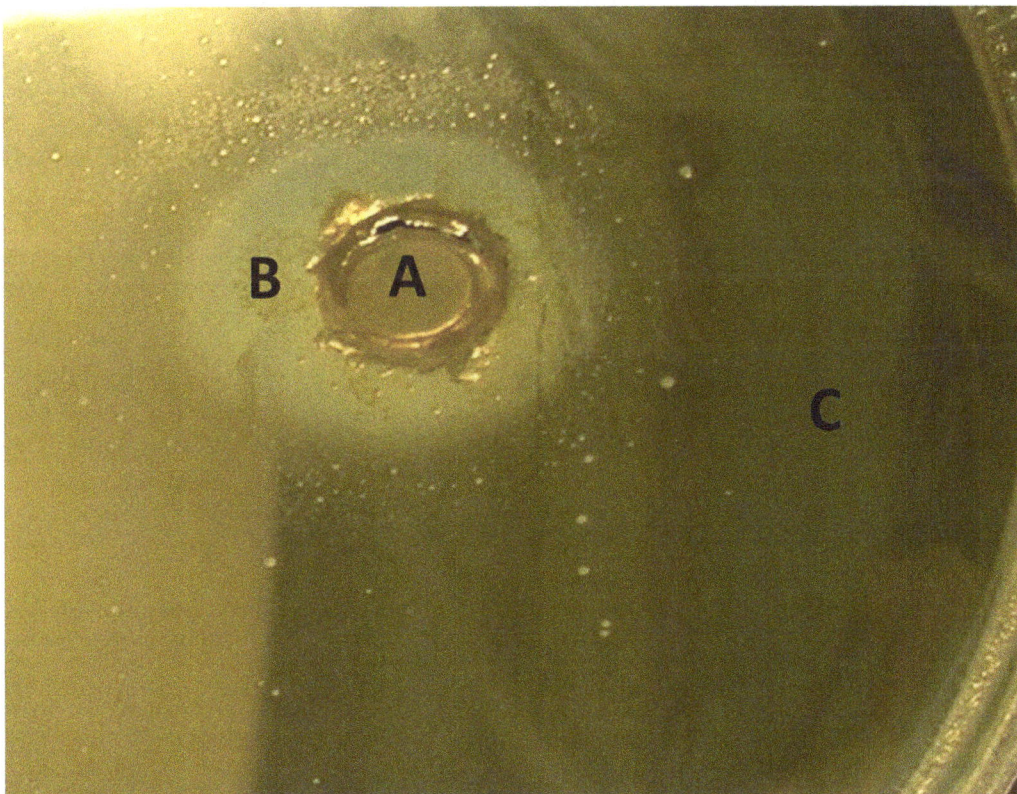

Figure 5. Enzyme inhibition assay using trimethoprim (20 µg/ml) and trimethoprim susceptible *S. aureus* ATCC 25923. An agar plate section containing trimethoprim (20 µg/ml), was inoculated with trimethoprim susceptible *S. aureus* ATCC 25923. An agar plug containing the trimethoprim inhibiting soil bacteria were placed into the inoculated trimethoprim agar. The soil bacteria producing the enzyme inhibiting trimethoprim is located in the center circle (A). The ring of growth of trimethoprim susceptible *S. aureus*, which has been protected from the inhibitory effects of trimethoprim by the soil bacteria enzyme (B). There is no *S. aureus* growth outside the zone of trimethoprim inhibition by the soil bacteria (C).

Broth. The solution was shaken and allowed to settle for 5 minutes. Aliquots of 500 µl from supernatants of each soil suspension were inoculated into 4.5 ml of LB broth in Ritter Riplates® (Ritter GmbH, Schwabmuenchen, Germany) and incubated at 22°C for 3 days. Aliquots of 200 µl from each soil sample culture was transferred to 1.8 ml LB broth containing 20 µg/ml of each antibiotic and incubated at 22°C for 4 days. The cultures were serially diluted on LB plates containing 20 µg/ml of the corresponding antibiotic and incubated at 22°C for 48 h. The individual colonies were plated onto antibiotic LB agar plates to obtain pure cultures. µg/ml.

Antibiotic Resistance Profiling

The antibiotic susceptibilities were investigated as previously described [5]. Four hundred and twelve isolates representing the antibiotic resistance populations of the ten soils were isolated. Each isolate was inoculated into 200 µl LB in 96 well plates from frozen glycerol stocks and incubated at 22°C for 3 days. LB plates, each containing 1 antibiotic at 20 µg/ml, were inoculated using a multipoint inoculator. LB agar plates without antibiotic were used as controls. The inoculated plates were incubated at 22°C for 4 days. Bacteria which presented visible growth on agar plates containing 20 µg/ml of antibiotic were defined as resistant [5].

Twenty three different antibiotics were tested. The ESBL phenotype was determined by resistance to cefotaxime and susceptibility to cefotaxime (20 µg/ml) with clavulanic acid (4 µg/ml). Phenotypic efflux resistance was defined as resistance to the antibiotic and susceptibility to the antibiotic when carbonyl cyanide 3-chlorophenylhydrazone (CCCP) at 20 µg/ml was added to the agar. Growth of 152 isolates were inhibited by CCCP at 20 µg/ml. Inhibition of these isolates by CCCP were tested at doubling dilution concentrations of 0.5 µg/ml to 10 µg/ml. These isolates were not inhibited by CCCP alone at a final concentration of 1 µg/ml. Therefore, efflux in these isolates was defined as resistance to an antibiotic and susceptibility to the antibiotic when 1 µg/ml CCCP was added.

Total Soil DNA Extraction

The total soil DNA from each of the ten soils was extracted using the Mo Bio PowerSoil® DNA Isolation Kit, Mo Bio Laboratories Inc. (Süd-Laborbedarf GmbH, Gauting, Germany).

Antibiotic Resistance Gene Screening

Phenotypically ESBL positive isolates and the extracted total DNA from the ten soils were screened by PCR for the bla_{TEM}, bla_{SHV}, bla_{OXA}, bla_{CTX-M} Groups 1, 2, 8, 9 and 25, bla_{VEB}, bla_{PER}, bla_{GES} ESBL resistance genes [30–32]. The fluoroquinolone resistant isolates and the extracted total DNA from the ten soils were screened by PCR for the $aac(6')Ib$-cr, qnrA, qnrB, qnrS and qep resistance genes using previously described primers and parameters [28,33,34]. $aac(6')Ib$-cr PCR products were sequenced using the PCR primers. The DNA samples extracted directly from the soils were additionally screened for the presence of bla_{NDM-1}. The bla_{NDM-1} PCR primers were NDMF 5′ GAAGCTGAGCACCG-CATTAG 3′ and NDMR 5′ TGCGGGCCGTATGAGTGATT 3′. The annealing temperature was 55°C and the PCR product of the positive control was approximately 800 bp. All resulting PCR products were sequenced. Positive controls were included in each PCR run except for bla_{VEB}, bla_{PER}, bla_{GES}.

All new data has been deposited in GenBank with the accession numbers KC916938, KC916939, KC916940.

Phylogenetic Profiling

The 16S ribosomal RNA (rRNA) genes of all isolates were amplified using bacterial 16S primers: Bact_63f 5′ CAGGCC-TAACACATGCAAGTC 3′ and Bact_1389r 5′ ACGGGCGGTGTGTACAAG 3′ [35]. The resulting PCR product was on average 1363 bp. For the bacterial isolates which produced no PCR product with these primers a second primer set was utilised [36]: 16SF 5′ TCCTACGGGAGGCAGCAGT 3′ and 16SR 5′ GGACTACCAGGGTATCTAATCCTGTT 3′. 16S rRNA gene amplicons were sequenced and classified using Greengenes [37]. Phylogenetic trees were constructed using the neighbor-joining algorithm in ARB using the Greengenes aligned 16S rRNA gene database [38].

Enzyme Inhibition Assays

The enzyme inhibition assays were performed using a plug assay. Escherichia coli ATCC 25922 or Staphylococcus aureus ATCC 25923 inocula were prepared to a 0.5 McFarland standard and 80 µl was spread on LB agar plates containing 20 µg/ml of penicillin, cefotaxime, kanamycin, tetracycline, gentamicin, ami-kacin, colistin, ciprofloxacin, levofloxacin, nalidixic acid, trimeth-oprim, rifampicin, erythromycin, vancomycin, streptomycin or chloramphenicol. Wells were cut aseptically in the antibiotic agar. Plugs corresponding to the same diameter as the wells were cut aseptically from the resistant bacteria plates and inserted into the wells of antibiotic containing plates. The plates were incubated at 37°C for 1–2 days. A circle of growth around the plugs, which reduced in cell density in relation to its distance from the plug, indicated antibiotic inactivation.

Supporting Information

Figure S1   The antibiotic resistance profiles segregated according to soil for all antibiotics. The antibiotics are color coded according to class.
(TIF)

Figure S2   The numbers of antibiotics each soil bacterial community is resistant to as a percentage of the total bacterial popoulation within the given soil.
(TIF)

Figure S3   The quinolone antibiotic resistance profiles of the soil bacteria separated according to soil sample.
(TIF)

Figure S4   The aminoglycosides antibiotic resistance profiles of the soil bacteria separated according to soil sample.
(TIF)

Figure S5   The tetracycline antibiotic resistance profiles of the soil bacteria separated according to soil sample.
(TIF)

Figure S6   The colistin antibiotic resistance profiles of the soil bacteria separated according to soil sample.
(TIF)

Acknowledgments

Work was conducted within the European research network COST TD0803 Detecting evolutionary hotspots of antibiotic resistances in Europe (DARE). The authors thank Dr Mark Toleman (Cardiff University, Wales), Prof Patrice Nordmann (South Paris University, France) and Dr Neil Woodford (HPA, England) for providing reference strains.

Author Contributions

Conceived and designed the experiments: FW BD. Performed the experiments: FW. Analyzed the data: FW. Contributed reagents/materials/analysis tools: BD. Wrote the paper: FW BD.

References

1. Pruden A, Pei RT, Storteboom H, Carlson KH (2006) Antibiotic resistance genes as emerging contaminants: Studies in northern Colorado. Environ Sci Technol 40: 7445–7450.
2. Rosenblatt-Farrell N (2009) Antibiotics in the Environment. Environ. Health Perspect. 117: A248–A250.
3. American Academy of Microbiology (2009) Antibiotic resistance: an ecological perspective on an old problem (American Academy of Microbiology, Washington DC).
4. Aminov RI (2009) The role of antibiotics and antibiotic resistance in nature. Environ Microbiol 11: 2970–2988.
5. D'Costa VM, McGrann KM, Hughes DW, Wright GD (2006) Sampling the antibiotic resistome. Science 311: 374–377.
6. Bhullar K, Waglechner N, Pawlowski A, Koteva K, Banks ED, et al. (2012) Antibiotic resistance is prevalent in an isolated cave microbiome. PLoS One 7: e34953.
7. Clinical and Laboratory Standards Institute (2006) Methods for dilution antimicrobial susceptibility tests for bacteria that grow aerobically; approved standard. 7th ed. CLSI document M7–A7. Clinical and Laboratory Standards Institute, Wayne, PA.
8. Staley JT, Konopka A (1985) Measurement of in situ activities of nonphotosynthetic microorganisms in aquatic and terrestrial habitats. Annu Rev Microbiol 39: 321–346.
9. Amann R, Fuchs BM, Behrens S (2001) The identification of microorganisms by uorescence in situ hybridisation. Curr Opin Biotech 12: 231–236.
10. Benveniste R, Davies J (1973) Aminoglycoside antibiotic-inactivating enzymes in Actinomycetes similar to those present in clinical isolates of antibiotic-resistant bacteria. Proc. Natl Acad Sci USA 5170: 2276–2280.
11. Nordmann P, Poirel L (2005) Emergence of plasmid-mediated resistance to quinolones in Enterobacteriaceae. J Antimicrob Chemother 56: 463–469.
12. Oliver A, Pérez-Díaz JC, Coque TM, Bacquero F, Cantón R (2001) Nucleotide sequence and characterization of a novel cefotaxime-hydrolyzing beta-lactamase (CTX-M-10) isolated in Spain. Antimicrob Agents Chemother 45: 616–620.
13. Zheng B, Tan S, Gao J, Han H, Liu J, et al. (2011) An unexpected similarity between antibiotic-resistant NDM-1 and beta-lactamase II from *Erythrobacter litoralis*. Protein Cell 2: 250–258.
14. Cases I, de Lorenzo V (2005) Promoters in the environment: transcriptional regulation in its natural context. Nat Rev Microbiol 3: 105–118.
15. Davies J, Davies D (2010) Origins and evolution of antibiotic resistance. Microbiology and Molecular Biology Reviews. 74: 417–433.
16. Martinez JL (2008) Antibiotics and antibiotic resistance genes in natural environments. Science 321: 365–367.
17. Leclercq R, Canton R, Brown DFJ, Giske CG, Heisig P, et al. (2011) EUCAST expert rules in antimicrobial susceptibility testing. Clin. Microbiol. and Infect. doi: 10.1111/j.1469-0691.2011.03703.x.
18. Fajardo A, Martínez-Martín N, Mercadillo M, Galán JC, Ghysels B, et al. (2008) The neglected intrinsic resistome of bacterial pathogens. PLoS One 3: e1619.
19. Alonso A, Sanchez P, Matinez LJ (2001) Environmental selection of antibiotic resistance genes. Environ Microbiol 3: 1–9.
20. Lee HH, Molla MN, Cantor CR, Collins JJ (2010) Bacterial charity work leads to population-wide resistance. Nature 467: 82–85.
21. Lister PD, Wolter DJ, Hanson ND (2009) Antibacterial-resistant Pseudomonas aeruginosa: clinical impact and complex regulation of chromosomally encoded resistance mechanisms. Clin Microbiol Rev. 22: 582–610.
22. Walsh TR, Weeks J, Livermore DM, Toleman MA (2011). Dissemination of NDM-1 positive bacteria in the New Delhi environment and its implications for human health: an environmental point prevalence study. Lancet Infect Dis 11: 355–362.
23. Poirel L, Schrenzel J, Cherkaoui A, Bernabeu S, Renzi G, et al. (2011) Molecular analysis of NDM-1 producing enterobacterial isolates from Geneva, Switzerland. J Antimicrob Chemother 66: 1730–1733.
24. Fernández L, Gooderham WJ, Bains M, McPhee JB, Wiegand I, et al. (2010) Adaptive resistance to the "last hope" antibiotics polymyxin B and colistin in Pseudomonas aeruginosa is mediated by the novel two-component regulatory system ParR-ParS. Antimicrob Agents Chemother 54: 3372–82.
25. Bengoechea JA, Skurnik M (2000) Temperature-regulated efflux pump/potassium antiporter system mediates resistance to cationic antimicrobial peptides in Yersinia. Mol Microbiol. 37: 67–80.
26. Martinez JL (2008) Antibiotics and antibiotic resistance genes in natural environments. Science 321: 365–367.
27. Allen HK, Donato J, Wang HH, Cloud-Hansen KA, Davies J, et al. (2010) Call of the wild: antibiotic resistance genes in natural environments. Nat Rev Microbiol 8: 251–259.
28. Ma JY, Zeng Z, Chen Z, Xu X, Wang X, et al. (2009) High prevalence of plasmid-mediated quinolone resistance determinants qnr, aac(6')-Ib-cr, and qepA among ceftiofur-resistant enterobacteriaceae isolates from companion and food-producing animals. Antimicrob Agents Chemother 53: 519–524.
29. Dolejska M, Villa L, Poirel L, Nordmann P, Carattoli A (2012) Complete sequencing of an IncHI1 plasmid encoding the carbapenemase NDM-1, the ArmA 16S RNA methylase and a resistance-nodulation-cell division/multidrug efflux pump. J Antimicrob Chemother. 2013 Jan;68(1): 34–9. doi: 10.1093/jac/dks357.
30. Colom K, Pérez J, Alonso R, Fernández-Aranguiz A, Lariño E, et al. (2003) Simple and reliable multiplex PCR assay for detection of bla(TEM), bla(SHV) and bla(OXA-1) genes in Enterobacteriaceae. FEMS Microbiol Lett 223: 147–151.
31. Wang CX, Cai PQ, Chang D, Mi ZH (2006) A Pseudomonas aeruginosa isolate producing the GES-5 extended-spectrum beta-lactamase. J Antimicrob Chemother 57: 1261–1262.
32. Woodford N, Fagan EJ, Ellington MJ (2006) Multiplex PCR for rapid detection of genes encoding CTX-M extended-spectrum beta-lactamases. J Antimicrob Chemother 57: 154–155.
33. Gay K, Robicsek A, Strahilevitz J, Park CH, Jacoby G, et al. (2006) Plasmid-mediated quinolone resistance in non-Typhi serotypes of Salmonella enterica. Clin Infect Dis 43: 297–304.
34. Robicsek A, Strahilevitz J, Jacoby GA, Macielag M, Abbanat D, et al. (2006) Fluoroquinolone-modifying enzyme: a new adaptation of a common aminoglycoside acetyltransferase. Nature Medicine 12: 83–88.
35. Dantas G, Sommer MOA, Oluwasegun RD, Church GM (2008) Bacteria subsisting on antibiotics. Science 320: 100–103.
36. Nadkarni MA, Martin FE, Jacques NA, Hunter N (2002) Determination of bacterial load by real-time PCR using a broad-range (universal) probe and primers set. Microbiology 148: 257–266.
37. DeSantis TZ, Hugenholtz P, Larsen N, Rojas M, Brodie EL, et al. (2006) Greengenes, a chimera-checked 16S rRNA gene database and workbench compatible with ARB. Appl Environ Microbiol 72: 5069–5072.
38. Ludwig W, Strunk O, Westram R, Richter L, Meier H, et al. (2004) ARB : a software environment for sequence data. Nucleic Acids Res 32: 1363–1371.

7

Factors Influencing Bank Geomorphology and Erosion of the Haw River, a High Order River in North Carolina, since European Settlement

Janet Macfall*, Paul Robinette, David Welch

Center for Environmental Studies, Elon University, Elon, North Carolina, United States of America

Abstract

The Haw River, a high order river in the southeastern United States, is characterized by severe bank erosion and geomorphic change from historical conditions of clear waters and connected floodplains. In 2014 it was named one of the 10 most threatened rivers in the United States by American Rivers. Like many developed areas, the region has a history of disturbance including extensive upland soil loss from agriculture, dams, and upstream urbanization. The primary objective of this study was to identify the mechanisms controlling channel form and erosion of the Haw River. Field measurements including bank height, bankfull height, bank angle, root depth and density, riparian land cover and slope, surface protection, river width, and bank retreat were collected at 87 sites along 43.5 km of river. A Bank Erosion Hazard Index (BEHI) was calculated for each study site. Mean bank height was 11.8 m, mean width was 84.3 m, and bank retreat for 2005/2007-2011/2013 was 2.3 m. The greatest bank heights, BEHI values, and bank retreat were adjacent to riparian areas with low slope (<2). This is in contrast to previous studies which identify high slope as a risk factor for erosion. Most of the soils in low slope riparian areas were alluvial, suggesting sediment deposition from upland row crop agriculture and/or flooding. Bank retreat was not correlated to bank heights or BEHI values. Historical dams (1.2–3 m height) were not a significant factor. Erosion of the Haw River in the study section of the river (25% of the river length) contributed 205,320 m^3 of sediment and 3759 kg of P annually. Concentration of suspended solids in the river increased with discharge. In conclusion, the Haw River is an unstable system, with river bank erosion and geomodification potential influenced by riparian slope and varied flows.

Editor: Xiaoyan Yang, Chinese Academy of Sciences, China

Funding: This work was supported by the North Carolina Clean Water Management Trust Fund (JM). Haw River Assembly was the grant recipient, with funds passed to JM through Elon University to financially support all of this research. The funders had no role in study design, data collection and analysis, decision to publish, or preparation of the manuscript.

Competing Interests: The authors have declared that no competing interests exist.

* Email: macfallj@elon.edu

Introduction

Streams and rivers are considered to be in a state of dynamic equilibrium when the sediment delivered to the channel is in balance with the capacity of the stream to transport and discharge that sediment [1]. Stream channels alternatively experience periods of alluvial deposition, followed by erosional downcutting of the alluvium, followed by periods of additional deposition. These cycles have created a landscape of terraces and floodplains, sculpted by the streams and rivers flowing through them [2]. Globally, changes in land use, climate and other factors have altered the historic patterns of transport and discharge, with significant changes to river shape, processes, sediment dynamics and water quality [2]. With these changes, soil erosion has been identified as a significant challenge in both developing and developed countries.

A landscape perspective of rivers and their watersheds demonstrates the influence of land use and disturbance on river structure and ecology at multiple scales [5]. European settlement in the southeastern United States began a period of forest clearing in the 1700's, followed by row crop agriculture [3]. These practices had deleterious ecological consequences to surface waters in the form of increased sediment loads and habitat degradation [3,4,6–9].

Before urbanization and agricultural clearing, streams in the Piedmont region of the southeastern United States, had low levels of suspended solids and high connectivity between the stream and surrounding floodplains [3,4]. It has been estimated that 25 km^3 of soil have eroded from agricultural lands in the Piedmont region between the coastal plain and the Appalachian Mountains, with an average of 14 cm of topsoil lost from North Carolina since the early 1700's [3]. This erosion from agriculture has left a legacy of upland gullies and sediment deposition near and in streams and rivers [3].

Another anthropogenic disturbance to streams and rivers is the proliferation of dams and artificial water bodies. There are over 2 million artificial surface water impoundments including 82,000 dams in the continental United States, with some dating from the 17th century [10,11]. Both natural and man-made dams can have profound impacts on the ecology and geomorphology of rivers, altering patterns of sediment transport and deposition, water and energy flow, and aquatic habitat [10,12,13]. Downstream channel degradation due to dams has been documented for more than 85 years and in many cases has been extreme [14–17]. Downstream changes often include channel incision, channel pattern change (braided to single-thread or vice-versa), loss or encroachment of vegetation, and bank collapse [13,18,19]. Upstream of the dam there may be sediment deposition within the impoundment, leading to incision when the dam is removed [20].

Upstream urbanization, with the increase in impervious surface, has been shown to alter the flow of streams, increasing the frequency of "flashiness" and bank incision [2]. Susceptibility to erosion and hydromodification from urbanization varies with stream bed composition, armoring, bank height, bed and bank materials, precipitation patterns, and other factors [21,22].

Sediment loading from bank erosion is a land management problem of global importance [23,24]. Sediment is one of the most common pollutants from non-point sources, with over 6,000 water bodies across the United States showing significant suspended sediments [25–27]. Different bank materials, aerial and subaerial weathering, variations in grain size, shear strength of the bank materials, bank angle, and water potential can influence river bank mass wasting, failure and fluvial entrainment [23–25]. The ability to predict bank failure and erosion, however, is often uncertain, especially for stream banks with a varied depositional history such as from anthropogenic disturbance [26].

Processes that determine channel geomorphology differ between first and second order streams and larger rivers with larger watersheds [27–29]. Few high order rivers have been studied [27,30]. In large rivers, conditions leading to river widening often are nonlinear, with energy adjustment resulting in different and sometimes opposite adjustment processes. This was seen in the North Fork Toutle River system in Washington State in the NW United States, where following the eruption of Mount St. Helens, one river was dominated by aggradation and widening, while another similar river was dominated by degradation [31]. Unstable channels continue to adjust following major disturbances, both anthropogenic and natural, until a stable floodplain is established with a progressive armoring of the channel bed [32,33].

Sediment and nutrient loading are a global water quality concern, and are significant issues in the Haw River, a high order river in the North Carolina Piedmont. Jordan Reservoir, a major drinking water supply, is formed by a dam on the Haw [34]. However, water in the reservoir is considered impaired due to algal blooms from excess nutrients. To improve water quality, nutrient reduction goals have been established. Understanding the patterns of geomorphic change of the river and contributions of the river bank to sediment and nutrient load may provide a model for water quality improvement in this and similar systems.

Similarly to other developing areas, the Haw River watershed has a history of profound disturbance through forest conversion to row crop agriculture, the construction of dams, and upstream urbanization. In 2014, it was identified as the 9th most threatened river in the United States by the organization American Rivers. The river today has little resemblance to the clear water and low banks from historical descriptions [35].

The primary objective of this study was to identify factors influencing bank geomorphic change and erosion of the Haw River using field measurements. Few reports on erosion of high order rivers have been published, with most based on model estimations. River traits included river slope, riparian soil type and slope, land cover, bank angle, surface protection, bank height, bankfull height and root density and depth. The Haw River has highly incised, unstable banks exhibiting extensive mass wasting, undercutting with bank collapse and fluvial entrainment. Understanding alluvial channel behavior and the channel response to disturbance will provide insight into understanding the factors controlling erosion patterns, shape and balance of the Haw and other high order rivers.

Materials and Methods

For study sites that were located on public lands, the field sites were located in local parks (town parks). Local government agencies were project partners and did not require permits for access. For sites that were located on private land, we obtained landowner permission for access. However, most landowners in our region are reluctant to allow access to their lands by strangers and requests for access may result in alienation of landowners. This research is being used as the foundation for development of the Haw River Trail - a project which uses recreation to achieve regional conservation goals. Because of the trust and contacts established with the research project described in the submitted paper, landowners have been willing to work with us and with local governments on this conservation/recreation project. Alienation of landowners would compromise progress on the conservation work we are developing on the Haw. When we made contact for this study, some of the landowners requested that their identities not be made public. Landowners highly value the personal connections that were established with this project, which is contributing to the success of the conservation work. However, landowners in this region also highly value their privacy and private property rights.

We will share an excel spreadsheet with the GPS coordinates through requests to the senior author. The authors will assist landowner contacts if requested.

The Haw River is located in the north central Piedmont region of North Carolina (Figure 1). This area is located between the coastal plain to the east, and the Appalachian Mountains to the west. North Carolina borders the Atlantic Ocean in the southeastern United States. The river is approximately 177 km long with a watershed of about 3952 km^2, including agricultural, forested, urban and suburban land cover. Nearly one million people live within the Haw River watershed, including the urbanized areas of Greensboro, Graham, Elon, and Burlington, North Carolina which are all upstream of the study reach.

The Haw River ends at the confluence with the Deep River near Moncure, North Carolina, forming the headwaters of the Cape Fear River. For most of its length, including the section studied, the river is relatively straight with few meanders. The watershed is characterized by low, gently rolling hills. Low naturally formed levees (typically ≤1.5 m in height) are frequently found on the river bank. Behind the river bank is a backswamp of varied width. Behind the backswamp is the toe slope, extending to the upland areas. The river channel bottom ranges from bedrock to mixed sand/silt/cobbles with large woody debris. The river is a riffle – pool system [36]. Mean discharge at the United States Geological Survey monitoring station near Bynum, North Carolina (1973–2012) immediately downstream of the study area was 34 m^3/s, with discharge ranging from 1642 m^3/s (1996) to 0.005 m^3/s (1983) [37]. Elevation ranged from 97m at the lower end of the study area to about 304 m at the top of the watershed upriver of the study section.

The study area extended 43.5 km, from the intersection of the Haw River and Interstate 40/85 in Burlington, North Carolina to the intersection of the river and US 15-501 in Bynum, North Carolina, flowing through the counties of Alamance, Orange, and

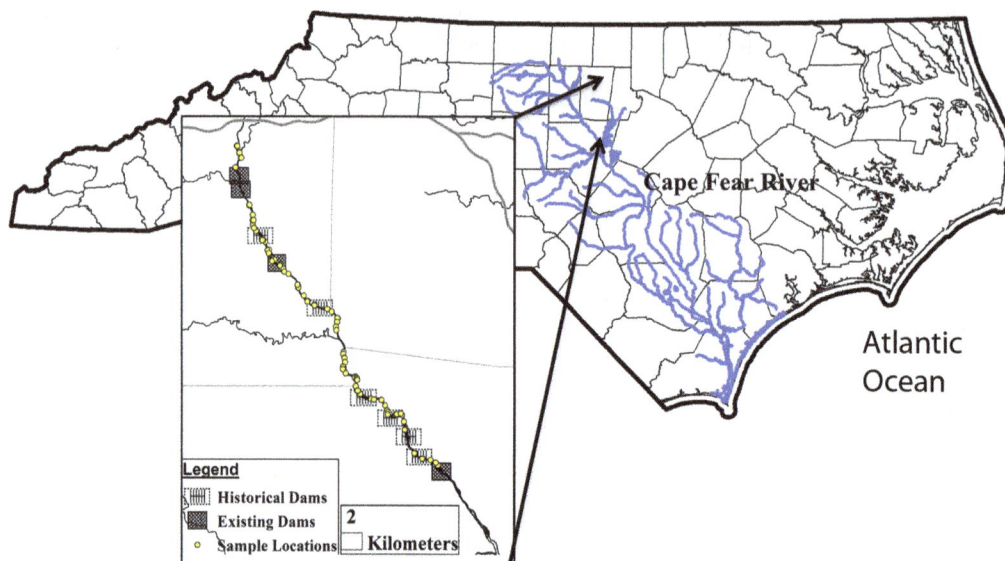

Figure 1. Location of the study area on the Haw River.

Chatham. The upriver location beginning the study area was approximately 112 km downstream from the headwaters and 65 km upstream from the confluence of the Haw River and the Deep River. The studied portion of the river is a fourth order river and higher [38]. Data were collected from 87 bank sites within the study area. Sites were accessed by kayak and land. (Figure 1).

An opportunistic sampling strategy for study site selection which was contingent upon landowner permission was used. Most study sites were on privately owned land. Study sites that were on public land were located in local parks where the local governments were project partners and no permits were required. Eighty-seven river sections that had more than 6 m in length of relatively homogenous slope, bank angle, bank height, and vegetation cover were selected for study. There were no endangered or protected species present at any of the study sites. Measurements were collected during 2006–2007, primarily during spring, summer and fall.

The geographic locations of the bank sites were documented using a Magellan handheld GPS. The lengths of the banks with homogenous characteristics were measured using a Leica Laser Distometer and heights were determined using a telescoping measuring rod and the distometer. The sites were also photographed and sketched to record the appearance of toppled trees and bare soil.

Bank Erosion Hazard Index Methods

A Bank Erosion Hazard Index (BEHI) was used to assess the potential stability of the Haw River's banks [39,40]. This method estimates and predicts erosion potential of rivers and streams, providing documentation of geomorphic field parameters and channel adjustment predictions. While the BEHI methodology is often a small component of more extensive river geomorphic assessments, it is a convenient and useful tool to rapidly create an inventory of river bank characteristics and to assess river bank and river adjustments. Rosgen's BEHI method is normally used to assess lower order rivers and is not often used on large rivers, such as the Haw River [39].

The BEHI is based on five measurements related to bank erosion potential (Table 1). Measurements were conducted

according to methods outlined in Rosgen [39,40]. The bank height was measured at the top of the vertical or sloped area rising from the channel bed. Rooting depth was measured across the bank surface. Root density was determined from a visual estimation of the surface coverage across the bank surface. Surface protection was determined from the percentage of the bank with root, vegetation or hardened structure coverage. From these field measures, bank height ratio (bank height/bankfull height) and root depth ratio (root depth/bank height) were calculated [39].

Table 1 lists criteria used for calculation of the BEHI measurements (Table 1). In the Very Low erosion risk category, a value of between 1–2 is assigned. For example in a low risk stream, the Bank Height Ratio is low, there is extensive rooting in the bank (80–100% of the bank), there is high root density, low bank angle, and substantial surface protection, Summing the BEHI index values for the physical and biological traits gives a total of 5–10, indicating a low potential for erosion. In contrast, attributes that would suggest extreme risk of erosion (the bottom row) include high banks, few roots extending down the bank with low root density, an undercut riverbank with high angle, and little surface protection.

A BEHI value was calculated at each studied site on the river based on Rosgen's BEHI method [40]. A value of 1 (very low) to 10 (extreme) was assigned to each of the bank metrics, as indicated in Table 1. Numbers for all traits were summed for each site. A total value of <10 was considered a very low bank erosion hazard index. A value> 45 was considered an extreme bank erosion hazard index.

Identification of the bankfull stage can be difficult to determine for rivers with an unstable channel. Rather than presenting a stable state, rivers are open systems which respond to variations in energy and materials, making bankfull assessment sometimes uncertain [31]. In cases where the bank height exceeded the bankfull height, the bankfull height was measured at the place on the river bank with visible vegetation changes indicating plants experienced flooding, such as a change in root morphology or root washout [26,27,41]. Where roots were not present, the bankfull height was measured from the point on the bank where bank

Table 1. River bank erosion metric ranking scores for calculation of the BEHI.

Category		Bank Ht. Ratio (m/m)	Root Depth Ratio (%)	Root Density (%)	Bank Angle (Degrees)	Surface Protection (%)	Total Index
Very Low	Value	1.0-1.1	100-80	100-80	0-20	100-90	
	Index	1-2	1-2	1-2	1-2	1-2	≤10
Low	Value	1.1-1.2	80-55	80-55	20-60	90-50	
	Index	2-4	2-4	2-4	2-4	2-4	10-20
Moderate	Value	1.2-1.5	55-30	55-30	60-80	50-30	
	Index	4-6	4-6	4-6	4-6	4-6	20-30
High	Value	1.5-2.0	30-15	30-15	80-90	30-15	
	Index	6-8	6-8	6-8	6-8	6-8	30-40
Very High	Value	2.0-2.8	15-5	15-5	90-120	15-5	
	Index	8-9	8-9	8-9	8-9	8-9	40-45
Extreme	Value	>2.8	<5	<5	>120	<5	
	Index	10	10	10	10	10	>45

materials showed evidence of saturation or shear force stress from flowing water.

Soils were sampled for nutrient analysis by bulk density sampler from the top 20 cm of the river bank at 31 locations throughout the study area. Duff was removed and soils were air dried and analyzed for bulk density and nutrient content by the NC Department of Agriculture and Consumer Services.

River Width and Slope Calculation Methods

The river width was measured using a geographic information system (ArcGIS, ESRI, Inc.) to analyze aerial photographs of the study area. Aerial photographs for Alamance, Orange (2005) and Chatham counties (2007) were obtained from the county GIS offices. Aerial photographs for Alamance and Orange (2011) were from DigitalGlobe (supplied through ESRI, Inc., Redlands, CA) and for Chatham Co. (2013) through NCOneMap. Widths were determined from the aerial photographs by measuring a line placed perpendicular to the center line of the river. The river width measurements were taken of the visible water surface, representing the river width at base flow. The river was at base flow at all dates of image acquisition. Digital resolution was 0.27 m. Locations of dams which are still present in the river were recorded from these aerial photographs. To confirm consistency of measurements, lengths of 10 hardened structures (dams, bridges, buildings) were measured in 2005/2007 and 2011/2013 images. There were no differences in measurements between the two image sets for hardened structures.

The slope of the river was estimated from Digital Elevation Models obtained from the NC Floodplain Mapping Program, 2013. Digital vertical resolution was 0.25 m and horizontal resolution was 6 m. A center line was established on the river image and points were placed every 500 meters. The river slope was measured for the entire length of the study section of the river.

Historical Dam Locations

The locations of historical dams, which are no longer present on the river, were determined by comparing the GIS data layers with photographs and topographic maps. Locations of historical dams were estimated from U.S. Geological Survey topographic quad-rangles (scale 1:24,000) and historical records [42–44](Figure 1).

Soil and Land Cover Methods

Aerial photographs from 2005/2007 were used to digitize different types of land cover within 153 meters (500 ft.) of both of the river banks for evaluation of the riparian corridor over the entire river length studied. Within the GIS, polygons were drawn around each type of designated land cover. The areas for these polygons were then calculated and summed for each land cover category. The land cover classes were:

Forest – areas with evident canopy coverage of mature trees
Open - areas lacking trees or shrubs
Shrubland – areas dominated with small-canopied vegetation
Impervious – areas such as roads and buildings
Water – streams, ponds and other water features

Soil type and traits at each study site were determined from the Alamance, Orange and Chatham Co. Soil Survey. Soil maps (SSURGO) were obtained from the Natural Resource Conservation Service Web Soil Survey [45]. Locations for each study site were identified in the Soil Surveys using GIS. Soil types were determined for soils immediately adjacent to each study location, using the GIS shapefiles for soils in each county. Slopes which are characteristic for each soil type adjacent to the study sites were recorded based on the SSURGO soil type descriptions.

Statistical Analysis

Statistical analyses were conducted with SAS Enterprise Guide 4.3 (SAS Institute, Cary, NC). A Pearson product moment correlation analysis was calculated to determine independence of the river attributes and erosion potential. Analysis of variance or a Whitney Mann Rank Sum Test was calculated to determine effect of riparian slope and historical dams. Data used in these analyses is freely available at http://www.elon.edu/e-web/academics/elon_college/environmental_studies/macfall-plos-one.xhtml.

Results

Bank Erosion Hazard Index

A majority, 84% percent, of the studied banks had a BEHI value of moderate to high erosion potential (Table 2). The mean bank height was 11.8±0.5 m and the mean bankfull height was 5±0.1 m. The mean bank angle was 53°, with a maximum of 90°. The river had widened (bank retreat) by 2.3 m over the six year period from 2005/2007 to 2011/2013, suggesting rapid change to the river bank and ongoing erosion (Table 2).

River Width and Slope

The slope of the river varied throughout the study area, with some sections showing little drop and others a much steeper gradient over a short run (Figure 2). Upstream of the Saxapahaw Dam (9.1 m in height), the highest dam in the river segment studied, the river was 300 m wide, its widest point in the study segment (Table 2).

A number of significant correlations between the river bank geomorphology and erosion potential with attributes of the studied sites were noted (Table 3). There was a significant negative correlation between river width with bank height and BEHI, suggesting that as the river widens, the bank height and BEHI decrease.

One of the most significant results was that the slope of the riparian soils adjacent to the river bank was negatively correlated with bank height, BEHI and bank retreat. The BEHI, bank retreat and the bank heights were significantly higher at study sites with low riparian slope (<2%) compared to riparian areas with greater slopes (Table 4), suggesting greater erosion and erosion potential in areas with low riparian slope. The flatter the land adjacent to the river banks, the more the river had widened, the greater the height of the river banks, the greater the erosion potential (BEHI) and the more erosion had occurred through river widening.

Rooting patterns were negatively correlated with bank heights. There are likely two reasons for this observation. First is the protective effect from erosion which has been described from vegetation. A second reason, however, is if bank collapse or slumping had occurred, trees on the river bank would have also moved downward with the soil, increasing both the root depth ratio and root density lower in the river bank face.

In contrast to the correlation with bank height, root density and root depth ratio were not correlated with bank retreat. Areas with measureable erosion and river widening were also independent of river physical attributes including bank height, bank angle and the BEHI (Table 3).

Based on river bank height and bank retreat measurements, the amount of soil lost through erosion can be estimated for the six year period between 2005/2007 and 2011/2013. The mean value for annual soil loss from the 43.5 km section of the river which was studied was 205,320±23,000 m^3 per year. The bulk density measurement of the riparian soil adjacent to the river was 1.01. Based on this bulk density measure, 2.3 * 10^8 kg of soil are lost annually from the studied river segment. The P concentration

from soil nutrient analyses was measured to be 17.6±1.0 mg/L. Since P generally does not leach through the soil profile and is primarily retained in the surface, if P loading to the river with erosion is only from the top 20 cm of soil; approximately 3759 kg (8286 lb.) of P enters the river annually with sediment [46].

Historical Dams

Historically, there have been ten dams along the river reach studied, three of which still exist. The only two dams still providing hydroelectric power are the Saxapahaw Dam and the Bynum Dam (Table 5). There were low BEHI values and bank heights upriver of and closest to the Bynum and Saxapahaw Dams, as the sites were adjacent to upper reaches of the pools formed behind the dams. BEHI values and bank heights indicating severe erosion were measured downstream of the Saxapahaw dam. No measurements were made downstream of the Bynum dam, the downstream limit of this study.

Dams which are no longer present on the Haw River had little effect on river bank height, BEHI estimates of erosion potential or bank retreat. While it is generally accepted that sediment is deposited upstream of dams in the impoundments, the presence of historical dams on the Haw River does not appear to be a strong factor influencing current conditions (Table 6). In addition, the heights of the dams (with the exception of Saxapahaw Dam) were all lower than the mean values for bank heights measured in the current study. Impacts will likely be much greater with the remaining dams in Saxapahaw and Bynum, however, which are larger.

Riparian Land Cover and Soil Types

Most of the riparian areas adjacent to the study sites were forested with mature trees, primarily hardwoods, at the time of this study (Table 7). There was no relationship between BEHI values or bank heights with land cover in 2005/2007, as most of the riparian lands were forested. However, much of the land had been in agriculture for the past two centuries, so current geomorphic patterns may be a legacy from past land use.

Agriculture was a major land use for the first half of the past century in the 3 study area counties within the watershed (Table 8). Both farm number and percentage of agricultural land decreased, with the greatest loss in the last half of the twentieth century. Counties in the Haw River watershed had 80–85% of the land area in agriculture at the beginning of the twentieth century. The 2007 agricultural census indicates this area had shrunk to 24–32% of each county [47–49].

There were twenty one different soil types at the 87 BEHI study sites (Table 9). The most common soil type was Riverview Silt Loam, followed by Buncomb loamy fine sand.

Discussion

The banks of the Haw River, located in the central Piedmont of North Carolina, are deeply incised (11.8 m mean bank height), with steep banks in the straight reaches, inside bends, and outside bends of the river channel. There is overhanging vegetation at the top of banks, with mass wasting, bank failures, and collapse of trees into the river. The high banks, BEHI indices and bank retreat indicate active erosion is occurring and the potential for future erosion is high. Over half of the study sites had a BEHI of moderate to high erosion potential, supporting observations that the banks along the Haw River in the study area are unstable. The average difference between bank height and bankfull height was 6.6 meters.

Table 2. Summary statistics for bank characteristics of the Haw River.

Attribute	Mean	SE	Min	Max	Median
River Width 2005/2007 (m)	84.3	5.7	27.4	300.5	64.8
River Width 2011/2013 (m)	86.7	5.6	27.4	300.2	68.6
Bank retreat (m)	2.3	0.3	−1.4	9.5	1.5
Riparian slope (%)	6.6	0.9	1	30	1
Bank angle	53.2	2.5	3	90	55
Elevation (m)	122.7	1.4	97	145.7	120.4
River bed slope (%)	0.08	0.01	0	0.37	0
Bank height (m)	11.8	0.5	1.8	29.8	12.1
Bankfull height (m)	5.0	0.1	1.8	9.2	5
Bank height ratio	2.4	0.09	1	4.2	2.4
Root depth ratio	0.59	0.04	0	1.0	0.6
Surface protection (%)	35	2.4	4.3	83.4	30
Root density (%)	43	2	0	100	45
BEHI	24.3	0.7	9.1	39.6	24.1
BEHI Category	**Very low**	**Low**	**Moderate**	**High**	**Very High**
Number of Sites	1	13	58	15	0
Percent of Sites	1.2	14.9	66.7	17.2	0
Total sites	87				

The slopes of the riparian areas were significantly correlated with the BEHI, bank height and bank retreat, primarily through an increase in bank height, BEHI and bank retreat as the slope of the riparian areas decreased (Tables 3 and 4). This suggests that river banks adjacent to riparian areas with low slope are more eroded and erodible than river reaches with more steeply sloped riparian areas.

The riparian areas with low slope are mostly alluvial (Table 9), likely including migrating soils from the surrounding agricultural uplands and deposited sediment from past floods. Gross floodplain sediment trapping potential has been shown to be a function of floodplain area, with larger floodplains having greater trapping potential [50]. If sediment migrating from the uplands was deposited in riparian areas with low slope, the deposited sediment would likely be more erodible than the original base. Riparian areas with high slope adjacent to the river would be less likely to have extensive sediment deposition as the waters carrying the sediment have comparatively more energy than areas which were more flattened.

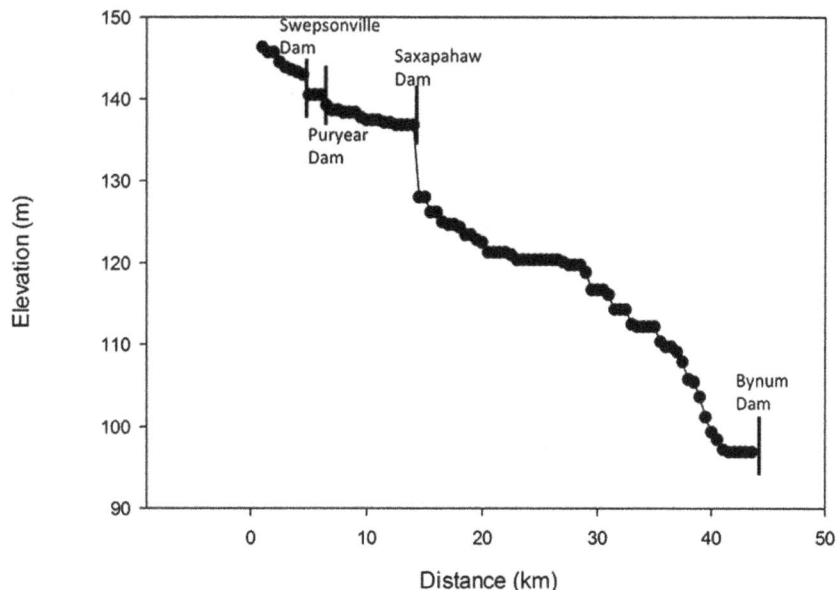

Figure 2. Longitudinal profile of the Haw River study site.

Table 3. Pearson Product-Moment Correlation Analysis of Erosion Factors on the Haw River.

Attribute		Bank height (m)	Bankfull height (m)	BEHI	Bank Retreat (m)
Bank angle	r	−0.04	−0.04		0.11
	p	0.7	0.7		0.29
Bank height (m)	r		**0.55**		−0.04
	p		**<0.001**		0.7
Bank height ratio	r	**0.76**	−0.03		−0.07
	p	**<0.001**	0.76		0.49
Bank retreat (m)	r	−0.04	0.10	0.11	
	p	0.7	0.38	0.34	
River width 2005/2007 (m)	r	**−0.68**	**−0.48**	**−0.40**	−0.09
	p	**<0.001**	**<0.001**	**<0.001**	0.37
Riparian slope (%)	r	**−0.341**	−0.15	**−0.50**	**−0.25**
	p	**<0.001**	0.16	**<0.001**	**0.02**
River bed slope (%)	r	0.02	**0.21**	−0.04	0.01
	p	0.9	**0.05**	0.73	0.9
Surface protection (%)	r	**−0.19**	−0.10		−0.002
	p	**0.08**	0.35		0.98
Root density (%)	r	**−0.36**	−0.17		0.004
	p	**<0.001**	0.12		0.96
Root depth ratio	r	**−0.29**	**−0.26**		−0.16
	p	**0.006**	**0.02**		0.14

Riparian sediment deposition from agriculture on the Haw is consistent with descriptions of sediment deposition layers between 1 and 6 m in depth which have been found adjacent to other streams of the eastern United States. A history of row crop agriculture has caused a loss of 14 cm of topsoil from the North Carolina Piedmont, with sediments migrating from the upland fields to riparian areas and associated rivers and streams [3,4]. These sediment deposit layers are likely an outcome of land clearing for development, agriculture within the watershed, deposition behind dams, and deposits from past flood events, showing the potential for substantial sediment migration to storage areas along riparian river borders [17]. Recent and continued sediment deposition from current agriculture into riparian areas has been documented in the upper Midwest of the United States [51], similarly to patterns on the Haw River.

The riparian areas were mostly forested at the time of the study, with root density and root depth ratios negatively correlated with bank height, but not with bank retreat. Although mature trees were present in the riparian zone throughout the length of the study site, roots rarely extended to the base of the bank. Root

Table 4. Analysis of Variance comparing bank characteristics between areas with low (<2%) and high (≥ 2%) riparian slopes.

Traits/Location	n	Median values (m)	Mean values (m)	SE	P*
Bank retreat					
Low slope	61	2.1	2.7	0.31	0.03
High slope	26	0.8	1.5	0.45	
Bank height					
Low slope	61	12.7	12.8	0.42	0.002
High slope	26	10.4	9.6	1.23	
Bank angle					
Low slope	61	60.0	56.1	2.60	0.06
High slope	26	40.0	46.1	5.35	
BEHI					
Low slope	61	26.5	26.3	0.70	<0.001
High slope	26	20.6	19.5	1.08	

*Mann Whitney Rank Sum Test was used for Bank retreat and Bank height analyses. ANOVA was used for BEHI and Bank Angle analyses.

Table 5. Historical and current dams on the Haw River listed in order on the river.

Dam Name	Dam Height (m)	Years
Virginia Falls Dam	3.0	1874- present, removed 2013
Puryear Dam	2.4	1763 - present
Cedar Cliffs Dam	1.5	1860–1910
Saxapahaw Dam	9.1	1938 - present
Dark's Dam	-	1790–1875
Elliot's Falls	1.2	1778–1810
Love's Dam	-	1790–1920
Pace's Dam	2.4	1789–1924
Burnett-Powell Dam	-	1776–1880
Bynum Dam	3.0	1874 - Present

density in river banks has been shown to reduce scour through both mechanical root reinforcement and matric suction [52]. However, on the Haw, the bank height frequently exceeded rooting depth (Table 2). The absence of roots at the base of the river bank would mean that roots are not present to provide reinforcement and protection from shear. In contrast to the protective effect of roots on bank height, the lack of correlation with bank retreat suggests vegetation provided no protection from erosion of the bank toe slope at the river's edge, a significant concern in river management [53].

The influence of dams on river geomorphology has been well documented [10,12,13]. In the present study, dams were lower than the heights of banks measured in the study area and the eroded bank areas extended well upriver beyond dam impoundments. The highest dam currently on the Haw is the Saxapahaw Dam, with a height of 9.1 m. The mean height for all dams in the studied segment was 2.9 m, including both historical and present dams. Yet the mean bankfull measurement was 5 m, and the mean bank height was 11.8 m, higher than all the dams. If bank erosion was primarily through legacy sediments from dam impoundments, it would be expected that the bankfull and bank heights would not exceed the heights of the dams. The Haw River bed is rocky, lined with rocks from 0.2 m diameter to large boulders and bedrock, so the contribution to bank heights from erosion of the river bed is likely to be minimal. Comparison of the bank heights, BEHI values and bank retreat upstream of the

legacy dam sites with other study sites showed no difference between locations (Table 6). Although low dams can have significant ecological effects in many cases, the impact to higher order rivers such as the Haw appears to be small.

The lack of a correlation between bank height and the BEHI with bank retreat suggests that different but related processes are occurring. Bank retreat often occurs following bank failure and collapse, with the river bank collapsing downward towards the water's edge then eroding away as the river flows past. River bank stability can be highly variable, with many factors contributing to the structural integrity. The negative correlation between bank height and BEHI with river width suggests that many of the tall banks are stable, reflective of the wide range of conditions present in the river corridor. However, both high banks and bank retreat were correlated with low riparian slope, suggesting this condition increases risk from both patterns of erosion [54].

Geomorphic patterns of the Haw River are consistent with conceptual models describing changes in river geomorphology following disturbance such as the removal of a small dam. Following dam removal, the sequence would be: a) lowered water surface, b) degradation, c) degradation and widening d) aggradation and widening, ending with e) quasi-equilibrium [32,55,56]. This process usually happens quickly after removal of small dams, reaching quasi-equilibrium within a few years. A similar sequence of geomorphological changes following disturbance seems to be

Table 6. Effect of historical dams on river bank geomorphology.

Traits/location	n	Median values (m)	Mean values (m)	SE	p
Bank retreat					
Behind dam	8	1.8	2.3	0.87	0.94
Not behind dam	79	1.5	2.2	0.28	
Bank height					
Behind dam	8	11.7	11.4	1.5	0.83
Not behind dam	79	12.1	11.9	0.52	
BEHI					
Behind dam	8	23.2	24.1	2.5	0.65
Not behind dam	79	24.4	27.2	0.69	

*Mann Whitney Rank Sum Test was used for Bank retreat and Bank height analyses. ANOVA was used for BEHI analysis.

Table 7. Land cover within the 153 m zone adjacent to BEHI study sites.

Land Use	Percentage
Forest	78.4
Open	15.3
Shrubland	5.5
Impervious	0.8

occurring on the Haw River, with the river currently at stages b, c and d.

Disturbances on the Haw River include historical dams which are no longer in place, three dams which are currently in place, a history of row crop agriculture throughout the watershed and upstream urbanization. This is similar to the history of dams and land use on other rivers of the Piedmont and around the world [18,57]. Erosion of legacy sediments from agriculture followed by urbanization is has also been documented on a smaller stream in the Piedmont region, leaving similarly incised banks [36].

Reaching dynamic equilibrium does not always occur quickly. Following an eruption of Mount St. Helens in 1857, stable floodplains and re-vegetation of riparian zones had not yet been re-established at the time of the 1980 eruption [33]. In this case, dynamic equilibrium had not been reached over a century after the earlier disturbance. Similar patterns have been observed in other regions [5,58,59]. For the Haw River, re-connection with flood plains and equilibrium will likely take centuries or longer, if it all, compared with the decadal time scale described with dam removal.

In addition to the effects from dams and past agriculture, changes in flow can affect soil loss from erosion and sediment load. Data from a U.S.G.S. monitoring station just downstream of the study area were examined from 2005–2010, with flow and sediment measures collected on the same day [37,60]. As flow increased, so did sediment loading to the river, as reflected in the concentration of suspended solids (Figure 3). River discharge ranged from 2 m^3/sec to 659 m^3/sec, averaging 31 m^3/sec. The suspended solids in the river ranged from 1 mg/L to 187 mg/L, averaging 15 mg/L. As the energy in the river increased with flow, the concentration of suspended solids also increased exponentially, a relationship well documented in the literature [2]. As the region upstream of the study area continues to urbanize with increasing impervious surface, high flow events are likely to also increase with increasing erosion, a significant management concern.

Sediment loading from bank erosion to the Haw River will also contribute P. The analyses of riparian soil indicated approximately 3759 kg (8286 lb.) of P enters the river annually from sediment. This is about 1.5% of the total target P load from non-point sources in the Haw River flowing to Jordan Reservoir (of a targeted 106,884 kg), and potentially 6% of the total P load to the Haw River from river bank erosion (the study area was about 25% of the river length). Currently this is an unaccounted non-point source of P, with no Best Management Practices or nutrient loading targets assigned to this P source [34].

Like many other high order rivers globally, the geomorphology of the Haw River in North Carolina is changing following agricultural and other disturbances, with significant sediment and P loading to the river. The river is undergoing a process of reshaping with river widening and the formation of very high river banks. Occasional high flow events further contribute to erosion with the concentration of suspended solids in the water column increasing with discharge (Figure 3).

Future study of the Haw River and similar systems could be directed at understanding the patterns of erosion seen on this river. One major area of investigation would be to determine the origin and age of sediments deposited along the river, especially those in low slope areas which are experiencing the most erosion. Another study would be to evaluate the impacts of past land use and land cover on contemporary geomorphology, flow and erosion. Comparison with commonly used models for erosion estimates would also be valuable.

Conclusion

Few studies on the patterns of bank erosion and hydrogeo-morphic change following disturbance have been published for high order rivers such as the Haw River. However, differences in geomorphology have been described between high and low order segments of rivers, with factors impacting the river attributes changing throughout the river's length. These studies show processes and attributes of low order rivers may not be applicable to larger river systems [52].

For most studies estimating erosion and erosion potential of larger rivers, a modeling approach has been used. Common methods are the Universal Soil Loss Equation and the Soil and Water Assessment Tool [61,62]. In contrast, field measurements of large systems are seldom the major approach used to identify factors influencing erosion and geomorphic change.

For the Haw River, regions of high erosion potential (as indicated by the BEHI and bank height) are negatively correlated with river width, suggesting regions high BEHI values and bank heights are have narrow river width and are relatively stable. On this river, erosion as measured by bank retreat is independent of most physical features such as bank height, BEHI, bank angle and

Table 8. History of farming within the studied counties of the Haw River watershed.

Year		Alamance	Chatham	Orange
1910	Number of farms	2508	3640	1957
	% of county in agriculture	80	85	84
1950	Number of farms	2940	2977	2038
	% of county in agriculture	79	66	70
2007	Number of farms	753	1089	604
	% of county in agriculture	32	24	24

Table 9. Soil types at BEHI study sites.

Soil ID	Soil Type	% slope	# sites
AdE	Appling, sandy loam, steep phase	20	2
Ba	Buncomb loamy fine sand, frequently flooded	1	14
BaE	Badin Nanford Complex	23	6
CbE	Cecil fine sandy loam, moderately steep phase	17	2
Cg	Congaree fine sandy loam, frequently flooded	1	2
ChA	Chewacla and Wehadkee soils, frequently flooded	1	1
Cp	Congaree fine sandy loam, frequently flooded	1	4
GaE	Georgeville silt loam, moderately steep phase	17	1
GbE3	Georgeville silt loam, severely eroded, moderately steep phase	20	1
GcE	Goldston slaty silt loam, moderately steep phase	17	5
GkE	Georgeville Badin complex	23	2
LbE	Lloyd loam, moderately steep phase	17	1
Mc	Mixed alluvial land, poorly drained	1	6
Md	Mixed alluvial land, well drained	1	6
NaD	Nanford Badin complex	5	1
RvA	Riverview silt loam, frequently flooded	1	28
TaD	Tirza silt loam, strongly sloping phase	30	1
WcE	Wilkes stony soils, moderately steep phase	17	4

bankfull height, suggesting different erosional processes are occurring. Although these physical features are often used to predict erosion potential, such as through use of the BEHI, that does not appear to be the case in this study. Independence of bank physical features with bank retreat has been observed in a few other systems [65].

Areas with low riparian slope (<2%) appear to have the highest erosion risk, experiencing high bank heights, BEHI values and

bank retreat. This relationship is in contrast to some common models used for erosion estimation, which rank high slope as an erosion risk factor [61,62]. In the Haw River system, a history of extensive soil loss from upland agriculture suggests erosion of agricultural legacy sediments deposited on areas with low slope beside the river has occurred, with high erodibility. Increasing sediment concentration with flow suggests the river is still changing, reshaping as an outcome of past and present distur-

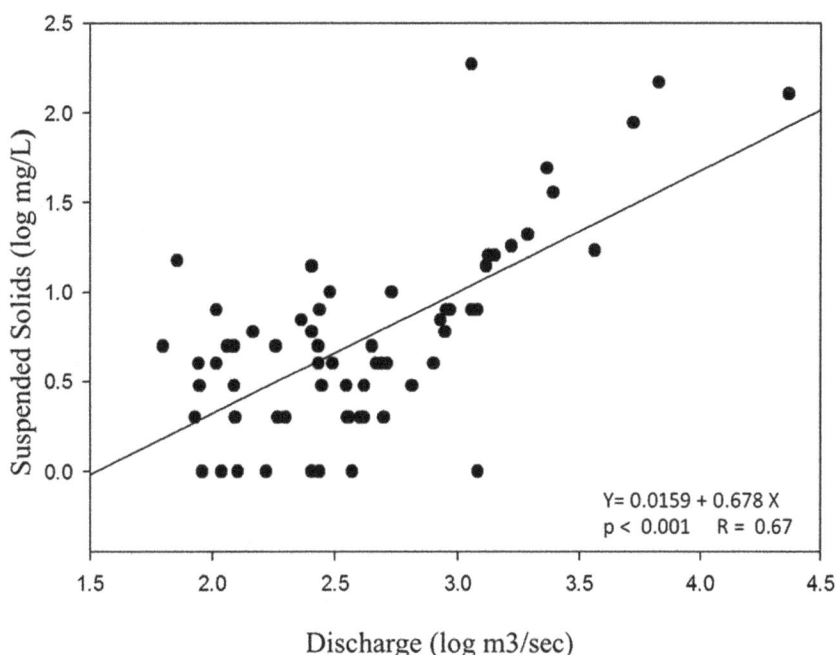

Y= 0.0159 + 0.678 X
p < 0.001   R = 0.67

Figure 3. Relationship between log of river discharge and log of total suspended solids in the Haw River.

bance. The potential for further impairment as upstream urban development increases should be a management concern for the river health and water quality.

Rate of bank retreat and river widening is similar to the rate reported for other high order rivers. In rivers of southern Minnesota, in the Midwest of the United States, LIDAR studies have indicated widening rate of 0.57–5 m/yr since European settlement [63]. In our study, the Haw has widened 0.38 m/yr, slightly less than in Minnesota. But both studies indicate the rivers are not at equilibrium.

The Haw River is an unstable system, with river bank erosion and geomodification potential influenced by disturbance, riparian slope and episodes of high flow. The greatest erosion, measured by bank height and bank retreat occurred in regions with low riparian slope, usually with alluvial soils, suggesting erosion of deposited sediments. Historical dams were not a significant factor in influencing current conditions on the Haw River. This study

provides a model for high order rivers, identifying factors driving erosion and changes to channel morphology which will help in management of these and other high order river systems.

Acknowledgments

We would like to thank Dr. Greg Jennings for his assistance and guidance throughout development of this project. We would also like to sincerely thank the reviewers of this manuscript for their thoughtful and insightful comments.

Author Contributions

Conceived and designed the experiments: JM PR DW. Performed the experiments: JM PR DW. Analyzed the data: JM PR DW. Contributed reagents/materials/analysis tools: JM PR DW. Wrote the paper: JM PR DW. Obtained landowner permission for access: JM PR DW.

References

1. Zaimes G, Emanuel R (2006) Stream processes Part I - Basic. Arizona Watershed Stewardship Guide.
2. Leopold LB (1994) A View of the River. Cambridge, Ma. Harvard University Press.
3. Trimble SW (1974) Man-induced soil erosion on the southern Piedmont, 1790–1970. Ankeny, Iowa: Soil and Water Conservation Society.
4. Trimble SW (1975) A volumetric estimate of soil eroson on the southern Piedmont. USDA Agricultural Research Service Publication S-40: 142–154.
5. Allen JD (2004) Landscapes and Riverscapes: The Influence of Land Use on Stream Ecosystems. Annual review of ecological and evolutionary systematics 35: 257–284.
6. Bennett HH (1928) The geographic relation of soil erosion to land productivity. Geographical Review 18: 579–605.
7. Jacobson RB, Coleman DJ (1986) Stratigraphy and recent evolution of Maryland Piedmont flood plains. American Journal of Science 286: 617–637.
8. Richter BD, Baumgartner JV, Powell J, Braun DP (1996) A method for assessing hydrologic alteration within ecosystems. Conservation Biology 10: 1163–1174.
9. Florsheim JL, Pellerin BA, Oh NH, Ohara N, Bachand PAM (2011) From deposition to erosion: Spatial and temporal variability of sediment sources, storage, and transport in a small agricultural watershed. Geomorphology 132: 272–286.
10. Smith SV, Renwick WH, Bartley JD, Buddenmeier RW (2002) Distribution and significance of small articial water bodies across the United States landscape. Science of the Total Environment 299: 21–36.
11. U.S. Army Corps of Engineers (2013) National Inventory of dams. http://geo.usace.army.mil/pgis/f?p=397:1:479164877021486. Accessed November, 2013
12. Beyer PJ (2005) Introduction to the special issue: Dams and geomorphology. Geomorphology 71: 1–2.
13. Walter RC, Merritts DJ (2008) Natural streams and the legacy of water-powered mills. Science 319: 299–304.
14. Lawson JM (1925) Effect of Rio Grande storage on river erosion and deposition. Engineering news Record: 327–334.
15. Williams GP, Wolman MG (1984) Downstream effects of dams on alluvial rivers. Geological Survey Professional Paper 1286: 1–83.
16. Petts G, Gurnell A (2005) Dams and geomorphology: Research progress and future directions. Geomorphology 71: 27–47.
17. Hupp CR, Schenk ER, Richter JM, Peet RK, Townsend PA (2009) Bank erosion along the dam-regulated lower Roanoke River, North Carolina. The Geological Society of America Special Paper 451: 97–108.
18. Schenk ER, Hupp CR (2009) Legacy effects of colonial millponds on floodplain sedimentation, bank erosion and channel morphology, Mid-Atlantic, USA. Journal of the American Water Resources Association 45: 596–606.
19. Ligon FK, Dietrick WE, Thrush WJ (1995) Downstream ecological effects of dams: a geomorphic perspective. BioScience 45: 183–192.
20. Doyle MW, Stanley EH, Orr CH, Selle AR, Sethi SA, et al. (2005) Stream ecosystem response to small dam removal: Lessons from the heartland. Geomorphology 71: 227–244.
21. Bledsoe BP, Stein E, Hawkley RD, Booth D (2012) Framework and tool for rapid assessment of stream susceptibility to hydromodofication. Journal of the American Water Resources Association 48: 788–808.
22. Paul MJ, Meyer JL (2001) Streams in the urban landscape. Annual review of ecological and evolutionary systematics 32: 333–365.
23. Thorne CR, American Society of Civil Engineers Task Committee on Hydraulics (1998) River width adjustment I: processes and mechanisms. Journal of Hydraulic Engineering 124: 881–902.
24. Thorne CR, American Society of Civil Engineers Task Committee on Hydraulics (1998) River width adjustment II: modeling. Journal of Hydraulic Engineering 124: 903–917.
25. Sass CK, Deane TD (2012) Application of rosgen's bancs model for NE Kansas and the development of predictive streambank erosion curves. Journal of the American Water Resources Association 48: 774–778.
26. Jennings GD, Harman WH. Measurement and stabilization of stream bank erosion in North Carolina. ASAE. In: Ascough JC, Flanagam DC, editors; 2001; Honolulu, Hawaii.
27. Rosgen D (1996) Applied River Morphology. Pagosa Springs, Colorado: Wildland Hydrology Books.
28. Abernethy B, Rutherford ID (1998) Where along a rivers length will vegetation most effectively stabilize stream banks? Geomorphology 23: 55–75.
29. Anderson RJ, Bledsoe BP, Hession WC (2004) Width of streams and rivers in response to vegetation, bank material and other factors. Journal of the American Water Resources Association 40: 1159–1172.
30. Rosgen D (1996) The use of infrared photography for the determination of sediment production. Fluvial Process and Sedimentation. Edmonton, Alberta: Canadian National Research Council. pp. 381–402.
31. Simon A, Doyle MW, Kondolf M, Shields FD, Rhoads B, et al. (2007) Critical evaluation of how the Rosgen classification and associated "natural channel design" methods fail to integrate and quantify fluvial processes and channel response. Journal of the American Water Resources Association 43: 1117–1131.
32. Doyle MW, Stanley EH, Orr CH, Selle AR, Sethi SA, et al. (2005) Stream ecosystem response to small dam removal: Lessons from the Heartland. Geomorphology 71: 227–244.
33. Simon A, Thorne CR (1996) Channel adjustment of an unstable coarse-grained stream: opposing trends of boundary and critical shear stress, and the applicability of extremal hypotheses. Earth Surface Processes and Landforms 21: 155–180.
34. General Assembly of North Carolina (2009) Jordan Water Supply Nutrient Strategy Rules. Session Law 2009-216, House Bill 239, 15A NCAC 02B . 0263 -. 0273, .0311(p).
35. Cassebaum AM (2011) Down along the Haw: A history of a North Carolina river. Raleigh, NC: McFarland and Company.
36. Hupp CR, Noe GB, Schenk ER, Bentham AJ (2013) Recent and historic sediment dynamics along Difficult Run, a suburban Virginia Piedmont Stream. Geomorphology 180–181: 156–169.
37. U.S. Geological Survey (2013) U.S.G.S. Water Information System. Available: http://waterdata.usgs.gov/nc/nwis/uv/?site_no=02096960&PARAmeter_cd=00065,00060. Accessed 2013 May 12.
38. Strahler AN (1952) Hypsometric (area altitude) analysis of erosional topology. Geological Society of America Bulletin 63: 1117–1142.
39. Rosgen DL (2001) A practical method of computing streambank erosion rate. pp. 9–15.
40. U.S. Dept. of Agriculture (2007) Rosgen Geomorphic Channel Design. Part 654 Stream Restoration Design National Engineering Handbook: Natural Resources Conservation Service. pp. 1–76.
41. Harman WH, Jennings GD, Patterson JM, Clinton DR, Slate LO (1999) Bankfull hydraulic geometry relationships for North Carolina Streams. In: Olsen DS, Potyondy JP, editors; Bozeman, Mt.
42. Spoon WL (1893) Map of Alamance County, NC. Cleveland, OH: H.B. Granahan and Co. Engravers.
43. Spoon WL (1928) Map of Alamance Co., NC. Cleveland, Ohio: H.B. Stranahan and Co. Engravers.
44. Chilton M (2008) A historical atlas of the Haw River. Carrboro, NC: Private printing.
45. U.S. Dept. of Agriculture (2014) Web Soil Survey. Natural Resource Conservation Service. Available: http://websoilsurvey.sc.egov.usda.gov/App/HomePage.htm. Accessed 2014 Apr.

46. Jobbagy EG, Jackson RB (2001) The distribution of soil nutrients with depth: Global patterns and the imprint of plants. Biogeochemistry 53: 51–77.
47. U.S. Dept. of Agriculture (1913) Thirteenth census of the United States taken in the year 1910. Agriculture 1909 and 1910, Reports by States with statistics for counties.
48. U.S. Dept. of Agriculture (1952) United States Census of Agriculture: 1950. Counts and State Economic Areas. North Carolina and South Carolina.
49. U.S. Dept. of Agriculture (2009) Census of Agriculture 2007, North Carolina State and County Data. Geographic Area Series, Part 33.
50. Schenk ER, Hupp CR, Gellis A, Noe G (2013) Developing a new stream metric for comparing stream function using a bank-floodplain sediment budget: a case study of three Piedmont streams. Earth Surface Processes and Landforms 38: 771–784.
51. Knox JC (2001) Agricultural influence on landscape sensitivity in the Upper Mississippi River Valley. Catena 42: 193–224.
52. Pollen-Bankhead N, Simon A (2010) Hydrologic and hydraulic effects of riparian root networks on streambank stability: Is mechanical root-reinforcement the whole story? Geomorphology 116: 353–362.
53. Docker BB, Hubble TCT (2008) Quantifying root-reinforcement of river bank soils by four Australian tree species. Geomorphology 100: 401–418.
54. Atkinson PM, German SE, Sear DA, Clark MJ (2003) Exploring the Relations between River bank Erosion and Geomorphological Controls using Geographically Weighted Logistic Regression. Geographical Analysis 35: 58–82.
55. Schumm A, Watson C, Harvey M (1986) Incised channels, morphology, dynamics and controls. Littleton, Co: Water Resources Publications.
56. Simon A, Hupp CR (1986) Channel evolution in modified Tennessee channels; Las Vegas, Nv. Federal Interagency Sedimentation Conference. pp. 71–82.
57. Cruise JF, Laymon CA, Al-Hamdan OZ (2010) Impact of 20 Years of Land-Cover Change on the Hydrology of Streams in the Southeastern United States1. Journal of the American Water Resources Association 46: 1159–1170.
58. Gregory S, Swanson F, McKee A, Cummins K (1991) An ecosystem perspective of riparian zones. BioScience 42: 540–551.
59. Parker C, Simon C, Thorne CR (2008) The effects of variability in bank material properties on river bank stability: Goodwin Creek, Mississippi. Geomorphology 101: 533–543.
60. (2011) Cape Fear River Basins Water Monitoring Coalition. Available: http://www.cormp.org/CFP/. Accessed 2011 Jul 11.
61. University T&M (2014) Soil and Water Assessment Tool. Available: http://swat.tamu.edu/. Accessed 2014 Jun.
62. Soupios M, Vallianaatos F (2009) Soil erosion prediction using the Revised Universal Soil Loss Equation (RUSLE) in a GIS Framework. Chania, Northwestern Crete, Greece. Environmental Geology 57: 483–497.
63. Kessler AC, Gupta SC, Brown MK (2013) Assessment of river bank erosion in Southern Minnesota rivers post European settlement. Geomorphology 201: 312–322.

Digital Mapping of Soil Organic Carbon Contents and Stocks in Denmark

Kabindra Adhikari[1]*, Alfred E. Hartemink[1], Budiman Minasny[2], Rania Bou Kheir[3†], Mette B. Greve[3], Mogens H. Greve[3]

1 Department of Soil Science, University of Wisconsin–Madison, Madison, Wisconsin, United States of America, 2 Department of Environmental Sciences, The University of Sydney, Sydney, New South Wales, Australia, 3 Department of Agro-ecology, Aarhus University, Tjele, Denmark

Abstract

Estimation of carbon contents and stocks are important for carbon sequestration, greenhouse gas emissions and national carbon balance inventories. For Denmark, we modeled the vertical distribution of soil organic carbon (SOC) and bulk density, and mapped its spatial distribution at five standard soil depth intervals (0–5, 5–15, 15–30, 30–60 and 60–100 cm) using 18 environmental variables as predictors. SOC distribution was influenced by precipitation, land use, soil type, wetland, elevation, wetness index, and multi-resolution index of valley bottom flatness. The highest average SOC content of 20 g kg^{-1} was reported for 0–5 cm soil, whereas there was on average 2.2 g SOC kg^{-1} at 60–100 cm depth. For SOC and bulk density prediction precision decreased with soil depth, and a standard error of 2.8 g kg^{-1} was found at 60–100 cm soil depth. Average SOC stock for 0–30 cm was 72 t ha^{-1} and in the top 1 m there was 120 t SOC ha^{-1}. In total, the soils stored approximately 570 Tg C within the top 1 m. The soils under agriculture had the highest amount of carbon (444 Tg) followed by forest and semi-natural vegetation that contributed 11% of the total SOC stock. More than 60% of the total SOC stock was present in Podzols and Luvisols. Compared to previous estimates, our approach is more reliable as we adopted a robust quantification technique and mapped the spatial distribution of SOC stock and prediction uncertainty. The estimation was validated using common statistical indices and the data and high-resolution maps could be used for future soil carbon assessment and inventories.

Editor: Dafeng Hui, Tennessee State University, United States of America

Funding: The study was funded by the Danish Ministry of Climate and Energy through SINKS Project (2009–2012). The funders had no role in study design, data collection and analysis, decision to publish, or preparation of the manuscript.

Competing Interests: The authors have declared that no competing interests exist.

* Email: kadhikari@wisc.edu

† Deceased.

Introduction

Digital Soil Mapping uses statistical tools to quantify the spatial relationship between soil property values to its environmental covariates [1]. The digital mapping of soil organic carbon (SOC) at fine resolution is a challenging task [2] and the mapping is also a high priority for SOC assessment and monitoring [3]. Spatial models on SOC prediction has a long history (e.g., [4]). A range of techniques have been used to predict and map SOC from landscape to national or continental levels and Minasny et al. [5] provided a comprehensive review. Several researchers applied splines to model the vertical distribution of SOC in the soil profiles and predicted SOC at a landscape scale using data-mining tools and environmental variables as predictors [6–8]. Mishra et al. [9] calculated SOC pool in each soil horizon and applied geographic weighted regression to map SOC pool at a regional scale in mid-west USA. Odgers et al. [10] used splines to derive SOC content from soil map units and predicted SOC at six standard soil depths for the entire USA. Arrouays et al. [11], Bou Kheir et al. [12], Chaplot et al. [13], Doblas-Miranda et al. [14], mapped SOC at

national level using different statistical tools ranging from statistical aggregation to advanced regression tools such as regression trees.

The distribution of SOC changes across the landscape and it also varies by depth. In most soils, SOC is higher in the surface horizons and it decreases with depth. Such depth-wise variability is mostly continuous [15–17] except in soils with a strong human impact like some soils in the Netherlands [18]. Although most SOC studies and inventories are confined to 30 cm soil depth [19–21], the amount of SOC stored below 30 cm is of relevant in many ecosystems [22,23]. For accurate quantification of SOC stocks, a depth function needs to be modeled. Several tools exist: spline function [15,16], exponential decay function [24] or soil-type specific or profile depth functions [18,25]. For modeling SOC with depth, the equal-area spline function has been proven to be useful in several studies [7,8,10]. Spline predicted SOC values with depth act as a geo-referenced point data to which environmental variables are joined and prediction models are generated using digital soil mapping techniques.

In Denmark, studies on SOC dynamics and its quantification has started after a national wide soil database was established between the years 1975–1985. Based on simple statistical scaling-up techniques, Krogh et al. [26] calculated a total stock of about 579 Tg and reported that 69% of it was stored in the soils under agriculture. Greve et al. [27] estimated topsoil SOC contents (g 100 g^{-1}) for the whole country but have not assessed the SOC stocks. Bou Kheir et al. [12] predicted the spatial extent of organic soils across Denmark. Similarly, Olesen [28] and Taghizadeh-Toosi et al. [29] estimated the stock but did not map its distribution. Most of these studies have estimated the SOC content and stock but they have not explored the spatial distribution of SOC stock nor quantified the uncertainty of the SOC predictions.

We applied digital soil mapping techniques to quantify the SOC content and stocks for Denmark. The main objectives of this study were: to model the vertical distribution of SOC content and bulk density in soil profiles, to predict and map their spatial distribution using environmental variables, to identify major environmental variables responsible for SOC distribution, to estimate the SOC

stock for the soils of Denmark, and to assess the uncertainties in the SOC predictions.

Materials and Methods

Study area

Denmark is situated in Northern Europe covering an area of approximately 43,000 km^2. The temperate climate is characterized by a mild winters with annual mean winter and summer temperatures of about 0°C and 16°C [30]. Precipitation is well distributed throughout the year with an average annual rainfall of 800 mm in the west to 500 mm to the east. The country is relatively flat with a mean elevation above mean sea level is about 31 m. Denmark is divided into 10 physiographic regions—referred in this paper as geo-regions—based on geographical, climatic and soil-formation criteria.

The soils are coarse sandy to clayey to heavy clays as defined in the Danish Soil Classification System [31]. Soil in the western part

Table 1. List of environmental variables used to predict the distribution of soil organic carbon and its stock in Denmark.

Environmental variables	Scorpan factor	Type of variable	Description	Range of values	Scale or resolution	Reference
Soil map	S	Categorical	Map of Soil types based on soil texture (8 classes)	-	1:50,000	[31]
Precipitation	C	Continuous	Average annual rainfall (mm) (1961–1990)	525 to 905	30.4 m	[31]
Geo-regions	C	Categorical	Scanned geographical regions map (10 classes)	-	1:100,000	[31]
Insolation	C	Continuous	Potential incoming solar radiation (2011)	254 to 698	30.4 m	[61]
Mid-slope position	C, N	Continuous	Covers the warmer zones of slopes	0 to 1	30.4 m	[62]
Land use	O	Categorical	CORINE Land cover data adopted in Denmark (31 classes)	-	1:100,000	[45]
Elevation	R	Continuous	Elevation of the land surface derived from LiDAR (m)	0 to 170	30.4 m	
Slope gradient	R	Continuous	Maximum rate of change between the cells and neighbors (degree)	0 to 90	30.4 m	[63]
Slope aspect	R	Continuous	Direction of the steepest slope from the North (degree)	0 to 360	30.4 m	[63]
Flow accumulation	R	Continuous	Number of upslope cells	1 to 73645	30.4 m	
SAGA wetness index	R	Continuous	Wetness Index. WI = ln (A$_s$ / tan β): where A$_s$ is modified catchment area and β is the slope gradient	7.2 to 19	30.4 m	[64]
Multi-resolution valley bottom flatness	R	Continuous	Possible depositional areas	0 to 11	30.4 m	[65]
Valley depth	R	Continuous	Extent of the valley depth (m)	0 to 90	30.4 m	
Wetlands	S, R	Categorical	Map showing the presence or absence of wetlands	-	1:20,000	[31]
Landscape	R	Categorical	Landform types (10 classes)	-	1:100,000	[31]
Altitude above channel network	R	Continuous	Vertical distance to channel network base level (m)	0 to 56	30.4 m	
Slope length factor	R	Continuous	LS-factor of Universal Soil Loss Equation (m)	0 to 47	30.4 m	[66]
Geology	P	Categorical	Scanned and registered geological map (86 classes)	-	1:100,000	[31]

S-soil types; C-climate, O-organisms; R-relief; P-parent material; N-spatial position.

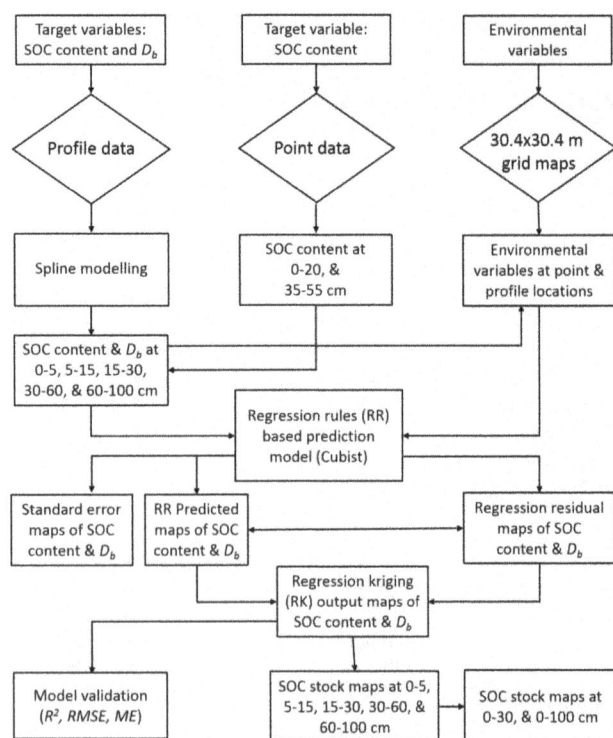

Figure 1. Schematic representation of overall prediction scenario.

of the country are developed in non-glaciated sandy parent material along the glacial flood-plains and Saalian moraine, whereas the soils from the central and eastern region have been developed on relatively young basal till high in finer materials [31]. Most of the soils in the north have been formed in sand mixed with uplifted marine sediments. More than 66% of the soils are classified as Podzols (Spodosols) and Luvisols (Alfisols). Podzols occupy a major portion in the west and Luvisols and Cambisols (Inceptisols, Entisols) in the central and eastern part of the country [32]. Peat soils (Histosols) occur in poorly drained basins throughout the country. About 66% of the total land area is used for agriculture with grain and potato as the main crops. Forest areas include spruce, pine and beach and these cover more than 12%.

Point data

Point (pedon) data on soil organic carbon (SOC) (g kg^{-1}) and bulk density (D_b) (Mg m^{-3}) were derived from two databases: Danish Soil Classification database (DSC) and Danish Soil Profile

database (DSP). DSC consists of about 36,000 point observations from the topsoil (0−20 cm) sampled randomly from agricultural fields in the period 1975−1980. From about 6,000 same pedons, soils from the subsoil (35−55 cm) was also sampled. SOC content was determined by dry combustion using a LECO IR-12 furnace. DSP consists of a grid based data (7×7 km spacing) established during the 1990s for improve fertilizer recommendation in Denmark [33]. At each 850 grid intersection, soil samples were collected based on genetic horizons and were analyzed for SOC by dry combustion. For some selected horizons, samples were taken to determine D_b. In addition, soil data from about 1100 profiles were used, and these were collected during the establishment of main gas pipeline system and other research activities across Denmark [34].

In total 40,250 topsoil point samples and 1,994 soil profiles with horizon based SOC data were used in this study. About 1,133 soil profiles included D_b measurements.

Environmental covariates

The environmental covariates data used in this study are terrain parameters from the Digital Elevation Model (DEM) of Denmark derived using Light Detection and Ranging (LiDAR) technology. The LiDAR points were interpolated using Delaunay Triangulation and the output Triangular Irregular Network (TIN) surface was converted to a grid DEM with a 1.6×1.6 m spacing. This study resampled the original 1.6 m DEM to 30.4 m using simple aggregation considering the mean value. This 30.4 m grid size was also used in the previous studies in Denmark (e.g., [32,35]). Before aggregating to a coarser grid, the DEM was corrected by removing the pits and peaks of about 50 cm dimensions in order to ensure a regular flow on the surface. Once the DEM was processed, a number of terrain parameters (e.g., slope aspect, slope gradient, elevation, SAGA wetness index, multi-resolution index of valley bottom flatness (MrVBF), altitude above channel network, slope-length factor, over-land flow distance, and valley depth) were derived. Tools and algorithms incorporated in SAGA GIS [36] and Arc GIS V10.2 [37] were used to process the DEM and to derive these parameters.

Other covariates used were six choropleth maps which were compiled at different cartographic scales including: soil map, landscape types, geo-regions, geology, land use, and wetlands – see Table 1 and for more detail Adhikari et al. [35].

Key environmental variables affecting the spatial distribution of SOC and D_b in Denmark were identified. The relative usage of the environmental variables used during SOC and D_b prediction was calculated for each depth and their importance was expressed in percentage. This was done with *Cubist* software which determines the relative importance of variables based on their usage in the prediction rules. The prediction method adopted in this study is summarized in Figure 1.

Table 2. Average soil bulk density (Mg m^{-3}) for different soil organic carbon levels (g 100 g^{-1}) within the central wetlands [Source: [44]].

SOC content	Soil depth (cm)		
	0−30	30−60	60−90
<6	1.15	0.56	0.76
6−12	0.77	0.61	0.44
>12	0.39	0.25	0.19

Table 3. Descriptive statistics of soil organic carbon content (g kg^{-1}) and bulk density (D_b) (Mg m^{-3}) data used in this study.

Parameters	Spline predicted data										Point observation	
	Soil depth (cm)											
	0–5		5–15		15–30		30–60		60–100		0–20	35–55
	SOC	D_b	SOC	D_b	SOC	D_b	SOC	D_b	SOC	D_b	SOC	SOC
Minimum	0.07	0.47	0.09	0.47	0.02	0.47	0.02	0.47	0.01	0.47	0.10	0.10
Maximum	562.31	1.84	562.1	1.84	562.22	1.99	564.01	2.01	570.01	1.96	562.21	559.22
Interquartile Range	15.92	0.21	14.11	0.20	10.22	0.19	6.12	0.17	2.33	0.16	9.16	7.61
Mean	35.22	1.44	30.71	1.44	23.81	1.46	15.61	1.52	9.91	1.59	19.71	15.01
Std. error of mean	1.45	0.00	1.18	0.00	1.11	0.00	1.14	0.00	1.03	0.00	0.07	0.48
Std. deviation	64.81	0.17	52.9	0.17	49.74	0.15	51.2	0.15	46.41	0.15	15.38	44.85
Coef. variation	184	12.02	175.91	11.71	208.9	10.71	328.31	9.81	465.42	9.51	78.12	298.61
Skewness	4.72	-1.01	5.61	-0.96	6.31	-1.03	7.11	-1.32	7.91	-2.02	15.22	7.91

Spline predicted data represents soil profile data from the Danish Soil Profile Database, whereas point observation represents data from the Danish Soil Classification.

Modeling SOC and D_b distribution

The vertical distribution of SOC and D_b in the soil profiles was modeled with mass preserving equal-area quadratic splines [15] in R [38]. The mathematical derivation of the spline has been described by Malone et al. [8]. As the fitting quality of splines to profile attribute data depends on a smoothing parameter–lambda (λ), we tested seven λ values (0.00001, 0.0001, 0.001, 0.01, 0.1, 1, 10) for SOC and D_b data from all the profiles and selected a λ value that showed the best fit for all the profiles using the root mean square. With increasing λ value, the fit becomes more rough. During the fit, we pegged the spline by introducing a 1 cm thick slice with a same SOC value at the topmost layer to prevent unnecessary extrapolation on the surface horizon.

Once the depth function of SOC and D_b were modeled, a weighted-average value of these properties were derived for five soil depths (0–5, 5–15, 15–30, 30–60, and 60–100 cm) based on the GlobalSoilMap specifications [39]. To these new values for the standard depths from all soil profiles, environmental variables were intersected and used for statistical analysis and spatial prediction.

Mapping to the spatial domain

Spatial prediction of SOC content and D_b at five depths was based on Regression kriging (RK) [40]. RK assumed that the spatial prediction function consists of a deterministic model formed by a regression, and the residuals of the regression (unexplained variation) are spatially correlated. The general principle of RK includes (1) regression, and (2) simple kriging of the residuals from the regression, where outputs from these two steps are added to obtain the final prediction. For the regression step, we adopted Regression-rules (RR) derived using *Cubist* software [41]. This tool generates a set of condition–based rules where each rule comes with a multiple regression prediction function that only operates once the conditions specified by the rule are met [42].

To build the SOC and D_b prediction model in *Cubist*, the data set was split randomly into two sets: 75% for calibration and 25% for model validation. Before the data split, SOC content from the topsoil observations (0–20 cm) were joined to the spline predicted SOC content from 0–5 and 5–15 cm soil depths. Similarly, SOC content from subsoil observations (35–55 cm) were attached to the spline predicted SOC at 30–60 cm soil depth. This way, a larger number of point SOC observations were incorporated to the splined SOC data from the 7×7 km grid profiles. The *Cubist* tool was run for log transformed SOC [log SOC g kg^{-1}] and D_b data from each depth interval and the output was converted to a regular grid map using a program written in FORTRAN. For each calibration location, the difference between measured and RR predicted value was calculated and its spatial distribution over the study area was generated using local variogram and point kriging in VESPER program [43]. This continuous residual surface was added to the corresponding RR output to get a final prediction of SOC and D_b for all five depths. Together with the prediction, a map showing the uncertainty of the prediction was generated. Both SOC prediction and uncertainty maps at each depth were then back-transformed to SOC in g kg^{-1}.

Bulk density in the peat areas

Peat lands are mostly present along the central part of the wetlands across Denmark. Bulk density in those areas were adjusted according to Greve et al. [44]. For the three surface layers (i.e., 0–5, 5–15, and 15–30 cm soil depths), D_b from 0–30 cm was used, and for the 30–60 and 60–100 cm, D_b from 30–60 cm and 60–90 cm were used (Table 2).

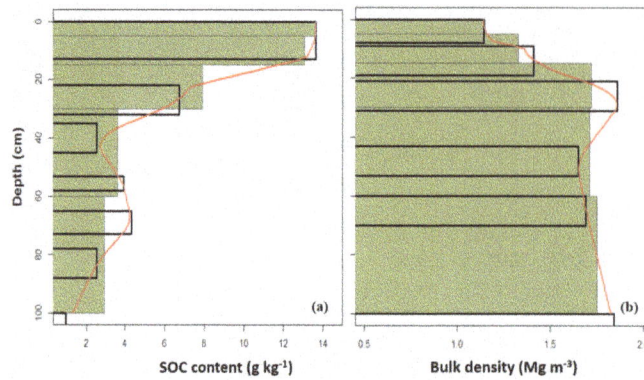

Figure 2. Example of a fitted spline for soil organic carbon content (a), and for bulk density (b). Horizontal bars represent measured soil organic carbon and bulk density at different soil horizons, continuous line through horizons represents a fitted spline, and horizontal olive-green bars give an weighted-average values of these properties at five standard soil depth intervals (i.e., 0−5, 5−15, 15−30, 30−60 and 60−100 cm).

SOC stocks

SOC stock for each soil depth was calculated according to Eq. (1) using SOC content and D_b data. SOC stocks from the five layers were summed to obtain a SOC stock to 1 m soil depth. D_b in Eq. (1) was corrected for gravel content.

$$SOC_{stock} = [SOC_{content} \times D_b \times D]/10 \qquad (1)$$

Where SOC_{stock} is the SOC stock (t ha^{-1}), $SOC_{content}$ the SOC content (g kg^{-1}), D_b the soil bulk density (Mg m^{-3}) and D the given soil layer thickness (cm).

Model validation

Model performance in predicting SOC content and D_b was evaluated on 25% of the point data. The following three indices that were calculated:

$$R^2 = \frac{\sum\limits_{i=1}^{n} \left(pred_i - \overline{obs}\right)^2}{\sum\limits_{i=1}^{n} \left(obs_i - \overline{obs}\right)^2} \qquad (2)$$

$$ME = \frac{1}{n}\sum\limits_{i=1}^{n}(obs_i - pred_i) \qquad (3)$$

$$RMSE = \sqrt{\frac{1}{n}\sum\limits_{i=1}^{n}(obs_i - pred_i)^2} \qquad (4)$$

Where *obs* and *pred* are observed and predicted SOC and D_b values from *n* number of observations at i^{th} locations, *ME* is mean error, and *RMSE* is the root mean square error.

Predicted SOC stock was determined for all five depth intervals and then aggregated to 0−30 and 0−100 cm soil depth. These two C stock maps were stratified based on soil and land use types. The soil class map of Denmark based on the FAO legend [32] and a land use map (CORINE data) were used. The CORINE database for Denmark has 31 classes [45] but in this study the legend was reduced to five major land uses types: artificial surface (urban areas, industrial areas, road network and ports etc.), agricultural areas, forest and semi-natural areas,

wetlands, and others (e.g., coastal lagoon, estuaries etc.) for ease of comparison to other studies (e.g., [11,22,26]). We also stratified our stock maps based on Danish geo-regions. We first derived the area of each soil, land use and geo-regions class based on the number of predicted pixels within those classes and then calculated an average and total SOC stock for each class at 0−30 and 0−100 cm soil depth. The 0−30 cm depth represents the plough depth from agriculture areas and estimation of carbon for this depth is of interest to farm management. The top 1 m soil depth mostly represents a rooting depth of many field crops and may act as an important soil depth section for carbon balance and accounting studies.

Results

Summary of SOC and bulk density

SOC content was highly variable and ranged from 0 to 562 g kg^{-1} for the topsoil (0−20 cm) and from 0 to 569 g kg^{-1} in the subsoil (Table 3). Mean SOC decreased with soil depth and SOC at 60−100 cm was about four times lower than the SOC in the 0−5 cm layer. With depth, the coefficient of variation (CV) of the SOC content increased. The CV at 0−5 cm was 184% and that for the 60−100 cm was about 466%. The SOC data was positively skewed at all soil depths with a maximum skewness coefficient at 60−100 cm. The equal-area splines modeled the depth-wise distribution and generated a continuous SOC profile to 1 m depth. The best λ value to fit all soil profiles for both SOC and D_b data was 0.1. Also average D_b was found to be increased with soil depth. Up to 30 cm depth the D_b was on average 1.44 Mg m^{-3}, whereas it increased to 1.52 Mg m^{-3} below 60 cm depth. Bulk density appeared to be less variable with depth (Table 3).

Figure 2 shows a measured and spline predicted SOC and D_b for a coarse sandy soil under agriculture area from the Saalian moraine soilscape in western Denmark (West Jutland). Measured SOC from different horizons decreased with depth except at 53−73 cm where it increased due to podsolization. Although a spline should pass through the mid-point of each measured horizon for this soil profile, in this case the spline slightly extrapolated the SOC value at 35−55 cm due to the selected λ at 0.1.

Prediction rules for SOC and D_b

Depending on the soil depth, 17 to 54 condition-based regression rules were generated while predicting SOC and D_b.

Table 4. Relative usage (%) of the environmental variables to predict soil organic carbon at different soil depths in Denmark.

Environmental variables	Soil depth (cm)									
	0–5		5–15		15–30		30–60		60–100	
	CR[+]	PF[+]	CR	PF	CR	PF	CR	PF	CR	PF
Wetlands	72	-	55	-	10	-	-	-	-	-
Multi-resolution valley bottom flatness index	64	67	9	81	-	52	-	33	-	5
Geo-regions	62	-	65	-	17	-	73	-	10	-
Soil map	55	-	38	-	28	-	60	-	5	-
Precipitation	54	74	98	98	4	76	66	62	20	25
Landscape	53	-	26	-	15	-	48	-	5	-
Land use	45	-	60	-	5	-	-	-	-	-
Altitude above channel network	37	85	11	87	82	28	7	4	-	3
Elevation	36	88	31	88	50	15	32	21	-	5
Geology	27	-	23	-	30	-	62	-	100	-
SAGA wetness index	22	94	67	94	-	70	5	56	-	2
Valley depth	9	41	16	47	-	26	-	59	-	15
Slope gradient	2	78	8	93	-	54	-	63	-	10
Mid-slope position	2	50	2	52	-	54	4	15	-	2
Flow accumulation	-	30	8	35	-	26	-	-	-	-
Slope length factor	-	85	-	89	-	31	1	52	-	-
Insolation	-	34	-	16	-	-	-	17	5	-
Slope aspect	-	22	-	2	-	15	2	26	-	-

[+]CR-Variable usage in setting the rule conditions; PF-Variable usage in the linear prediction function.

Table 5. Predicted soil organic carbon content (g kg^{-1}) at five soil depths for each FAO-UNESCO soil groups.

FAO soil groups	Soil depth (cm)									
	0–5		5–15		15–30		30–60		60–100	
	Mean	Stdev.	Mean	Stdev.	Mean	Stdev.	Mean	Stdev.	Mean	Stdev.
Alisols	20.8	10.4	19.7	10.9	15.4	19.0	9.8	20.3	2.1	0.7
Arenosols	12.5	10.2	11.9	8.8	11.8	12.4	7.8	15.9	1.9	1.1
Cambisols	17.9	8.2	17.0	6.4	12.2	7.5	7.3	8.8	2.2	0.6
Luvisols	18.0	7.1	16.4	5.7	15.7	8.3	6.8	7.2	2.2	0.7
Podzols	21.9	10.9	21.4	14.1	16.6	25.3	9.1	17.4	2.1	1.2
Fluvisols	24.1	12.0	22.7	9.0	16.7	15.3	12.0	27.5	2.6	0.9
Gleysols	22.8	15.2	22.3	15.0	21.4	26.7	11.9	29.5	2.4	0.8
Podzoluvisols	20.8	6.3	20.8	6.6	14.7	10.2	8.7	8.3	2.0	0.7
Histosols	38.9	22.8	37.8	27.1	52.6	52.5	37.5	71.5	2.5	0.7
Unmapped areas	13.1	6.5	16.5	5.9	12.7	11.5	8.0	13.9	2.0	0.6

This paper only included one of the 54 rules produced during SOC prediction at 0–5 cm soil depth, as an example.

Rule 3: [400 cases, mean 3.1, range 2.2 to 4.5, est err 0.21]

if

Georegions in (1, 2, 3, 6)

Landscape in (5, 6, 7)

MrVBF>6.6

Soil map in (3, 4, 6, 7)

then

log SOC g kg^{-1} = 1.527+0.124*MrVBF+0.0067*Elevation+0.084*SAGA wetness index–0.0011*precipitation+0.06*slope gradient.

This rule used elevation, SAGA wetness index, MrVBF, precipitation and slope gradient to predict SOC in the areas where the MrVBF index is higher than 6.6 and consisted of fine sandy soils from a Moraine landscapes from the geo-regions (e.g. Himmerland). This rule was only valid for 400 locations where the mean SOC was 3.1 [log SOC g kg^{-1}] and the prediction error was about 0.2 [log SOC g kg^{-1}].

Identifying predictors

Variables were identified for SOC and D_b prediction based on their relative usages in the model (Table 4). In all models, precipitation appeared to be the most dominant variable followed by altitude above channel network and SAGA wetness index to predict SOC. As an example, to predict SOC at 5–15 cm, precipitation had a usage of 98% for both rule setting and developing a linear prediction model followed by SAGA wetness index which had a relative usage of 67% and 94%. For this model, insolation and slope aspect had the lowest contribution. Geology became robust with increasing soil depth, whereas land use was important for SOC prediction of the surface layers. Geology, soil map, wetlands, land use, precipitation, MrVBF, SAGA wetness index, elevation, slope gradient, slope-length factor, and altitude above channel network were among the predictors that had a relative importance of >60%. Similarly, land use, soil map, geology, slope gradient, SAGA wetness index, MrVBF, and elevation appeared to be the most important variables for predicting D_b at all soil depths.

Predicted maps

Predicted maps of SOC content (Figure 3) and D_b (map not shown) at five soil depths were produced at a resolution of 30.4×30.4 m. The highest mean SOC content was in the 0–5 cm layer (mean 20 g kg^{-1}; sd 11 g kg^{-1}). Predicted SOC content decreased with soil depth and at 60–100 cm, it was on average 2.2 g kg^{-1}. The soils of western Denmark have relatively higher SOC content than the rest of the country. The northern part has a moderate amount of SOC with two large raised bogs with high SOC contents. Along the coastline, especially in the west, soils with lower SOC concentration were mapped.

The prediction errors were higher towards the west and along the coastline. The prediction error increased with soil depth. For the 0–5 cm layer, the mean error was 1.1 g kg^{-1} and it increased to 1.8 g kg^{-1} at 60–100 cm soil depth.

Predicted SOC content for the FAO−UNESCO soil groups is shown in Table 5. Average SOC content ranged between 11.8 to 52.6 g kg^{-1} at 0–30 cm and between 1.9 to 37.5 g kg^{-1} at 30–100 cm. The highest SOC was observed in Histosols and the lowest in Arenosols at most soil depths.

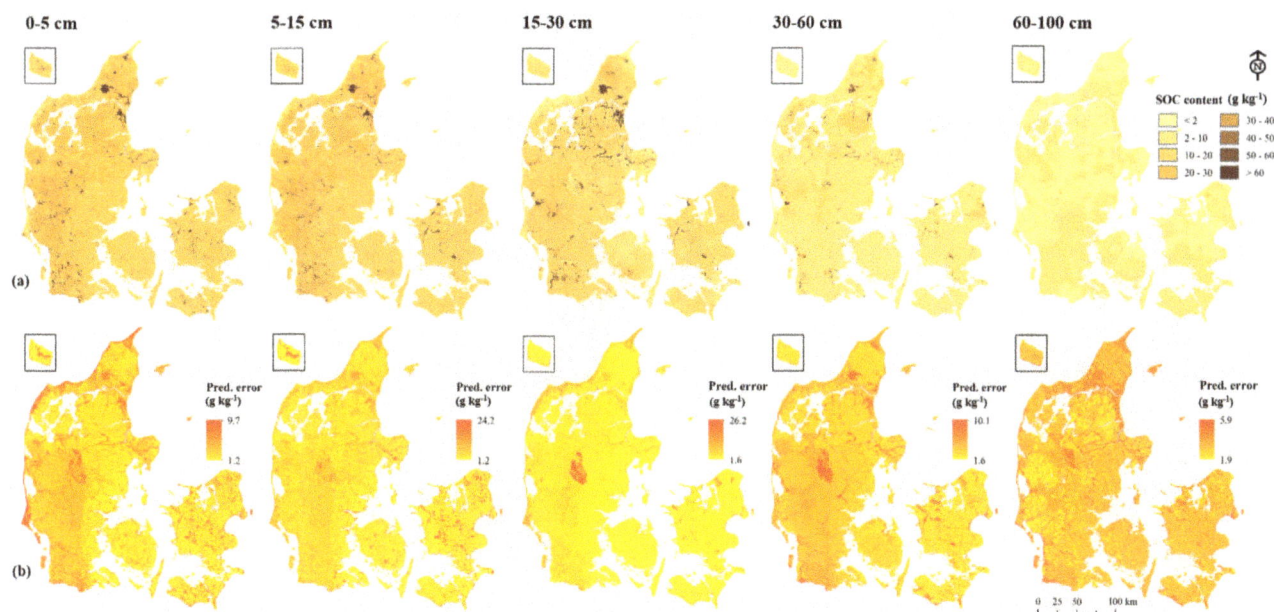

Figure 3. Predicted soil organic carbon content (a), and standard error maps (b) at five soil depths in Denmark.

Model validation

SOC prediction models performance is summarized in Table 6. The best prediction was found at $5-15$ cm soil depth for both training and test data sets. The model performance at $60-100$ cm was relatively poor compared to the rest of the soil depths. Negative ME values suggested that almost all the prediction models were negatively biased suggesting some under prediction of the mapping of SOC distribution.

SOC stocks

SOC stock maps were made for two soil depths (Figure 4). For $0-30$ cm, average SOC stocks were about 72 t ha^{-1} and for the top 1 m depth, it was about 120 t SOC ha^{-1}. Most of the western and northern parts of the country have more than 80 t SOC ha^{-1} in the top 30 cm whereas the average stock in the eastern part of the country was less than 80 t SOC ha^{-1}. Total SOC stock was calculated for each geo-region (Figure 5) and Himmerland and West Jutland had an average stock of about 135 t SOC ha^{-1} followed by North Jutland and Thy both having a mean stock of $>$ 120 t SOC ha^{-1}. The soils of West Jutland and East Denmark contain almost 50% of the total SOC stock in Denmark.

Luvisols and Podzols contain about 60% of the total SOC stock (Table 7). Other soil groups that contain significant amounts of

SOC were Cambisols (6%), Gleysols (9%), and Arenosols (9%). Although Histosols had SOC stock of 176 t ha^{-1}, its total content was 20 Tg SOC. For all soil groups, more than 58% of the total stock was in the top 30 cm. Unmapped areas in the soil map representing major Danish cities that may also contain a substantial amount of carbon [46].

Of the total stock of 570 Tg SOC about 59% is in the upper 30 cm. Soils under agriculture have an average stock of 121 t ha^{-1} and contain about 444 Tg which is almost 78% of the total estimated stock (Table 8). Another large fraction of SOC stock is found in the soils of the forest and semi-natural areas, and these had a stock of 39 Tg in the top 30 cm and about 67 Tg up to 1 m soil depth. Wetland areas contain large amounts of SOC, and average SOC stock within 1 m soil depth was about 152 t ha^{-1} which is nearly 2% of the total stock. Almost 90% of the total SOC stock within the top 1 m soil depth is found in the soils under agriculture, forest and natural areas.

Discussion

Here we have predicted the distribution of SOC contents and stocks across Denmark including an assessment of the prediction error. We have also estimated the SOC contents and stocks for different land uses and soil types. This discussion focuses on the

Table 6. Model performance to predict soil organic carbon content [log SOC g kg^{-1}] based on Training and Validation datasets.

Soil depth (cm)	Training data			Validation data		
	R^2	RMSE	ME	R^2	RMSE	ME
0−5	0.61	0.22	−0.008	0.41	0.24	−0.08
5−15	0.63	0.22	−0.006	0.42	0.24	−0.02
15−30	0.51	0.62	−0.03	0.43	0.66	−0.22
30−60	0.50	0.53	−0.05	0.29	0.56	0.02
60−100	0.28	0.47	−0.06	0.23	0.48	0.12

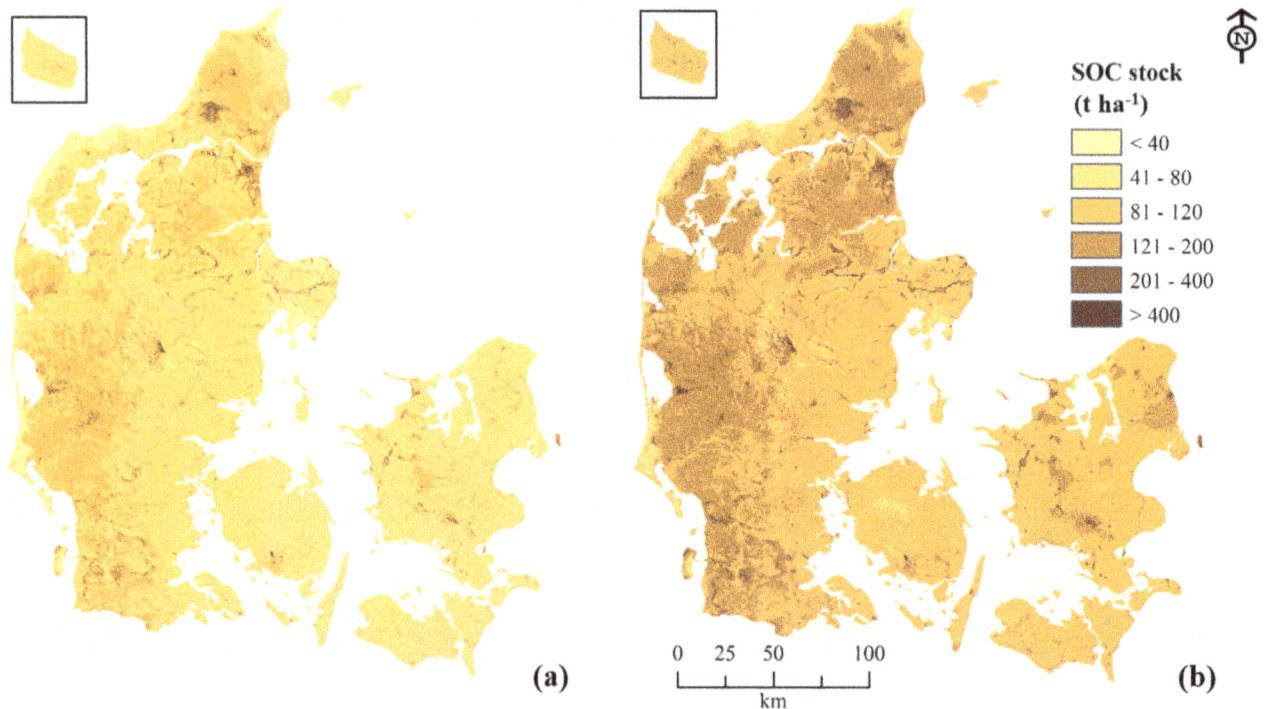

Figure 4. Predicted soil organic carbon stock maps at 0–30 cm (a), and 0–100 cm (b) soil depths for Denmark.

prediction model, the importance of the variables, the uncertainty, and the SOC contents and stocks.

Prediction model

The equal-area spline fit the discrete horizon data but it also harmonized the profile by disaggregating the horizon bulk data and generated a continuous function of SOC distribution. Several other researchers have advocated the usefulness of such splines in depth-wise mapping of SOC in different parts of the world applied from local to regional extents (e.g., [7,8,10]). Pegging of spline by introducing an artificial horizon on the surface benefitted our splines that restricted biased extrapolation of SOC on the surface horizon.

The spatial prediction method (i.e. rule-based regression using the *Cubist* software) was capable of exploring the relationship of SOC to its environment predictors. The prediction rules were conditioned to a given environmental settings such that each rule is valid only to that specific boundary within which SOC distribution presumably less heterogeneous compared to the areas outside where other conditions and rules prevail. For example, SOC content in forests or clayey soils might be different than the SOC from agriculture or sandy soils hence different prediction models operated in these two specific areas. Such a beneficial and advanced function has been used by Lacoste et al. [7] who found regression-rules combined with the spline depth function useful for producing a detailed pseudo-3D map of SOC content in heterogeneous agricultural landscape in France. Minasny and McBratney [42] found that regression rules provided a greater accuracy compared to partial least squares while predicting total carbon content. Several other studies have applied this tool in digital soil mapping (e.g., [35,47,48]).

Variable importance. The environmental variables used to predict SOC content showed a varying level of importance in the model. There was a large influence of precipitation, land use, soil type and some terrain parameters such as elevation, slope gradient, SAGA wetness index, and MrVBF on the spatial distribution of SOC content. The influence of topographic parameters on SOC distribution has been documented in other studies [6,8,9,49–51]. Similarly, land use, precipitation, soil types, wetlands were found important while mapping SOC distribution [11,14,18,52–55]. The influence of terrain parameters on SOC composition and distribution can be linked to its behavior on soil re-distribution through erosion and deposition, in the maintenance of vegetation cover and rooting depths, and in soil drainage that affects SOC decomposition as well as vegetation. In Denmark, the influence of elevation, soil types, geology, and slope gradient was also documented by Bou Kheir et al. [12] when predicting SOC in wet cultivated lands.

The categorical variables such as soil types, geo-regions, and land use were used in defining the rule conditions and continuous variables (terrain parameters) in regression functions. Some terrain parameters like elevation, SAGA wetness index, MrVBF, altitude above channel network etc. were utilized in setting rule conditions. This combined approach in defining the conditions partitioned the study area into several possible strata where SOC distribution was supposed to be more homogeneous and in each stratum several terrain parameters were again used to make sure that the within-stratum SOC variability was well captured by the model. This has made the model robust in predicting SOC content.

Uncertainty. Based on validation indices, the prediction models showed a higher performance (i.e., higher R^2 and a lower $RMSE$) in calibration data (75%) than in validation data (25%). The uncertainty of the SOC prediction increased with depth. The R^2 value ranged between 0.23–0.63 - the higher values for the surface layers. It suggests that our prediction was able to capture up to 63% of total SOC variability. The range of R^2 values was comparable to similar SOC mapping studies where internal validation was applied [24,50,56,57]. Values higher than 0.7 are unusual and values <0.5 are quite common in soil attribute

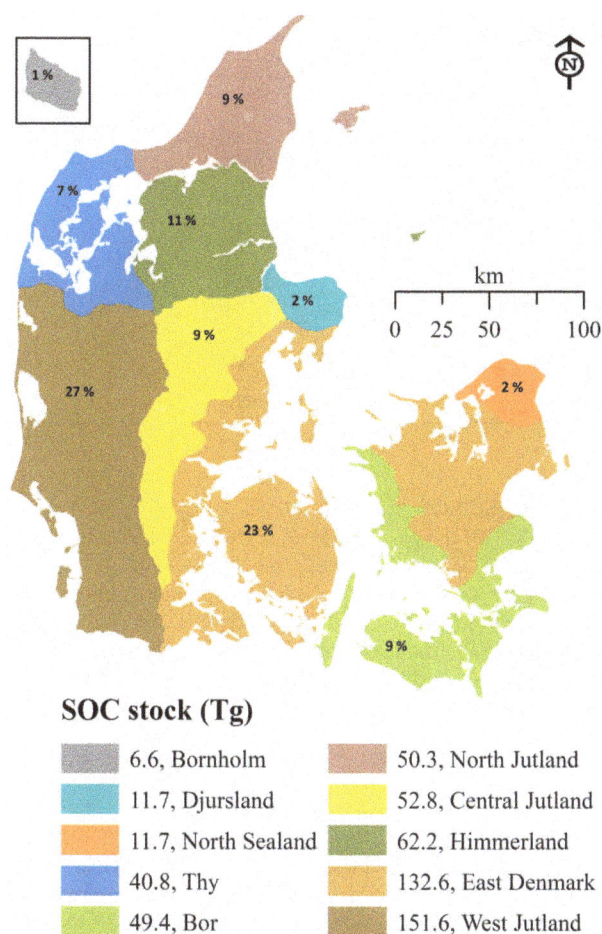

SOC stock (Tg)

6.6, Bornholm	50.3, North Jutland
11.7, Djursland	52.8, Central Jutland
11.7, North Sealand	62.2, Himmerland
40.8, Thy	132.6, East Denmark
49.4, Bor	151.6, West Jutland

Figure 5. Soil organic carbon stock (1 m depth) for the geo-regions in Denmark. Percentage values represent the fraction of the total soil organic carbon content stock (570 Tg).

predictions [58]. The difference in R^2 value between the two datasets could probably be attributed to the different data density used in prediction. Minasny et al. [5] reviewed several previous SOC mapping studies, and reported that with increasing data density the R^2 of prediction was larger. Similarly, SOC maps from the upper soil depths were associated with a low prediction error compared to the maps with depth. This could be linked to the terrain parameters used as predictors because most of these parameters explain soil surface phenomena and the uncertainty increases with depth [24]. Moreover, higher data density from the surface layers (e.g., $0-5$ and $5-5$ cm) might have a positive influence on prediction performance.

SOC content and stock. SOC data from two main sources were used in this study: high-density (about 1 observation per km^2) topsoil and subsoil samples, and from soil profiles at each 7-km grid-intersections together with the data from profiles along the pipe lines across Denmark. The grid and point data covered the entire country.

The SOC data were all from dry combustion analysis and from the period $1975-85$ so we have modeled and predicted here the SOC contents and stocks for that period. Assuming a steady-state condition at that period, i.e. the carbon levels represent the equilibrium with its physical environment and landuse. Likely, SOC contents have changed since that time, thus the map can be used as a baseline to indicate spatio-temporal changes.

A higher SOC content was found in surface soils, and the lowest average SOC content of 3 g kg^{-1} was found at $60-100$ cm soil depth. The western part of the country mostly consists of glacial floodplains and old Saalian moraine landscapes where soils are predominantly sandy. To increase the soil fertility in those areas, farmers has been adding for decades large amounts of manure in the form of pig slurry. This may have led to increased SOC content and gave the highest SOC stock in those areas compared to other geo-regions in Denmark as also reported in Taghizadeh-Toosi et al. [29].

Our results are in line with the previous SOC estimations in Denmark. According to Krogh et al. [26], total stock in Denmark to 1 m soil depth ranged between $563-598$ Tg with 579 Tg as the average which was comparable to our prediction of 570 Tg. Similarly, SOC stock from the topsoil ($0-28$ cm) was about 230 Tg while our prediction showed about 266 Tg. Small differences

Table 7. Soil organic carbon stock in the top 1 m soil depth according to FAO−UNESCO soil groups.

FAO Soil groups	Area (km^2)	Average SOC stock (t ha^{-1})		Total stock (Tg)		Relative stock (%)
		0−30 cm	0−100 cm	0−30 cm	0−100 cm	0−30 cm:0−100 cm
Alisols	921.9	71.3	118.3	7.3 (2%)	12.1 (2%)	60
Arenosols	3,585.9	60.3	105.0	29.5 (9%)	51.3 (9%)	57
Cambisols	2,910.2	64.0	109.9	20.8 (6%)	35.4 (6%)	58
Luvisols	14,499.4	62.3	107.6	100.1 (29%)	172.9 (30%)	58
Podzols	13,745.0	79.6	129.8	115.4 (34%)	189.0 (33%)	61
Fluvisols	879.6	80.2	144.5	7.7 (2%)	13.7 (2%)	56
Gleysols	3,310.0	85.3	140.5	30.3 (9%)	49.7 (9%)	61
Podzoluvisols	698.3	75.7	126.0	5.8 (2%)	9.7 (2%)	60
Histosols	1,039.6	120.8	176.1	14.0 (4%)	21.1 (4%)	69
Unmapped areas	1,320.45	63.1	109.8	9.0 (3%)	15.7 (3%)	57
Sum	42,910.6			340 (100%)	570 (100%)	

Table 8. Predicted soil organic carbon stock in different land use types derived for two different soil depths.

Major Land use types	Area (km²)	Soil depth (cm)	
		0−30	0−100
		SOC Stock in Tg	
Artificial surface (Urban, Industry, Roads, etc.)	3169.0	22.3 (6%)	38.7 (7%)
Agricultural areas	32942.3	266.4 (78%)	443.9 (78%)
Forest and semi-natural areas	5547.1	38.9 (13%)	66.7 (11%)
Wetlands	860.0	8.3 (2%)	13.5 (3%)
Other (Coastal lagoons, Estuaries, etc.)	391.9	4.1 (1%)	7.2 (1%)
Sum	42910.3	340 (100%)	570 (100%)

in stocks estimates between the two studies were noticed. For example, SOC stock in agricultural areas as predicted by Krogh et al. [26] was 404 Tg, whereas the estimate in the current study is 444 Tg. This could be due to a difference in areas estimated for agricultural lands in two studies. Our prediction used CORINE data that suggested an agricultural area of nearly 32,942 km^2, whereas this area was 28,883 km^2 in the previous study based on AIS (Area Information System). In a separate study, Olesen [28] reported a total stock of 604 Tg from 0−60 cm soil depth calculated for an area of 34,000 km^2 based on AIS. Unlike in our study where bulk density from the peat lands were adjusted, Olesen applied a standard bulk density of 1 Mg m^{-3} for all peats or organic soils and that might have increased the stock leading to over estimation.

Previous estimates of SOC stocks had not quantified the spatial distribution of SOC stock nor validated their prediction. Our approach seemed to be more reliable and the data generated could be useful for future SOC content and stock assessments.

A large amount of SOC was present in the soils under agriculture. Our estimation of 121 t ha^{-1} is in a comparable to the SOC stock of 111 t ha^{-1} estimated for arable lands in Scotland [59] but it was slightly higher than the arable stock in Southeast Germany [60]. Similarly, our predicted average SOC stock of 121 t ha^{-1} from the agricultural areas was lower than the findings of Taghizadeh-Toosi et al. [29]. The later study used a soil-type dependent standard D_b established for 7×7 km grid and perhaps overestimated the SOC stock.

The estimation of average SOC stock in different soil groups within 0−30 cm depth was comparable to Arrouays et al. [11], but the total stock was approximately nine times less than reported for France (which is almost 13 times larger than Denmark). Contrary to our study, the soils under forest soils contained more SOC than the arable soils in France. Similar results were also reported by Chaplot et al. [13] from Laos. A slightly lower average stock in the soils under forest in our study might appear due to the inclusion of non-forested areas (semi-natural areas) in the same class and exclusion of litter layer that possibly lowered its value. Decade long intensive soil management practices such as addition of large amount of manure to the agricultural soils might have increased SOC contents and consequently the SOC stock from those areas [29]. Likewise, our estimated average stock for 1 m soil depth from different soil groups, for example, Cambisols (110 t ha^{-1}), and Gleysols (140 t ha^{-1}) were in a agreement with the global estimate of Batjes [22] using the same soil depth. We also observed >50% stock stored within the top 30 cm soil depth for almost all soil groups which also corresponds to the finding of Batjes [22].

Although the SOC stock in Denmark might have changed substantially over the past few decades [29], our estimation based on the available data has provided a baseline SOC. Together with the estimation of uncertainty, the maps are more reliable and could be useful in environmental research in Denmark. It could support national carbon accounting and carbon balance studies and also act as a back ground information for monitoring temporal SOC changes in Denmark.

Conclusions

This study predicted the spatial distribution of SOC content (g kg^{-1}) at five soil depths intervals (0−5, 5−15, 15−30, 30−60, and 60−100 cm) and quantified its stock (t ha^{-1}) to 1 m soil depth for Denmark. DEM derivatives and soil maps were used as predictors where condition-based regression rules were applied to quantify the spatial relationship between measured SOC and D_b with the predictors. The following can be concluded from this study:

- Equal area spline modeled the continuous depth function of SOC and D_b data from discrete soil horizons in soil profiles from Denmark.

- The most important variables that influenced SOC distribution across Denmark were precipitation, wetlands, land use, soil types, elevation, and saga wetness index.

- Model performance was better for surface soil layers and almost all prediction models suffered from underpredictions.

- The total estimated SOC stock at 0−30 cm soil depth was about 340 Tg and that for 0−100 cm was 570 Tg.

- Almost 60% of the total SOC stock was found in Luvisols and Podzols.

- About 90% of SOC was held in soils under agriculture, forest and semi-natural vegetation. For the soils under agriculture, 60% of the SOC was found in the top 30 cm.

- West Jutland and east Denmark contained almost 50% of the total SOC stock.

- This article is an example for a national level SOC mapping based on GlobalSoilMap procedures and the methods applied can be tested and used in other part of the world.

Acknowledgments

The authors thank all soil surveyors in Denmark for compiling a national Danish soil database used in this study. We are grateful to the GlobalSoilMap consortium for providing the specifications that were used in this paper to predict soil carbon. We also thank the academic editor and two reviewers for their comments on this paper. During the course of writing, we were extremely distressed as our friend and co-author Prof. Rania Bou Kheir suddenly passed away. We dedicate this paper to her.

Author Contributions

Conceived and designed the experiments: KA MHG. Performed the experiments: KA MHG. Analyzed the data: KA AEH BM MHG. Contributed reagents/materials/analysis tools: KA AEH BM RBK MBG MHG. Contributed to the writing of the manuscript: KA.

References

1. McBratney AB, Mendonça-Santos ML, Minasny B (2003) On digital soil mapping. Geoderma 117: 3−52.
2. Hartemink AE, McSweeney K (2014) Soil Carbon. Dordrecht: Springer.
3. Hartemink AE, Lal R, Gerzabek MH, Jama B, McBratney AB, et al. (2014) Soil carbon research and global environmental challenges. PeerJ PrePrints 2: e366v361.
4. Jenny H, Salem AE, Wallis JR (1968) Interplay of soil organic matter and soil fertility with state factors and soil properties. "Study Week on Organic Matter and Soil Fertility", Pontificiae Academiae Scientiarvm Scripta Varia 32: 5−36.
5. Minasny B, McBratney AB, Malone BP, Wheeler I (2013) Digital mapping of soil carbon. Advances in Agronomy 118: 1−47.
6. Dorji T, Odeh IOA, Field DJ, Baillie IC (2014) Digital soil mapping of soil organic carbon stocks under different land use and land cover types in montane ecosystems, Eastern Himalayas. Forest Ecology and Management 318: 91−102.
7. Lacoste M, Minasny B, McBratney AB, Michot D, Viaud V, et al. (2014) High resolution 3D mapping of soil organic carbon in a heterogeneous agricultural landscape. Geoderma 213: 296−311.
8. Malone BP, McBratney AB, Minasny B, Laslett GM (2009) Mapping continuous depth functions of soil carbon storage and available water capacity. Geoderma 154: 138−152.
9. Mishra U, Lal R, Liu D, van Meirvenne M (2010) Predicting the spatial variation of the soil organic carbon pool at a regional scale. Soil Science Society of America Journal 74: 906−914.
10. Odgers NP, Libohova Z, Thompson JA (2012) Equal-area spline functions applied to a legacy soil database to create weighted-means maps of soil organic carbon at a continental scale. Geoderma 189−190: 153−163.
11. Arrouays D, Deslais W, Badeau V (2001) The carbon content of topsoil and its geographical distribution in France. Soil use and Management 17: 7−11.
12. Bou Kheir R, Greve MH, Bøcher PK, Greve MB, Larsen R, et al. (2010) Predictive mapping of soil organic carbon in wet cultivated lands using classification-tree based models: The case study of Denmark. Journal of Environmental Management 91: 1150−1160.
13. Chaplot V, Bouahom B, Valentin C (2010) Soil organic carbon stocks in Laos: spatial variations and controlling factors. Global Change Biology 16: 1380−1393.
14. Doblas-Miranda E, Rovira P, Brotons L, Martínez-Vilalta J, Retana J, et al. (2013) Soil carbon stocks and their variability across the forests, shrublands and grasslands of peninsular Spain. Biogeosciences 10: 8353−8361.
15. Bishop TFA, McBratney AB, Laslett GM (1999) Modelling soil attribute depth functions with equal-area quadratic smoothing splines. Geoderma 91: 27−45.
16. Poncehernandez R, Marriott FHC, Beckett PHT (1986) An improved method for reconstructing a soil-profile from analyses of a small number of samples. Journal of Soil Science 37: 455−467.
17. Hartemink AE, Minasny B (2014) Towards digital soil morphometrics. Geoderma 230−231: 305−317.
18. Kempen B, Brus DJ, Stoorvogel JJ (2011) Three-dimensional mapping of soil organic matter content using soil type-specific depth functions. Geoderma 162: 107−123.
19. IPCC (2003) Good practice guidance for land use, land-use change and forestry. Institute for Global Environmental Strategies.
20. Janssens IA, Freibauer A, Schlamadinger B, Ceulemans R, Ciais P, et al. (2005) The carbon budget of terrestrial ecosystems at country-scale–a European case study. Biogeosciences 2: 15−26.
21. Smith J, Smith P, Wattenbach M, Zaehle S, Hiederer R, et al. (2005) Projected changes in mineral soil carbon of European croplands and grasslands, 1990–2080. Global Change Biology 11: 2141−2152.
22. Batjes NH (1996) Total carbon and nitrogen in the soils of the world. European Journal of Soil Science 47: 151−163.
23. Jobbágy EG, Jackson RB (2000) The vertical distribution of soil organic carbon and its relation to climate and vegetation. Ecological applications 10: 423−436.
24. Minasny B, McBratney AB, Mendonça-Santos ML, Odeh IOA, Guyon B (2006) Prediction and digital mapping of soil carbon storage in the Lower Namoi Valley. Australian Journal of Soil Research 44: 233−244.
25. Mishra U, Lal R, Slater B, Calhoun F, Liu DS, et al. (2009) Predicting soil organic carbon stock using profile depth distribution functions and ordinary kriging. Soil Science Society of America Journal 73: 614−621.
26. Krogh L, Noergaard A, Hermansen M, Greve MH, Balstroem T, et al. (2003) Preliminary estimates of contemporary soil organic carbon stocks in Denmark using multiple datasets and four scaling-up methods. Agriculture, Ecosystems & Environment 96: 19−28.
27. Greve MH, Greve MB, Bøcher PK, Balstrom T, Madsen HB, et al. (2007) Generating a Danish raster-based topsoil property map combining choropleth maps and point information. Geografisk Tidsskrift-Danish Journal of Geography 107: 1−12.
28. Olesen JE (1991) Forelobig beregning af CO_2 emission fra danske landbrugsjord. AJMET Notat no 24 Afdeling for Jordbrugsmeteorologi. Tjele, Denmark: Research Center Foulum, Danish Institute of Agricultural Sciences.
29. Taghizadeh-Toosi A, Olesen JE, Kristensen K, Elsgaard L, Østergaard HS, et al. (2014) Changes in carbon stocks of Danish agricultural mineral soils during 1986−2009. European Journal of Soil Science.
30. Danmarks Meteorologiske Institut (1998) Danmarks Klima 1997. Copenhagen: Danmarks Meteorologiske Institut.
31. Madsen HB, Nørr AH, Holst KA (1992) The Danish soil classification: Atlas over Denmark Copenhagen, Denmark: The Royal Danish Geographical Society.
32. Adhikari K, Minasny B, Greve MB, Greve MH (2014) Constructing a soil class map of Denmark based on the FAO legend using digital techniques. Geoderma 214−215: 101−113.
33. Østergaard HS (1990) Kvadratnettet for nitratundersøgelser i Denmark 1986−89. Skejby, Århus: Landbrugets Rådgivningscenter, Landskontoret for Planteavl.
34. Madsen HB, Jensen NH (1985) The establishment of pedological soil databases in Denmark. Geografisk Tidsskrift-Danish Journal of Geography 85: 1−8.
35. Adhikari K, Kheir RB, Greve MB, Bøcher PK, Malone BP, et al. (2013) High-resolution 3-D mapping of soil texture in Denmark. Soil Science Society of America Journal 77: 860−876.
36. SAGA GIS System for Automated Geoscientific Analyses. Available: http://www.saga-gis.org.
37. ESRI (2012) ArcGIS Desktop: Release 10.1. Redlands, CA: Environmental Systems Research Institute.
38. R Development Core Team (2008) R: A language and environment for statistical computing. Vienna, Austria: R Foundation for Statistical Computing.
39. Arrouays D, Grundy MG, Hartemink AE, Hempel JW, Heuvelink GBM, et al. (2014) GlobalSoilMap: Toward a fine-resolution global grid of soil properties. Advances in Agronomy 125: 93−134.
40. Odeh IOA, McBratney AB, Chittleborough DJ (1995) Further results on prediction of soil properties from terrain attributes - Heterotopic cokriging and Regression-kriging. Geoderma 67: 215−226.
41. Quinlan JR (1993) C4. 5: Programs for machine learning. San Mateo, CA, USA: Morgan Kaufmann.
42. Minasny B, McBratney AB (2008) Regression rules as a tool for predicting soil properties from infrared reflectance spectroscopy. Chemometrics and intelligent laboratory systems 94: 72−79.
43. Minasny B, McBratney AB, Whelan BM (2005) VESPER version 1.62. Australian Centre for Precision Agriculture, McMillan Building A05, The University of Sydney. NSW.
44. Greve MH, Christiansen OF, Greve MB, Bou Kheir R (2014) Change in peat coverage in Danish cultivated soils during the past 35 Years. Soil Science (Accepted).
45. Stjernholm M, Kjeldgaard A (2004) CORINE landcover update in Denmark-Final report. Denmark: National Environment Research Institute (NERI).
46. Vasenev VI, Stoorvogel JJ, Vasenev II, Valentini R (2014) How to map soil organic carbon stocks in highly urbanized regions? Geoderma 226−227: 103−115.
47. Lacoste M, Lemercier B, Walter C (2011) Regional mapping of soil parent material by machine learning based on point data. Geomorphology 133: 90−99.
48. Bui EN, Henderson BL, Viergever K (2006) Knowledge discovery from models of soil properties developed through data mining. Ecological Modelling 191: 431−446.
49. Razakamanarivo RH, Grinand C, Razafindrakoto MA, Bernoux M, Albrecht A (2011) Mapping organic carbon stocks in eucalyptus plantations of the central highlands of Madagascar: A multiple regression approach. Geoderma 162: 335−346.
50. Mueller TG, Pierce FJ (2003) Soil carbon maps: Enhancing spatial estimates with simple terrain attributes at multiple scales. Soil Science Society of America Journal 67: 258−267.
51. Mendonça-Santos ML, Dart RO, Santos HG, Coelho MR, Barbara RLL, et al. (2010) Digital soil mapping of topsoil organic carbon content of Rio de Janeiro State, Brazil. In: J. L Boettinger, D. W Howell, A. C Moore, A. E Hartemink, S Kienast-Brown, editors. Digital Soil Mapping. Springer Netherlands. 255−266.
52. Ross CW, Grunwald S, Myers DB (2013) Spatiotemporal modeling of soil organic carbon stocks across a subtropical region. Science of The Total Environment 461−462: 149−157.
53. Martin MP, Wattenbach M, Smith P, Meersmans J, Jolivet C, et al. (2011) Spatial distribution of soil organic carbon stocks in France. Biogeosciences 8: 1053−1065.
54. Wiesmeier M, Barthold F, Blank B, Kögel-Knabner I (2011) Digital mapping of soil organic matter stocks using Random Forest modeling in a semi-arid steppe ecosystem. Plant Soil 340: 7−24.

55. Meersmans J, van Wesemael B, Goidts E, van Molle M, De Baets S, et al. (2011) Spatial analysis of soil organic carbon evolution in Belgian croplands and grasslands, 1960–2006. Global Change Biology 17: 466–479.

56. Bui EN, Henderson BL, Viergever K (2009) Using knowledge discovery with data mining from the Australian Soil Resource Information System database to inform soil carbon mapping in Australia. Global biogeochemical cycles 23: 1–15.

57. Zhao Y-C, Shi X-Z (2010) Spatial prediction and uncertainty assessment of soil organic carbon in Hebei province, China. Digital Soil Mapping. Springer. 227–239.

58. Beckett P, Webster R (1971) Soil variability: a review. Soils and Fertilizers 34: 1–15.

59. Chapman SJ, Bell JS, Campbell CD, Hudson G, Lilly A, et al. (2013) Comparison of soil carbon stocks in Scottish soils between 1978 and 2009. European Journal of Soil Science 64: 455–465.

60. Wiesmeier M, Spörlein P, Geuβ U, Hangen E, Haug S, et al. (2012) Soil organic carbon stocks in southeast Germany (Bavaria) as affected by land use, soil type and sampling depth. Global Change Biology 18: 2233–2245.

61. Böhner J, Antonić O (2009) Land-surface parameters specific to topo-climatology. In: T Hengl and H. I Reuter, editors. Geomorphometry- Concepts, Software, Applications. New York: Elsevier. 195–226.

62. Bendix J (2004) Geländeklimatologie – Gebrüder Borntraeger. Berlin, Stuttgart.

63. Zevenbergen LW, Thorne CR (1987) Quantitative analysis of land surface topography. Earth Surface Processes and Landforms 12: 47–56.

64. Moore ID, Gessler PE, Nielsen GA, Peterson GA (1993) Soil attribute prediction using terrain analysis. Soil Science Society of America Journal 57: 443–452.

65. Gallant JC, Dowling TI (2003) A multiresolution index of valley bottom flatness for mapping depositional areas. Water Resources Research 39: 1347–1359.

66. Desmet PJJ, Govers G (1996) A GIS procedure for automatically calculating the USLE LS factor on topographically complex landscape units. Journal of Soil and Water Conservation 51: 427–433.

Alfalfa (*Medicago sativa* L.)/Maize (*Zea mays* L.) Intercropping Provides a Feasible Way to Improve Yield and Economic Incomes in Farming and Pastoral Areas of Northeast China

Baoru Sun, Yi Peng, Hongyu Yang, Zhijian Li*, Yingzhi Gao*, Chao Wang, Yuli Yan, Yanmei Liu

Key Laboratory of Vegetation Ecology, Northeast Normal University, Changchun, China

Abstract

Given the growing challenges to food and eco-environmental security as well as sustainable development of animal husbandry in the farming and pastoral areas of northeast China, it is crucial to identify advantageous intercropping modes and some constraints limiting its popularization. In order to assess the performance of various intercropping modes of maize and alfalfa, a field experiment was conducted in a completely randomized block design with five treatments: maize monoculture in even rows, maize monoculture in alternating wide and narrow rows, alfalfa monoculture, maize intercropped with one row of alfalfa in wide rows and maize intercropped with two rows of alfalfa in wide rows. Results demonstrate that maize monoculture in alternating wide and narrow rows performed best for light transmission, grain yield and output value, compared to in even rows. When intercropped, maize intercropped with one row of alfalfa in wide rows was identified as the optimal strategy and the largely complementary ecological niches of alfalfa and maize were shown to account for the intercropping advantages, optimizing resource utilization and improving yield and economic incomes. These findings suggest that alfalfa/maize intercropping has obvious advantages over monoculture and is applicable to the farming and pastoral areas of northeast China.

Editor: Wen-Xiong Lin, Agroecological Institute, China

Funding: This work was supported by the National Natural Science Foundation of China (NSFC, 31072080 and 31270444), the program for New Century Excellent Talents in University (NCET-13-0717), Key Science and Technology Program of Jilin Province (20100212, 2012ZDGG008), National Key Technology R&D Program (2011BAD17B04-3-2), and the National Program on Key Basic Research Project (2012CB722202). The funders had no role in study design, data collection and analysis, decision to publish, or preparation of the manuscript.

Competing Interests: The authors have declared that no competing interests exist.

* Email: lizj004@nenu.edu.cn (ZL); gaoyz108@nenu.edu.cn (YG)

Introduction

The farming and pastoral area (FPA) of northeast China (NEC) is an agriculture-based ecozone combining forestry and animal husbandry, and it is an important grain commodity and animal husbandry base. However, it is also a vulnerable eco-environmental zone owing to low vegetation cover, fine sandy soil and strong wind [1–2]. Wind erosion, water erosion and unsustainable production activities (e.g., single cropping and multiple-year continuous cropping) have made the land subject to dust storms in winter and spring. Moreover, cropland soil is being increasingly eroded, causing low soil fertility and reduced crop productivity and quality [3–5]. Together with the growing use of agricultural chemicals, such as fertilizers and herbicides, sustainable agricultural development is now facing many serious challenges [6].

In addition to the cropland, grassland has been degraded due to overgrazing and excessive agricultural reclamation, and grassland degradation is reflected by the reduced grassland production and forb quality and low carrying capacity [7]. "Grain for green" initiatives including reseeding of forage grasses have long been considered as effective ways to restore grassland vegetation and help to balance the ecological system [8]. Moreover, with increasing demand for meat products and high quality forage

grass, China continues to give substantial support and increased financial investment to the development of the forage industry. Additionally, the Chinese government has strongly endorsed a proposal for boosting the development of the alfalfa industry so as to ensure the production, processing and sustainable supply of high quality forage grass [9]. Therefore, in the context of food and eco-environmental security and animal husbandry sustainable development, traditional farming patterns in the northeast FPA are being altered; with a tendency to adjust agricultural structure, introduce forage grass into the main crop farming system and establish an intercropping pattern between crops and forage grass, resulting in efficient resource utilization, a friendly ecological environment and good economic benefits [10–11].

Intercropping, the practice of growing two or more crops in proximity, is advantageous due to the differences in ecological characteristics and growth of the intercropped varieties. This can establish a composite population, producing complementary effects and increasing yield and economic incomes per unit area [12]. Furthermore, intercropping can improve soil fertility, alleviate disease and insect harm, and inhibit the growth of weeds [13–16]. Maize, as a principal crop of the northeast FPA, is an important food and forage crop. Its grain is an important fodder with high energy, known as "queen feed" [17]. Alfalfa is a

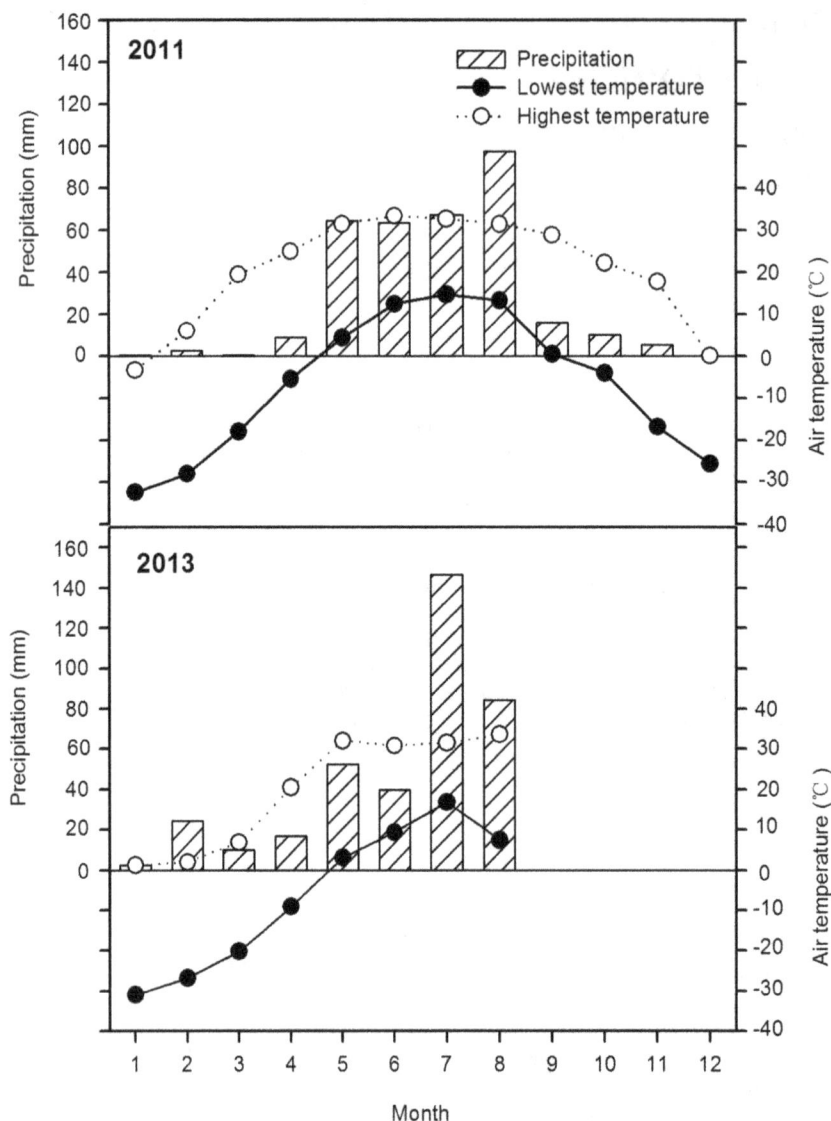

Figure 1. Monthly precipitation (bar) and air temperature (curve) of the experimental site in 2011 and 2013.

leguminous forb that has been prioritized in the development of the PFA of NEC; suitable because of its high yield, rich content and high quality of protein, abundant vitamins and minerals, good palatability and high digestibility [18]. Moreover, as a perennial, alfalfa supplies soil cover throughout the year; providing wind resistance, fixing soil and improving the environment of the planting area. Its large root system can significantly improve soil fertility and physico-chemical properties, leading to the win-win relationship between utilization and conservation [19]. Therefore, it is predictable that intercropping alfalfa with maize can not only guarantee regional food security and meet the nutritional requirements of forage industry, but also provide eco-environmental protections, and it is a promising cropping pattern in the future development of this region.

In China, maize has been traditionally cultivated in even rows. More recently, in attempts to improve maize yield agricultural scientists have experimented with alternative cultivation strategies. Indeed, a study has shown that in the main agricultural region of Jilin Province, planting maize in alternating wide and narrow rows can achieve up to 10% improved yield over the conventional

pattern [20]. Previous intercropping studies focused mainly on food crop combinations [21–22], but few studies have investigated the possibility of intercropping alfalfa with maize, a strategy combining annual food crop with perennial forage crop together. Those studies that have addressed alfalfa/maize intercropping focused on improving group yield, forage quality, soil fertility and the environment [23–24]. To our knowledge, no study has investigated whether intercropping alfalfa with maize in alternating wide and narrow rows could provide sustained high yield and economic incomes while taking investment and environmental factors into account, and whether there are some constraints limiting its popularization such as crop management and the acceptability of local farmers. Hence, a field experiment was conducted to explore the following questions: (1) Compared to an even row approach, does an alternating wide and narrow row planting approach in maize improve yield in the FPA of NEC? (2) Can intercropping alfalfa with maize provide sustained high yield and economic incomes? Which intercropping mode is best, considering farmer incomes and land management?

Materials and Methods

Experimental site

The study was conducted between 2007 and 2013 at the Grassland Ecosystem Field Station of the Northeast Normal University at Songnen Grassland (123° 44′ E and 44° 40′ N, 137.8–144.8 masl), a typical FPA of NEC. This area is characterized by a semi-arid and temperate continental monsoon climate with a mean annual temperature of 4.6–6.4°C, an annual accumulated temperature ($\geq 10°C$) of 2546–3375°C, mean annual precipitation of 300–400 mm (86% of precipitation occurring from May to September) and mean annual evaporation of 1500–2000 mm. The frost-free period lasts approximately 140 days, from the end of April to early October. The two experimental years contrasted each other in terms of precipitation. In 2011, the annual total precipitation was 335 mm and mostly occurred in the growing season (308 mm), whereas the year 2013 was a year with a higher precipitation amount (376 mm from January to August), better seasonal distribution and pronounced peak in July. Air temperature in 2011 and 2013 showed a similar dynamic with the maximum and minimum air temperatures of 33°C and −32°C, respectively (Figure 1). The soil type at the site is light chernozem with deep soil layers. The plough layer consists of organic C (17.24 ± 1.76 g kg^{-1}), total N (0.98 ± 0.15 g kg^{-1}), rapidly available P (5.88 ± 0.65 mg kg^{-1}) and rapidly available K (140.70 ± 11.75 mg kg^{-1}), with an initial soil pH of 7.46 ± 0.04.

Experiment materials and design

Alfalfa variety *Medicago sativa* L. cv. Dongmu No. 1 was used throughout this study. This variety was bred by Northeast Normal University to adapt readily to drought and cold and is now the principally cultivated variety in the study area. Alfalfa generally turns green in mid-April, continues to grow until end of October and can be harvested three times per year. The hay yield can reach up to 7500–10,000 kg ha^{-1} in the second year under the rainfed condition. *Zea mays* L. cv. Zhengdan 958 was chosen as the maize test variety. This variety matures at about 128 days and is widely cultivated by local farmers due to the high and stable yield and a great resistance to lodging and disease. The differences of alfalfa and maize in growth dynamics (Figure S4) make them easy to form temporal and spatial complementarity and promote the efficient utilization of light, water and nutrients.

At the beginning of the experiment, the whole field was fully ploughed to ensure uniform soil conditions. The experiment was conducted in a completely randomized block design with four blocks each containing five cropping patterns: maize monoculture in even (65 cm) rows (MME); maize monoculture in alternating wide (90 cm) and narrow rows (40 cm) (MMW); alfalfa monoculture in even (30 cm) rows (MA); MMW intercropped with alfalfa, with one row of alfalfa in the wide rows (23.1% alfalfa in intercropping area) (IMA1); MMW intercropped with alfalfa, with two rows of alfalfa in the wide rows (46.2% alfalfa in intercropping area) (IMA2). In the maize planting patterns, every three rows of maize were defined as one belt (1.3 m wide) with three belts in each treatment. Each plot had an area of 46.8 m^2 (3.9 m × 12 m), with 50 cm spacing between each plot and 1 m separating each block.

To establish the intercropping system, alfalfa was sown in early July 2007 at a seeding rate of 15 kg ha^{-1} and had been allowed to grow for 4 years before data collection in 2011. Every year maize was sown in early May with 26 cm separating each plant, and was irrigated with 75 mm before its sowing to ensure good germination. In all planting patterns, 135 kg P ha^{-1} and 90 kg K ha^{-1} were applied for alfalfa, and 225 kg N ha^{-1}, 120 kg P ha^{-1} and 60

kg K ha^{-1} were applied for maize. The commercial fertilizer used were: nitrogen, urea (46% nitrogen content); phosphate, diammonium phosphate (46% phosphorus content, 18% nitrogen content); potash, potassium chloride (60% potassium content). All fertilizers required by alfalfa were spread in the soil at the time of sowing. For maize, all of the phosphate and potash, and half of the nitrogen fertilizers were spread in the soil at the time of sowing and the remaining nitrogen fertilizers applied during the big flare opening period. Weeds were regularly controlled using a hand hoe, and pest and disease of alfalfa or maize were separately controlled timely with the idea of minimizing the pesticide application effects on the non-target crop.

Data collection

The data was collected from 2011. Unfortunately, there was considerably small snowfall and the air temperature was relatively high in the winter of 2011 (Figure 1), which weakened alfalfa resistance to cold and freezing [25]. In the early spring of 2012, the sprout of alfalfa was promoted due to the continuously high air temperature from 21th to 30th in March with the maximum 16.8°C, whereas the air temperature dramatically decreased from 31th March to 6th in April with the minimum −9.0°C (Figure S1). This unexpected cold snap made a serious freezing injury to the sprouting alfalfa due to its lowest cold resistence at that time [26]. Consequently, alfalfa achieved a low turning green rate. In order to ensure the sustainable production of alfalfa, we stopped data collection in 2012 and restarted sampling in 2013 when alfalfa turned a good recovery in its growth and development.

Light intensity was determined using a ST-80C illuminometer (Photoelectric Instrument Factory of Beijing Normal University, China) in 2011. Four layers for maize or alfalfa were selected in each treatment: (1) the reference layer, above the canopy; (2) the bottom layer, close to the soil surface; (3) the intermediate layer, at the point of 1/2 plant height and (4) the top layer, 10 cm below the top of the plant. Based on the light intensity, light transmission was calculated for each of the layers. Maize and alfalfa leaf area index (LAI) was measured using a LAI-2000 plant canopy analyzer (LI-COR, Inc., USA) in 2011 and 2013. From the first flowering stage of alfalfa, both light intensity and LAI were tested from 10:00–11:00 on a sunny day. The measurement was repeated every 15 days at three different positions within each testing belt.

Using time domain reflectometry (TDR 100, Campbell Scientific Inc., Logan, Utah, USA), soil water content (SWC) at depth of 0–20 cm was measured four times for all the treatments in each year (2011: 10th June, 24th July, 10th September and 4th October; 2013: 6th June, 12th July, 26th August and 30th September). The measuring time was corresponding to different developing stages of crop, that was the first, second and third flowering stage of alfalfa and the maturity stage of maize, respectively. Specifically, SWC was tested at ten different representative positions between the alfalfa or maize rows in the monoculture treatments for each plot. As to intercropping treatments, SWC was measured between the alfalfa rows (for IMA2), between the alfalfa and maize rows and between the maize rows with ten different representative positions in each belt of plots and then all measured data were averaged as the SWC condition of the intercropping system.

The final harvest of maize was taken in early October (2011: 5th October; 2013: 1th October). In each planting pattern, the second belt was selected for grain yield determination. Fresh weight was recorded before maize grains were oven-dried at 65°C to a constant weight and dry weight recorded. Water content and grain yield were calculated based on dry weight. Alfalfa was cut three times (2011: 11th June, 25th July and 11th September; 2013: 7th

June, 13th July and 27th August). For each alfalfa cut, fresh weight and dry weight were determined, and water content and hay yield calculated.

Data calculations

Light transmission. Light transmission (LT) was calculated using equation (1);

$$LT_m = \frac{LI_m}{LI_a} \qquad (1)$$

where m is the canopy layer (including the top, intermediate and bottom layers) of either alfalfa or maize, LT_m is the light transmission at layer m, LI_m is the light intensity at layer m, and LI_a is the light intensity above the canopy [27].

Output value per unit area. The output value per unit area (OVPUA) was calculated according to equation (2);

$$OVPUA = P_m \times Y_m + P_a \times Y_a \qquad (2)$$

where for each planting pattern, P_m and P_a denote the price of maize grain and alfalfa hay respectively. The price expressed in USD is based on the exchange rate of 630 ¥ 100 USD[-1] in 2011 and 613 ¥ 100 USD[-1] in 2013, thus P_m and P_a are respectively 361.90 USD t[-1] and 380.95 USD t[-1] in 2011 and 342.58 USD t[-1] and 358.89 USD t[-1] in 2013 based on local market values; Y_m and Y_a denote the yield of maize grain and alfalfa hay respectively [28].

Land equivalent ratio. Land equivalent ratio (LER) is an index that adopts yield as a comparison parameter to evaluate land use efficiency of different cultivated patterns relative to monoculture. However, the value of LER is not necessarily related to yield. The equation is defined as follows;

$$LER = (Y_{ia}/Y_{ma}) + (Y_{im}/Y_{mm}) \qquad (3)$$

where Y_{ia} and Y_{im} are the respective yields of alfalfa and maize in the total intercropped area, and Y_{ma} and Y_{mm} are the yields of monocultured alfalfa and maize. An LER greater than 1.0 reveals an intercropping advantage and the favors of intercropping on crops growth and yield, while an LER less than 1.0 indicates an intercropping disadvantage and the negative affections of inter-cropping on crops growth and yield [29–31].

Aggressivity. Aggressivity (A_{ac}) measures the relative re-source competitiveness of two intercropped species;

$$A_{ac} = Y_{ia}/(Y_{ma} \times P_a) - Y_{im}/(Y_{mm} \times P_c) \qquad (4)$$

where A_{ac} is the aggressivity of alfalfa relative to maize in the intercropping system, P_a and P_c are the intercropping area proportions occupied by alfalfa and maize respectively, while the meanings of other symbols are the same as those used for the LER equation. If A_{ac} is greater than 0, the competitive ability of alfalfa exceeds that of maize in intercropping; otherwise, maize has greater competitiveness [12,23].

Statistical analysis

Normal distribution and homogeneous variances were tested for all the data with Shapiro-Wilk test [32] using SPSS 17.0 software

(SPSS Inc., Chicago, IL, USA), and light transmission was biquadrate after reciprocal transformed to achieve normal distribution. One-way ANOVA was performed to examine the effects of cropping patterns on light transmission and soil water content (SWC). Repeated measures ANOVA in a general linear model (GLM) were conducted to assess the effects of planting modes on LAI, yield and output value per unit area (OVPUA), with year and alfalfa flowering stage as the repeated measures. The results were reported using the Greenhouse-Geisser correction when Mauchly's test of sphericity was violated. If the interaction between factors was significant, one-way ANOVA was conducted to evaluate the effects of cropping pattern or alfalfa flowering stage and significant differences of means were compared with Duncan's multiple-compare range test; while the effects of year were tested by independent-samples t test. Significant level was set at $P < 0.05$.

Results

Maize light transmission

With the exception of the first flowering stage of alfalfa (early June), no significant difference was found in light transmission at the top of maize among treatments (Figure 2A). At the stage of the first flowering of alfalfa, light transmission of monoculture maize (MME and MMW) was significantly higher than that of intercropped maize (IMA1 and IMA2) ($P < 0.0001$), whereas there was no significant difference between monoculture modes or between intercropping modes. This can be accounted for by the fact that during flowering, alfalfa grew taller than maize, which had an overshadowing effect. On the whole, however, no significant difference was observed for average light transmission at the top of maize, regardless of cropping pattern (Figure 2A inset).

During continuous growth of maize, there was a tendency for a reduction in both intermediate and bottom light transmission for all treatments. This was particularly evident before maize entered the big flare opening stage (24th July), with little change in light transmission occurring after this point (Figure 2B and C). In the first flowering stage of alfalfa, both intermediate and bottom light transmission of monoculture maize (MME and MMW) were higher than for intercropped maize (IMA1 and IMA2); however, the opposite pattern was observed in the resting period, when light transmission of IMA1 and IMA2 maize was higher than monoculture maize (Figure 2B and C). In addition, there were significant differences for intermediate and bottom light transmis-sion between intercropping and monoculture maize at the second alfalfa flowering stage (24th July) (intermediate: $P < 0.0001$; bottom: $P = 0.003$) and after the second cutting of alfalfa (8th August) (intermediate: $P = 0.002$; bottom: $P = 0.005$). Significant differences were also found for intermediate and bottom light transmission between intercropped maize (IMA1 and IMA2) at the second flowering stage of alfalfa, but differences between monocultures of maize were never significant. However, for maize intermediate or bottom canopy layers, no significant difference in average light transmission was observed, regardless of planting strategy (Figure 2A inset).

Alfalfa light transmission

Throughout the alfalfa growing season, there was no evident variance in top light transmission in any treatment, whereas the intermediate and bottom light transmission dynamics of all treatments showed a bimodal curve (Figure 2D, E and F), which can be attributed to alfalfa being cut twice (11th June and 25th July). Before cutting, alfalfa was flowering and therefore the canopy had high closure and lower light transmission. Upon

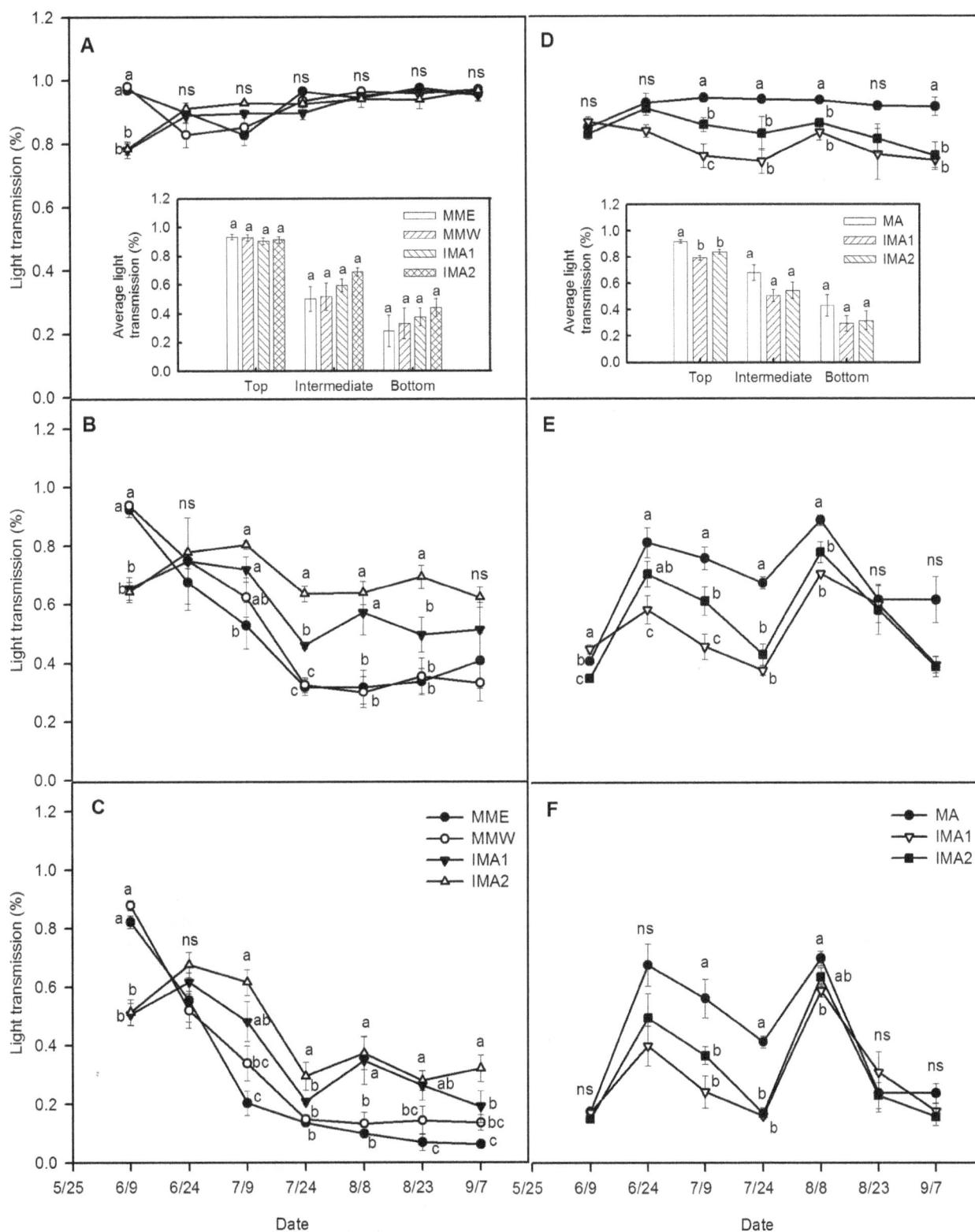

Figure 2. Light transmission dynamics of maize and alfalfa at different layers under monoculture and intercropping. The inset figures show the average light transmission of maize and alfalfa in different planting patterns during the vegetation period. A = top light transmission of maize, B = intermediate light transmission of maize, C = bottom light transmission of maize, D = top light transmission of alfalfa, E = intermediate light transmission of alfalfa, F = bottom light transmission of alfalfa. MME = monoculture maize in even rows, MMW = monoculture maize in alternating wide and narrow rows, MA = alfalfa monoculture, IMA1 = maize intercropped with one row of alfalfa in the wide rows, IMA2 = maize intercropped with two rows of alfalfa in the wide rows. Different letters for the same date indicate significant difference at $P < 0.05$ probability level, and ns represents no difference between treatments. Values = means ± SE.

cutting, light transmission increased. Subsequent alfalfa regrowth produced new canopy closure and decreased light transmission.

With the exception of the first, second and sixth measuring times, the top layer light transmission of monoculture alfalfa (MA) was significantly higher than that of intercropped alfalfa at all recorded time points (third: $P = 0.003$; forth: $P < 0.0001$; fifth: $P = 0.004$; seventh: $P = 0.010$). The season average light transmission of MA was significantly higher than that of intercropped alfalfa. There was no significant difference between the two intercropping modes (Figure 2D and inset). In the intermediate and bottom layers, light transmission showed complex patterns with time. In the intermediate layer, light transmission of MA was significantly higher than in the two intercropping models at the third, fourth and fifth measuring times (third: $P = 0.003$; forth: $P < 0.0001$; fifth: $P = 0.003$). With the exception for the first and third measuring times, there was no significant difference between the light transmissions of the two intercropping modes (Figure 2E). In the bottom layer, for the third and fourth measuring times, light transmission of MA was significantly higher than the two intercropping modes (third: $P = 0.006$; forth: $P < 0.0001$) and light transmission of the two intercropping modes showed no significant difference (Figure 2F). For intermediate and bottom layers, there were no significant differences in the season average light transmission among treatments (Figure 2D inset).

Soil water content (SWC)

SWC of all cropping patterns displayed a strong seasonal dynamic, with a peak in July and August (Figure 3). Irrespective of the growing stage, the difference of SWC between MME and MMW was not significant in both 2011 and 2013. Compared to monoculture, intercropping significantly reduced the SWC, and it was more evident in 2013 than 2011 (Figure 3). In 2011, with the exception of 24th July and 4th October, the SWC of IMA1 and IMA2 was significantly decreased compared to MMW, but with no significant difference compared with MA (both: $P < 0.0001$) (Figure 3A); while in 2013, the SWC of IMA1 and IMA2 was significantly lower than that of MMW as well as MA except for 12th July (all: $P < 0.0001$) (Figure 3B). The differences between treatments were also reflected by seasonal average SWC: the values of IMA1 and IMA2 were significantly lower than that of MMW and MME in both 2011 and 2013, while there was no significant difference between IMA1 and IMA2 in both years. For the MA treatment, the seasonal average SWC was significantly lower than that of MMW and MME in 2011, while there was no significant difference among treatments MMW, MME and MA in 2013 (Figure 3 3A inset and 3B inset).

Leaf area index (LAI)

There was no significant difference in maize LAI between MME and MMW (Figure 4). Compared to MMW, the LAI of intercropped maize (IMA1 and IMA2) was significantly reduced ($P < 0.0001$), and no significant difference was observed between the two intercropping patterns (Table 1; Figure 4). Regarding alfalfa LAI, the values of IMA1 and IMA2 were significantly higher than that of MA ($P = 0.009$), but with no significant difference between IMA1 and IMA2 (Table 1; Figure 5A). Meanwhile, alfalfa LAI was significantly affected by its flowering stage ($F = 32.648$, $P < 0.0001$), LAI in the first flowering stage was significantly higher than that in the second and third flowering stages ($P < 0.0001$) (Figure 5B).

Yield and output value per unit area (OVPUA)

Although both grain yield and OVPUA of MMW were higher (6.8% in 2011 and 6.5% in 2013) than the corresponding values of

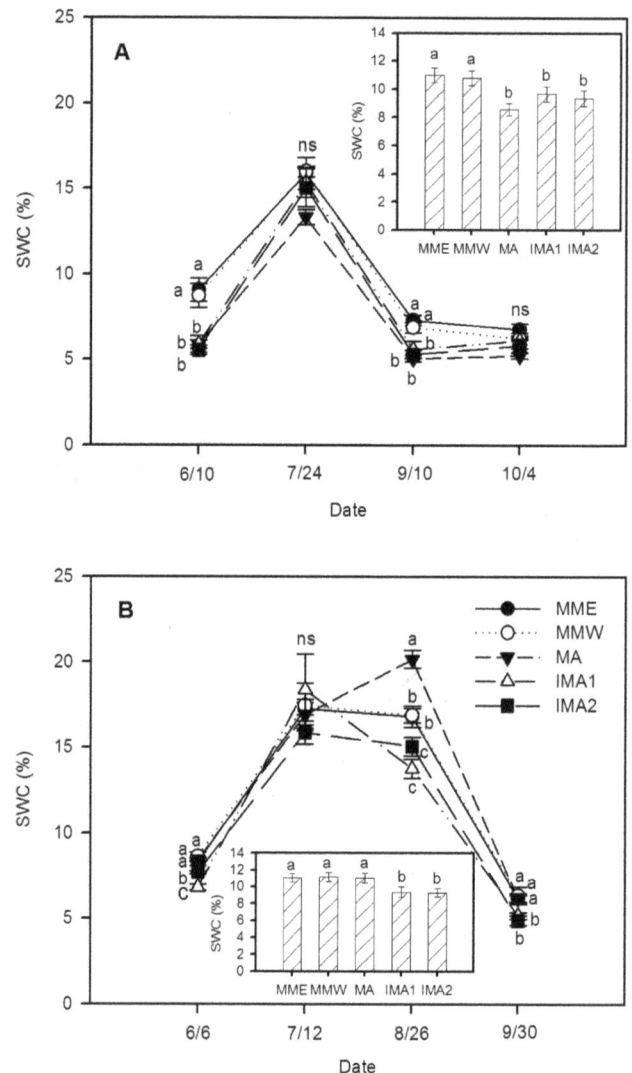

Figure 3. Soil water content comparisons of different cropping patterns in 2011 (A) and 2013 (B). The inset figures show the average soil water content in different planting patterns during the vegetation period. SWC = soil water content. The other symbols are the same as for Figure 2.

MME, the difference between the two cultivated patterns was not statistically significant (Table 2). Compared to monoculture, alfalfa hay yield in the intercropping treatments increased significantly, while maize grain yield in the same treatments was reduced dramatically in both years (all: $P < 0.0001$). The corresponding parameters in the IMA1 cropping pattern were altered in a greater extent than that in IMA2. Additionally, maize grain yield of IMA1 in 2013 was significantly higher than that in 2011 ($P = 0.028$) (Table 1 and 2).

The comprehensive benefits for total yield and OVPUA were significantly affected by the interaction between cropping pattern and year (Table 1). In 2011, both total yield and OVPUA of IMA1 and IMA2 were significantly enhanced compared to MA (both: $P < 0.0001$), while no significant increase was found compared to MMW. In 2013, both total yield and OVPUA of the two intercropping patterns were significantly higher than that of MA as well as MMW (both: $P < 0.0001$). In terms of year effects, both total yield and OVPUA of MA in 2013 was significantly

Alfalfa (Medicago sativa L.)/Maize (Zea mays L.) Intercropping Provides a Feasible Way to Improve Yield...

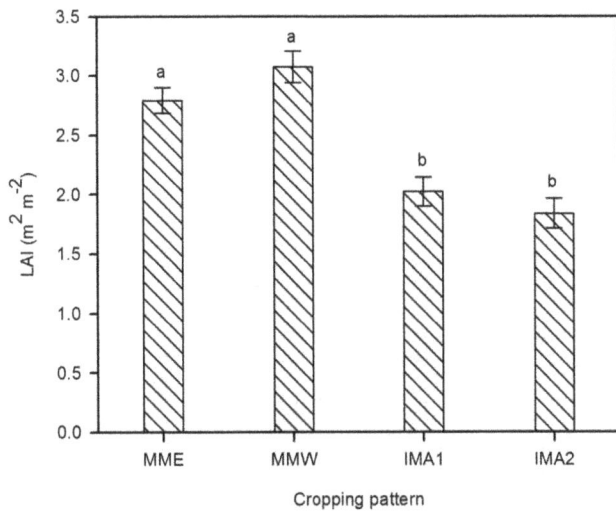

Figure 4. Leaf area index comparisons of maize at the harvest stage under monoculture and intercropping. Significant differences between different cropping patterns are indicated by lower case letters ($P < 0.05$). The other symbols are the same as for Figure 2.

higher than that in 2011 (total yield: $P = 0.038$, OVPUA: $P < 0.0001$), and there was also a significant increase in the total yield of IMA1 in 2013, as compared to 2011 ($P < 0.0001$) (Table 2).

Furthermore, land equivalent ratios (LERs) of IMA1 and IMA2 were 1.27 and 1.23 in 2011 and 1.12 and 1.08 in 2013, respectively, demonstrating that both intercropping strategies were advantageous, and that the IMA1 pattern was superior to IMA2. The calculated aggressivity (A_{ac}) values for IMA1 and IMA2 were 2.71 and 1.40 in 2011 and 1.31 and 0.65 in 2013, respectively, demonstrating that the resource competitiveness of alfalfa was greater than that of maize in the two intercropping systems (Table 2).

Discussion

Maize monoculture in wide and narrow rows vs. in even rows

Liu et al. [20] reported that in the main agricultural region of Jilin Province, compared to even rows, maize planted in alternating wide and narrow rows increased group light transmission, photosynthetic potential, leaf area and grain yield by more than 10%. The results presented here are in agreement with these findings; both grain yield and output value of MMW were enhanced by 6.8% in 2011 and 6.5% in 2013 relative to MME (Table 2). This was mainly attributable to the improved spatial structure of the group, which increased light transmission (Figure 2), improved maize growth conditions and promoted the formation of edge effect [33]. At harvest time, maize achieved a greater LAI and dry matter accumulation (Figure 4 and S3). Therefore, in the PFA of NEC, maize cropped in alternating wide and narrow rows also had a superior economic benefit compared to maize planted in even rows. We recommend that this approach should be popularized and put into widespread use.

The advantages of intercropping alfalfa with maize

Intercropping plays an important part in traditionally intensive agriculture and has captured attention for its efficient utilization of limited resources [21,34]. Among numerous agricultural intercropping modes, legume/cereal intercropping has been most

Table 1. Results of repeated measures ANOVA on maize leaf area index (LAI) and yield, alfalfa yield and comprehensive benefit analysis of total yield and output value per unit area (OVPUA), with cropping pattern (CP) as the independent variable and year (Y) as the repeated measure.

Factors	Maize			Alfalfa		Comprehensive benefit analysis		
	Df	LAI	Yield	Df	Yield	Df	Total yield	OVPUA
CP	3	26.691*	49.157**	2	430.879**	4	16.343**	15.063**
Y	1	0.349 ns	21.174**	1	0.248 ns	1	32.515**	8.052*
CP × Y	3	0.846 ns	9.024**	2	3.985 ns	4	10.850**	10.699**

Df = degrees of freedom, ns = no significant difference, * $p < 0.05$, ** $p < 0.01$

successful, with a long history and several apparent advantages [12,35–36]. Common patterns include intercropping peanuts [37], soybeans [38] or cowpeas [39] with maize. However, few studies have investigated the potential advantages of intercropping alfalfa with maize [24,40–41], and there has been no systematic study attempting to identify whether it can provide sustained high yield and economic incomes while taking investments and environmental factors into consideration or analyze the constraints limiting its popularization. Furthermore, previous intercropping studies have planted their maize in even rows [23,28]. To our knowledge, no field data are available for intercropping alfalfa with maize in alternating wide and narrow rows.

Owing to differences in the traits, growth and development characteristics of alfalfa and maize (Figure S4), the alfalfa/maize intercropping system resulted in temporal and spatial complementarity, which optimized resource utilization and promoted intercropping advantages [27]. The details are presented as follows:

First, as a perennial forb, alfalfa turns green in early spring, grows fast and mainly covers the soil by late April, whereas maize is sown in early or mid-May. We found that when alfalfa was entering the first flowering stage, maize was still a seedling with canopy lower than that of alfalfa. Thus alfalfa and maize occupy complementary spatial and temporal niches, resulting in complementarity in light interception (Figure 2) and facilitating the circulation and diffusion of air (especially CO_2) in the composite population. This result is consistent with many other studies [42–43]. The increase of light transmission of intercropped alfalfa produces an edge effect and enhances the LAI of alfalfa (Figure 5), thereby significantly improving the hay yield of alfalfa in the first cut, especially when using the IMA1 strategy (Figure S2). Furthermore, the hay yield of alfalfa in the first cut accounted for more than 50% of the total hay yield [19]; therefore, the increase in hay yield in the first cut made a great contribution to total hay yield (Figure S2).

Second, between alfalfa first cut and third flowering stage, maize achieves a period of peak growth, became taller than alfalfa; occupying a more advantageous position in the intercropping system so as to make full use of light, heat and other resources. This effect is particularly evident immediately after the second cutting of alfalfa, when intermediate and bottom light transmission of intercropped maize is dramatically enhanced (Figure 2B and C); producing better growing conditions for intercropped maize and accelerating dry matter accumulation (Figure S3). As a result, intercropped maize has an opportunity for recovery [44], partly compensating for the reduction in the maximum dry matter and grain yield caused by competition with alfalfa. In addition, many studies have demonstrated that there is nitrogen transfer from legumes to cereals in intercropping systems [45–46]. Especially, alfalfa, a perennial leguminous forb, has a strong ability to fix nitrogen, and its fixed and transferred nitrogen contributed as much as 30% to the total N accumulated in the associated grass [47–48]. Therefore, we speculate that there would be nitrogen transfer from alfalfa to maize in the alfalfa/maize intercropping system, which can enhance soil nitrogen availability, improve soil physical and chemical properties and be responsible for the improved growth and development of intercropped maize. It should be noted that the relatively favorable precipitation and allocation in the growing season in 2011 and 2013 was also beneficial in alleviating the growth inhibition of intercropped maize caused by water deficit (Figure 1). Consequently, the combined effects of these factors narrowed the grain yield gap between intercropped maize and monoculture maize and promoted the formation of intercropping advantages.

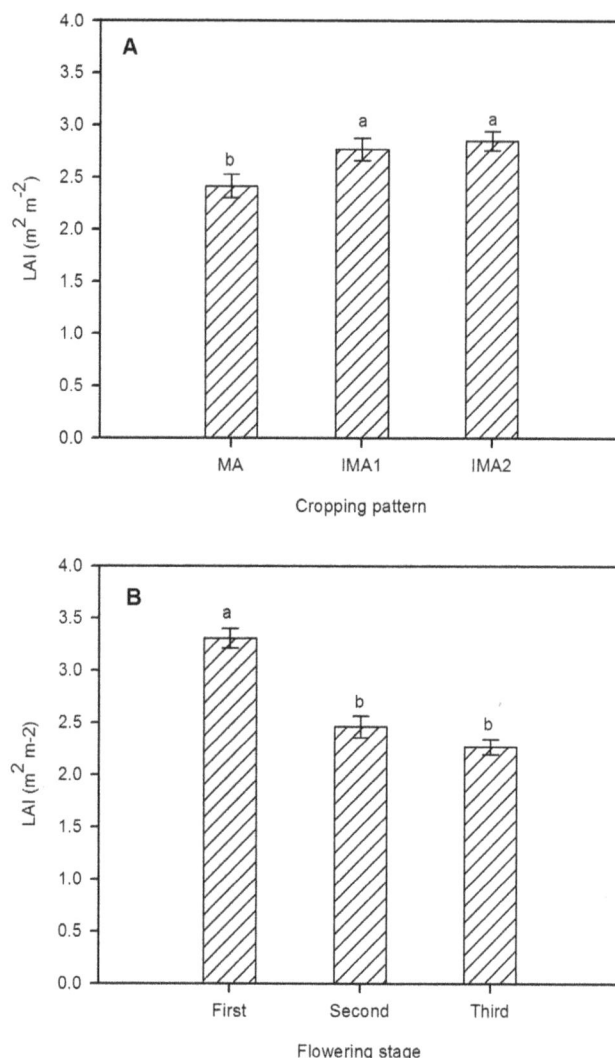

Figure 5. Leaf area index comparisons of alfalfa under monoculture and intercropping (A) and at different flowering stages (B). Significant differences between different alfalfa flowering stages are indicated by lower case letters (P <0.05). The other symbols are the same as for Figure 2 and 4.

Third, alfalfa can continue to grow for around one month after harvesting of maize (Figure S4), providing enough time for alfalfa roots to store adequate carbohydrates to overwinter [23]. Thus, the temporal and spatial differentiation in alfalfa/maize intercropping system avoided wasting light, heat, water, air and other natural resources, prolonged photosynthetic effective time, increased photosynthetic effective area, and ultimately enhanced group yield (Table 2).

Although intercropping alfalfa with maize occupies largely complementary aboveground ecological niches, there is belowground competition. Alfalfa has deep strong roots, that can penetrate>10 m into the soil but mainly proliferate at a soil depth of 0–60 cm. The roots of maize are shallow and mostly distributed at a soil depth of 0–40 cm [49]. Despite some differentiation in root distribution, alfalfa and maize will compete in the shallow soil layers where most water and nutrients are distributed. Moreover, alfalfa has a higher evaporation coefficient and requires more water than maize [18,50]. Therefore, when water is limited, intercropped alfalfa competes with maize for water in the shallow

Table 2. Comparisons of total yield and OVPUA of different planting patterns.

Treatment	Yield (t ha^{-1})			OVPUA (USD ha^{-1})			LER	A$_{ac}$
	Maize	Alfalfa	Total	Maize	Alfalfa	Total		
2011								
MME	11.00±0.33a	–	11.00±0.33a	3982.31±120.85	–	3982.31±120.85a	–	–
MMW	11.75±0.33a	–	11.75±0.33a	4253.31±120.38	–	4253.31±120.38a	–	–
MA	–	7.85±0.31a	7.85±0.31b	–	2991.92±118.56	2991.92±118.56b	–	–
IMA1	5.87±0.34 (7.63±0.44b)	6.08±0.53 (26.35±2.29b)	11.95 ± 0.30a	2124.17±122.92	2316.91±201.23	4441.08±116.95a	1.27	2.70
IMA2	3.71±0.64 (6.89 ± 1.19b)	7.21±0.34 (15.63±0.74c)	10.92±0.67a	1342.17±231.66	2747.93±129.75	4090.10± 246.77a	1.23	1.40
2013								
MME	10.68±0.29a	–	10.68±0.29a	3656.77±100.28	–	3656.77±100.28a	–	–
MMW	11.37±0.23a	–	11.37±0.23ab	3894.69±79.08	–	3894.69±79.08a	–	–
MA	–	11.76±0.05a	11.76 ± 0.05b*	–	4221.72±18.90	4221.72±18.90b*	–	–
IMA1	7.18±0.18 (9.34±0.23b*)	5.79±0.27 (25.06±1.17b)	12.97±0.19c*	2460.38 ± 61.59	2075.89±97.10	4536.26±67.64c	1.12	1.31
IMA2	4.77±0.15 (8.87 ± 0.27b)	7.79±0.11 (16.88±0.23c)	12.56 ± 0.05c	1633.92±50.89	2795.73±38.22	4429.65±14.12c	1.08	0.65

MME = monoculture maize in even rows, MMW = monoculture maize in alternating wide and narrow rows, MA = alfalfa monoculture, IMA1 = maize intercropped with two rows of alfalfa in the wide rows. Values in the parentheses are yields based on the whole of the intercropping area, including the areas occupied by both maize and alfalfa, and are equal to the yields of maize or alfalfa divided by their respective area proportion. The intercropping area proportions of maize and alfalfa were respectively 76.9% and 23.1% in the IMA1 treatment, while the intercropping area ratios occupied by alfalfa and maize were 53.8% and 46.2% in the IMA2 treatment. Different letters in the same column following the values indicate significant difference between different cropping patterns, and * denotes significant difference between years (P <0.05). Value = mean ± S.

layers to meet its growth and development besides utilizing deep groundwater [28], which contributed to the promotion of LAI and hay yield of intercropped alfalfa at the flowering stage (Figure 5 and S1; Table 2).

However, the strong competition of alfalfa with maize significantly reduced SWC of the intercropping system, especially at the first flowering stage of alfalfa and between its second and third flowering stage when maize was in the period of seedling and peak growth with high water demand (Figure 3). The water stress restrained the growth and development of intercropped maize, particularly in the early growth stage (first 80 days post-emergence) (Figure S3), and significantly declined the LAI and grain yield at harvest time (Figure 4; Table 2).

The competitiveness for resources of intercropped species differs due to competition and complementary in the intercropping system. Generally, cereals are considered to have a competitive advantage over legumes and have a decisive influence on the total yield in annual legume/cereal intercropping systems [31,51–52]. However, Zhang et al. [23] studied an alfalfa/maize intercropping system and found that alfalfa had much greater competitiveness than maize, and that alfalfa yield dramatically influenced the total yield of the intercropping system. Our results are consistent with this study. In our two intercropping modes (IMA1 and IMA2), the competitiveness for resources of alfalfa was much stronger than that of maize, and alfalfa hay yield was significantly improved whereas maize grain yield was significantly reduced. In addition, any reduction in maize yield was more than offset by increased alfalfa yield (Table 2). Thus both intercropping systems have consistently accomplished a successful tradeoff between complementarity and competition, enhanced land use capability and achieved a significant yield and output value advantage over monoculture, except that in 2011 the two intercropping patterns achieved a similar yield and OVPUA compared to MMW (Table 2).

When investment is considered, both IMA1 and IMA2 can improve economic benefits in a great extent relative to MMW in both years. This is because alfalfa is a perennial legume that does not require ploughing, sowing and fertilization after the establishment. Compared to maize, planting alfalfa can reduce costs by at least 584 USD ha^{-1} (based on the fertilizer level and seeding rate in our study and in which fertilizer, seed and farming labor savings were 376, 113 and 95 USD ha^{-1}). In this way, relative to MMW, the IMA1 mode not only increased output value (+187.77 USD ha^{-1} in 2011 and +641.57 USD ha^{-1} in 2013), but also reduced investment (−135 USD ha^{-1}), thus enhanced economic benefits by 322.77 USD ha^{-1} and 776.57 USD ha^{-1} in 2011 and 2013, respectively. Similarly, the final economic benefit of IMA2 was also enhanced by 106.79 USD ha^{-1} in 2011 and 804.96 USD ha^{-1} in 2013, as compared to MMW.

The higher total yield and output value of alfalfa/maize intercropping in 2013 than 2011 (Table 2) could be attributed to the following two aspects. First, the favorable higher amount and better distribution of precipitation in 2013, especially in the growing season, improved the SWC (Figure 3). Thus water competition between alfalfa and maize was minimized, nutrient availability to crops was enhanced [53] and crops obtained a superior condition for growth and development. Meanwhile, the improved growth conditions made alfalfa reach the flowering stages earlier, especially for the second and third times. Thus the co-growth time of alfalfa and maize with strong competition was shortened, and maize obtained a more unconstrained condition to utilize resource and grow. Second, with the intercropping year's increase, alfalfa roots descend into deeper soil layers [49] and could extract water at greater depths when the upper soil horizons

get drier, which would improve the growth conditions of intercropped maize; at the same time, it has been proved that alfalfa can fix and transfer more nitrogen to the associated grass as time goes on [54], thus, it is likely that alfalfa could transfer more nitrogen to intercropped maize and improve its growth and development.

Therefore, the comprehensive benefit of alfalfa intercropped with maize was not completely stable and had a certain variation with the changes of rainfall and planting years. It should be noted that the economic benefit of intercropping is also strongly correlated with local market prices of alfalfa and maize [28]. In addition, environmental stress (e.g. pest, disease and the freezing injury to alfalfa in early spring) should also be taken into account to evaluate the valorization of the comprehensive benefits of the intercropping system, and it could severely reduce crops productivity and the economic returns. However, compared to monoculture, intercropping enhanced crops resistance to stress and reduced the management costs and economic losses per unit area [14–16]. Overall, both of the two intercropping strategies have potential to improve economic incomes and are superior to monoculture. This result is in accordance with many other studies and manifests the advantages of intercropping [16,28,55]. Moreover, the intercropping mode of IMA1 was superior to that of IMA2, regardless of land use efficiency, total yield or output value.

Nevertheless, the maize/alfalfa intercropping system still remains relatively unpopular for a number of reasons. First, because of the distinct cultural requirements of each crop, it is difficult to manage the two crops together with existing farm machinery. In this study, all crop management was performed using manual labor. Sowing, fertilizer application, weed control and harvest were conducted separately due to growth and management differences between the two crops. It was also required to manage pest and disease control independently for each crop - when either crop suffered from pests or disease, appropriate pesticides and safeguards were selected to minimize influence on the other crop. It is clear that managing the intercropping system is more complicated and inconvenient than management of a monoculture system. In order to simplify management of this intercropping system, it will be necessary to integrate a multidisciplinary body of knowledge to develop efficient machinery and cautiously advocate the popularization of this intercropping system on a moderate scale to realize yield and economic advantages [56]. Secondly, although the 383,000 km^2 FPA and NEC areas [1] are well suited for the popularization of this intercropping system, local farmers have traditionally utilized a single cropping system and are likely unaware of intercropping systems, especially an intercropping system mixing a food crop and forage legume. Under current conditions, these farmers are more likely to select traditional planting strategies with simple and convenient management schedules and are little concerned about sustainable food production, animal husbandry or eco-environmental security (local investigation and personal communications). Therefore, the dissemination of efficient intercropping technologies and expert technical guidance, as well as financial support of government will necessarily play an important role in putting alfalfa/maize intercropping into practice in the FPA of NEC [56]. Finally, further research is required to assess the long-term benefits of the composite crop population and its responses to rainfall, planting years and environmental stresses (pest, disease and freezing injury), in order to avoid agronomic risks and economic loss.

Conclusions

Based on the above analyses, we conclude that alfalfa/maize intercropping has obvious advantages in grain yield and economic incomes; it guarantees regional food security and provides superior forage. Therefore, this intercropping strategy can help maximize use of limited land resources and promote sustainable development of agriculture and animal husbandry. Moreover, nitrogen fixation and transfer from alfalfa to maize can improve soil fertility and reduce fertilizer investment [19]. Furthermore, alfalfa hay yield increases continuously throughout the first five years [18], providing sustainable economic benefits.

With rising demand for meat, egg, milk and nutrient balance, China is giving increasing importance to the development of animal husbandry and investing the forage industry [9]. Therefore, alfalfa market prices are likely to increase. In this way, planting alfalfa will not only meet the demands of the animal husbandry industry, but also promote the rapid development of local economies. In addition, the multiple-year coverage of alfalfa on the soil can alleviate wind erosion and water erosion, improving the environment of planting areas [19]. In conclusion, there are clear and significant economic, social and ecological benefits in alfalfa/maize intercropping, and maize intercropped with one row of alfalfa was identified as the optimal strategy.

Supporting Information

Figure S1 Daily variation dynamics of air temperature from February to April in 2012. T-highest = highest temperature in a day, T-lowest = lowest temperature in a day.
(TIF)

Figure S2 Alfalfa hay yield at three different flowering stages under monoculture and intercropping. MA = alfalfa monoculture, IMA1 = maize intercropped with one row of alfalfa in the wide rows, IMA2 = maize intercropped with two rows of alfalfa in the wide rows. Different lower case letters for the same flowering stage in one year indicate significant difference, and significant differences of alfalfa total hay yield for one year between different cropping patterns are indicated by different capital letters ($P < 0.05$). Values = means ± SE.
(TIF)

Figure S3 Accumulation dynamics of maize above-ground dry matter under monoculture and intercropping. MME = monoculture maize in even rows, MMW = monoculture maize in alternating wide and narrow rows. The other symbols are the same as for Figure S2.
(TIF)

Figure S4 Growth dynamics of alfalfa and maize in the intercropping system.
(TIF)

Acknowledgments

Special thanks to Baotian Zhang for field help and Scott Diloreto for English check.

Author Contributions

Conceived and designed the experiments: ZL YG. Performed the experiments: BS CW YY YL. Analyzed the data: BS YP HY YG. Contributed reagents/materials/analysis tools: ZL. Wrote the paper: BS YG.

References

1. Zhang HX, Shao MA, Zhang XC (2004) The resuming of weak ecological environment and sustainable development in farming-pasture zone of north-eastern China. J Arid Land Resour Env 18: 129–134.
2. Luo CP, Xue JY (1995) Ecologically vulnerable characteristics of the farming-pastoral zigzag zone in northern China. J Arid Land Resour Env 9: 1–7.
3. Zhou DW, Lu WX, Xia LH, Wu FZ, Li JD, et al. (1999) Grassland degradation and soil erosion in the eastern ecotone between agriculture and animal husbandry in northern China. Resour Sci 21: 57–61.
4. Zhao LP, Wang HB, Liu HQ, Wang YL, Liu SX, et al. (2006) Mechanism of fertility degradation of black soil in corn belt of Song Liao Plain. Acta Pedol Sinica 43: 79–84.
5. He LN, Liang YL, Gao J, Xiong YM, Zhou MJ, et al. (2008) The effect of continuous cropping on yield, quality and soil enzymes activities in solar green house. J Northwest A F Univ (Nat. Sci. Ed.) 36: 155–159.
6. Zhen Z, Bo WJ, Wu GL, Luo XC, Zheng YH (2012) Important effect of the organic fertilizer on soil fertility and yield of crop: a case study in Zhende organic farm, Henan, China. J Eng Stud 4: 19–25.
7. Lin JX, Wang JF, Li XY, Zhang YT, Xu JT, et al. (2011) Effects of saline and alkaline stresses in varying temperature regimes on seed germination of Leymus chinensis from the Songnen Grassland of China. Grass Forage Sci 66: 578–584.
8. Tian X, Yang YF (2009) Current situation of grassland degradation and its management options in farming-pasturing ecotone in western Jilin Province and eastern Inner Mongolia. Chinese J Ecol 28: 152–157.
9. Lu XS (2012) The opportunity, connotation and future of Chinese grassland agriculture. The Second China Grassland Agriculture Conference. pp. 7–10.
10. Ren JZ (2002) Establishment of on agro-grassland systems for grain storage—A thought on restructure of agricultural framework in Western China. Acta Prataculturae Sinica 11: 1–3.
11. Zhu TC, Li ZJ, Zhang WZ, Liang CZ, Yang HJ, et al. (2002) A preliminary report on the cereal-forage rotation system in the plain of Northeast China. Acta Prataculturae Sinica 12: 34–43.
12. Li L, Sun JH, Zhang FS, Li XL, Yang SC, et al. (2001) Wheat/maize or wheat/soybean strip intercropping I. Yield advantage and interspecific interactions on nutrients. Field Crop Res 71: 123–137.
13. Hauggaard-Nielsen H, Ambus P, Jensen ES (2001) Interspecific competition, N use and interference with weeds in pea-barley intercropping. Field Crop Res 70: 101–109.
14. Miriti JM, Kironchi G, Esilaba AO, Heng LK, Gachene CKK, et al. (2012) Yield and water use efficiencies of maize and cowpea as affected by tillage and cropping systems in semi-arid Eastern Kenya. Agr Water Manage 115: 148–155.
15. Trenbath BR (1993) Intercropping for the management of pests and diseases. Field Crops Res 34: 381–405.
16. Rusinamhodzi L, Corbeels M, Nyamangara J, Giller KE (2012) Maize-grain legume intercropping is an attractive option for ecological intensification that reduces climatic risk for smallholder farmers in central Mozambique. Field Crop Res 136: 12–22.
17. Chen YX, Zhang X, Chen J, Zhou DW (2009) The maize proper harvesting methods in ecotone between agriculture and animal husbandry in Northeast China. J Agr Mech Res 31: 113–117.
18. Wang X, Ma YX, Li J (2003) The nutrient content and main biological characteristics of alfalfa. Pratacultural Sci 10: 39–41.
19. Li ZJ, Guo JX, Zhang YS, Wu ZY (2003) The role and status of alfalfa industry in the structure adjustment of planting industry in Jilin province. J Jilin Agr Sci 28: 40–46.
20. Liu WR, Feng YC, Zheng JY, Liu FC, Zhu XL, et al. (2003) Yield and benefit analysis of maize planted in wide and narrow rows. J Maize Sci 11: 63–65.
21. Lesoing GW, Francis CA (1999) Strip intercropping effects on yield and yield components of corn, grain sorghum, and soybean. Agron J 91: 807–813.
22. Gilbert RA, Heilman JL, Juo ASR (2003) Diurnal and seasonal light transmission to cowpea in sorghum-cowpea intercrops in Mali. Agron Crop Sci 189: 21–29.
23. Zhang GG, Yang ZB, Dong ST (2011) Interspecific competitiveness affects the total biomass yield in an alfalfa and corn intercropping system. Field Crop Res 124: 66–73.
24. Wang T, Zhu B, Xia LZ (2012) Effects of contour hedgerow intercropping on nutrient losses from the sloping farmland in the Three Gorges Area, China. J Mt Sci 9: 105–114.
25. Yin XJ, Cui GW (2006) Causes and preventing techniques of alfalfa freezing injury in northern cold regions. Feed Review 4:31–33.
26. Sun QZ, Wang YQ, Hou XY (2004) Alfalfa winter survival research summary. Pratacultural Sci 21:21–25.
27. Bedoussac L, Justes E (2010) Dynamic analysis of competition and complementarity for light and N use to understand the yield and the protein content of a durum wheat-winter pea intercrop. Plant Soil 330: 37–54.
28. Smith MA, Carter PR (1989) Strip intercropping corn and alfalfa. J Prod Agric 11: 345–353.
29. Anil L, Park J, Phipps RH, Miller FA (1998) Temperate intercropping of cereals for forage: a review of the potential for growth and utilization with particular reference to the UK. Grass Forage Sci 53: 301–317.

30. Chu GX, Shen QR, Gao JL (2004) Nitrogen fixation and N transfer from peanut to rice cultivated in aerobic soil in an intercropping system and its effect on soil N fertility. Plant Soil 263: 17–27.
31. Lithourgidis AS, Viachostergios DN, Dordas CA, Damalas CA (2011) Dry matter yield, nitrogen content, and competition in pea-cereal intercropping systems. Eur J Agron 34: 287–294.
32. Shapiro SS, Wilk MB (1965) An analysis of variance test for normality (complete samples). Biometrika 52:591–611.
33. Liu AN, Liu ZG, Zhou XG, Meng ZJ, Chen JP (2005) Study on edge effect and ecological effect in system of winter wheat intercropping with cotton. J Mt Agr Biol 24: 471–476.
34. Trenbath BR (1976) Plant interactions in mixed crop communities. In: Papendick RI, Sanchez PA, Triplett GB, editors. Multiple cropping. ASSA, CSSA, and SSSA. Madison Wis. pp. 129–170.
35. Willey RW (1990) Resource use in intercropping systems. Agr Water Manage 17: 215–231.
36. Mandal BK, Das D, Saha A, Mohasin M (1996) Yield advantage of wheat (Triticum aestivum) and chickpea (Cicer arietinum) under different spatial arrangements in intercropping. Ind J Agron 41: 17–21.
37. Inal A, Gunes A, Zhang F, Cakmak I (2007) Peanut/maize intercropping induced changes in rhizosphere and nutrient concentrations in shoots. Plant Physiol Biochem 45: 350–356.
38. Prasad RB, Brook RM (2005) Effect of varying maize densities on intercropped maize and soybean in Nepal. Exp Agr 41: 365–382.
39. Ghanbari A, Dahmardeh M, Siahsar BA, Ramroudi M (2010) Effect of maize (Zea mays L.)-cowpea (Vigna unguiculata L.) intercropping on light distribution, soil temperature and soil moisture in arid environment. J Food Agr Environ 8: 102–108.
40. Liebman M, Graef RL, Nettleton D, Cambardella CA (2011) Use of legume green manures as nitrogen sources for corn production. Renew Agr Food Syst 27: 180–191.
41. Guldan SJ, May T, Martin CA, Steiner RL (1998) Yield and forage quality of interseeded legumes in a high-desert environment. J Sustain Agr 12: 85–97.
42. Tsubo M, Walker S (2002) A model of radiation interception and use by a maize-bean intercrop canopy. Agr Forest Meteor 110: 203–215.
43. Awal MA, Koshi H, Ikeda T (2006) Radiation interception and use by maize/peanut intercrop canopy. Agr For Meteorol 139: 74–83.
44. Li L, Sun JH, Zhang FS, Li XL, Rengel Z, et al. (2001) Wheat/maize or wheat/soybean strip intercropping II. Recovery or compensation of maize and soybean after wheat harvesting. Field Crop Res 71: 173–181.
45. Shen QR, Chu GX (2004) Bi-directional nitrogen transfer in an intercropping system of peanut with rice cultivated in aerobic soil. Biol Fertil Soils 40: 81–87.
46. Li YF, Ran W, Zhang RP, Sun SB, Xu GH (2009) Facilitated legume nodulation, phosphate uptake and nitrogen transfer by arbuscular inoculation in an upland rice and mung bean intercropping system. Plant Soil 315: 285–296.
47. Ta TC, Faris MA (1987) Species variation in the fixation and transfer of nitrogen from legumes to associated grasses. Plant Soil 98: 265–274.
48. Yang SX, Yang ZZ (1992) A study on superiorities in mixed cropping of alfalfa and Siberian wildrye. Sci Agr Sin 25: 63–68.
49. Zhang GG, Zhang CY, Yang ZB, Dong ST (2013) Root distribution and N acquisition in an alfalfa and corn intercropping system. J Agr Sci 5: 128–142.
50. Sun HR (2003) Alfalfa's pre-blossoming transpiration coefficients and comparison of the water consumption coefficients between alfalfa and maize in terms of economic yield. Acta Agrestia Sinica 4: 346–349
51. Misra AK, Acharya CL, Rao AS (2006) Interspecific interaction and nutrient use in soybean/sorghum intercropping system. Agron J 98: 1097–1108.
52. Wahla IH, Ahmad R, Ehsanullah AA, Jabbar A (2009) Competitive functions of components crops in some barley based intercropping systems. Intl J Agr Biol (Pakistan) 11: 69–71.
53. Jensen JR, Bernhard RH, Hansen S, McDonagh J, MØberg JP, et al. (2003) Productivity in maize based cropping systems under various soil-water-nutrient management strategies in semi-arid Alfisol environment in East Africa. Agr Water Manage 59: 217–237.
54. Goodman PJ (1988) Nitrogen fixation transfer and turnover in upland and lowland grass-clover awards, using ^{15}N isotope dislution. Plant Soil 112: 247–254.
55. Mucheru-Muna M, Pypers P, Mugendi D, Kung'u J, Mugwe J, et al. (2009) A staggered maize-legume intercrop arrangement robustly increases crop yields and economic returns in the highlands of Central Kenya. Field Crop Res 115: 132–139.
56. Yu Y, He Y (2009) The investigation and popularization of the efficient mode of grain-economic intercropping. Shanxi J Agr Sci 5: 90–93.

10

Determination of Critical Nitrogen Dilution Curve Based on Stem Dry Matter in Rice

Syed Tahir Ata-Ul-Karim, Xia Yao, Xiaojun Liu, Weixing Cao, Yan Zhu*

National Engineering and Technology Center for Information Agriculture, Jiangsu Key Laboratory for Information Agriculture, Nanjing Agricultural University, Nanjing, Jiangsu, P. R. China

Abstract

Plant analysis is a very promising diagnostic tool for assessment of crop nitrogen (N) requirements in perspectives of cost effective and environment friendly agriculture. Diagnosing N nutritional status of rice crop through plant analysis will give insights into optimizing N requirements of future crops. The present study was aimed to develop a new methodology for determining the critical nitrogen (N_c) dilution curve based on stem dry matter (S_{DM}) and to assess its suitability to estimate the level of N nutrition for rice (*Oryza sativa* L.) in east China. Three field experiments with varied N rates (0–360 kg N ha^{-1}) using three Japonica rice hybrids, Lingxiangyou-18, Wuxiangjing-14 and Wuyunjing were conducted in Jiangsu province of east China. S_{DM} and stem N concentration (SNC) were determined during vegetative stage for growth analysis. A N_c dilution curve based on S_{DM} was described by the equation ($N_c = 2.17W^{-0.27}$ with W being S_{DM} in t ha^{-1}), when S_{DM} ranged from 0.88 to 7.94 t ha^{-1}. However, for $S_{DM} < 0.88$ t ha^{-1}, the constant critical value $N_c = 1.76\%$ S_{DM} was applied. The curve was dually validated for N-limiting and non-N-limiting growth conditions. The N nutrition index (NNI) and accumulated N deficit (N_{and}) of stem ranged from 0.57 to 1.06 and 51.1 to −7.07 kg N ha^{-1}, respectively, during key growth stages under varied N rates in 2010 and 2011. The values of ΔN derived from either NNI or N_{and} could be used as references for N dressing management during rice growth. Our results demonstrated that the present curve well differentiated the conditions of limiting and non-limiting N nutrition in rice crop. The S_{DM} based N_c dilution curve can be adopted as an alternate and novel approach for evaluating plant N status to support N fertilization decision during the vegetative growth of Japonica rice in east China.

Editor: Guoping Zhang, Zhejiang University, China

Funding: This work was supported by grants from the National High-Tech Research and Development Program of China (863 Program) (2011AA100703), Special Program for Agriculture Science and Technology from Ministry of Agriculture in China (201303109), Priority Academic Program Development of Jiangsu Higher Education Institutions (PAPD), and Science and Technology Support Plan of Jiangsu Province (BE2011351, BE2012302). The funders had no role in study design, data collection and analysis, decision to publish, or preparation of the manuscript.

Competing Interests: The authors have declared that no competing interests exist.

* Email: yanzhu@njau.edu.cn

Introduction

Estimating nitrogen (N) nutritional status is a key to investigating, monitoring, and managing cropping systems [1]. Conventional farming has led to extensive use of N as a tool for ensuring profitability in the soils with uncertain fertility levels, which has raised the concerns about environmental sustainability. A reliable diagnosis of crop N requirement and nutritional status give insight into optimization of qualitative and quantitative aspects of crop production. It also improve N use efficiency and add to environmental protection [2]. Soil and plant-based strategies are two principle approaches, extensively used to derive information about the N nutrition status of crops, for satisfying their demand for N and to minimize N losses [3]. The former rarely describes the intensity of N release over a longer period, so the latter are widely accepted and adopted. Therefore, the present study investigates a plant-based strategy for an in-season assessment of N nutrition status for rice crop.

In plant-based approaches, the N nutrition status is generally monitored to determine the requirement for top dressing in crops [3]. For this purpose, several plant-based diagnostic tools, such as critical N concentration (N_c) approach, chlorophyll meter, hyper-

spectral reflectance and remote sensing, have been successfully used for in-season N management [4]. They differ in scope, in context of reference spatial scale, in terms of monetary and time resources, as well as skills and expertise required for their implementation at field [5]. Despite being simple, chlorophyll meter readings are affected by leaf thickness, abiotic stress and nutrient variability [6]. Canopy reflectance method's accuracy is affected by solar illumination, soil background effects and sensor viewing geometry [4]. However, the concept of N_c can be used as a potential alternate to these techniques, and it can give insight into relative N status of a crop. The present study utilizes this concept for an in-season N fertilizer management in rice crop.

The concept of N_c is crop specific, precise, simple and biologically sound, because it is based on actual crop growth. Whole plant dry matter based N_c approach was successfully applied for N management in winter wheat [7,8], corn [9] and spring wheat [10]. This approach was successfully applied for a Indica rice in tropics and Japonica rice in subtropical temperate region [11,12]. Dry matter partitioning among different plant organs affects the weight/N concentration relationship, and changes the shape of the dilution curve, thus limits its acceptance as a reliable method [13,14]. The concept of N_c for specific plant

Table 1. Changes of stem dry matter (S_{DM}) with time (days after transplantation) under different N rates in two rice cultivars in experiments conducted during 2010 and 2011.

Year	Cultivar	DAT	Sampling date	Stem dry matter/Applied N (kg ha^{-1})					F prob.	LSD
				0	80	160	240	320		
2010	LXY-18	16	07-Jul	0.23	0.27	0.38	0.48	0.49	*	0.028
	LXY-18	26	17-Jul	0.63	0.78	0.95	1.11	1.12	*	0.055
	LXY-18	36	27-Jul	1.04	1.28	1.55	1.77	1.81	*	0.075
	LXY-18	48	08-Aug	2.23	2.73	3.25	3.61	3.51	*	0.226
	LXY-18	60	20-Aug	3.87	4.47	4.94	5.23	5.29	*	0.146
	LXY-18	70	30-Aug	4.72	5.56	6.7	7.01	7.22	*	0.279
	WXJ-14	16	07-Jul	0.22	0.27	0.32	0.36	0.35	*	0.019
	WXJ-14	26	17-Jul	0.39	0.54	0.73	0.9	0.91	*	0.045
	WXJ-14	36	27-Aug	0.55	0.8	1.13	1.38	1.46	*	0.063
	WXJ-14	48	08-Aug	1.22	1.65	1.99	2.18	2.23	*	0.11
	WXJ-14	60	20-Aug	2.77	3.46	3.72	4.19	3.97	*	0.233
	WXJ-14	70	30-Sep	3.69	4.4	5.04	5.81	5.7	*	0.205

Year	Cultivar	DAT	Sampling date	Stem dry matter/Applied N (kg ha^{-1})					F prob.	LSD
				0	90	180	270	360		
2011	LXY-18	18	09-Jul	0.22	0.33	0.4	0.56	0.59	*	0.042
	LXY-18	30	21-Jul	0.67	0.76	0.92	1.19	1.19	*	0.053
	LXY-18	42	02-Aug	1.12	1.28	1.42	1.78	1.76	*	0.132
	LXY-18	54	15-Aug	2.24	2.41	2.76	3.43	3.48	*	0.127
	LXY-18	64	25-Aug	3.59	4.02	4.37	4.75	4.85	*	0.164
	LXY-18	74	04-Sep	5.68	5.91	6.41	7.84	8.04	*	0.172
	WXJ-14	18	09-Jul	0.19	0.28	0.32	0.34	0.36	*	0.02
	WXJ-14	30	21-Jul	0.37	0.6	0.71	0.86	0.9	*	0.06
	WXJ-14	42	02-Aug	0.54	0.96	1.2	1.37	1.47	*	0.128
	WXJ-14	54	15-Aug	1.51	1.84	2.24	2.52	2.68	*	0.137
	WXJ-14	64	25-Aug	2.49	3.07	3.36	4.06	4.04	*	0.169
	WXJ-14	74	04-Sep	4.41	5.04	5.75	6.2	6.27	*	0.178

*: F statistic significant at 0.01 probability level.

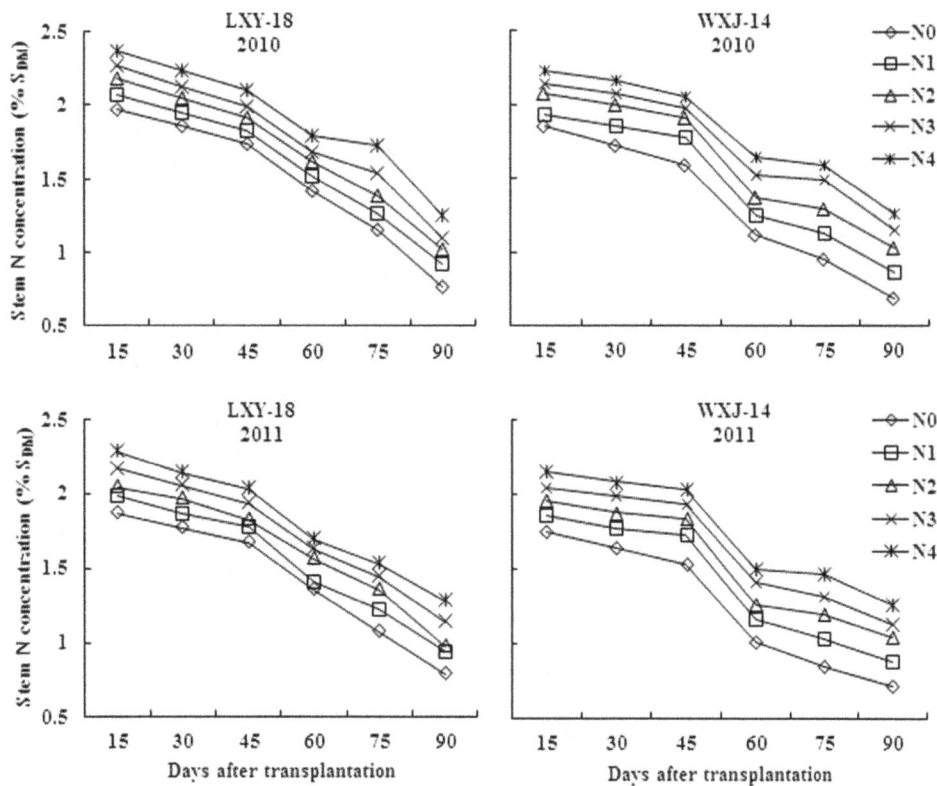

Figure 1. Changes of stem nitrogen concentration (% S_{DM}) with time (days after transplantation) for rice under different N rates in experiments conducted during 2010 and 2011.

organs (e.g., leaves and stem) is similar to that on whole plant basis. Leaf based diagnosis of N status in crops is affected by progressive shading by newer leaves, decline of leaf N concentration due to aging, pest attack, abiotic stresses and increase in the proportion of structural tissues [15]. Stem sap nitrate concentration is influenced by phenological phase, cultivar, temperature and solar radiation [16]. During vegetative phase, the contribution of stem dry matter (S_{DM}) towards total plant dry matter is significantly higher than that of leaf dry matter (L_{DM}), hence it is the most determining factor for N dilution of the whole plant [17]. Thus, the idea of using N_c curve based on S_{DM} over whole plant dry matter and L_{DM} based methods, can be used as an alternate approach for determination of N_c dilution curve.

The objectives of this work were to develop a N_c dilution curve based on S_{DM} and to assess the plausibility of this curve to estimate N nutrition status of Japonica rice. The estimation based on this approach will be more reliable than existing methods due to consistency at different growth stages.

Materials and Methods

Ethics statement

The experiments land is owned and managed by Nanjing Agricultural University, Nanjing, China. Nanjing Agricultural University permits and approvals obtained for the work and study. The field studies did not involve wildlife or any endangered or protected species.

Experimental details

Three field experiments with multiple N rates (0–360 kg N ha^{-1}) were conducted using three contrasting Japonica

rice hybrids, Lingxiangyou-18 (LXY-18), Wuxiangjing-14 (WXJ-14) and Wuyunjing (WYJ), at Yizheng (32°16′N, 119°10′E) and Jiangning (31°56′N, 118°59′E) located in lower Yangtze River Reaches of east China. The soil was clay loam and was classified as Ultisoles. The rice-wheat cropping system is practiced in the region. The applied N rates varied significantly among different farmers. The average rate of N fertilizer reached 387 kg ha^{-1} during the period of 2004–2008 [18].

The whole experimental area was ploughed and subsequently harrowed before transplanting. All bunds were compacted to prevent seepage into and from adjacent plots. A plastic lining was installed to a depth of 40 cm between drain and the bund of each plot to minimize seepage across the bunds towards the drains. To further minimize seepage of water from control plot (N_0), double bunds were constructed separating them and the adjacent plots. Experiments were arranged in a randomized complete block design with three replications. The size of each experimental plot was 8 m by 4.5 m, with planting density of approximately 22.2 hills per m^2. At site 1, soil pH, organic matter, total N, available phosphorous (P), and available potassium (K) were 6.2, 17.5 g kg^{-1}, 1.6 g kg^{-1} 43 mg kg^{-1}, 90 mg kg^{-1}, and 6.4, 15.5 g kg^{-1}, 1.3 g kg^{-1} 38 mg kg^{-1}, and 85 mg kg^{-1} in 2010 and 2011, respectively. The corresponding soil properties were 6.5, 13.5 g kg^{-1}, 1.13 g kg^{-1} 45 mg kg^{-1}, 91 mg kg^{-1} in 2007 at site 2. For experiments conducted at site 1 in 2010 and 2011, treatment consisted of five N rates as 0, 80, 160, 240, and 320 kg N ha^{-1}, and 0, 90, 180, 270, and 360 kg N ha^{-1}, respectively, while for experiment conducted at site 2 in 2007, treatment consisted of three N rates as 110, 220, and 330 kg N ha^{-1}. N in all experiments was distributed as 50% at pre planting, 10% at tillering, 20% at jointing, and 20% at booting, with urea as the N source. Aside from N fertilizer, phosphorus (135 kg ha^{-1}) and potassium (190 kg ha^{-1}) fertilizers were basally incorporated at the last harrowing and leveling in all plots before transplanting as monocalcium phosphate Ca(H$_2$PO$_4$)$_2$ and potassium chloride (KCl). Rice seedlings at five leaves stage were transplanted in experimental fields on June 20 (site 1) in 2010 and 2011, and on 29 June (site 2) in 2007, respectively. Pre-emergence herbicides were used to control weeds at early growth stages. Also plots were regularly hand-weeded until canopy was closed to prevent weed damage. Insecticides were used to prevent insect damage. All other agronomic practices were used according to local recommendations to avoid yield loss.

Sample collection and measurement

Rice plants were sampled from each plot at the intervals of 10–12 days from 0.23 m^2 area (5 hills) at active tillering, mid tillering, stem elongation, panicle initiation, booting and heading stages during the period of each experiment for growth analysis. The plants were manually severed at ground level on each sampling date. Fresh plants were divided into green leaf blades and culm plus sheath. Samples were oven-dried at 105°C for half an hour to rapidly stop metabolism and then at 70°C until constant weight to obtain stem dry matter (S$_{DM}$, t ha^{-1}). The dried stem samples were ground and analyzed for total stem N concentration (SNC, %) by Kjeldahl method. Stem N accumulation (SNA, kg N ha^{-1}) was obtained as summed product of the S$_{DM}$ by the SNC. The SNC of whole-plant stem was calculated as SNA divided by S$_{DM}$.

Statistical analysis

The S$_{DM}$ and SNC data for each sampling date, year and variety was separated and subjected to analysis of variance (ANOVA) using GLM procedures in SPSS-16 software package (SPSS Inc., Chicago. IL, USA). The differences among treatment means were measured by using the least significant difference (LSD) test at 90% level of significance, instead of classically used 95% in order to reduce the occurrence of Type II errors that could be high in such field experiments. For each measurement date, year and variety, the variation in the SNC versus S$_{DM}$ across the different N levels was combined into a bilinear relation composed of a linear regression representing the joint increase in SNC and S$_{DM}$ and a vertical line corresponding to an increase in SNC without significant variation in S$_{DM}$. The theoretical N$_c$ points corresponds to the ordinate of the breakout of the bilinear regression. Regression analysis was performed using Microsoft Excel (Microsoft Cooperation, Redmond, WA, USA).

Construction and validation of critical, maximum and minimum N dilution curves

For determination of N$_c$ dilution curve it is necessary to determine the N concentration that did not limit the S$_{DM}$ production either by its excess or deficiency. The data used to construct the N$_c$ dilution curve came from two experiments conducted in 2010 and 2011 by distinguishing the data points for N-limiting and non-N-limiting growth. The N-limiting growth treatment is defined as a treatment for which an additional N application leads to a significant increase in S$_{DM}$. The non-N-

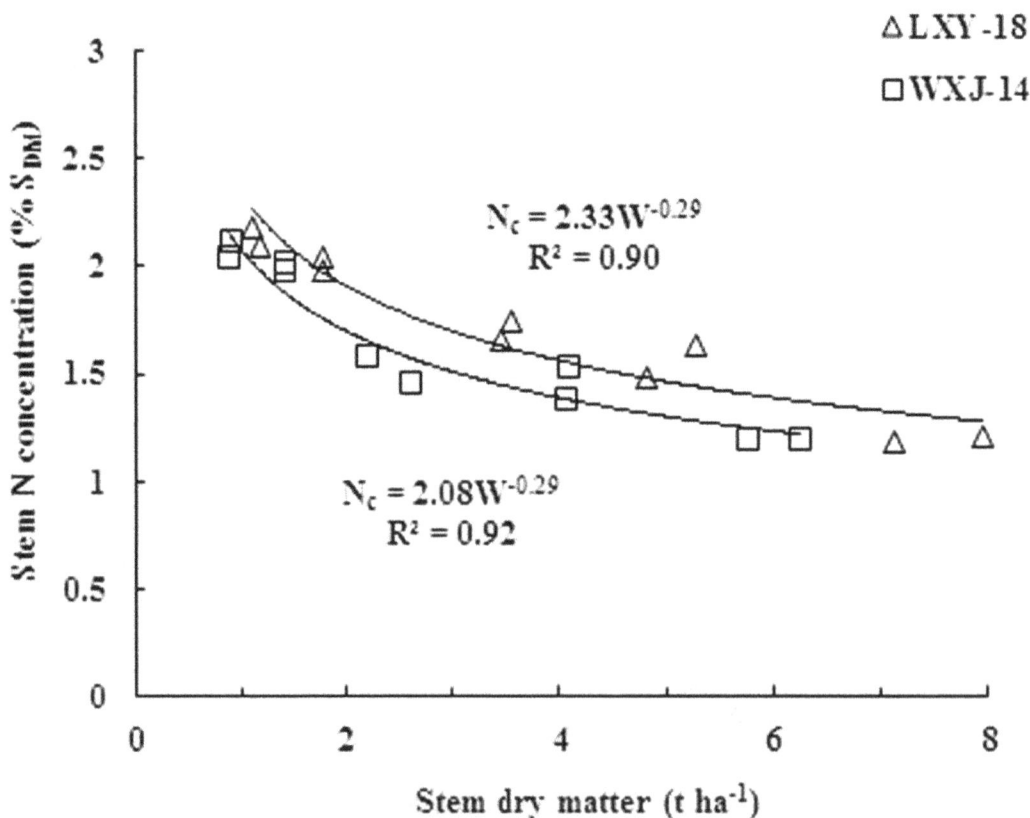

Figure 2. Critical nitrogen data points and N$_c$ dilution curves in stem obtained by non-linear fitting for two rice cultivars (LXY-18, N$_c$ = 2.33W$^{-0.29}$ and WXJ-14, N$_c$ = 2.08W$^{-0.29}$) under different N rates in experiments conducted during 2010 and 2011.

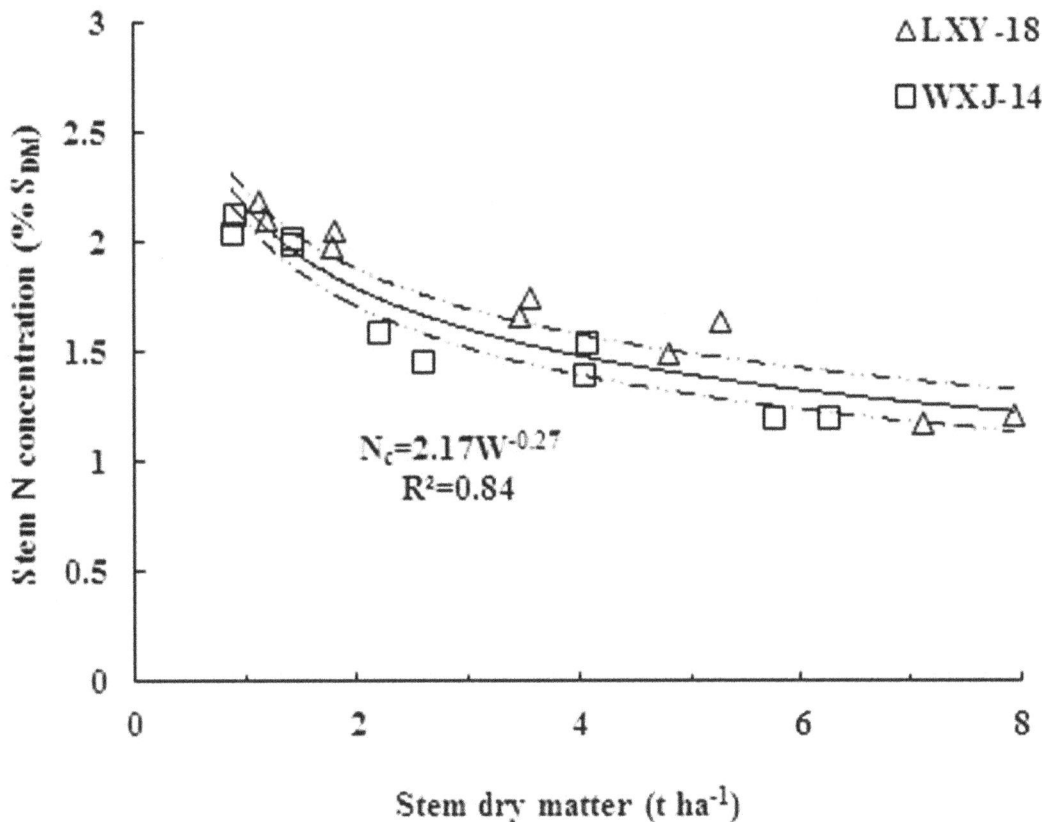

Figure 3. Critical nitrogen data points used to define the N_c dilution curve when data were pooled over for two rice cultivars (LXY-18 and WXJ-14). The solid line represents the N_c dilution curve ($N_c = 2.17W^{-0.27}$; $R^2 = 0.84$) describing the relationship between the N_c and stem dry matter of rice. The dotted lines represent the confidence band ($P = 0.95$).

limiting growth treatment is defined as a treatment, for which a supplement of N application does not lead to an increase in S_{DM} and, at the same time, exhibits a significant increase in SNC. If at the same measurement date, statistical analysis distinguished at least one set of N-limiting and non-N limiting data point, these data points were used either for construction of the N_c dilution curve or to validate it [7]. Consistent with earlier studies, an allometric function based on power regression (Freundlich model) was used to determine the relationship between the observed decreases in N_c with increasing S_{DM}. The N_c dilution curve was validated first by using the data points not retained for establishing the parameters of the allometric function in 2010 and 2011, and then with independent data set from experiment conducted in 2007.

The data points (n = 13) from most plethoric N treatments (N_4 plots) was assumed to represent the maximum N dilution curve (N_{max}) while the minimum N dilution curve (N_{min}) was determined by using the data points (n = 13) from the most N-limiting treatments for which N application was zero (N_0 check plots).

Calculation of critical N dilution curve based diagnostic tools

To identify the N status in the S_{DM} of rice during vegetative growth, the nitrogen nutrition index (NNI) and accumulated nitrogen deficit (N_{and}) were established for each sampling date, experiment and variety. The NNI value was obtained by dividing the total N concentration of S_{DM} by N_c value determined by critical dilution curve, [9]. The N_{and} value for rice crop on each

sampling date was obtained by subtracting the N accumulation under the N_c condition (N_{cna}) from actual N accumulation (N_{na}) under different N rates [12]. For in-season recommendation of supplemental N application, the difference value of NNI (ΔNNI), N_{and} (ΔN_{and}) and difference value of N application rate (ΔN) between different N treatments was calculated according to the method proposed by Ata-Ul-Karim et al. [12].

Results

Stem dry matter and nitrogen concentration

The S_{DM} production was significantly affected by N fertilization during the growth period of rice. The increase in S_{DM} followed a continuous increasing trend along with sampling dates for both the varieties during each year with increasing N rates from N_0 to N_4; however, there was no significant difference between N_3 and N_4 in all the cases (Table 1). This increase in the S_{DM} production with N fertilization may be linked to a higher absorption of N fertilizer. S_{DM} ranged from minimum 0.22 t ha^{-1} and 0.19 t ha^{-1} (N_0) in WXJ-14 to a maximum of 7.22 t ha^{-1} and 8.04 t ha^{-1} (N_4) in LXY-18 during 2010 and 2011, respectively. The results showed that there was no positive correlation between S_{DM} and N rates, as the S_{DM} tend to decrease when N rate exceeded a critical level. During each experimental year, S_{DM} conferred with the following inequality under different N ratess.

$$S_{DM0} < S_{DM1} < S_{DM2} < S_{DM3} = S_{DM4} \qquad (1)$$

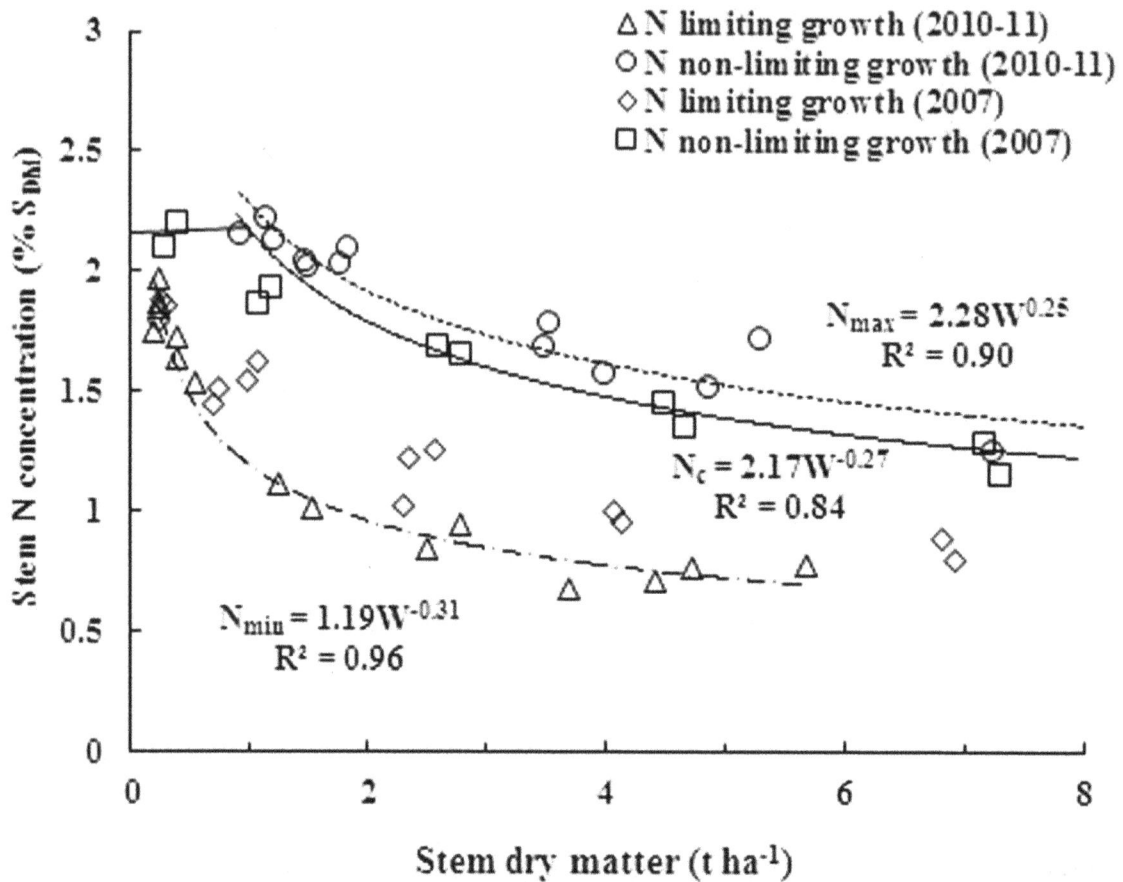

Figure 4. Comprehensive validation of N_c dilution curve using independent data set from experiment conducted in 2007. Data points ($\diamond$) represent N limiting growth conditions, while ($\square$) represent N non-limiting conditions. The solid line in the middle represents the N_c curve ($N_c = 2.17W^{-0.27}$) describing the relationship between the N_c and stem dry matter of rice. The data points ($\triangle$) and ($\bigcirc$) not engaged for establishing the parameters of allometric function (2010 and 2011) were used to develop two boundary curves, (–•–•–•) minimum limit curve ($N_{min} = 1.19\ W^{-0.31}$) and (------) maximum limit curve ($N_{max} = 2.27W^{-0.25}$).

where S_{DM0}, S_{DM1}, S_{DM2}, S_{DM3} and S_{DM4} stands for S_{DM} of N_0, N_1, N_2, N_3 and N_4, respectively.

Stem N concentration response to N fertilizer rates was usually linear and a higher rate of N mostly resulted in a higher SNC, hitherto a decline in SNC was observed with increasing S_{DM} from active tillering to heading. Maximum variation in SNC of both cultivars was observed on 16 and 18 DAT, while minimum on 70 and 74 DAT, in years 2010 and 2011, respectively. The SNC ranged from 2.28 to 0.78 for LXY-18 and 2.16 to 0.71 for WXJ-14 during 2010, while 2.36 to 0.77 for LXY-18 and 2.23 to 0.68 for WXJ-14 during 2011 (Fig. 1).

Determination of critical nitrogen dilution curves based on stem dry matter

A set of twenty theoretical data points for both cultivar, obtained from two experiments (10 data points for each cultivar) from active tillering to heading were used to calculate the N_c for a given level of S_{DM}. The S_{DM} data that fit the statistical criteria for establishing N_c dilution curve varied from 0.88 t ha^{-1} to 7.94 t ha^{-1}. A power functions were fitted to the calculated N_c points as equations (2) and (3), the coefficient for which were 0.90 and 0.92 for LXY-18 and WXJ-14, respectively (Fig. 2).

$$N_c = 2.33\,W^{-0.29} \qquad \left(W \geq 0.88\ t\ ha^{-1}, R^2 = 0.90, n = 10 \right) \quad (2)$$

$$N_c = 2.08\,W^{-0.29} \qquad \left(W \geq 0.88\ t\ ha^{-1}, R^2 = 0.92, n = 10 \right) \quad (3)$$

where W is the S_{DM} expressed in t ha^{-1}; N_c is the critical N concentration in stem expressed in % S_{DM}; a and b are estimated parameters. The parameter a represents the N concentration in the S_{DM} when W = 1 t ha^{-1}, and b represents the coefficient of dilution describing the relationship between N concentration and S_{DM}.

The F-value (0.72) of two curves was less than the critical value of $F_{(1-18)} = 4.41$ at 5% probability level, showing non-significant difference between the curves [19], thus the data for the two varietal groups were united, and a unified dilution curve was determined as equation 4.

$$N_c = 2.17\,W^{-0.27} \qquad \left(W \geq 0.88\ t\ ha^{-1}, R^2 = 0.84, n = 20 \right) \quad (4)$$

The model accounted for 84% of the total variance. At early growth stages of rice crop, the N_c varied between 2.24% S_{DM} to 2.10% S_{DM} (95% confidence interval) for a S_{DM} of 0.88 t ha^{-1} at the lower end while 7.94 t ha^{-1} at the higher end, respectively (Fig. 3).

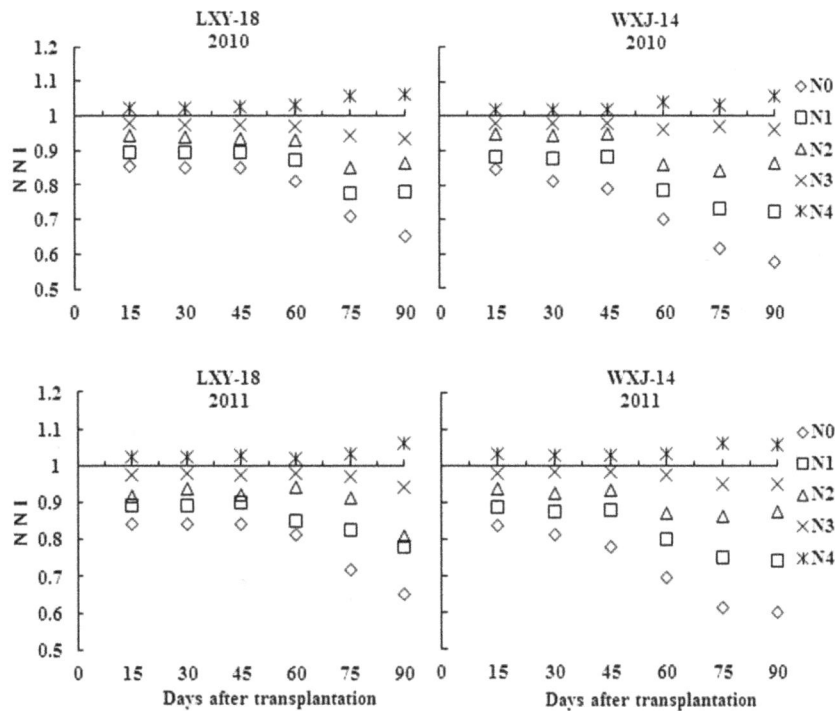

Figure 5. Changes of nitrogen nutrition index (NNI) with time (days after transplantation) for rice stem under different N rates in experiments conducted during 2010 and 2011.

For the S_{DM} range of 0.1 to 0.88 t ha^{-1}, corresponding to early growth stages, increasing N rates at sowing did not significantly affect S_{DM}, because N requirement is relatively low during these early stages. Therefore, the N_c dilution curve cannot be applied to the low S_{DM}<0.88 t ha^{-1} at early growth stages due to relatively smaller decline of N_c with increasing S_{DM}. For these S_{DM}, the N_c could not been determined by the same statistical method because the very high slope of the linear regression resulted in a highly variable estimate [7]. Hence, for the data points of S_{DM} ranging from 0.37 to 0.88 t ha^{-1} a constant N_c (1.76% S_{DM}) was calculated as the mean value between the minimum N concentration of non-limiting N points (2.26% S_{DM}) and the maximum N concentration of limiting N points (1.25% S_{DM}), based on extrapolation of equation 4.

The above S_{DM} based N_c dilution curve was dually validated for N-limiting and non-N-limiting situations within the range for which it was developed. First, the curve was partially validated by combining the data points not engaged for establishing the parameters of the allometric function. In addition, the comprehensive validation of the curve was performed by using the data points from an independent experiment conducted in 2007. The results revealed that the N concentration data that led to the highest significant yields in S_{DM} were positioned close to or above the N_c dilution curve and considered to be non-N-limiting concentrations, whereas the data for the lowest significant S_{DM} yields, were positioned close to or under the N_c dilution curve and classified as N limiting values (Fig. 4). To determine N_{max}, data points were selected only from non-N-limiting treatments (n = 13), and for N_{min}, data points were selected from the treatment without N application (n = 13). Thus, the present N_c dilution curve could well discriminate the N limiting and non-N-limiting growing conditions in this study

Changes of NNI and N_{and}

Nitrogen nutrition index and N_{and} are helpful in determining the crop nutrition status i.e. deficient, optimal or excess of N nutrition. N nutrition is considered as optimum when NNI = 1 and N_{and} = 0, while NNI>1 and N_{and}<0 indicates luxury consumption of N nutrition, values of NNI<1 and N_{and}>0 represents N shortage. NNI and N_{and} can be used to quantify the intensity of the N stress after the onset of N deficiency. Our results of significant differences in NNI and N_{and} across the growing seasons, N rates, and phenological stages in rice are in agreement with earlier reports for maize and wheat [10]. As seen in Figure 5 and 6, during 2010 and 2011 the NNI ranged from 0.65 to 1.06 for LXY-18 and 0.57 to 1.06 for WXJ-14, while the N_{and} ranged from 51.1 kg ha^{-1} to −7.07 kg ha^{-1} for LXY-18 and 43.3 kg ha^{-1} to −4.5 kg ha^{-1} for WXJ-14. The results showed that NNI amplified while N_{and} declined with increasing N rates, while both intensified steadily with growth of rice crop and reached to peaks at heading stage for N_0, N_1, N_2 and N_3 (N limiting treatments), nevertheless, for N_3 this intensification was minor. In contrast, surplus N nutrition existed till heading stage for N_4 (non-N-limiting treatment). The estimates based on NNI and N_{and} can be used to identify the N nutritional status at any stage of rice growth, allowing us to assess whether the N fertilizer dosage was ample enough to obtain higher yield in practice. These results confirmed the plausibility of using NNI and N_{and} to assess the status of N nutrition in rice plants growing under various conditions and stages.

Figure 7 and 8 showed that ΔN had a positive correlation with ΔNNI and ΔN_{and}. The simple linear regression equation showed non-significant differences between two varieties, although noticeable differences were observed among different phenological stages. Therefore, ΔN during growth period for both varieties could be derived from ΔNNI and ΔN_{and}, respectively, according

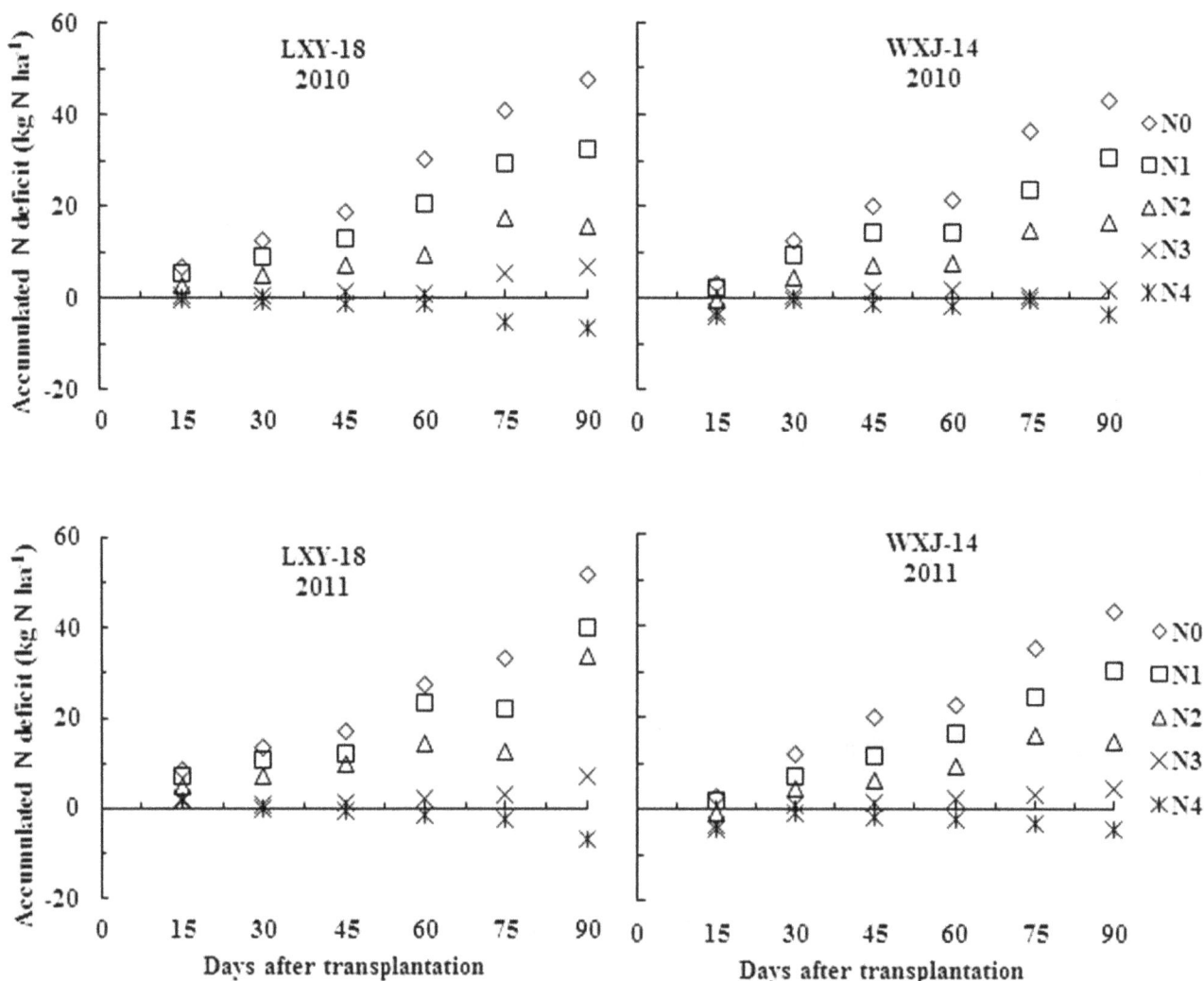

Figure 6. Changes of accumulated N deficit (N$_{and}$) with time (days after transplantation) for rice stem under different N rates in experiments conducted during 2010 and 2011.

to the equations 5 & 6 as follows:

$$\Delta N = A \times \Delta NNI + B \quad (5)$$

$$\Delta N = C \times \Delta N_{and} + D \quad (6)$$

The parameters A, B, C, and D could be calculated from days after transplanting (DAT) using the equations as:

$$A = -16.60 \times DAT + 2101 \quad (R^2 = 0.95) \ (7)$$

$$B = -0.024 \times DAT^2 + 2.57 \times DAT - 40.07 \quad (R^2 = 0.62) \ (8)$$

$$C = 18.97 ln(DAT) - 89.22 \quad (R^2 = 0.98) \ (9)$$

$$D = -9.98 ln(DAT) + 52.76 \quad (R^2 = 0.19) \ (10)$$

The ΔN obtained on the basis of relationship between ΔNNI, ΔN_{and} and ΔN, allowed us to make corrective decisions of N dressing recommendation for the precise N management during or even before the period of highest demand of the rice crop.

Discussion

Application of N fertilizer for crop production is an economically viable option in terms of low cost as compared to the value of the marketable agricultural products themselves; however, N usage cannot assure a significant increase in crop productivity due to diminishing returns after certain levels. There is an increasing demand by strategy makers for simple-to-use, technically established and economically viable N indicators, which may allow monitoring and assessment of policy measures and offer tools for farm N management. With the advent of technology, more emphasis should be put on plant-based indicators, which simultaneously reflect the interactions between the plant and the

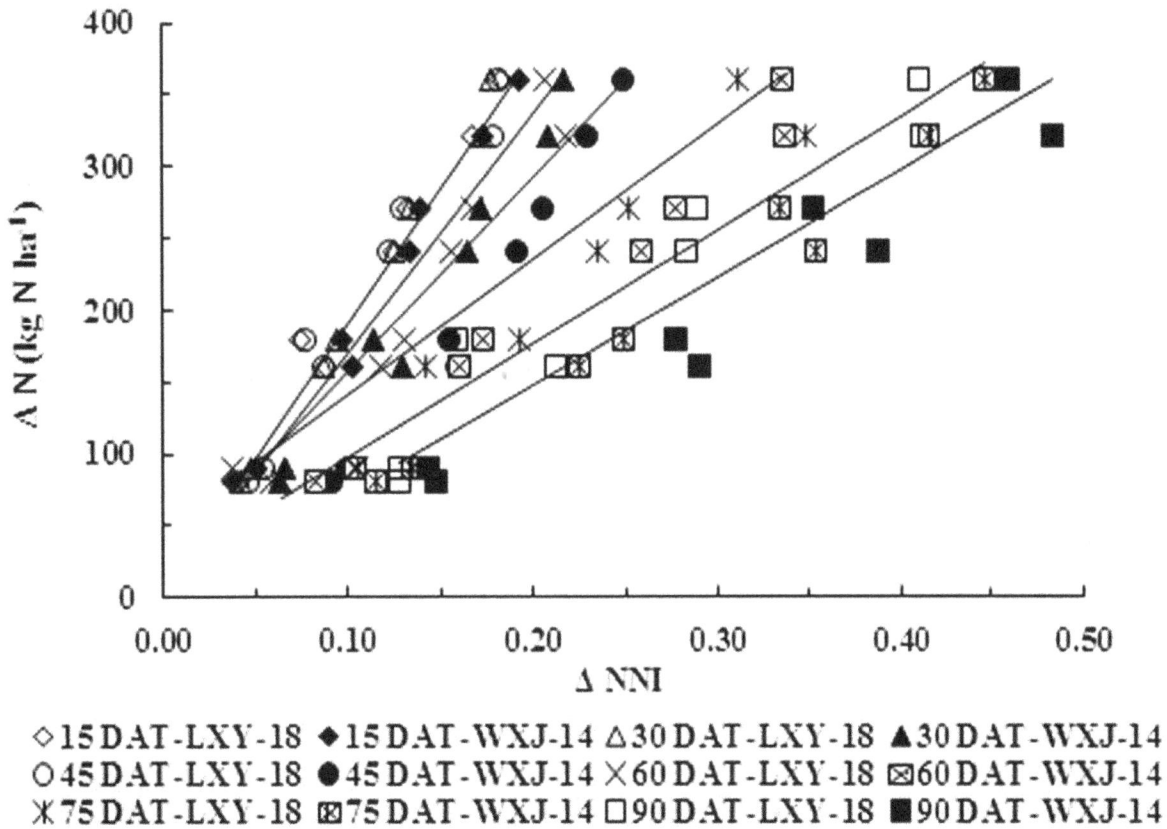

Figure 7. Relationship between changes of nitrogen nutrition index (ΔNNI) and changes of nitrogen application rates (ΔN, kg N ha^{-1}) at different growth stages in experiments conducted during 2010 and 2011. The open symbols represent different growth stages for LXY-18 while filled symbols represent different growth stages for WXJ-14. (ΔN = A × ΔNNI + B; A = −16.60 × DAT + 2101, R^2 = 0.95; B = −0.024 × DAT^2 + 2.57 × DAT − 40.07, R^2 = 0.62).

soil. So far, there have been several reports on estimating the N_c concentration on whole plant dry matter basis in various crops, including rice [11,12], and on L_{DM} basis in rice [20], yet no attempt was made to determine the N_c dilution curve on S_{DM} basis for any crop including rice. The current study has developed a S_{DM} based N_c dilution curve for rice in east China, thus providing a new approach for diagnosing and regulating N in crop species.

Minimum and maximum nitrogen dilution curves

An obvious variability in SNC for a given range of S_{DM} was observed when all the data from three year experiments were analyzed for interpretation. This variability in SNC towards maturity of rice crop in present study was in agreement with earlier studies on winter wheat [7] and Japonica rice [12], and this variability could be attributed to a decline in the fraction of total plant N associated with photosynthesis [21], change in leaf/shoot ratio and self-shading of leaves [8].

Two boundary curves for N maximum (N_{max}) and minimum (N_{min}) have been determined by using maximum and minimum N concentration in S_{DM} and can be represented as equations:

$$N_{max} = 2.28 W^{-0.25} \quad (11)$$

$$N_{min} = 1.19 W^{-0.31} \quad (12)$$

The N_{max} curve corresponding to the maximum N uptake in the S_{DM} without interfering with productivity and it can be considered as the first assessment of a maximum N dilution on S_{DM} basis in crops, and can be obtained with increasing N rates for maximum growth and N accumulation. This curve is an estimate of the maximum N accumulation capacity of stem which is regulated by mechanism associated with the growth and availability of soil N directly or indirectly via N metabolism [22]. The N_{max} curve in the present study shows a luxury consumption of N under N_4 treatment, when N concentration exceeds N_c dilution curve and S_{DM} does not increase with increasing N rate. In contrast, the N_{min} curve is considered as a lower limit at which the N metabolism would soon stop to function. It corresponds to the minimum N taken up by rice plants under N_0 treatment in present study. Thus, the N_{min} were used as the threshold concentration for proper metabolic functionality of the plant.

Moreover, the value of parameter b for the N_{max} was not significantly different from that of N_c dilution curve, which indicate that the partitioning of dry matter remains relatively constant when N uptake exceeds the N_c dilution curve. This is consistent with the concept of N_c dilution curve, which represents the lowest N at which maximum dry matter accumulation occurs. This implies that under luxury consumption of N, when N exceeds N_c dilution curve, dry matter accumulations does not increase with N and hence, dry matter partitioning will have similar value of parameter b. In contrast, for N_{min} curve under N stress, the value for parameter b tended to be slightly lower than the dilution curve.

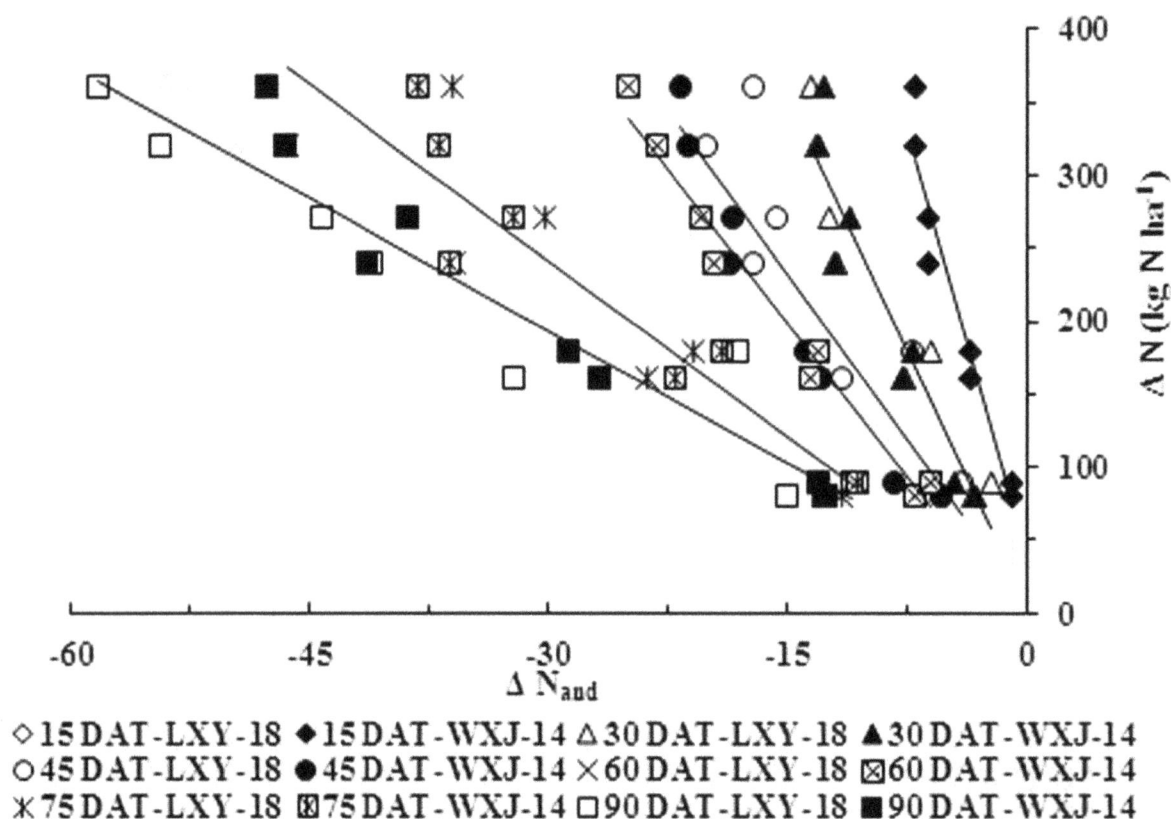

Figure 8. Relationship between changes of accumulated N deficit (ΔN_{and}) and changes of nitrogen application rates (ΔN, kg N ha^{-1}) at different growth stages in experiments conducted during 2010 and 2011. The open symbols represent different growth stages for LXY-18 while filled symbols represent different growth stages for WXJ-14 ($\Delta N = C \times \Delta N_{and} + D$; $C = 18.97$ ln(DAT)-89.22, $R^2 = 0.98$ and $D = -9.98$ ln(DAT)+ 52.76, $R^2 = 0.19$), respectively.

The relatively low value for b was associated with a change in dry matter partitioning.

Comparison with other critical nitrogen dilution curves

The concept of N_c dilution curve on whole plant dry matter and L_{DM} basis have already been successfully implicated for several crops including rice, yet no attempt was made to construct a S_{DM} based N_c dilution curve in any crop including rice. Figure 9 showed that the parameter a of N_c dilution curve on S_{DM} basis with Japonica rice developed in present study (2.17) was lower than the reference curve on whole plant dry matter basis of Indica rice in tropics (5.20) by Sheehy et al. [11] as well as lower than the curves developed with Japonica rice on whole plant dry matter basis (3.53) by Ata-Ul-Karim et al. [12], and L_{DM} basis (3.76) by Yao et al. [20].

The differences observed between the parameter a of dilution curve developed in present study and the curves on whole plant dry matter basis [11,12] were due to morphological aggregation of structural components, which relates to the weight/N concentration in the whole plant [13]. Stress responses may cause differences in the partitioning of dry matter among various plant organs, and thereby affect the shape of the dilution curves. Moreover, dissimilarities in climatic conditions and genetic differences of Indica and Japonica rice contributed to the differences between the curves. The ability of Indica to hold higher plant N content and total N uptake [23–25] and faster growth rate [26], compared with those of Japonica rice, also lead to the differences between N_c dilution curve of Sheehy et al. [11] and that described in the

present study. The differences of S_{DM} based curve with that of L_{DM} based one [20] are mainly attributed to leaf/stem ratio, because decrease in stem N during vegetative phase is related to decline in the metabolic biomass with high N contents, and increase in proportion of structural and non-photosynthetic biomass with low N contents [8]. Thus, higher proportion of structural biomass in stem than in leaves is responsible for the differences between the L_{DM} and S_{DM} based curves of Japonica rice.

The parameter b of the dilution curve indicates the dilution intensity of N during growth and the higher values of b indicate lower N dilutions [17]. The coefficients b were (-0.50, -0.28, -0.22 and -0.274) for N_c dilution curve of Indica rice and for Japonica rice based on whole plant dry matter, L_{DM}, and on S_{DM}, respectively. The observed differences between the coefficients b of Indica rice and current S_{DM} based dilution curve might be explained by the differences in duration of vegetative phase in tropical and subtropical climates, while the differences between coefficients b of the curves of Japonica rice based on L_{DM} and S_{DM} were directly related to the distribution of dry matter between green leaves and the stem [17]. In contrast, the differences between coefficients b of the curves of Japonica rice on whole plant dry matter basis compared with that of S_{DM} basis, are negligible due to the reason that stem have a dilution effect on the N in the above ground tissues, because of their higher weight percentage in the total dry matter [27]. Therefore, the S_{DM} based dilution curve can be used as a potential alternative for in-season estimation of

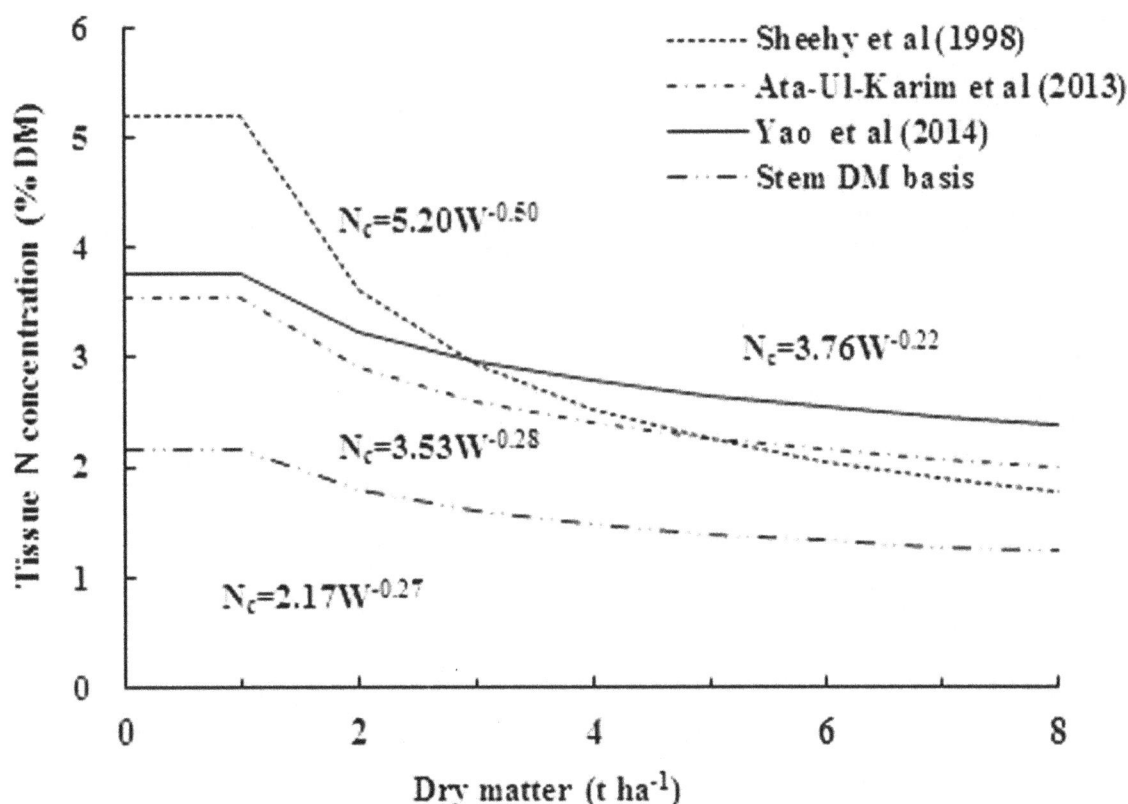

Figure 9. Comparison of different N_c dilution curves. The (------) represents the N_c dilution curve of Sheehy et al. (1998) ($N_c = 5.20W^{-0.50}$) on plant dry matter basis in Indica rice under tropic environment. The (–●–●–) represents the N_c dilution curve of Ata-Ul-Karim et al. (2013) ($N_c = 3.53W^{-0.28}$) on plant dry matter basis in Japonica rice in Yangtze River Reaches. The (——) line represents N_c dilution curve of Yao et al. (2014) ($N_c = 3.76W^{-0.22}$) on leaf dry matter basis in Japonica rice in Yangtze River Reaches, and the (–●●–●●–) line represents N_c dilution curve on stem dry matter basis in present study ($N_c = 2.17W^{-0.27}$).

plant N nutrition status, instead of existing whole plant dry matter and L_{DM} based approaches.

Implication for nitrogen diagnosis

The application of the present N_c dilution curve as a diagnostic tool for accurate N management to make corrective decisions of N dressing recommendation during rice production is very interesting. The N_c dilution curve can be used for a priori analysis intended to optimize fertilizer N management or for a posteriori diagnosis intended to detect N limiting nutrition for rice within experimental trials or fields in production. The a priori diagnosis of plant N status consists of timely detection of plant N deficiency during the crop growth cycle to determine the necessity of applying additional N fertilizer. Present study showed that the N_c dilution curve, resulting NNI and N_{and} effectively distinguished conditions of deficient, optimal and surplus N nutrition in rice. The values of ΔN in present study obtained on the basis of relationship between ΔNNI, ΔN_{and} and ΔN, permitted us to make corrective decisions of N dressing recommendation for precise N management during or even before the period of peak demand of the rice crop. The main limitation in using the present NNI and N_{and} directly as diagnostic tools is the need to determine the actual dry matter and N concentration, which can be monitored by the non-destructive means including remote sensing [28–30]. Moreover, a good correlation between these analytical tools and chlorophyll meter readings was previously reported by [9]. These indirect methods could possibly be a substitute for assessing NNI

and N_{and} and portray crops and environments in conditions where they cannot be measured directly [31]. Thus, the models of NNI and N_{and}, based on N_c dilution curve in relation to actual growth status, can be exploited directly for the estimation of crop N status to recommend the necessities of further N application during plant growth. These novel algorithms can also be combined into crop growth and management models to forecast crop N status and quantify N dressing plan. Although, NNI and N_{and} calculated in present study distinguished well the N-limiting and non-N-limiting growth conditions, a more comprehensive validation using different N management practices, N availabilities and cultivars is mandatory to robustly confirm the reliability of NNI and N_{and} usage as an investigative indicators for different ecological regions and rice production systems.

Conclusions

In conclusion, we found that N fertilization endorses increase in the S_{DM}, which was influenced by variations in SNC. A higher rate of N fertilizer generally increased SNC in Japonica rice; however, towards advancing maturity this increase followed a declining trend under different N levels, sampling dates and growing seasons. S_{DM} during vegetative growth period ranged from minimum value of 0.19 (N_0) in WXJ-14 to a maximum value of 8.04 (N_4) in LXY-18, whereas SNC varied from 0.68% in WXJ-14 to 2.36% in LXY-18 on S_{DM} basis under different N rates and growth stages. A new N_c dilution curve on S_{DM} basis for Japonica rice grown in east China was developed and can be described by equation, $N_c = 2.17W^{-0.274}$, when S_{DM} ranges from 0.88 and 7.94 t ha^{-1}, however for $S_{DM} < 0.88$ t ha^{-1}, the constant critical value $N_c = 1.76\%$ S_{DM} was applied, which was independent of S_{DM}. Additionally, the values of NNI and N_{and} at different sampling dates for N limiting condition were generally <1 and >0, while >1 and <0, respectively for non-N-limiting supply. The values of ΔN derived on the basis of relationship between ΔNNI, ΔN_{and} and ΔN, can be used to make corrective decisions of N dressing recommendation for precise N management, prior to or on the onset of the period of highest demand of the rice crop. We conclude that the S_{DM} based dilution curve developed in the present study offers a new vision into plant N status and can possibly be adopted as an alternate practical tool for reliable diagnosis of plant N status to correct N fertilization decision during the vegetative growth of rice in east China.

Author Contributions

Conceived and designed the experiments: ST AUK XY XL WC YZ. Performed the experiments: ST AUK XY XL. Analyzed the data: ST AUK XY YZ. Wrote the paper: ST AUK YZ.

References

1. Jaggard K, Qi A, Armstrong M (2009) A meta-analysis of sugarbeet yield responses to nitrogen fertilizer measured in England since 1980. J Agric Sci-(Camb) 147: 287–301.
2. Ghosh M, Mandal B, Mandal B, Lodh S, Dash A (2004) The effect of planting date and nitrogen management on yield and quality of aromatic rice (*Oryza sativa*). J Agric Sci-(Camb) 142: 183–191.
3. Cabangon R, Castillo E, Tuong T (2011) Chlorophyll meter-based nitrogen management of rice grown under alternate wetting and drying irrigation. Field Crops Res 121: 136–146.
4. Lin FF, Qiu LF, Deng JS, Shi YY, Chen LS, et al. (2010) Investigation of SPAD meter-based indices for estimating rice nitrogen status. Compu Electron Agric 71: S60–S65.
5. Confalonieri R, Debellini C, Pirondini M, Possenti P, Bergamini L, et al. (2011) A new approach for determining rice critical nitrogen concentration. J Agric Sci-(Camb) 149: 633–638.
6. Smeal D, Zhang H (1994) Chlorophyll meter evaluation for nitrogen management in corn. Commun Soil Sci Plant Anal 25: 1495–1503.
7. Justes E, Mary B, Meynard JM, Machet JM, Thelier-Huche L (1994) Determination of a critical nitrogen dilution curve for winter wheat crops. Ann Bot 74: 397–407.
8. Yue S, Meng Q, Zhao R, Li F, Chen X, et al. (2012) Critical nitrogen dilution curve for optimizing nitrogen management of winter wheat production in the North China Plain. Agron J 104: 523–529.
9. Ziadi N, Brassard M, Bélanger G, Cambouris AN, Tremblay N, et al. (2008) Critical nitrogen curve and nitrogen nutrition index for corn in eastern Canada. Agron J 100: 271–276.
10. Ziadi N, Belanger G, Claessens A, Lefebvre L, Cambouris AN, et al. (2010) Determination of a critical nitrogen dilution curve for spring wheat. Agron J 102: 241–250.
11. Sheehy JE, Dionora MJA, Mitchell PL, Peng S, Cassman KG, et al. (1998) Critical nitrogen concentrations: implications for high-yielding rice (*Oryza sativa* L.) cultivars in the tropics. Field Crops Res 59: 31–41.
12. Ata-Ul-Karim ST, Yao X, Liu X, Cao W, Zhu Y (2013) Development of critical nitrogen dilution curve of Japonica rice in Yangtze River Reaches. Field Crops Res 149: 149–158.
13. Kage H, Alt C, Stützel H (2002) Nitrogen concentration of cauliflower organs as determined by organ size, N supply, and radiation environment. Plant Soil 246: 201–209.
14. Vouillot MO, Huet P, Boissard P (1998) Early detection of N deficiency in a wheat crop using physiological and radiometric methods. Agronomie 18: 117–130.
15. Ziadi N, Bélanger G, Gastal F, Claessens A, Lemaire G, et al. (2009) Leaf nitrogen concentration as an indicator of corn nitrogen status. Agron J 101: 947–957.
16. Lemaire G, Jeuffroy MH, Gastal F (2008) Diagnosis tool for plant and crop N status in vegetative stage: Theory and practices for crop N management. Eur J Agron 28: 614–624.
17. Oliveira ECAd, de Castro Gava GJ, Trivelin PCO, Otto R, Franco HCJ (2013) Determining a critical nitrogen dilution curve for sugarcane. J Plant Nutr Soil Sci 176: 712–723.
18. Chen J, Huang Y, Tang Y (2011) Quantifying economically and ecologically optimum nitrogen rates for rice production in south-eastern China. Agric Ecosyst Environ 142: 195–204.
19. Hahn WS (1997) Statistical Methods for Agriculture and Life Science. Seol: Free Academy Publishing Co. 747 p.
20. Yao X, Ata-Ul-Karim ST, Zhu Y, Tian Y, Liu X, et al. (2014) Development of critical nitrogen dilution curve in rice based on leaf dry matter. Eur J Agron 55: 20–28.
21. Bélanger G, Richards JE (2000) Dynamics of biomass and N accumulation of alfalfa under three N fertilization rates. Plant Soil 219: 177–185.
22. Gayler S, Wang E, Priesack E, Schaaf T, Maidl FX (2002) Modeling biomass growth, N-uptake and phenological development of potato crop. Geoderma 105: 367–383.
23. Islam M, Islam M, Sarker A (2008) Effect of phosphorus on nutrient uptake of Japonica and Indica rice. J Agric Rural Dev 6: 7–12.
24. Shan Y, Wang Y, Yamamoto Y, Huang J, Yang L, et al. (2001) Study on the differences of nitrogen uptake and use efficiency in different types of rice. J Yangzhou Univ (Nat Sci Ed) 4: 42.
25. Yoshida H, Horie T, Shiraiwa T (2006) A model explaining genotypic and environmental variation of rice spikelet number per unit area measured by cross-locational experiments in Asia. Field Crops Res 97: 337–343.
26. Ying J, Peng S, He Q, Yang H, Yang C, et al. (1998) Comparison of high-yield rice in tropical and subtropical environments: I. Determinants of grain and dry matter yields. Field Crops Res 57: 71–84.
27. Oliveira ECAd, Freire FJ, Oliveira RId, Freire M, Simoes Neto DE, et al. (2010) Extração e exportação de nutrientes por variedades de cana-de-açúcar cultivadas sob irrigação plena. Rev Bras de Ciênc Solo 34: 1343–1352.
28. Wang W, Yao X, Tian Y, Liu X, Ni J, et al. (2012) Common spectral bands and optimum vegetation indices for monitoring leaf nitrogen accumulation in rice and wheat. J Integr Agric 11: 2001–2012.
29. Zhao B, Yao X, Tian Y, Liu X, Ata-Ul-Karim ST, et al. (2014) New critical nitrogen curve based on leaf area index for winter wheat. Agron J 106: 379–389.
30. Ata-Ul-Karim ST, Zhu Y, Yao X, Cao W (2014) Determination of critical nitrogen dilution curve based on leaf area index in rice. Field Crops Res: In press.
31. Debaeke P, Rouet P, Justes E (2006) Relationship between the normalized SPAD index and the nitrogen nutrition index: Application to durum wheat. J Plant Nutr 29: 75–92.

11

Emissions of CH_4 and N_2O under Different Tillage Systems from Double-Cropped Paddy Fields in Southern China

Hai-Lin Zhang[1]*, Xiao-Lin Bai[1,2], Jian-Fu Xue[1], Zhong-Du Chen[1], Hai-Ming Tang[3], Fu Chen[1]*

1 College of Agronomy and Biotechnology, China Agricultural University, Key Laboratory of Farming System, Ministry of Agriculture, Beijing, China, **2** Patent Examination Cooperation Center of the Patent Office, SIPO, Beijing, China, **3** Soil and Fertilizer Institute of Hunan Province, Changsha, China

Abstract

Understanding greenhouse gases (GHG) emissions is becoming increasingly important with the climate change. Most previous studies have focused on the assessment of soil organic carbon (SOC) sequestration potential and GHG emissions from agriculture. However, specific experiments assessing tillage impacts on GHG emission from double-cropped paddy fields in Southern China are relatively scarce. Therefore, the objective of this study was to assess the effects of tillage systems on methane (CH_4) and nitrous oxide (N_2O) emission in a double rice (*Oryza sativa* L.) cropping system. The experiment was established in 2005 in Hunan Province, China. Three tillage treatments were laid out in a randomized complete block design: conventional tillage (CT), rotary tillage (RT) and no-till (NT). Fluxes of CH_4 from different tillage treatments followed a similar trend during the two years, with a single peak emission for the early rice season and a double peak emission for the late rice season. Compared with other treatments, NT significantly reduced CH_4 emission among the rice growing seasons ($P<0.05$). However, much higher variations in N_2O emission were observed across the rice growing seasons due to the vulnerability of N_2O to external influences. The amount of CH_4 emission in paddy fields was much higher relative to N_2O emission. Conversion of CT to NT significantly reduced the cumulative CH_4 emission for both rice seasons compared with other treatments ($P<0.05$). The mean value of global warming potentials (GWPs) of CH_4 and N_2O emissions over 100 years was in the order of NT<RT<CT, which indicated NT was significantly lower than both CT and RT ($P<0.05$). This suggests that adoption of NT would be beneficial for GHG mitigation and could be a good option for carbon-smart agriculture in double rice cropped regions.

Editor: Ben Bond-Lamberty, DOE Pacific Northwest National Laboratory, United States of America

Funding: This research was supported by Special Fund for Agro-scientific Research in the Public Interest Grant (200903003), Ministry of Agriculture of China. The funders had no role in study design, data collection and analysis, decision to publish, or preparation of the manuscript.

Competing Interests: The authors have declared that no competing interests exist.

* E-mail: hailin@cau.edu.cn (HLZ); chenfu@cau.edu.cn (FC)

Introduction

With the current rise in global temperatures, numerous studies have focused on greenhouse gases (GHG) emissions [1–3]. Agriculture production is an important source of GHG [4]. In addition to carbon dioxide (CO_2), methane (CH_4) and nitrous oxide (N_2O) also play an important role in global warming. The global warming potentials (GWPs) of CH_4 and N_2O are 25 and 298 times that of CO_2 in a time horizon of 100 years, respectively [5]. In addition to industrial emissions, farmland is another important source of atmospheric GHG [6–9]. Numerous results indicate rice (*Oryza sativa* L.) paddy field is a significant source of CH_4 [9,10]. The anaerobic conditions in wetland rice field are favorable for fostering CH_4 emission [11].

A considerable number of studies have shown that some farm operations can influence CH_4 and N_2O emission. For example, water/nitrogen (N) management, organic matter application and tillage can regulate CH_4 and N_2O emission [12–14]. Tillage and crop residues retention have a great influence on CH_4 and N_2O emission through the changes of soil properties (e.g., soil porosity, soil temperature and soil moisture, etc.) [15,16]. In some experiments, conversion of conventional tillage (CT) to no-till (NT) can significantly reduce CH_4 and N_2O emission [17,18]. However, tillage effects on CH_4 and N_2O emission are not always consistent among different studies. Dendooven et al. reported that CH_4 emission were not significantly affected by tillage [19]. In addition, some studies show that crop residues retention can increase CH_4 and N_2O emission from paddy fields [20–22].

Most previous studies of CH_4 and N_2O emissions in paddy field have focused on the effects of water and N management on GHG emission [23–26]. However, tillage can result in changes to GHG emission through the alteration of soil properties and biochemical processes. Although CT is widely adopted around the world, it strongly disturbs the soil, consumes more energy, and even leads to disaster (i.e., the 1930s Dust Bowl in the U.S.). Conservation tillage is increasingly being adopted in the world because of the numerous benefits (e.g., saving time/energy/fuel, controlling soil erosion and increasing water use efficiency). Presently, more and more countries in Asia are facing the problem of labor shortages and high labor cost in planting rice. Conservation tillage in paddy fields (e.g., NT, direct seeding) has increasingly been adopted in Asia, especially in Southern China. Currently, the labor shortage in agriculture has been a major constraint confronting rural

China. Because of energy and labor savings, NT has been widely adopted as a principal conservation technology in China. Furthermore, it is estimated that about 2.18×10^8 Mg yr^{-1} of rice crop residues are generated in China, accounting for 27.51% of the gross crop residue production [27]. Xiao et al. [28] reported that only 9.81% of crop residue was returned to croplands as fertilizer, but >20% of crop residue was burned directly in the field or thrown away, thus increasing environmental pollution and threatening public safety. Therefore, rational use of tillage and crop residues is of great importance for GHG emission mitigation in China.

Until now, most studies on GHG emissions in paddy fields have been based on single rice (one rice cropping in one year) or rice–wheat (*Triticum aestivum* L.) cropped fields and very few studies have involved tillage impacts on emissions of CH_4 and N_2O in double rice (two rice crops in one year, early rice and late rice) cropped fields [4,12,29]. The lower Yangtze region is a typical double rice cropped area in China, accounting for 40–60% of total arable land in this region [30]. Due to the important role of rice paddies in global agriculture, adopting reasonable agricultural management is of great importance in the mitigation of global GHG emissions. Therefore, it is valuable to examine GHG emissions in paddy fields under different tillage systems and to improve reasonable practices for mitigation of GHG emissions. The objective of this paper was to assess tillage effects on emissions of CH_4 and N_2O, and to identify the influencing factors controlling CH_4 and N_2O emission under different tillage methods.

Materials and Methods

Ethics Statement

This experiment was established in a long-term experiment site (Ningxiang, 112°18′E, 28°07′N, Hunan province, China), which belongs to Soil and Fertilizer Institute of Hunan Province. This research was performed in cooperation with China Agricultural University and Soil and Fertilizer Institute of Hunan Province. The farm operations of this experiment were similar to rural farmers' operations and did not involve endangered or protected species. The experiment was approved by the Key Laboratory of Farming System, China Agricultural University and Soil and Fertilizer Institute of Hunan Province.

Site Description

The experimental area has a subtropical monsoonal humid climate, with an annual average precipitation of 1358.3 mm and annual average temperature of 16.8°C. The typical cropping system in this area is double rice cropping in a year (i.e., early rice and late rice). Normally, rotary tillage is conducted one or two days before rice seedling transplanting. Principal properties of the surface soil (0–20 cm) are presented in Table 1. The experimental site had been cultivated with rice under rotary tillage (RT) without crop residue retention for ~30 years before the initiation of the experiment. Generally, early rice is transplanted in early April and harvested in early July and late rice is immediately transplanted

after the early rice harvest and is subsequently harvested in middle October.

Experimental Design and Treatments

The field experiment was established in 2005 with three tillage treatments: conventional tillage (CT), rotary tillage (RT) and no-till (NT). The treatments were laid out in a randomized complete block design with three replications and the area of each plot was 66.7 m^2. For all treatments, rice residue was retained on the soil surface after rice harvest until tillage operations were conducted. No-till operation was conducted in NT and the rice residue was retained on the soil surface throughout the entire study period. The CT plots were plowed once to a depth of ~15 cm using a moldboard plow and rotavated twice to a depth of ~8 cm on the day of rice seedling transplanting. The RT plots were rotavated four times to a depth of ~8 cm on the day of rice seedling transplanting.

Early rice (*Zhongjiazao 32#*) was transplanted on April 7, 2007 and April 10, 2008. Late rice (*Xiangwanshan 13#*) was transplanted on July 10 both in 2007 and 2008. All plots received 375 kg ha^{-1} compound fertilizer($N:P_2O_5:K_2O = 20:12:14$)as basal fertilizer at seedling transplanting. One week after seedling transplanting, the plots were top-dressed with urea (46% of N), 150 kg ha^{-1} for the early rice and 75 kg ha^{-1} for the late rice. Selective herbicides (34% Quinclorac, 4% Bensulfuron-methyl) were applied prior to rice transplanting in all treatments. The planting density was ~803 640 strains ha^{-1} and ~12 500 kg ha^{-1} yr^{-1} of rice residue was retained to the soil during the experimental years.

Data Collection

Soil temperature was measured by thermometers (DF-201A, Beijing Dongfang Mingguang Electronic Science And Technology Co., Ltd) with a measuring range of −30°C to +100°C. The thermometers were inserted into the 5 cm and 10 cm soil depth and data were recorded at 10-day intervals after rice seedling transplanting. Soil bulk densities (ρ_b) at 0–5 cm, 5–10 cm and 10–20 cm depth were determined by the core method.

Soil porosity (SP, m^3 m^{-3}) was calculated by using the formula below:

$$SP = 1 - \rho_b / \rho_s \qquad (1)$$

Where, ρ_s is soil particle density, Mg m^{-3}.

Soil samples were collected from each treatment plot prior to rice seedling transplanting and at the rice harvest.

Fluxes of CH_4 and N_2O were measured with the closed chamber method [31]. For each plot, three chamber bases were inserted into the soil (5 cm depth) after tillage operations. To avoid soil disturbance, every chamber base was placed at a fixed position until rice harvest. A removable wooden bridge (2 m long and 0.5 m wide) was placed near the chamber base for convenience of sampling. The chamber base had a 5 cm deep groove for installation. A chamber made with polymethyl methacrylate was placed at the chamber base. The cross-sectional area of each

Table 1. Principal soil properties of the test soil.

Soil layer (cm)	Bulk density (g cm^{-3})	Soil organic matter (g kg^{-1})	Available N (mg kg^{-1})	Available P (mg kg^{-1})	Available K (mg kg^{-1})	pH (H$_2$O)
0–20	1.21	34.90	224.10	4.38	97.10	6.26

Table 2. Mean monthly precipitation and air temperature from April to October between 2005 and 2008 at the experimental site.

Month	Precipitation (mm)				Air temperature (°C)			
	2005	2006	2007	2008	2005	2006	2007	2008
April	92.2	235.0	38.0	26.3	20.6	19.9	25.8	18.7
May	400.8	125.0	119.0	27.3	22.6	23.6	26.6	24.5
June	272.1	201.0	119.0	25.6	27.2	27.0	26.6	26.6
July	66.7	133.0	44.0	30.9	30.2	30.1	30.8	30.0
August	80.4	154.0	126.0	58.1	27.0	29.5	29.6	28.7
September	47.5	18.0	121.0	43.2	24.6	24.0	23.5	25.6
October	64.4	40.0	3.0	18.2	18.2	21.3	19.4	20.2
Mean	146.3	129.4	81.4	32.8	24.3	25.1	26.0	24.9

Source: China Meteorological Data Sharing Service System. These data represent the mean monthly precipitation and temperature. The early and late rice growing period was April to October.

chamber was 0.36 m^2 (0.6 m$\times 0.6$ m) and the height was 0.8 m. Chambers were closed by filling the groove of the base with water during gas sampling, and the chamber was equipped with a small fan to mix air inside the chamber. Gas samples were collected with vacuum vials. In order to minimize the underestimation of gas fluxes with the closed chamber method, the time-course of each gas sampling was kept within 10 min [32]. Measurements were conducted every 4 hours on each sampling day. Gas samples were collected at least three times per month. During the tillage period (~1 week) and the field drainage period (~10 days), gas collection was conducted daily. The gas samples were analyzed for CH_4 and N_2O using a gas chromatography with FID and ECD (model 6890N, Agilent Technologies, CA).

The fluxes of CH_4 and N_2O emissions were calculated by using the formula below [33]:

$$F = \frac{Mw}{Mv} \times \frac{Tst}{Tst + T} \times \frac{dc}{dt} \times h \qquad (2)$$

Where F is the emission fluxes (mg m^{-2} min^{-1}); M_w is the molar mass of trace gas (g mol^{-1}); Mv is the molar volume of trace gas (L mol^{-1}); T_{st} is the absolute temperature (273.2 K); T is the air temperature at sampling (°C); dc/dt is the change in the rate of CO_2 or CH_4 concentrations (ppbv min^{-1}); and h is the height of the chamber (m).

The cumulative emissions within one year were calculated assuming the existence of linear changes in gas fluxes between two successive sampling dates. Meteorological data were obtained from China National Meteorological Bureau.

GWPs is defined as the cumulative radiative forcing both direct and indirect effects integrated over a period of time from the emission of a unit mass of gas relative to some reference gas [34]. Carbon dioxide was chosen as this reference gas. The GWPs conversion parameters of CH_4 and N_2O (over 100 years) were adopted with 25 and 298 kg ha^{-1} CO_2-equivalent [35].

Statistical Analyses

Statistical analyses were performed with SPSS 11.0 analytical software package (SPSS Inc., Chicago, IL, US). Statistical analysis was performed with ANOVA to analyze the effects of tillage on ρ_b, SP, CH_4 and N_2O flux among the treatments. The Tukey-HSD was calculated for comparison of the treatment means. With regard to CH_4 and N_2O fluxes, data for each sampling day were

analyzed separately. Differences among treatments were declared to be significant at $P<0.05$.

Results

Air Temperature and Precipitation

In general, air temperature during May and September ranges from 22 to 30°C in this region. April and October are the coldest months during the rice growing period, with mean air temperature ~20°C. The mean air temperature in 2007 was higher than that of other years, but the air temperatures were slightly lower than the average of other years in September and October of 2007 (Table 2). Mean precipitation changed dramatically compared with the two years, 81.4 mm in 2007 and 32.8 mm in 2008. The precipitation is mainly distributed between May and August, especially during May and June in this region. However, the precipitation in August and September of 2007 was more than the average and these months had the highest precipitation in 2007. Precipitation in 2008 was much less compared to that of other years (Table 2).

Soil Bulk Density

Regardless of tillage practice, ρ_b increased with soil depth, but ρ_b increased more in NT than the other tillage treatments. Among the tillage treatments, ρ_b varied in the order of RT>CT>NT at 0–5 cm depth (Fig. 1), but varied in the order of NT>CT>RT at 5–10 cm depth for both the early and the late growing season. Compared with NT, ρ_b was lower at 5–10 cm and 10–20 cm depth under RT and CT. Figure 1 indicated that ρ_b under RT changed dramatically during the rice growing season, especially at 0–10 cm depth. At 0–5 cm and 5–10 cm depth, ρ_b under RT were higher in the early rice season than in the late rice season (0.23 vs. 0.13 g cm^{-3}). In both the early and the late rice growing season, ρ_b under RT was significantly different from that under NT (Tukey HSD. early rice season: 0–5 cm, df = 8 F = 31.907 $P<0.05$; 5–10 cm, df = 8 F = 20.100 $P<0.05$; 10–20 cm, df = 8 F = 10.323 $P<0.05$. Late rice season: 0–5 cm, df = 8 F = 35.083 $P<0.05$; 5–10 cm, df = 8 F = 43.017; $P<0.05$; 10–20 cm df = 8 F = 8.089 $P<0.05$). Because of minimal soil disturbance, ρ_b under NT increased greatly in the deeper soil layers (Fig. 1). The significant change of ρ_b in RT may be due to soil disturbance and crop residue incorporation, whereas NT had the crop residue remaining on the soil surface.

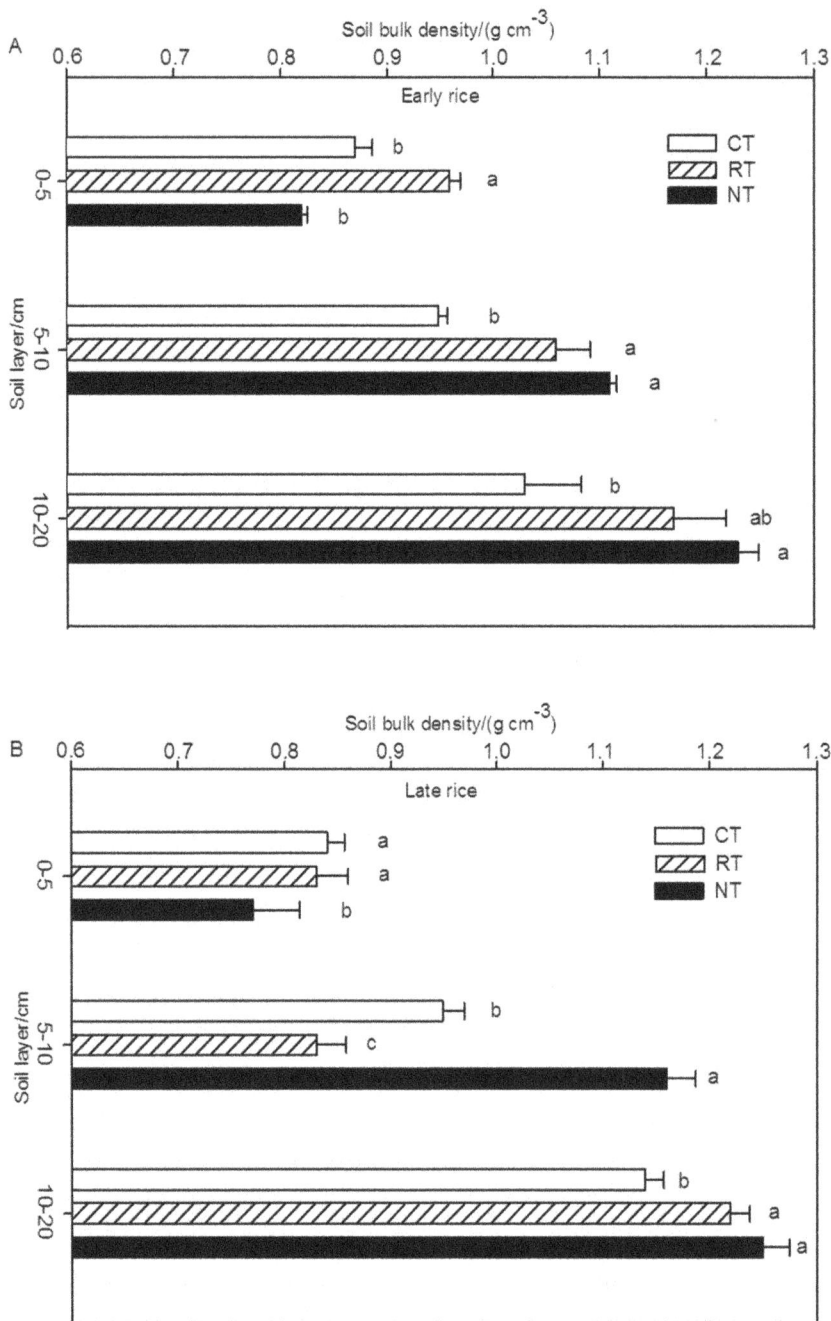

Figure 1. Soil bulk density of different tillage treatments in 2008 (A for the early rice season and B for the late rice season). Data are means of three replications; means followed by different letters are significantly different at $P<0.05$. Sampling was done during the harvest of the early and late rice in 2008.

Soil Porosity

Soil porosity decreased with soil depth among all the treatments (Fig. 2). For the early rice season, SP at $0-5$ cm depth was 68.48%, 63.18% and 61.03% for NT, CT and RT, respectively. Tukey HSD statistical test showed that SP for NT and CT significantly differed with that of RT ($0-5$ cm, df = 8 F = 69.651 $P<0.05$; $5-10$ cm, df = 8 F = 18.589 $P<0.05$; $10-20$ cm, df = 8 F = 10.393 $P<0.05$). The order of SP at depths of $5-10$ cm and $10-20$ cm varied with CT>RT> NT; and SP for CT and RT were 11.5% and 8.9% higher than that of NT, respectively. The

trend of SP in the late rice season varied similarly with that of the early rice season ($0-5$ cm, df = 8 F = 30.167 $P<0.05$; $5-10$ cm, df = 8 F = 195.166 $P<0.05$; $10-20$ cm df = 8 F = 6.957 $P<0.05$). Conversion of traditional tillage to NT, SP at $5-10$ cm depth was higher 1.83% and 7.27% than that for CT and RT, respectively. Compared with NT, SP for CT significantly increased at $10-20$ cm depth in the early rice season. During the early rice growing season, SP at $5-10$ cm depth varied in the order of CT>RT>NT ($P<0.05$). However, during the late rice season, SP at $5-10$ cm depth followed in the order of NT>RT>CT

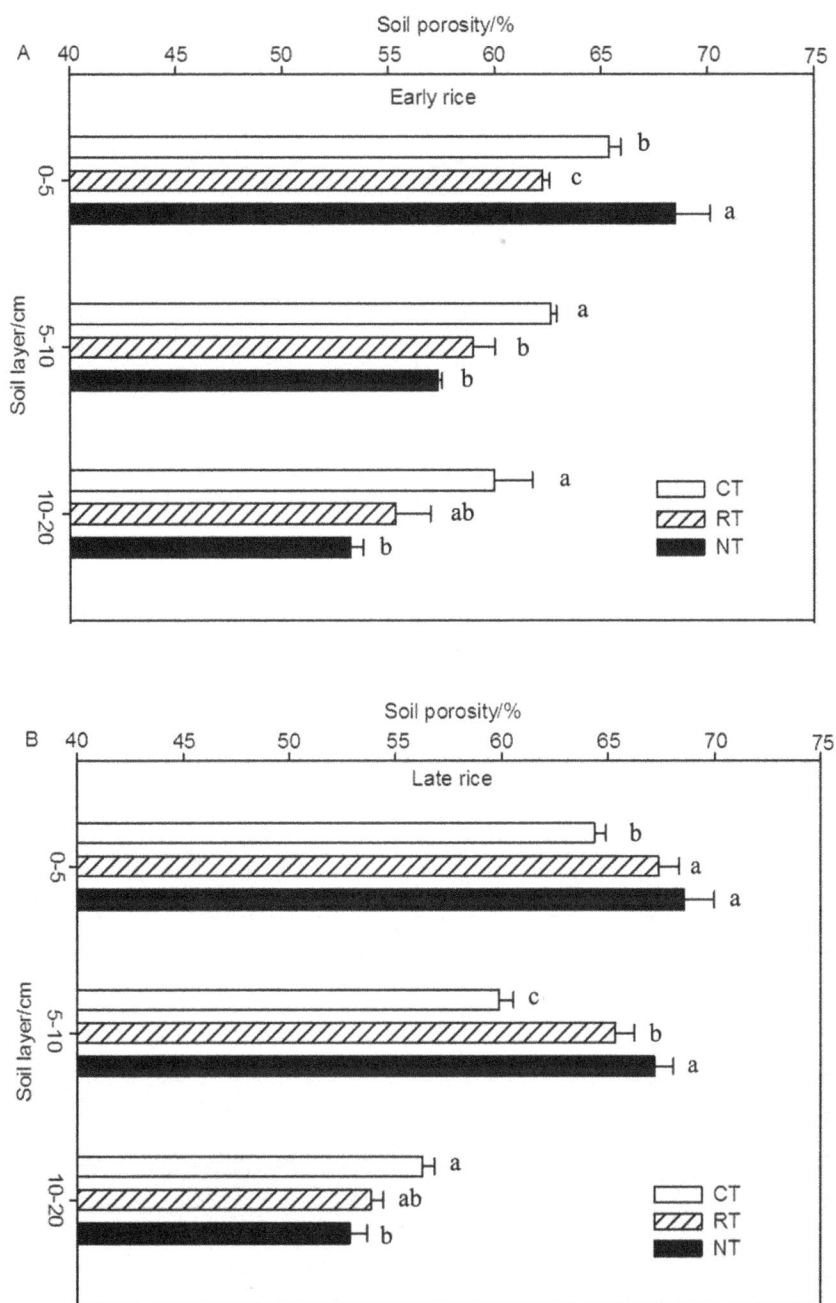

Figure 2. Soil porosity of different tillage treatments in 2008 (A for the early rice season and B for the late rice season). Data are means of three replications; means followed by different letters are significantly different at $P<0.05$.

($P<0.05$) and 9.84% and 6.35% higher for NT and RT than for CT, respectively.

CH₄ Emission

For the early rice season, paddy soil was the atmospheric source of CH₄ under all treatments in both years. The flux of CH₄ showed a single peak pattern characterized by three stages (Fig. 3−a, b). The first stage was the increasing stage of CH₄ emission. The flux of CH₄ showed a continuous increase under all the treatments and attained the highest fluxes during the aeration stage. The CH₄ emissions from both CT and RT displayed similar trends and were higher than that from NT (Fig. 3−a, b). The

second stage was the decreasing stage of CH₄ emission. The flux of CH₄ decreased rapidly from the aeration stage to the flooding stage during the early rice season. The emission fluxes in 2007 and 2008 were in the same order of RT>CT>NT and significant differences among the treatments were observed in 2008 ($P<0.05$). The third stage was characterized by stable CH₄ emission. The flux of CH₄ remained at a low level and tended to be stable from the flooding stage to the harvest stage. In 2008, the cumulative emissions were 228.3, 276.3 and 188.1 kg ha⁻¹ for CT, RT and NT, respectively and were 17.9%, −1.7% and 16.2% lower in 2007, respectively. The difference between 2007 and 2008 was possibly due to weather differences.

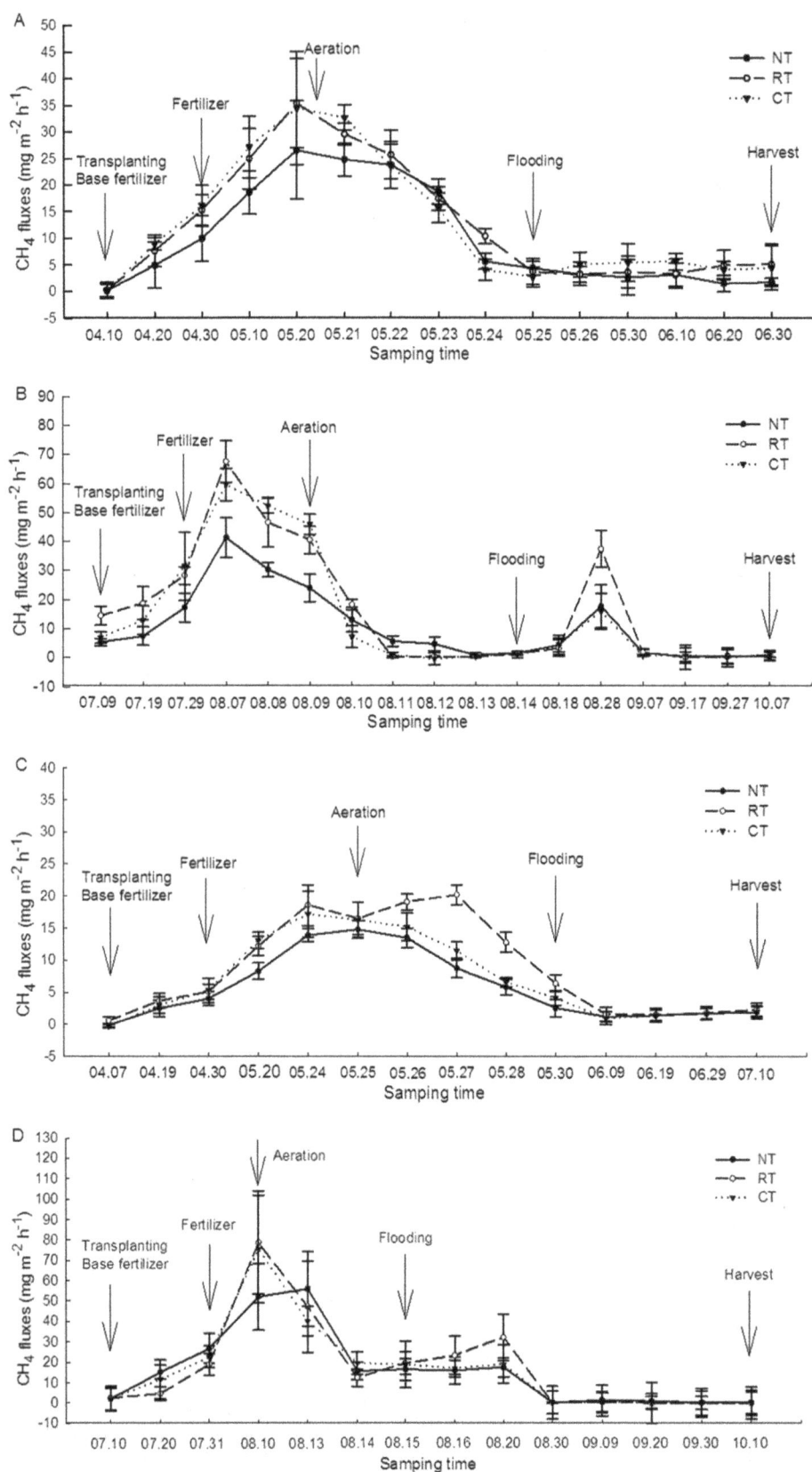

Figure 3. CH$_4$ flux under different tillage during the rice growing seasons (A, B for the early rice season and the late rice season in 2007; C, D for the early rice season and the late rice season in 2008, respectively). Vertical bars represent standard errors of the mean (n = 3).The arrows in the figures indicate the time of field operation.

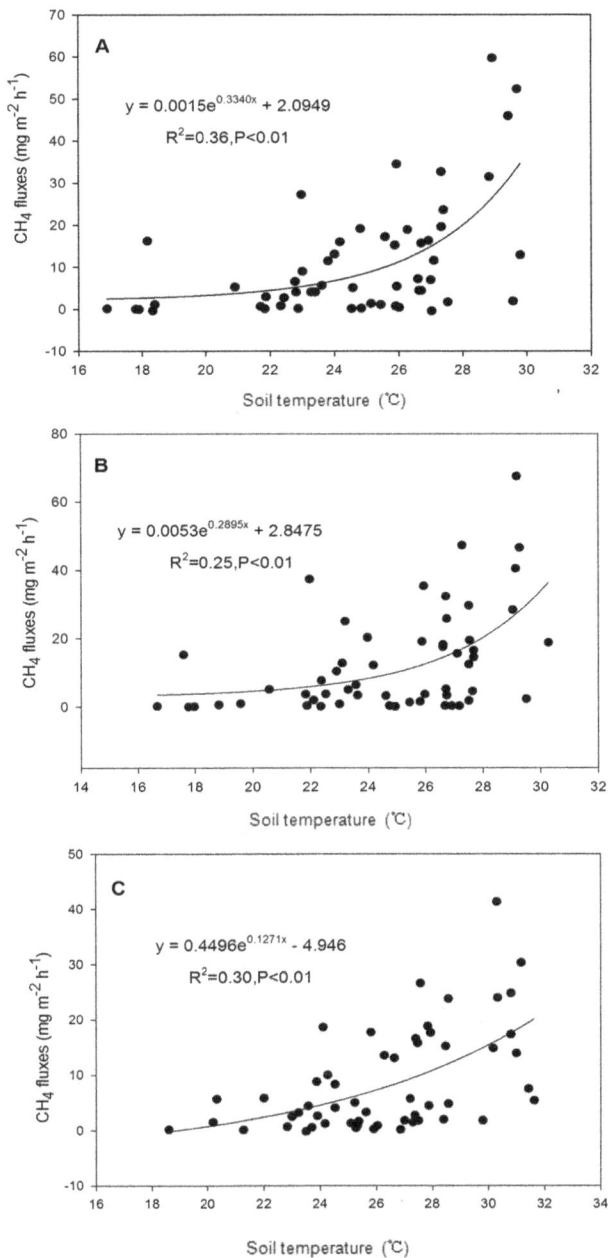

Figure 4. Relationship between soil temperature and CH₄ emission from paddy fields (A for CT at 5 cm depth soil, B for RT at 5 cm depth soil, and C for NT at surface soil). R^2: coefficient of determination.

The flux of CH_4 for the late rice season (Fig. 3-a, b) showed a double emission peak. Before flooding, the CH_4 emission flux exhibited similar trends to that of the early rice season. However, there was another small peak emission after the flooding stage which was lower than the first peak emission. For both years, CT had higher CH_4 emission in the second peak fluxes than that of RT and NT. The cumulative emissions of CH_4 for the late rice season in 2008 were 526.2, 565.5 and 506.2 kg ha^{-1} for CT, RT and NT, respectively; and 68.5%, 39.3% and 140.8% higher than in 2007, respectively.

The cumulative CH_4 emission under NT was lower than that under CT and RT (Fig. 3-a, b), and the difference was significant

at the peak emission ($P<0.05$). In contrast, CT emitted more CH_4 during the early and the late rice growing seasons, with a higher peak emission than that of NT and RT (Fig. 3-a, b).

The emission of CH_4 was greatly correlated with soil temperature (Fig. 4). There were significant correlations between CH_4 emission and soil temperature among the treatments. There was a significant correlation between CH_4 emission and soil temperature at 5 cm depth for CT and RT, while significant correlation for NT was at the soil surface.

Compared with the GWPs of CH_4 emission (over 100 years), the mean value of 2007 and 2008 for NT was significantly lower than for CT and RT ($P<0.05$) with 16814, 18988 and 14112 kg ha^{-1} CO_2-equivalent for NT, RT and CT, respectively.

N₂O Emission

The N_2O emission exhibited an impulse type for both the early and the late rice season in 2007 and 2008 (Fig. 5-a, b). Regardless of tillage methods, the N_2O emission exhibited an emission peak after tillage, aeration and flooding. The first peak of N_2O emission appeared ~10 days after tillage. The emission varied in the order of RT>CT>NT in 2008, and RT was significantly higher than NT ($P<0.05$). The emission order was NT>CT>RT in 2007, but no significant differences among treatments ($P<0.05$) were observed. The N_2O emission fluxes decreased after fertilizer application, but aeration and flooding triggered emission peaks.

All the three tillage treatments were weak sources of N_2O (Table 3). In 2008, the cumulative N_2O emission was 0.01, 0.30 and 0.30 kg ha^{-1} for CT, RT and NT, respectively. However, the emissions in 2008 were nearly 60% lower than those in 2007 for all the treatments. The annual difference of the cumulative emission was possibly due to influences from meteorological factors (i.e., temperature, precipitation). Regardless of the year, the N_2O emission fluxes for NT was more stable than that for CT and RT, ranging from 13.1−33.0 μg m^{-2} h^{-1} in the late rice season. On the other hand, the emission fluxes for RT and CT changed greatly from day to day. However, aeration strongly influenced N_2O emissions for all the treatments. In general, about 68%−81% of the cumulative N_2O emissions occurred from aeration to harvest in 2007. Compared with CT and RT, NT significantly increased the N_2O emission from aeration to harvest in both 2007 and 2008 ($P<0.05$).

Compared with the GWPs of N_2O emission (over 100 years), the mean value of 2007 and 2008 for CT was significantly lower than that for RT and NT ($P<0.05$). The values were 126.7, 166.9 and 152.0 kg ha^{-1} CO_2-equivalent for NT, RT and CT, respectively.

Discussion

CH₄ Emission

Large variations in CH_4 emission were observed during the rice growing seasons, which may be attributed to differences in meteorological conditions. However, soil tillage had significant effects on CH_4 emission across the entire rice growing seasons. In this study, NT had a lower CH_4 emission compared with other treatments ($P<0.05$), which is consistent with the results of Zhang et al. [36]. Gregorich et al. attributed the differences in gas fluxes between NT and CT to differences in the physical environment [37]. Wang et al. indicated that the major differences in CH_4 production zone resulted from the disturbed depth by the different tillage methods [38]. Therefore, the CH_4 production zone may vary according to the adopted tillage method. Wang et al. also reported that the main oxidation zone of CH_4 was the root surface and the interface between soil and water [38]. The rice residues

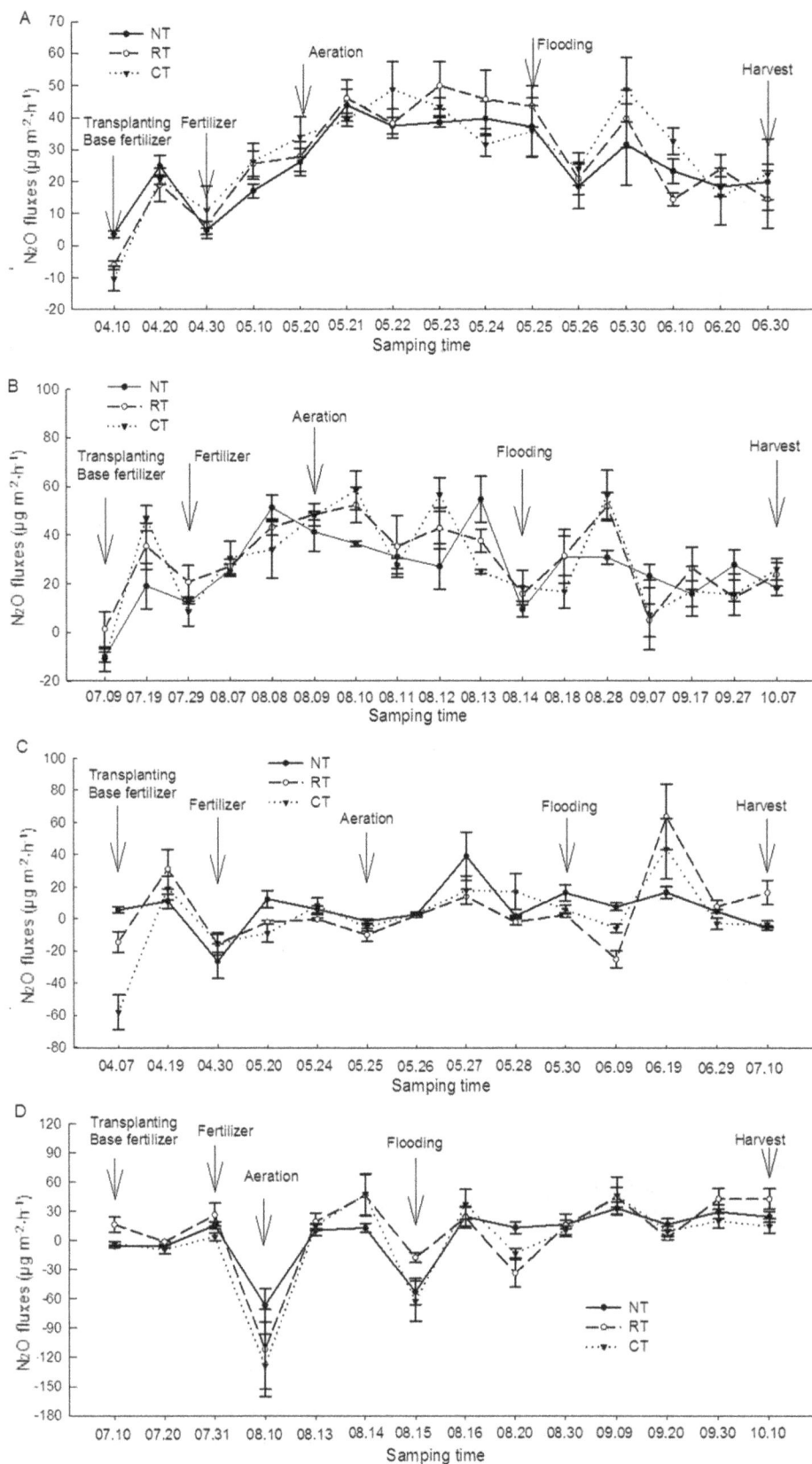

Figure 5. N₂O flux under different tillage during the rice growing seasons (A, B for the early rice season and the late rice season in 2007; C, D for the early rice season and the late rice season in 2008, respectively). Vertical bars represent standard errors of the mean (n = 3).The arrows in the figures indicate the time of field operations.

Table 3. Cumulative N_2O emissions of each farm operation phase during the rice growing period.

Year			CT (kg ha^{-1})	RT (kg ha^{-1})	NT (kg ha^{-1})
			Treatments		
2007	Early rice	Before aeration	0.09b	0.08c	0.10a
		During aeration	0.13a	0.12a	0.10b
		After aeration	0.24a	0.19b	0.19b
	Late rice	Before aeration	0.12b	0.13a	0.06c
		During aeration	0.10b	0.11a	0.10b
		After aeration	0.16c	0.18b	0.18a
Total emission			0.84a	0.82b	0.72c
2008	Early rice	Before aeration	−0.11c	0a	−0.03b
		During aeration	0b	0b	0.02a
		After aeration	0.09b	0.13a	0.09b
	Late rice	Before aeration	−0.16b	−0.07a	−0.05a
		During aeration	−0.04c	−0.02a	−0.03b
		After aeration	0.22c	0.26b	0.29a
Total emission			0.01c	0.30a	0.30b

Values are means of three replications for each treatment; means followed by different letters are significantly different at $P<0.05$.

retention may have increased the soil oxide layer. In this study, NT significantly increased the SP at $0-5$ cm depth (Fig. 2), and thus had a larger oxide layer than other treatments, which may be beneficial to the oxidization of CH_4. Regina et al. indicated that CH_4 oxidation rate was higher when there were more macropores or fewer micro-pores in the soil [39]. In addition, CH_4 emission was influenced by soil temperature and soil redox potential (Eh). Yu et al. [40] reported that CH_4 emission showed an exponential decrease by an Eh increase.

In this study, the crop residues were distributed on the soil surface under NT. Furthermore, the decomposition of residues consumed limited soil dissolved oxygen. All these factors discussed above resulted in Eh decrease and consequently a reduction of CH_4 emission under NT. Khalil et al. [41] observed an increase in CH_4 emissions from paddy fields with increasing soil temperature. In this study, temperature was another major factor affecting CH_4 emission (Fig. 4). In general, NT decreased soil temperature especially during the hotter days. Therefore, low temperatures also reduced the CH_4 emission when compared with other treatments.

In this study, CH_4 emission from the late rice season was 65% higher than that from the early rice season, which indicates that the late rice paddy is the principal CH_4 source in double paddy fields. Temperature was the major reason for the differences in the CH_4 emission pattern between the early and the late rice season. The soil temperature had a predictive functional relationship with CH_4 emission. Zhu et al. [42] and Bossio et al. [43] reported a strong correlation between CH_4 emission and soil temperature. Furthermore, Whalen and Reeburgh [44] reported that temperature had important influence on CH_4 emission from soils and the combination of high soil moisture and low temperature was favorable to decrease CH_4 emission. In this study, an exponential model was used for fitting CH_4 emission and soil temperature. Our results showed that there was a significant correlation between CH_4 emission and soil temperature. But the coefficient of determination was not high, and this may be due to the

fluctuation of soil temperature influenced by the alternation of wetting and drying in paddy. In this experimental area, the late rice season was the hottest time of the summer. Therefore, high temperatures enhanced the decomposition rate of crop residues in the moist environment. During the decomposition process of crop residues, a large number of organic compounds are produced and oxygen is consumed, thus decreasing the soil Eh, leading to an increase in the possibility of CH_4 emission. In contrast to the warm temperatures of the late rice season, the air temperatures of the early rice season were lower, which resulted in slower crop residue decomposition and therefore little CH_4-substrate. Hence, these differences in weather factors (e.g., temperature) resulted in the different characteristics of CH_4 between the early and the late rice seasons.

N_2O Emission

In our study, the fluxes of N_2O emission show a great fluctuation during the rice growth seasons, but it remained at a low level. Indeed, the N_2O emission was strongly influenced by external factors and many emission peaks occurred during the rice growing season. The emission of N_2O was dramatically different between the two years. This difference is possibly due to the variations in weather. Some studies show that extreme precipitation and drying could increase N_2O emission [45,46]. Hao et al. [47] reported that aeration and water flooding led to outbreaks of emissions. The precipitation in 2007 was much higher than the precipitation in 2008. This precipitation difference may explain the fluctuations of N_2O emissions between the two years.

The N_2O emission differences among the treatments were possibly due to farm operations (e.g., tillage, drainage). Some results indicated that N_2O production and emission was greatly influenced by tillage because of the breaking of the soil uniformity [48]. Nitrogen (mainly as NO_3^--N or NH_4^+-N) can remain stable in homogeneous soil and thus may decrease N_2O production. Tillage practices change the soil nutrients and crop residue distribution. The distribution of soil nutrients was relatively even under CT and RT by cutting, mixing, overturning the soil and crop residues. However, the crop residues were well-distributed only in the 0–8 cm soil layer under RT because of the shallow tilled depth. High stratification ratio of soil nutrients (e.g., N, SOC) across different depths is observed in NT systems [48,49], which means that the soil nutrient distributions are not even among different depths. Therefore, the different distribution of crop residues and soil nutrients among the treatments influences the N_2O production and emission. In addition, similar to CH_4, N_2O emission is also influenced by soil Eh. Weier et al. reported that the rate of N_2O emission decreased with increasing soil reducibility [49]. Generally, crop residues in CT are mainly distributed within the plow layer (0–20 cm) and had a strong redox potential due to decomposition of crop residues. Therefore, N_2O produced from CT soils tended to be further deoxidized to N_2, which consequently decreased N_2O emission. Similar results were also reported by Steinbach and Alvarez [50] who observed NT increased N_2O emission.

Conclusion

Paddy fields with rice residues retention were a source of atmospheric CH_4, regardless of the tillage practice. Compared with other treatments, NT reduced CH_4 emission among the rice growing seasons. The GWPs (based on CH_4 emission) under NT was significantly $(P<0.05)$ lower than that of CT and RT. The N_2O emission was vulnerable to external influences and varied greatly during the rice growing seasons. Although the cumulative emission under NT was more than other treatments, GWPs of

N_2O was relative low compared to that of CH_4. Therefore, N_2O emission was a weak source of GHG in paddy fields. The GWPs (based on CH_4 and N_2O) of NT is lower than that of CT and RT. Thus, adoption of NT is beneficial in GHG mitigation and could be a good practice of carbon-smart agriculture in double rice cropped regions.

Acknowledgments

We would like to express our sincere thanks to Mr. Shadrack Dikgwatlhe and Mr. Jay Lytle for language assistance.

Author Contributions

Conceived and designed the experiments: HLZ FC. Performed the experiments: XLB HMT. Analyzed the data: HLZ XLB JFX ZDC. Contributed reagents/materials/analysis tools: XLB HMT FC. Wrote the paper: HLZ XLB.

References

1. Levy PE, Mobbs DC, Jones SK, Milne R, Campbell C, et al. (2007) Simulation of fluxes of greenhouse gases from European grasslands using the DNDC model. Agric Ecosyst Environ 121: 186–192.
2. Saggar S, Hedley CB, Giltrap DL, Lambie SM (2007) Measured and modelled estimates of nitrous oxide emission and methane consumption from a sheep-grazed pasture. Agric Ecosyst Environ 122: 357–365.
3. Hernandez-Ramirez G, Brouder SM, Smith DR, Van Scoyoc GE (2009) Greenhouse gas fluxes in an eastern corn belt soil: Weather, nitrogen source, and rotation. J Environ Qual 38: 841–854.
4. Wassmann R, Neue HU, Ladha JK, Aulakh MS (2004) Mitigating greenhouse gas emissions from rice-wheat cropping systems in Asia. Environ Devel Sustain 6: 65–90.
5. Forster P, Ramaswamy V, Artaxo P, Berntsen T, Betts R, et al. (2007) Changes in atmospheric constituents and in radiative forcing. In: Solomon S, Qin D, Manning M, Chen Z, Marquis M, et al. (Eds.) Climate Change 2007: The Physical Science Basis. Contribution of Working Group I to the Fourth Assessment Report of the Intergovernmental Panel on Climate Change, Cambridge University Press, Cambridge, United Kingdom and New York, NY, USA.
6. Lokupitiya E, Paustian K (2006) Agricultural soil greenhouse gas emissions: A review of national inventory methods. J Environ Qual 35: 1413–1427.
7. Verma A, Tyagi L, Yadav S, Singh SN (2006) Temporal changes in N_2O efflux from cropped and fallow agricultural fields. Agric Ecosyst Environ 116: 209–215.
8. Liu H, Zhao P, Lu P, Wang YS, Lin YB, et al. (2008) Greenhouse gas fluxes from soils of different land-use types in a hilly area of South China. Agric Ecosyst Environ 124: 125–135.
9. Tan Z, Liu S, Tieszen LL, Tachie-Obeng E (2009) Simulated dynamics of carbon stocks driven by changes in land use, management and climate in a tropical moist ecosystem of Ghana. Agric Ecosyst Environ 130: 171–176.
10. Wassmann R, Dobermann A (2006) Greenhouse Gas Emissions from Rice Fields: what do we know and where should we head for? Paper presented at: The 2nd Joint International Conference on "Sustainable Energy and Environment". Bangkok, Thailand. 21–23 November. Paper D-030 (O).
11. Pandey D, Agrawal M, Bohra JS (2012) Greenhouse gas emissions from rice crop with different tillage permutations in rice-wheat system. Agric Ecosyst Environ 159: 133–144.
12. Yagi K, Minami K (1990) Effect of organic matter application on methane emission from some Japanese paddy fields. Soil Sci Plant Nutr 36: 599–610.
13. Yagi K, Tsuruta H, Kanda KI, Minami K (1996) Effect of water management on methane emission from a Japanese rice paddy field: Automated methane monitoring. Global Biogeochem Cycles 10: 255–267.
14. Nishimura S, Sawamoto T, Akiyama H, Sudo S, Yagi K (2004) Methane and nitrous oxide emissions from a paddy field with Japanese conventional water management and fertilizer application. Global Biogeochem Cycles 18, GB2017, doi:10.1029/2003GB002207.
15. Al-Kaisi MM, Yin X (2005) Tillage and crop residue effects on soil carbon and carbon dioxide emission in corn-soybean rotations. J Environ Qual 34: 437–445.
16. Yao Z, Zheng X, Xie B, Mei B, Wang R, et al. (2009) Tillage and crop residue management significantly affects N-trace gas emissions during the non-rice season of a subtropical rice-wheat rotation. Soil Biol Biochem 41: 2131–2140.
17. Matthias AD, Blackmer AM, Bremmer JM (1980) A simple chamber technique for field measurements of emissions nitrous oxide from soils. J Environ Qual 9: 251–256.
18. Estavillo JM, Merino P, Pinto M, Yamulki S, Gebauer G, et al. (2002) Short term effect of ploughing a permanent pasture on N_2O production from nitrification and denitrification. Plant Soil 239: 253–265.
19. Dendooven L, Patiño-Zúñiga L, Verhulst N, Luna-Guido M, Marsch R, et al. (2012) Global warming potential of agricultural systems with contrasting tillage and residue management in the central highlands of Mexico. Agric Ecosyst Environ 152: 50–58.
20. Toma Y, Hatano R (2007) Effect of crop residue C: N ratio on N_2O emissions from Gray Lowland soil in Mikasa, Hokkaido, Japan. Soil Sci Plant Nutr 53: 198–205.
21. Lou Y, Ren L, Li Z, Zhang T, Inubushi K (2007) Effect of rice residues on carbon dioxide and nitrous oxide emissions from a paddy soil of subtropical China. Water Air Soil Pollut 178: 157–168.
22. Lu F, Wang X, Han B, Ouyang Z, Duan X, et al. (2010) Net mitigation potential of straw return to Chinese cropland: Estimation with a full greenhouse gas budget model. Ecol Appl 20: 634–647.
23. Sun W, Huang Y (2012) Synthetic fertilizer management for China's cereal crops has reduced N_2O emissions since the early 2000s. Environ Pollut 160: 24–27.
24. Leytem AB, Dungan RS, Bjorneberg DL, Koehn AC (2011) Emissions of ammonia, methane, carbon dioxide, and nitrous oxide from dairy cattle housing and manure management systems. J Environ Qual 40: 1383–1394.
25. Jiao Z, Hou A, Shi Y, Huang G, Wang Y, et al. (2006) Water management influencing methane and nitrous oxide emissions from rice field in relation to soil redox and microbial community. Commun Soil Sci Plant Anal 37: 1889–1903.
26. Li CF, Zhou DN, Kou ZK, Zhang ZS, Wang JP, et al. (2012) Effects of Tillage and Nitrogen Fertilizers on CH_4 and CO_2 Emissions and Soil Organic Carbon in Paddy Fields of Central China. PLoS ONE 7(5): e34642. doi:10.1371/journal.pone.0034642.
27. Zhong HP, Yue YZ, Fan JW (2003) Characteristics of crop straw resources in China and its Utilization. Res Sci 25(4): 62–67. (In Chinese).
28. Xiao T, He C, Ling X, Jin C. Wu C, et al. (2010) Comprehensive utilization, situation and countermeasure of crop straw resources in China. World Agric 12: 31–33. (In Chinese).
29. Hanaki M, Ito T, Saigusa M (2002) Effect of no-till rice (Oryza sativa L.) cultivation on methane emission in three paddy fields of different soil types with rice straw application. Jpn J Soil Sci Plant Nutr 73: 135–143.
30. Xiong YM, Huang GQ, Cao KW, Liu LW (2003) Review on Developing Ryegrass–Rice Rotation System in Double-cropping Rice Area of Middle and Lower Reaches of Changjiang River. Acta Agric Jiangxi 15: 47–51. (In Chinese).
31. Lapitan RL, Wanninkhof R, Mosier AR (1999) Methods for stable gas flux determination in aquatic and terrestrial systems. Develop Atmos Sci 24: 29–66.
32. Nakano T, Sawamoto T, Morishita T, Inoue G, Hatano R (2004) A comparison of regression methods for estimating soil-atmosphere diffusion gas fluxes by a closed-chamber technique. Soil Biol Biochem 36: 107–113.
33. Zheng X, Wang M, Wang Y, Shen R, Li J, et al. (1998) Comparison of manual and automatic methods for measurement of methane emission from rice paddy fields. Adv Atmos Sci 15: 569–579.
34. IPCC (1996) Climate Change 1995: The Science of Climate Change. In: Houghtom JT, Meira Filho LG, Callander BA, Harris N, Kattenberg A, et al. (Eds.) Intergovernmental Panel on Climate Change, Cambridge University Press, Cambridge, United Kingdom.
35. IPCC (2007) Climate change 2007: The physical science basis. Contribution of working group I to the fourth assessment report of the intergovernmental panel on climate change. Cambridge University Press, Cambridge, United Kingdom and New York, NY, USA.
36. Zhang JK, Jiang CS, Hao QJ, Tang QW, Cheng BH, et al. (2012) Effects of tillage-cropping systems on methane and nitrous oxide emissions from agro-ecosystems in a purple paddy soil. Environ Sci 33(6): 1980–1986. (In Chinese).
37. Gregorich EG, Rochette P, Hopkins DW, McKim UF, St-Georges P (2006) Tillage-induced environmental conditions in soil and substrate limitation determine biogenic gas production. Soil Biol Biochem 38: 2614–2628.
38. Wang M, Li J, Zhen X (1998) Methane Emission and Mechanisms of Methane Production, Oxidation, Transportation in the Rice Fields. Sci Atmos Sin 22: 600–612. (In Chinese).
39. Regina K, Pihlatie M, Esala M, Alakukku L (2007) Methane fluxes on boreal arable soils. Agric Ecosyst Environ 119: 346–352.
40. Yu K, Böhme F, Rinklebe J, Neue HU, DeLaune RD (2007) Major biogeochemical processes in soils - A microcosm incubation from reducing to oxidizing conditions. Soil Sci Soc Am J 71: 1406–1417.
41. Khalil MAK, Rasmussen RA, Shearer MJ, Chen ZL, Yao H, et al. (1998) Emissions of methane, nitrous oxide, and other trace gases from rice fields in China. J Geophys Res 103: 25241–25250.
42. Zhu R, Liu Y, Sun L, Xu H (2007) Methane emissions from two tundra wetlands in eastern Antarctica. Atmos Environ 41: 4711–4722.
43. Bossio DA, Horwath WR, Mutters RG, Van Kessel C (1999) Methane pool and flux dynamics in a rice field following straw incorporation. Soil Biol Biochem 31: 1313–1322.
44. Whalen SC, Reeburgh WS (1996) Moisture and temperature sensitivity of CH_4 oxidation in boreal soils. Soil Biol Biochem 28: 1271–1281.

45. Zona D, Janssens IA, Verlinden MS, Broeckx LS, Cools J, et al. (2011) Impact of extreme precipitation and water table change on N$_2$O fluxed in a bio-energy poplar plantation. Biogeosci Discuss 8: 2057–2092.

46. Xu W, Liu G, Liu W (2002) Effects of precipitation and soil moisture on N$_2$O emissions from upland soils in Guizhou. Chin J Appl Ecol 13(1): 67–70. (in Chinese).

47. Hao X, Chang C, Carefoot JM, Janzen HH, Ellert BH (2001) Nitrous oxide emissions from an irrigated soil as affected by fertilizer and straw management. Nutr Cy Agroecosyst 60: 1–8.

48. Kay BD, VandenBygaart AJ (2002) Conservation tillage and depth stratification of porosity and soil organic matter. Soil Till Res 66: 107–118.

49. Weier KL, Doran JW, Power JF, Walters DT (1993) Denitrification and the dinitrogen/nitrous oxide ratio as affected by soil water, available carbon, and nitrate. Soil Sci Soc Am J 57: 66–72.

50. Steinbach HS, Alvarez R (2006) Changes in soil organic carbon contents and nitrous oxide emissions after introduction of no-till in Pampean agroecosystems. J Environ Qual 35: 3–13.

Comparison of the Rhizosphere Bacterial Communities of Zigongdongdou Soybean and a High-Methionine Transgenic Line of This Cultivar

Jingang Liang[1,9], Shi Sun[2,9], Jun Ji[1], Haiying Wu[3], Fang Meng[1], Mingrong Zhang[3], Xiaobo Zheng[1], Cunxiang Wu[2]*, Zhengguang Zhang[1]*

1 Department of Plant Pathology, College of Plant Protection, Nanjing Agricultural University, and Key Laboratory of Integrated Management of Crop Diseases and Pests, Ministry of Education, Nanjing, China, 2 The National Key Facility for Crop Gene Resources and Genetic Improvement (NFCRI), MOA Key Laboratory of Soybean Biology (Beijing), Institute of Crop Science, The Chinese Academy of Agricultural Sciences, Beijing, China, 3 Nanchong Academy of Agricultural Science, Nanchong, China

Abstract

Previous studies have shown that methionine from root exudates affects the rhizosphere bacterial population involved in soil nitrogen fixation. A transgenic line of Zigongdongdou soybean cultivar (ZD91) that expresses *Arabidopsis* cystathionine γ-synthase resulting in an increased methionine production was examined for its influence to the rhizosphere bacterial population. Using 16S rRNA gene-based pyrosequencing analysis of the V4 region and DNA extracted from bacterial consortia collected from the rhizosphere of soybean plants grown in an agricultural field at the pod-setting stage, we characterized the populational structure of the bacterial community involved. In total, 87,267 sequences (approximately 10,908 per sample) were analyzed. We found that Acidobacteria, Proteobacteria, Bacteroidetes, Actinobacteria, Chloroflexi, Planctomycetes, Gemmatimonadetes, Firmicutes, and Verrucomicrobia constitute the dominant taxonomic groups in either the ZD91 transgenic line or parental cultivar ZD, and that there was no statistically significant difference in the rhizosphere bacterial community structure between the two cultivars.

Editor: Raffaella Balestrini, Institute for Sustainable Plant Protection, C.N.R., Italy

Funding: This research was supported by Genetically Modified Organisms Breeding Major Projects of China (Grant No. 2014ZX08011-003) and Jiangsu Provincial Coordination Panel for Recruitment and Selection of Specially-Appointed Professors. The funders had no role in study design, data collection and analysis, decision to publish, or preparation of the manuscript.

* Email: zhgzhang@njau.edu.cn (ZZ); wucx@mail.caas.net.cn (CW)

9 These authors contributed equally to this work.

Introduction

The global commercial cultivation of transgenic crops has increased from 1.7 million hectares in 1996 to 170.3 million hectares in 2012 [1]. Modern agricultural biotechnology and genetic engineering have allowed the development of crops with improved properties, such as the transgenic high-methionine soybean line Zigongdongdou 91 (ZD91) [2]. Most studies to date have suggested that the release of transgenic plants results in only minor changes to the microbial community structure and that these minor changes are often transient and short lived [3–15].

Soil microorganisms are affected by soil characteristics, environmental conditions, and crop management strategies [16–19]. Organic compounds in root exudates and rhizosphere microbial communities perform fundamental processes that contribute to nutrient cycling and plant health and root growth, which are governed by complex interactions driven by various factors such as soil types and plant species [3,20–23]. The type and amount of root exudation may be an inherent property of a plant, and the rhizosphere microbial community is also expected to be unique [3,24–26]. Thus, an altered exudate composition may result in a corresponding shift in the rhizosphere microorganism

community [27–30]. According to Faragova et al. [31], genetically modified plants release root exudates different from those of non-transgenic counterparts, therefore affecting the microbial community of the rhizosphere. This occurs even if the genetically modified plant is of different cultivars of the same plant species [32–34]. It was reported that methionine has the effect of inhibiting nitrification in soil, and transgenic alfalfa that produces high levels of methionine has a profound effect on bacteria population involved in the nitrogen cycle of the soil [31,35]. In a recent comparative study that employed Polymerase Chain Reaction-Denaturing Gradient Gel Electrophoresis (PCR-DGGE) technology to examine the bacterial communities in the rhizosphere of a transgenic high-methionine soybean, we found that the growth stage was the main factor affecting the community structure of the soil bacteria [36]. This finding is consistent with that of van Overbeek and van Elsas [25], Jin et al. [37] and Inceoglu et al. [38], in which bacterial community structures vary per growth stage changes. In addition, a study of rape (*Brassica napus*) indicated that the rhizosphere bacterial composition of a transgenic cultivar could be distinguished from that of its non-transgenic cultivar [39]. Conversely, in a study of potatoes, the microbial community structure in the rhizosphere was not significantly affected by genetic engineering [40,41].

The biological nitrogen cycle is one of the most important nutrient cycles in the terrestrial ecosystem. It includes four major processes: nitrogen fixation, mineralization (decay), nitrification and denitrification. Studies have shown that there are three functional genes: *nifH*, *amoA*, and *nosZ*, which encode the key enzymes involved in nitrogen fixation, ammonia oxidization and complete denitrification, respectively. Microorganisms capable of nitrogen fixation are called diazotrophs whose composition is determined by associated plant species. In addition, agricultural practices can also influence soil diazotrophs [42–45]. There are two types of diazotrophs: free-living and symbiotic [46,47]. The Rhizobia, *Frankia*, and Cyanobacteria are three important symbiotic diazotrophs. Rhizobia make up a paraphyletic group that falls into two classes in the phylum Proteobacteria: Alphaproteobacteria and Betaproteobacteria. Nitrification, which converts ammonium to nitrate, includes two steps: ammonia oxidation to nitrite, and nitrite oxidation to nitrate. The oxidation of ammonia to nitrite is performed by two groups of organisms, ammonia-oxidizing bacteria (AOB) and ammonia-oxidizing archaea (AOA) [48]. AOB is found among the Betaproteobacteria and Gammaproteobacteria [49,50]. Denitrification is a microbially-facilitated process of nitrate reduction. So far, more than 60 genera of denitrifying microorganisms have been identified [51].

High-throughput TAG-encoded FLX amplicon pyrosequencing (using 16S rRNA genes) is a great technical advance for microbial ecology studies [52]. This technique can produce ~400,000–1,000,000 reads in a single 4 h run, with an accuracy of approaching 99%, and the tag sequence contains adequate information for taxonomic assignment [53]. Several recent studies have used this technique with the 16S rRNA gene to determine the microbial diversity in various environmental samples [3,52–60]. Given the wealth of information available due to the deep sequencing of environmental DNA, it is possible to determine whether these lineages are ecologically coherent [3]. It also has been proposed that some taxa are extremely coherent ecologically. Such taxa can be said to follow an r- or K-type life strategy [61–63].

In the present study, we used 16S rRNA gene-based pyrosequencing to compare the community dynamics of bacteria inhabiting the rhizosphere of field-grown transgenic cultivar ZD91 during the pod-setting stage to those of the corresponding non-transgenic ZD plants. Our study showed a wide variety of rhizosphere bacterial population associated with these cultivars but there was no significant difference in the bacterial population between transgenic cultivar ZD91 and parental cultivar ZD.

Results

Quantification of *nifH*, *amoA*, and *nosZ* genes, and carbon, nitrogen concentrations in the rhizospheres

To examine the effect of transgenic soybean on the abundance of diazotrophs, ammonia-oxidizing bacteria, and denitrifiers, we used quantitative real-time PCR (qPCR) assays to quantify *nifH*, *amoA*, and *nosZ* genes to inform nitrogen-cycling related bacteria collected from the rhizosphere soil. The abundance of *nifH*, *amoA*, and *nosZ* genes between ZD and ZD91 showed no significant differences (Table 1).

We also estimated whether there are any differences in carbon and nitrogen levels in the rhizosphere of ZD and ZD91 and found no significant differences (Table 1).

Amino acids in root exudates

Amino acids included aspartic acid, threonine, serine, glutamic acid, glycine, alanine, valine, methionine, isoleucine, leucine, tyrosine, phenylalanine, lysine and histidine in root exudates were determined. However, there were no differences in amino acids contents between ZD and ZD91 (Table 2).

Pyrosequencing analysis of the bacterial communities of the rhizospheres

We performed 16S rRNA gene-based pyrosequencing analysis of the V4 region to characterize the bacterial community composition in the rhizospheres surrounding ZD and ZD91 cultivars. A total of 87,267 high-quality sequences was obtained, with an average of 10,908 sequences per sample. The results were summarized in Table 3 with detailed sample characteristics shown in Table S1. We used the grouping of cultivars based on the number of replicates (ZD, four replicates: ZD_1, ZD_2, ZD_3, and ZD_4; ZD91, four replicates: ZD91_1, ZD91_2, ZD91_3, and ZD91_4).

The rarefaction curves tended towards the saturation plateau while Good's coverage estimations revealed that 78–84% of the species were from eight samples, indicating that the sequence coverage was sufficient to capture the diversity of the bacterial communities in the samples (Fig. 1). The estimators of community richness (Chao and ACE) and diversity (Shannon and Simpson) were shown in Table 3, and there were no significant differences in these indices between ZD and ZD91. The null hypothesis was tested through AMOVA analysis (Table 4) and it did show that there were no significant differences between ZD and ZD91.

Taxonomic composition

Although probably present, this study did not attempt to characterize any Archaea. All sequences were classified from phylum to genus according to the RDP classifier using the default settings. 22 different phyla, including 1 archaeal phylum, were identified from the samples. The overall microbiota structure for ZD and ZD91 at the phylum level was shown in Figure 2 and Table S2. Acidobacteria, Proteobacteria, Bacteroidetes, Actinobacteria, Chloroflexi, Planctomycetes, Gemmatimonadetes, Firmicutes, and Verrucomicrobia accounted for >90% of the reads. Again, there were no observable differences among these categories between ZD and ZD91 at this taxonomic level (Fig. 3).

A total of 13,299 operational taxonomic units was found in the complete data set (Table S3). The ten most abundant OTUs within each sample were identified (Table S4). There were 12 dominant OTUs in total, including Gp4, Gp6, *Terrimonas, Sphingosinicella, Sphingomonas, Flavisolibacter, Sphingomonadaceae, Lysobacter, Chryseobacterium, Acinetobacter, Levilinea and Opitutus*. A total of 2047 common OTUs was determined in rhizosphere soil and there was no significant difference between ZD and ZD91 ($P<0.01$) (Fig. 4). The ten most abundant genera from each of the eight samples were classified as a percentage of the total sequences per sample (Table S5). This analysis showed that there were 17 dominant genera in total, including four members of the Acidobacteria, five of the Bacteroidetes, three Alphaproteobacteria, two Gammaproteobacteria, and one each of the Actinobacteria, Gemmatimonadetes and Verrucomicrobia.

To analyze whether the distribution of low-abundance genera differs between samples, sequences with abundances below 0.1% were analyzed. A total of 481 such low-abundance genera was found. Of the 481 genera, 214 (70.63%), 203 (68.35%), 200 (61.54%), and 219 (64.60%) were observed in ZD_1, ZD_2,

Table 1. Changes in abundance of *nifH*, *amoA*, and *nosZ* genes (Log_{10} copies/g soil), and carbon (C), nitrogen (N) contents in rhizosphere soil ($P<0.01$).

	ZD	ZD91
nifH	9.05±0.02 A	9.06±0.12 A
amoA	8.03±0.02 A	8.13±0.01 A
nosZ	7.70±0.08 A	7.90±0.02 A
Total C (%)	1.68±0.02 A	1.65±0.03 A
Total N (%)	0.06±0.01 A	0.07±0.01 A

ZD_3, and ZD_4, respectively. In comparison, 208 (67.97%), 205 (61.75%), 219 (68.65%), and 206 (65.19%) were detected in ZD91_1, ZD91_2, ZD91_3, and ZD91_4, respectively. There were no genera unique to cultivar ZD or cultivar ZD91. Rare genera (abundances below 0.1%) accounted for over 60% in all samples, and there was no significant difference between cultivar ZD (66.28±4.02%) and cultivar ZD91 (65.89±3.14%).

Root-colonizing pseudomonads are known to be involved in plant growth and health. These bacteria increase plant growth either directly through the production of phytohormones or other stimulants and increasing the bioavailability of nutrients in the soil, or indirectly by the suppression of plant diseases and the induction of systemic resistance in plants [10]. A total of 6 pseudomonad species was found in our samples and there was no significant difference between ZD (0.07±0.03%) and ZD91 (0.05±0.04%).

Nitrogen-fixing Bacteria

In the present study, *Frankia* was not found in any samples. Sequences related to Cyanobacteria, Alphaproteobacteria, and Betaproteobacteria were not significantly different between cultivar ZD (0.22±0.07%, 14.04±5.07%, and 1.23±0.30%) and cultivar ZD91 (0.14±0.08%, 13.56±4.88%, and 1.11±0.25%) samples.

Nitrifying bacteria

In the present study, there were 35 Gammaproteobacteria genera detected. There was no significant difference between the sequences for cultivar ZD (3.47±0.17%) and ZD91 (4.49±1.38%). Nitrifying bacteria, which oxidize nitrite, include *Nitrobacter*, *Nitrospina*, *Nitrococcus*, and *Nitrospira* [65], but only *Nitrospira* was detected in this study. In addition, there was no significant difference between the sequences for cultivar ZD (0.36±0.16%) and cultivar ZD91 (0.44±0.16%).

Denitrifying bacteria

In the present study, 36 such bacterial genera were found. The sequences related to denitrifying bacteria were not significantly different between cultivar ZD (3.13±1.31%) and cultivar ZD91 (2.80±0.28%).

Shared OTUs in ZD and ZD91 libraries

The bacterial OTUs in the cultivar ZD and cultivar ZD91 libraries were further investigated for shared OTUs. As shown in Table S6, ZD_1, ZD_2, ZD_3, and ZD_4 had 700 OTUs in common. A statistical analysis revealed that the OTUs common to the four libraries comprised 58.96%, 59.62%, 52.19%, and 46.82% of the reads in the ZD_1, ZD_2, ZD_3, and ZD_4 libraries, respectively. Acidobacteria, Proteobacteria, and Bacteroidetes comprised 531 of the shared OTUs (75.86% in propor-

Table 2. Amino acids secreted by soybean roots in the root box at pod-setting stage (μmol/ml) ($P<0.01$).

	ZD	ZD91
Asp	3.44±1.56 A	1.13±0.48 A
Thr	2.85±1.28 A	0.99±0.41 A
Ser	12.13±6.05 A	4.11±1.82 A
Glu	2.31±0.59 A	1.30±0.23 A
Gly	4.83±2.43 A	1.60±0.86 A
Ala	3.63±1.72 A	1.32±0.51 A
Val	2.14±0.83 A	1.01±0.25 A
Met	0.32±0.28 A	0.31±0.27 A
Ile	1.04±0.41 A	0.40±0.17 A
Leu	1.54±0.55 A	0.77±0.24 A
Tyr	2.03±0.58 A	1.33±0.20 A
Phe	1.22±0.47 A	0.65±0.17 A
Lys	1.03±0.57 A	0.33±0.20 A
His	2.12±1.36 A	0.68±0.46 A

Table 3. Pyrosequencing data summary.

Sample	Sequences	OTUs	ACE	Chao	Shannon	Simpson	Coverage
ZD	10534±1353A	3391±505A	8900±1019A	6381±780A	7.34±0.21A	0.002±0.001A	0.82±0.02A
ZD91	11283±1818A	3649±370A	9637±795A	6911±688A	7.47±0.17A	0.001±0.000A	0.82±0.03A

Values are the mean±standard deviation of four replicates. Within each vertical column, values followed by the same letter are not statistically different from each other ($P<0.01$). The number of OTUs, richness estimators Chao and ACE, diversity estimators Shannon and Simpson, and Good's coverage were calculated at 3% distance.

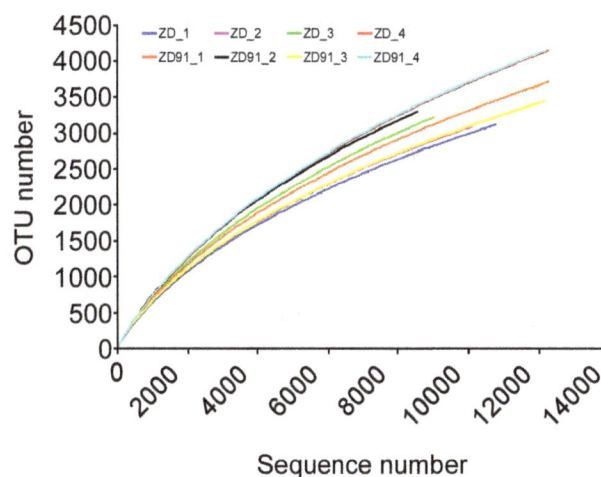

Figure 1. Rarefaction analysis. Rarefaction curves of OTUs clustered at 97% sequence identity across the samples. The sample labeled with ZD_1, ZD_2, ZD_3 and ZD_4 correspond to four replicates of cultivar ZD; ZD91_1, ZD91_2, ZD91_3 and ZD91_4 represent four replicates of transgenic cultivar ZD91.

tion). Further, the OTUs common to the four libraries of the transgenic cultivar comprised 54.04%, 49.01%, 52.17%, and 42.50% of the reads in the ZD91_1, ZD91_2, ZD91_3, and ZD91_4 libraries, respectively. Acidobacteria, Proteobacteria, and Bacteroidetes included 534 of the shared OTUs (76.29% in proportion).

Bacterial community dynamics

To examine the effect of cultivar on the total distribution of phyla and genera, we performed principal coordinate analysis (PCoA) of our data. The PCoA analysis of weighted UniFrac distances was used to calculate pairwise distances between the rhizosphere bacterial communities of the eight samples [64]. The first two axes of the PCoA represented 40.52% and 19.06% of the total variation, respectively (Fig. 5). AMOVA analysis (Table S7) was used to find out whether the separation of cultivar ZD and cultivar ZD91 in the PCoA was statistically significant. The result showed p-value was 0.859 indicating no significantly different between cultivar ZD and cultivar ZD91. Furthermore, we calculated ADONIS differences between cultivar ZD and cultivar ZD91 (Table 5), and found no significant difference between these two cultivars. Thus, the bacterial communities in the cultivars were not different from each other.

Discussion

In this study, we monitored the quantitative changes of the key genes involved in N-cycling as well as elements carbon, nitrogen concentrations in rhizosphere soil of ZD and ZD91, and analyzed the amino acids contents in the root exudates of different cultivars. We found that there were no significant differences between cultivar ZD and cultivar ZD91 in nitrogen-cycling genes and carbon, nitrogen concentrations. The results showed that the amino acids contents in the root exudates of ZD91 were lower than ZD, but they were not statistically significant.

We assessed the dynamics of the relative soil abundance of bacteria and the structure of soil bacterial communities as a function of cultivar type using the transgenic soybean cultivar ZD91 and its parental non-transgenic soybean cultivar ZD in a single experimental field. According to the pyrosequencing results,

Table 4. AMOVA analysis between ZD and ZD91.

	ZD91-ZD	Among	Within	Total
Based on OTU (Jclass distance coefficient)	SS	0.261484	1.603340	1.86483
	df	1.000000	6.000000	7.00000
	MS	0.261484	0.267224	
	Fs: 0.978522			
	p-value: 0.745			
Based on OTU (Thetayc distance coefficient)	SS	0.0321963	0.290100	0.322296
	df	1.0000000	6.000000	7.000000
	MS	0.0321963	0.048350	
	Fs: 0.665901			
	p-value: 0.565			
Based on weighted UniFrac distances	SS	0.0172751	0.1477470	0.165022
	df	1.0000000	6.0000000	7.000000
	MS	0.0172751	0.0246245	
	Fs: 0.701539			
	p-value: 0.861			

Experiment-wise error rate: 0.05.

genetic modification via the insertion and expression of *AtD-CGS* did not affect the composition of the bacterial community in the rhizosphere of soybean. However, the profiles of independent replicates differed in terms of the abundance of various phyla (Fig. 5 and Table S2), but this variation did not correlate with the experimental variables. All plots were on the same field, but due to the large size of the field, the replicates were located an average of 20 meters apart from each other. Thus, field heterogeneity was not considered when evaluating the impact of transgenic soybean on rhizosphere bacterial community. Furthermore, we analyzed the OTUs of high abundances and repeated well (occurred in more than three samples, defined as the common OTUs) in the soil.

Since the singletons and low abundance OTUs were detected randomly, and there was no significant difference within the treatment (among the 4 replicates) and between the treatments (Fig. 4).

Due to logistic and technical (e.g., low sequence numbers) constraints, studies of soil microbial diversity and function have focused mainly on the most dominant species [3]. Previous studies have demonstrated weaknesses in traditional molecular methods

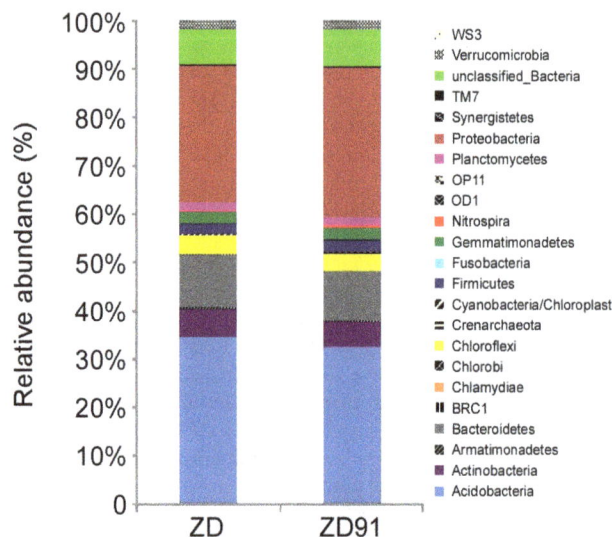

Figure 2. Bacterial composition at the phylum level. Relative read abundance of bacterial phyla within the communities. Sequences that could not be classified into any known group were assigned as unclassified_Bacteria.

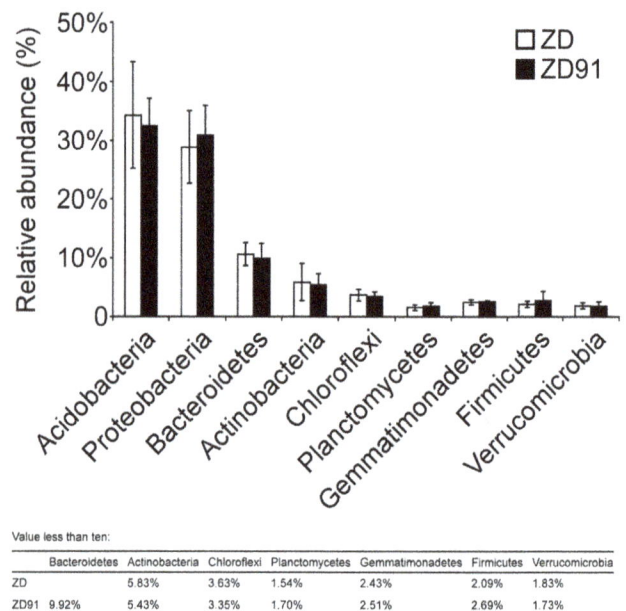

Figure 3. Relative abundance of main phyla in the rhizosphere of soybean cultivars. Error bars indicate standard errors. The p-values of the nine groups of bacteria, in order, were 0.734, 0.633, 0.662, 0.834, 0.687, 0.690, 0.751, 0.493, and 0.840 when testing for differences between cultivar ZD and transgenic cultivar ZD91.

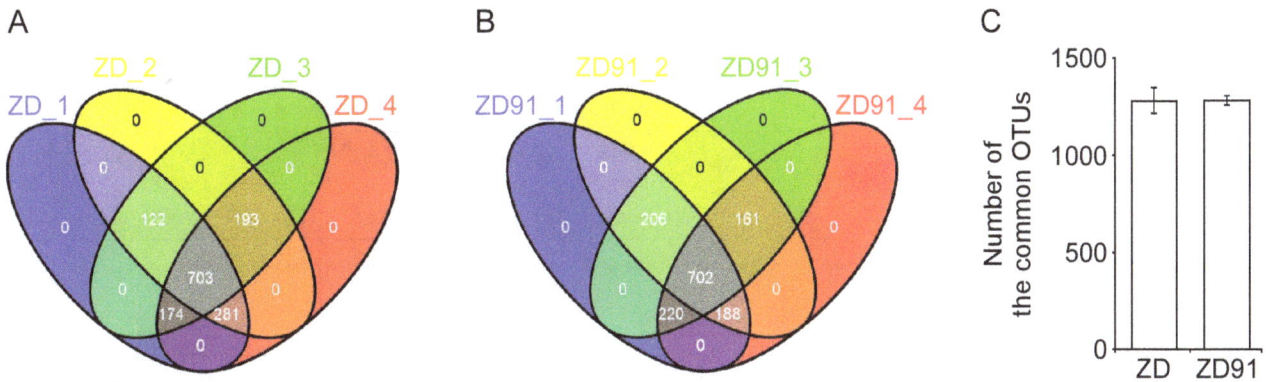

Figure 4. Common OTUs recovered from soil. (A) Venn diagram showing variable overlap between four replicates of ZD. (B) Venn diagram showing variable overlap between four replicates of ZD91. (C) Number of the common OTUs between ZD and ZD91.

(e.g., DGGE, RFLP, ARDRA, and SSCP) in terms of resolution; specifically, it is difficult to detect subtle changes in banding patterns or changes in closely related species, leading to the underestimation of bacterial diversity by one to two orders of magnitude [4,54,57,66]. Moreover, some rare species may have large effects on soil function, in spite of their low total biomass or rarity [67,68]. In this study, pyrosequencing provided a promising, fast, economical, and unique opportunity to access the less abundant bacterial taxa surrounding soybean plants, as well as the chance to compare their presence across cultivars in the pod-setting stage [54,55,57]. Similar to the results of Liu et al. [20] and Xu et al. [21,69], our sequencing results revealed that numerous bacterial phyla (e.g., Acidobacteria, Proteobacteria, Bacteroidetes, Actinobacteria, Chloroflexi, Planctomycetes, Gemmatimonadetes, Firmicutes, Nitrospirae, Verrucomicrobia, BRC1, Chlamydiae, Cyanobacteria, Fusobacteria, Chlorobi, OD1, OP11, Synergis-

tetes, TM7, Armatimonadetes, and WS3) commonly inhabit the soybean rhizosphere.

Moreover, dominant members of particular phyla or classes were used as indicators to reveal the prevailing ecological pattern (e.g., along the copiotroph/oligotroph scale) [3]. Members of the Betaproteobacteria and Bacteroidetes, as well as species belonging to the genus *Pseudomonas*, tend to be favored in r-selection/copiotrophic soils, which have higher carbon availability, while Acidobacteria are favored under K-selection/oligotrophic conditions [61–63]. Consequently, such groups may have higher abundances in soils with higher versus lower levels of easily available carbon, respectively. Furthermore, the ratio of Proteobacteria to Acidobacteria can be used as an indicator of the soil trophic level [70]. Although it is rather implausible that a whole phylum would have shared ecological characteristics [3], our indicators revealed variability (based on the ratio of Proteobacteria to Acidobacteria) occurs in the response of Proteobacteria versus Acidobacteria to shifting ecological conditions in the soil. The ratio is 0.16 in oligotrophic soil [71,72], 0.34 in low-input agricultural soil [73], 0.46 in low-nutrient systems [62], and 0.87 in high-input agricultural systems [74]. The values obtained in this study, 0.93 ± 0.46 in cultivar ZD, and 0.98 ± 0.31 in cultivar ZD91, indicating a high-nutrient system, but again there was no significant difference between the two cultivars.

The pyrosequencing protocol used in previous investigations and here carries certain limitations [3,52–60]. Any PCR primer set for the amplification of 16S rDNA genes may miss a considerable amount of the extant microbial diversity; thus, at present, no clear answer can be given with respect to the real extant community make-up [3]. Because PCR biases mostly affect the detection of rare sequences in a sample, it is important to view such data on the rare biosphere with proper cautions.

Of the ten most abundant bacterial OTUs in the samples, several were related to Acidobacteria. Of the ten most abundant bacterial genera, several were related to both Acidobacteria and Proteobacteria. The Acidobacteria are physiologically diverse and ubiquitous, especially in soil, but are underrepresented in culture [75–77]. Because most Acidobacteria have not been cultured, their ecology and metabolism are not well understood [76]. However, these bacteria may be important contributors to ecosystems because they are particularly abundant in soil [78]. The Proteobacteria are a major group of bacteria, and include many of the species responsible for nitrogen fixation.

We also monitored structural and quantitative changes in the bacterial communities of each soybean regarding nitrogen cycling.

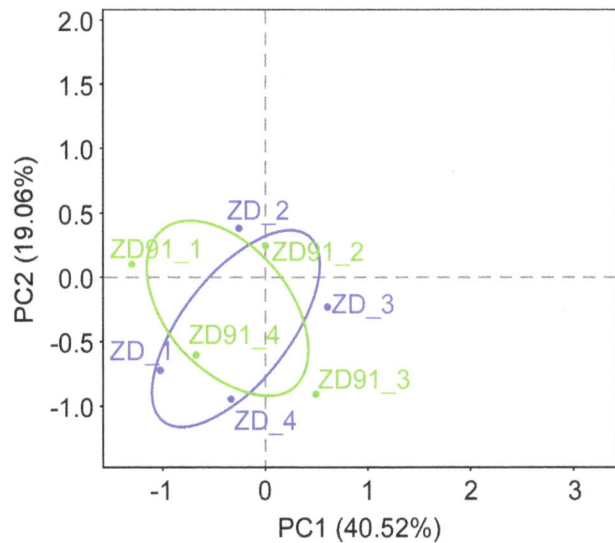

Figure 5. Principal coordinate analysis (PCoA) plot of samples using the weighted UniFrac distance metric. The variance explained by each principal coordinate axis is shown in parentheses. Datasets were subsampled to equal depth prior to the UniFrac distance computation. A circle was drawn with R around the samples of the same type (blue represents cultivar ZD, and green represents transgenic cultivar ZD91).

Table 5. ADONIS differences between ZD and ZD91.

	Df	SumsOfSqs	MeanSqs	F. Model	R^2	Pr(>F)
qiime.data$map [[opts$category]]	1	0.017275	0.017275	0.70124	0.10464	0.8492
Residuals	6	0.147814	0.024636		0.89536	
Total	7	0.165089			1.00000	

ADONIS differences is a non-parametric method to measure statistical significance of sample grouping. It was calculated based on weighted UniFrac distance metric.

The abundance of nitrogen-transforming bacteria were not significantly different between cultivars ZD and ZD91.

In conclusion, we found that 454-pyrosequencing can be a useful tool for risk assessment studies to analyze the immediate impact of plants on the diversity of soil bacteria in rhizospheres. Of all DNA reads, 7.56% belonged to unclassified bacteria, containing a suite of diverse taxa. To understand the ecological impact of unclassified sequences, they must be identified and ideally an organism should be cultured. This once again stresses that culture-independent techniques should go hand-in-hand with culture-dependent ones [3]. For ecological risk identification, however, the observation of shifts during one stage of plant development and in a one-year period is insufficient. The positive identification of risk requires that changes be demonstrated at all stages of soybean development and over an extended time frame. The fact that no effect between transgenic and non-transgenic plants was detected suggests only minor physiological changes caused by the insertion of *AtD-CGS* into the soybean genome.

Materials and Methods

Soybean Cultivars

Transgenic soybean cultivar (ZD91) contains the *Arabidopsis* cystathionine γ-synthase (*AtD-CGS*) gene which has been introduced artificially into the soybean cultivar Zigongdongdou (ZD) using *Agrobacterium*-mediated transformation, and which exhibits a high content of methionine in the seeds.

Field tests setup and sampling

An experimental field - Nanchong (30°48′N, 106°04′E), Sichuan Province, China, in which a completely randomized block design was set out in 2011, was used. The basic properties of the soil were 46.98 g/kg organic matter, 0.51 g/kg total nitrogen, 11.73 mg/kg available phosphorus, 220.40 mg/kg available potassium, and 52.81 mg/kg alkali-hydrolyzed nitrogen. The studied region was 2.8 acres. For each soybean cultivar, four replicate plots, which were randomly distributed over the field, were used. The field was under standard agricultural practice. The samples were collected at the pod-setting stage (80 days after seedling emergence) of the two soybean cultivars. Rhizosphere soil samples were collected as described previously [38]. Briefly, five plants were carefully sampled using five-point sampling method and the plants with adhering soil were immediately taken to the laboratory. The loosely adhering soil on the roots was shaken off, and the resulting roots (containing rhizosphere soil) were pooled per plot. Using the pooled sample, soil tightly adhering to the roots was brushed off and collected (constituting rhizosphere soil).

Ethics statements

This study was approved by the Ministry of Agriculture of the People's Republic of China and the genetically modified organisms safety team of Nanjing Agricultural University, China. The field studies did not involve endangered or protected species. The land was not privately owned or protected in any way.

Soil DNA extraction, PCR amplification and pyrosequencing

Soil DNA was extracted from rhizosphere soil samples by employing the FastDNA SPIN Kit for Soil (MP Biomedicals, USA) as recommended by the manufacturer. DNA concentrations were quantified by using a NanoDrop 1000 Spectrophotometer (Thermo Scientific, USA) according to the manufacturer's protocol. The 16S rDNA gene (V4 region) was amplified using the barcode primers 515F [79] and 926R [80]. The primers were designed as 5′-adapter+barcode+gene specific primer-3′ (Table S8). The PCR reaction mixture (20 μL) contained 2 μL 10-fold buffer, 1.6 μL dNTP mix (2.5 mM each), 1 μL BSA (20 mg/ml), 0.25 μL rTaq polymerase (2.5 U/μL), 0.2 μL of each of the primers (10 μM), 0.5–5 μL (1–50 ng) of isolated DNA as template. The mixtures were placed in a TaKaRa PCR Thermal Cycler Dice (TaKaRa, Japan) and thermal cycling was performed as follows: initial denaturation consisting of 5 min at 95°C; followed by 30 cycles consisting of 30 sec at 94°C, 30 sec at 57°C and 30 sec at 72°C; and final extension for 10 min at 72°C. Analysis of PCR products on 1% agarose gel revealed bands of the corresponding size. These bands were cut and purified using the QIAquick Gel Extraction Kit as recommended by the manufacturer (Qiagen, Netherlands), and then purified by Agencourt AMPure XP (Beckman Coulter, USA). Quantification of the purified PCR products was performed using the Quant-iT PicoGreen dsDNA Assay Kit (Invitrogen, Netherlands) and a TBS-380 Fluorometer (Turner Biosystems, USA) as recommended by the manufacturer. The PCR products from different samples were then mixed in equal ratios (7.2 ng per sample) for pyrosequencing with a Roche Genome Sequencer GS-FLX Titanium platform [56].

Real-time PCR

The abundance of *nifH*, *amoA*, and *nosZ* genes in all samples were quantified using real-time PCR. Quantitative real-time PCR was performed according to the methods from previous studies: *nifH* (as a measure of N-fixing bacteria) used primers nifH-F (5′-AAA GGY GGW ATC GGY AAR TCC ACC AC-3′) and nifH-R (5′-TTG TTS GCS GCR TAC ATS GCC ATC AT-3′) [81]; *amoA* (as a measure of ammonia-oxidizing bacteria) used primers amoA-1F (5′-GGG GTT TCT ACT GGT GGT-3′) and amoA-2R (5′-CCC CTC KGS AAA GCC TTC TTC-3′) [82]; and *nosZ* (as a measure of denitrification bacteria) used primers nosZ-F (5′-CGY TGT TCM TCG ACA GCC AG-3′) and nosZ 1622R (5′-CGS ACC TTS TTG CCS TYG CG-3′) [83]. Purified PCR products from a common DNA mixture (equal amounts of DNA from all samples) were used to prepare sample-derived quantification standards as previously described [84]. In comparison to using a clone (plasmid) as standard, this method avoids the difference of PCR amplification efficiency between standards and samples caused by the different sequence composition in the PCR templates (single sequence in a plasmid for the standard vs. mixture of thousands of sequences in a soil sample) [83]. The copy number of gene in each sample was calculated by http://cels.uri.edu/gsc/cndna.html and the data were Log_{10} transformed [84]. The 20 μL reaction mixture contained 10 μL SYBR *Premix Ex Taq* (Tli RNaseH Plus) (2×) (TaKaRa, Japan), 0.4 μL of each primer (10 μM), 0.4 μL ROX Reference Dye II (50×) (TaKaRa, Japan), 2 μL template, and 6.8 μL dH$_2$O (sterile distilled water). Real-time amplification was performed in an Applied Biosystems 7500 Fast Real-Time PCR System using the following program: 95°C for 30 s; 40 cycles of 95°C for 5 s, (58°C for *nifH* gene; 57°C for *amoA* gene; 64°C for *nosZ* gene) for 31 s, 72°C for 34 s. A dissociation step was added at the end of the qPCR to assess amplification quality. The R^2 of all these standard curves were higher than 0.99. All quantitative PCR reactions were performed in triplicate.

Root box design and determination of amino acids secreted by roots

The root box used in this study was designed as reported previously [85]. There were three replicates of root box for each genotype. The lower chamber of the root box was filled with 500 mL of sterile ultrapure water, which was used to collect the amino acids exuded by the roots during a 72-h period at the pod-setting stage. Samples of root exudate were collected, filtered through a 0.45 μm membrane, freeze-dried, and redissolved in 1 mL of deionized water. Amino acids contents in the root exudates were analyzed by L-8900 Amino Acid Analyzer (Hitachi, Japan).

Chemical analysis of carbon and nitrogen contents in rhizosphere soil

The concentrations of carbon and nitrogen in rhizosphere soil were determined by Vario MICRO cube (Elementar, Germany). The detection was based on JY/T 017–1996 General rules for elemental analyzer which was approved by Ministry of Education of the People's Republic of China.

Bioinformatic analysis

All sequences generated in this study can be downloaded from NCBI Sequence Read Archive, accession number: SRP026381. On the basis of several previous reports describing sources of errors in 454 sequencing runs, the valid reads should comply with the following rules: the sequences with low quality (length< 200 bp, with ambiguous base 'N', and average base quality score< 20) were removed; the sequences that matched the primer and one of the used barcode sequences and had at least an 80% match to a previously determined 16S rRNA gene sequence were retained [54,56,83,86]. The sequences were used for chimera check using UCHIME method in Mothur [87], and chimeric sequences were removed.

The high-quality sequences were assigned to samples according to barcodes. Sequences were aligned in accordance with RDP (Ribosomal Database Project) alignment [88] and clustered into operational taxonomic units (OTUs). OTUs were defined using a threshold of 97% identity, within and between groups, which was a criterion for species level delineation in previous studies [56]. OTUs that reached 97% similarity level were used for diversity (Shannon and Simpson), richness (Chao and ACE), Good's coverage, and Rarefaction curve analysis by using Mothur [87].

The OTUs which occurred in more than three samples from either ZD or ZD91 were defined as the common OTUs [89]. Taxonomic assignments of OTUs exhibiting 97% similarity were performed by using Mothur in accordance with RDP Classifier at 80% confidence threshold. The null hypothesis was tested using OTU and weighted UniFrac distances. The study objective was to test whether the transgenic soybean had an effect on rhizosphere bacterial communities. AMOVA analysis was done by using Mothur. ADONIS differences were calculated by using R.

To compare bacterial community structures across all samples based on the relative abundance of bacterial phyla, Principal coordinate analysis (PCoA) was performed by using Qiime [90]. AMOVA analysis was used to find out whether the separation of ZD and ZD91 in the PCoA was statistically significant.

Statistical analysis

One-way analysis of variance and Duncan pair-wise comparisons ($P<0.01$) were used to determine the minimum significant difference between soybean cultivars (rhizosphere soils) by employing SPSS version 17.0 for Windows.

Supporting Information

Table S1 Pyrosequencing data and estimator index of each sample. The number of OTUs, richness estimator Chao and ACE, diversity estimator Shannon and Simpson, and Good's coverage were calculated at 3% distance.
(DOC)

Table S2 Bacterial composition of the communities at the phylum level. The number in the table (except the first line and column) represent pyrosequencing reads.
(DOC)

Table S3 Different OTUs for each sample.
(XLS)

Table S4 Classification of the 10 most abundant bacterial OTUs in the eight samples. Relative abundance (%, a percentage of the total sequences per sample) of each OTU is under the genus name/group number. OTUs were identified using 97% cutoffs. All Gp4 OTU classify in the same class 'Acidobacteria_Gp4' and all Gp6 OTU classify in the same class 'Acidobacteria_Gp6'.
(DOC)

Table S5 The ten most abundant genera found in the eight samples. Relative abundance (%, a percentage of the total sequences per sample) of each genus is included in parentheses. All Gp6 classify in the same class 'Acidobacteria_Gp6', all Gp4 classify in the same class 'Acidobacteria_Gp4' and all Gp3 classify in the same class 'Acidobacteria_Gp3'.
(DOC)

Table S6 Shared OTUs in the samples. (A) Shared OTUs in the ZD libraries. (B) Shared OTUs in the ZD91 libraries.
(DOC)

Table S7 AMOVA analysis between ZD and ZD91 in PCoA. AMOVA analysis in PCoA (classfily method: UPGMA). AMOVA analysis was used to find out whether the separation of ZD and ZD91 in the PCoA is statistically significant.
(DOC)

Table S8 Primers with tags and adapters used in pyrosequencing.
(DOC)

Acknowledgments

We are grateful to Ping Wang of LSUHSC (USA), Ruifu Zhang and Zhongli Cui of Nanjing Agricultural University (China) for their helpful suggestions.

Author Contributions

Conceived and designed the experiments: ZZ CW SS XZ. Performed the experiments: JL JJ FM. Analyzed the data: JL. Contributed reagents/materials/analysis tools: CW SS HW MZ. Wrote the paper: JL ZZ.

References

1. James C (2012) Global Status of commercialized biotech/GM crops: 2012. Ithaca: The International Service for the Acquisition of Agri-biotech Applications. 1 p.

2. Song S, Hou W, Godo I, Wu C, Yu Y, et al. (2013) Soybean seeds expressing feedback-insensitive cystathionine gamma-synthase exhibit a higher content of methionine. J Exp Bot 64: 1917–1926.

3. Inceoglu O, Al-Soud WA, Salles JF, Semenov AV, van Elsas JD (2011) Comparative analysis of bacterial communities in a potato field as determined by pyrosequencing. PLoS One 6: e23321.

4. Liu B, Zeng Q, Yan FM, Xu HG, Xu CR (2005) Effects of transgenic plants on soil microorganisms. Plant Soil 271: 1–13.

5. Milling A, Smalla K, Maidl FX, Schloter M, Munch JC (2004) Effects of transgenic potatoes with an altered starch composition on the diversity of soil and rhizosphere bacteria and fungi. Plant Soil 266: 23–39.

6. Li XG, Zhang TL, Wang XX, Hua K, Zhao L, et al. (2013) The composition of root exudates from two different resistant peanut cultivars and their effects on the growth of soil-borne pathogen. Inter J Biol Sci 9: 164–173.

7. Hur M, Kim Y, Song HR, Kim JM, Choi YI, et al. (2011) Effect of genetically modified poplars on soil microbial communities during the phytoremediation of waste mine tailings. Appl Environ Microbiol 77: 7611–7619.

8. Hannula SE, de Boer W, van Veen J (2012) A 3-year study reveals that plant growth stage, season and field site affect soil fungal communities while cultivar and GM-trait have minor effects. PLoS One 7: e33819.

9. Duc C, Nentwig W, Lindfeld A (2011) No adverse effect of genetically modified antifungal wheat on decomposition dynamics and the soil fauna community - A Field Study. PLoS One 6: e25014.

10. Meyer JB, Song-Wilson Y, Foetzki A, Luginbuhl C, Winzeler M, et al. (2013) Does wheat genetically modified for disease resistance affect root-colonizing Pseudomonads and Arbuscular Mycorrhizal Fungi? PLoS One 8: e53825.

11. Di Giovanni GD, Watrud LS, Seidler RJ, Widmer F (1999) Comparison of parental and transgenic alfalfa rhizosphere bacterial communities using Biolog GN metabolic fingerprinting and enterobacterial repetitive intergenic consensus sequence-PCR (ERIC-PCR). Microb Ecol 37: 129–139.

12. Rasche F, Hodl V, Poll C, Kandeler E, Gerzabek MH, et al. (2006) Rhizosphere bacteria affected by transgenic potatoes with antibacterial activities compared with the effects of soil, wild-type potatoes, vegetation stage and pathogen exposure. FEMS Microbiol Ecol 56: 219–235.

13. Heuer H, Kroppenstedt RM, Lottmann J, Berg G, Smalla K (2002) Effects of T4 lysozyme release from transgenic potato roots on bacterial rhizosphere communities are negligible relative to natural factors. Appl Environ Microbiol 68: 1325–1335.

14. Duke SO, Lydon J, Koskinen WC, Moorman TB, Chaney RL, et al. (2012) Glyphosate effects on plant mineral nutrition, crop rhizosphere microbiota, and plant disease in glyphosate-resistant crops. J Agric Food Chem 60: 10375–10397.

15. Demaneche S, Sanguin H, Pote J, Navarro E, Bernillon D, et al. (2008) Antibiotic-resistant soil bacteria in transgenic plant fields. Proc Natl Acad Sci U S A 105: 3957–3962.

16. Govaerts B, Mezzalama M, Unno Y, Sayre KD, Luna-Guido M, et al. (2007) Influence of tillage, residue management, and crop rotation on soil microbial biomass and catabolic diversity. Appl Soil Ecol 37: 18–30.

17. McLaughlin A, Mineau P (1995) The impact of agricultural practices on biodiversity. Agric, Ecosyst & Environt 55: 201–212.

18. Garbeva P, Postma J, Van Veen J, Van Elsas J (2006) Effect of above-ground plant species on soil microbial community structure and its impact on suppression of Rhizoctonia solani AG3. Environ Microbiol 8: 233–246.

19. Salles J, Van Elsas J, Van Veen J (2006) Effect of agricultural management regime on Burkholderia community structure in soil. Micro Ecol 52: 267–279.

20. Liu J, Wang G, Jin J, Liu J, Liu X (2011) Effects of different concentrations of phosphorus on microbial communities in soybean rhizosphere grown in two types of soils. Ann Microbiol 61: 525–534.

21. Xu Y, Wang G, Jin J, Liu J, Zhang Q, et al. (2009) Bacterial communities in soybean rhizosphere in response to soil type, soybean genotype, and their growth stage. Soil Biol Biochem 41: 919–925.

22. Wang G-H, Jin J, Xu M-N, Pan X-W, Tang C (2007) Inoculation with phosphate-solubilizing fungi diversifies the bacterial community in rhizospheres of maize and soybean. Pedosphere 17: 191–199.

23. Albareda M, Dardanelli MS, Sousa C, Megías M, Temprano F, et al. (2006) Factors affecting the attachment of rhizospheric bacteria to bean and soybean roots. FEMS Microbiol Lett 259: 67–73.

24. Garbeva P, van Veen JA, van Elsas JD (2004) Microbial diversity in soil: selection microbial populations by plant and soil type and implications for disease suppressiveness. Annu Rev Phytopathol 42: 243–270.

25. van Overbeek L, van Elsas JD (2008) Effects of plant genotype and growth stage on the structure of bacterial communities associated with potato (Solanum tuberosum L.). FEMS Microbiol Ecol 64: 283–296.

26. Germida JJ, Siciliano SD (2001) Taxonomic diversity of bacteria associated with the roots of modern, recent and ancient wheat cultivars. Biol Fertility Soils 33: 410–415.

27. Duineveld BM, Rosado AS, van Elsas JD, van Veen JA (1998) Analysis of the dynamics of bacterial communities in the rhizosphere of the chrysanthemum via denaturing gradient gel electrophoresis and substrate utilization patterns. Appl Environ Microbiol 64: 4950–4957.

28. Gelsomino A, Keijzer-Wolters AC, Cacco G, van Elsas JD (1999) Assessment of bacterial community structure in soil by polymerase chain reaction and denaturing gradient gel electrophoresis. J Microbiol Methods 38: 1–15.

29. Schmalenberger A, Tebbe CC (2002) Bacterial community composition in the rhizosphere of a transgenic, herbicide-resistant maize (Zea mays) and comparison to its non-transgenic cultivar Bosphore. FEMS Microbiol Ecol 40: 29–37.

30. Nannipieri P, Ascher J, Ceccherini M, Landi L, Pietramellara G, et al. (2008) Effects of root exudates in microbial diversity and activity in rhizosphere soils. Molecular mechanisms of plant and microbe coexistence: Springer. 339–365.

31. Faragova N, Faragó J, Drabekova J (2005) Evaluation of abundance of aerobic bacteria in the rhizosphere of transgenic and non-transgenic alfalfa lines. Folia Microbiol 50: 509–514.

32. da Silva KRA, Salles JF, Seldin L, van Elsas JD (2003) Application of a novel Paenibacillus-specific PCR-DGGE method and sequence analysis to assess the diversity of Paenibacillus spp. in the maize rhizosphere. J Microbiol Methods 54: 213–231.

33. Miller HJ, Henken G, Vanveen JA (1989) Variation and composition of bacterial-populations in the rhizospheres of maize, wheat, and grass cultivars. Can J Microbiol 35: 656–660.

34. Rengel Z, Ross G, Hirsch P (1998) Plant genotype and micronutrient status influence colonization of wheat roots by soil bacteria. J Plant Nutrition 21: 99–113.

35. Vasantharajan V, Bhat J (1968) Interrelations of micro-organisms and mulberry. Plant Soil 29: 156–169.

36. Pu CL, Liang JG, Gao JY, Wu CX, Zhang MR, et al. (2012) Effects of high producing methionine soy-bean transferred cystathionine γ-synthase gene on community structure of bacteria in soil. J Nanjing Agric Univ 35: 8–14.

37. Jin J, Wang GH, Liu XB, Liu JD, Chen XL, et al. (2009) Temporal and spatial dynamics of bacterial community in the rhizosphere of soybean genotypes grown in a black Soil. Pedosphere 19: 808–816.

38. Inceoglu O, Salles JF, van Overbeek L, van Elsas JD (2010) Effects of plant genotype and growth stage on the Betaproteobacterial communities associated with different potato cultivars in two fields. Applied and Environ Microbiol 76: 3675–3684.

39. Siciliano SD, Germida JJ (1999) Taxonomic diversity of bacteria associated with the roots of field-grown transgenic Brassica napus cv. Quest, compared to the non-transgenic B-napus cv. Excel and B. rapa cv. Parkland. FEMS Microbiol Ecol 29: 263–272.

40. Lukow T, Dunfield PF, Liesack W (2000) Use of the T-RFLP technique to assess spatial and temporal changes in the bacterial community structure within an agricultural soil planted with transgenic and non-transgenic potato plants. FEMS Microbiol Ecol 32: 241–247.

41. Lottmann J, Berg G (2001) Phenotypic and genotypic characterization of antagonistic bacteria associated with roots of transgenic and non-transgenic potato plants. Microbiol Res 156: 75–82.

42. Tan XY, Hurek T, Reinhold-Hurek B (2003) Effect of N-fertilization, plant genotype and environmental conditions on nifH gene pools in roots of rice. Environ Microbiol 5: 1009–1015.

43. Shen JP, Zhang LM, Zhu YG, Zhang JB, He JZ (2008) Abundance and composition of ammonia-oxidizing bacteria and ammonia-oxidizing archaea communities of an alkaline sandy loam. Environ Microbiol 10: 1601–1611.

44. Briones AM, Okabe S, Umemiya Y, Ramsing NB, Reichardt W, et al. (2002) Influence of different cultivars on populations of ammonia-oxidizing bacteria in the root environment of rice. Appl Environ Microbiol 68: 3067–3075.

45. Wang YA, Ke XB, Wu LQ, Lu YH (2009) Community composition of ammonia-oxidizing bacteria and archaea in rice field soil as affected by nitrogen fertilization. Syst Appl Microbiol 32: 27–36.

46. Cleveland CC, Townsend AR, Schimel DS, Fisher H, Howarth RW, et al. (1999) Global patterns of terrestrial biological nitrogen (N-2) fixation in natural ecosystems. Global Biogeochem Cycles 13: 623–645.

47. Hsu SF, Buckley DH (2009) Evidence for the functional significance of diazotroph community structure in soil. Isme J 3: 124–136.

48. Treusch AH, Leininger S, Kletzin A, Schuster SC, Klenk HP, et al. (2005) Novel genes for nitrite reductase and Amo-related proteins indicate a role of

uncultivated mesophilic crenarchaeota in nitrogen cycling. Environ Microbiol 7: 1985–1995.

49. Purkhold U, Pommerening-Roser A, Juretschko S, Schmid MC, Koops HP, et al. (2000) Phylogeny of all recognized species of ammonia oxidizers based on comparative 16S rRNA and amoA sequence analysis: Implications for molecular diversity surveys. Appl Environ Microbiol 66: 5368–5382.

50. Kowalchuk GA, Stephen JR (2001) Ammonia-oxidizing bacteria: A model for molecular microbial ecology. Annu Rev Microbiol 55: 485–529.

51. Philippot L, Hallin S, Schloter M (2007) Ecology of denitrifying prokaryotes in agricultural soil. Adv Agronomy 96: 249–305.

52. Zhou HW, Li DF, Tam NFY, Jiang XT, Zhang H, et al. (2011) BIPES, a cost-effective high-throughput method for assessing microbial diversity. Isme J 5: 741–749.

53. Qiu MH, Zhang RF, Xue C, Zhang SS, Li SQ, et al. (2012) Application of bio-organic fertilizer can control Fusarium wilt of cucumber plants by regulating microbial community of rhizosphere soil. Biol Fertility Soils 48: 807–816.

54. Nacke H, Thurmer A, Wollherr A, Will C, Hodac L, et al. (2011) Pyrosequencing-based assessment of bacterial community structure along different management types in German forest and grassland soils. PLoS One 6: e17000.

55. Hirsch J, Strohmeier S, Pfannkuchen M, Reineke A (2012) Assessment of bacterial endosymbiont diversity in Otiorhynchus spp. (Coleoptera: Curculionidae) larvae using a multitag 454 pyrosequencing approach. Bmc Microbiol 12: S6.

56. Zhang CH, Zhang MH, Wang SY, Han RJ, Cao YF, et al. (2010) Interactions between gut microbiota, host genetics and diet relevant to development of metabolic syndromes in mice (vol 4, pg 232, 2010). Isme J 4: 312–313.

57. Lee OO, Wang Y, Yang JK, Lafi FF, Al-Suwailem A, et al. (2011) Pyrosequencing reveals highly diverse and species-specific microbial communities in sponges from the Red Sea. Isme J 5: 650–664.

58. Yu K, Zhang T (2012) Metagenomic and metatranscriptomic analysis of microbial community dtructure and gene expression of activated sludge. PLoS One 7: e38183.

59. Zhang T, Zhang XX, Ye L (2011) Plasmid metagenome reveals high levels of antibiotic resistance genes and mobile genetic elements in activated sludge. PLoS One 6: e26041.

60. Chen WG, Liu FL, Ling ZX, Tong XJ, Xiang C (2012) Human intestinal lumen and mucosa-associated microbiota in patients with colorectal cancer. PLoS One 7: e39743.

61. Fierer N, Bradford MA, Jackson RB (2007) Toward an ecological classification of soil bacteria. Ecology 88: 1354–1364.

62. Smit E, Leeflang P, Gommans S, van den Broek J, van Mil S, et al. (2001) Diversity and seasonal fluctuations of the dominant members of the bacterial soil community in a wheat field as determined by cultivation and molecular methods. Appl Environ Microbiol 67: 2284–2291.

63. Castro HF, Classen AT, Austin EE, Norby RJ, Schadt CW (2010) Soil microbial community responses to multiple experimental climate change drivers. Appl Environ Microbiol 76: 999–1007.

64. Yashiro E, McManus PS (2012) Effect of streptomycin treatment on bacterial community structure in the apple phyllosphere. PLoS One 7: e37131.

65. Gerardi MH (2002) Nitrification and denitrification in the activated sludge process. New York: John Wiley and Sons, Inc. 43–55.

66. Sogin ML, Morrison HG, Huber JA, Welch DM, Huse SM, et al. (2006) Microbial diversity in the deep sea and the underexplored "rare biosphere". Proc Natl Acad Sci U S A 103: 12115–12120.

67. Power ME, Tilman D, Estes JA, Menge BA, Bond WJ, et al. (1996) Challenges in the quest for keystones. BioSci 46: 609–620.

68. Hooper D, Chapin Iii F, Ewel J, Hector A, Inchausti P, et al. (2005) Effects of biodiversity on ecosystem functioning: a consensus of current knowledge. Ecol Monogr 75: 3–35.

69. Xu YX, Wang GH, Jin J, Liu JD, Zhang QY, et al. (2007) Comparison of bacterial community in rhizosphere of soybean by anaylysis of 16S rDNA obtained directly from soil and culture plate. Soybean Sci 26: 907–913.

70. Hartman WH, Richardson CJ, Vilgalys R, Bruland GL (2008) Environmental and anthropogenic controls over bacterial communities in wetland soils. Proc Natl Acad Sci U S A 105: 17842–17847.

71. Dunbar J, Takala S, Barns SM, Davis JA, Kuske CR (1999) Levels of bacterial community diversity in four arid soils compared by cultivation and 16S rRNA gene cloning. Appl Environ Microbiol 65: 1662–1669.

72. Kuske CR, Barns SM, Busch JD (1997) Diverse uncultivated bacterial groups from soils of the arid southwestern United States that are present in many geographic regions. Appl Environ Microbiol 63: 3614–3621.

73. Borneman J, Skroch PW, O'Sullivan KM, Palus JA, Rumjanek NG, et al. (1996) Molecular microbial diversity of an agricultural soil in Wisconsin. Appl Environ Microbiol 62: 1935–1943.

74. McCaig AE, Glover LA, Prosser JI (1999) Molecular analysis of bacterial community structure and diversity in unimproved and improved upland grass pastures. Appl Environ Microbiol 65: 1721–1730.

75. Barns SM, Cain EC, Sommerville L, Kuske CR (2007) Acidobacteria phylum sequences in uranium-contaminated subsurface sediments greatly expand the known diversity within the phylum. Appl Environ Microbiol 73: 3113–3116.

76. Quaiser A, Ochsenreiter T, Lanz C, Schuster SC, Treusch AH, et al. (2003) Acidobacteria form a coherent but highly diverse group within the bacterial domain: evidence from environmental genomics. Mol Microbiol 50: 563–575.

77. Rappé MS, Giovannoni SJ (2003) The uncultured microbial majority. Annu Rev Microbiol 57: 369–394.

78. Eichorst SA, Breznak JA, Schmidt TM (2007) Isolation and characterization of soil bacteria that define Terriglobus gen. nov., in the phylum Acidobacteria. Appl Environ Microbiol 73: 2708–2717.

79. Turner S, Pryer KM, Miao VP, Palmer JD (1999) Investigating deep phylogenetic relationships among Cyanobacteria and plastids by small subunit rRNA sequence analysis1. J Eukaryot Microbiol 46: 327–338.

80. Liu Z, Lozupone C, Hamady M, Bushman FD, Knight R (2007) Short pyrosequencing reads suffice for accurate microbial community analysis. Nucleic Acids Res 35: e120.

81. Rösch C, Mergel A, Bothe H (2002) Biodiversity of denitrifying and dinitrogen-fixing bacteria in an acid forest soil. Appl Environ Microbiol 68: 3818–3829.

82. Rotthauwe J-H, Witzel K-P, Liesack W (1997) The ammonia monooxygenase structural gene amoA as a functional marker: molecular fine-scale analysis of natural ammonia-oxidizing populations. Appl Environ Microbiol 63: 4704–4712.

83. Mao Y, Yannarell AC, Mackie RI (2011) Changes in N-transforming archaea and bacteria in soil during the establishment of bioenergy crops. PLoS One 6: e24750.

84. Chen J, Yu Z, Michel FC, Wittum T, Morrison M (2007) Development and application of real-time PCR assays for quantification of erm genes conferring resistance to macrolides-lincosamides-streptogramin B in livestock manure and manure management systems. Appl Environ Microbiol 73: 4407–4416.

85. Yang T, Liu G, Li Y, Zhu S, Zou A, et al. (2012) Rhizosphere microbial communities and organic acids secreted by aluminum-tolerant and aluminum-sensitive soybean in acid soil. Biol Fertility Soils 48: 97–108.

86. Wu S, Wang G, Angert ER, Wang W, Li W, et al. (2012) Composition, diversity, and origin of the bacterial community in grass carp intestine. PLoS One 7: e30440.

87. Schloss PD, Westcott SL, Ryabin T, Hall JR, Hartmann M, et al. (2009) Introducing mothur: open-source, platform-independent, community-supported software for describing and comparing microbial communities. Appl Environ Microbiol 75: 7537–7541.

88. Wang Q, Garrity GM, Tiedje JM, Cole JR (2007) Naive Bayesian classifier for rapid assignment of rRNA sequences into the new bacterial taxonomy. Appl Environ Microbiol 73: 5261–5267.

89. Yang W, Zheng Y, Gao C, He X, Ding Q, et al. (2013) The arbuscular mycorrhizal fungal community response to warming and grazing differs between soil and roots on the Qinghai-Tibetan Plateau. PLoS One 8: e76447.

90. Caporaso JG, Kuczynski J, Stombaugh J, Bittinger K, Bushman FD, et al. (2010) QIIME allows analysis of high-throughput community sequencing data. Nat methods 7: 335–336.

How Does Conversion of Natural Tropical Rainforest Ecosystems Affect Soil Bacterial and Fungal Communities in the Nile River Watershed of Uganda?

Peter O. Alele[1,2,3,7]*, **Douglas Sheil**[4,5,6,7], **Yann Surget-Groba**[1], **Shi Lingling**[1,2], **Charles H. Cannon**[1,8]

1 Key Laboratory of Tropical Forest Ecology, Xishuangbanna Tropical Botanical Garden (XTBG), Chinese Academy of Sciences, Kunming, Yunnan, P. R. China, **2** University of the Chinese Academy of Sciences, Beijing, P. R. China, **3** Great Nile Conservation Centre (GNCC), Lira, Uganda, **4** Department of Ecology and Natural Resource Management, Norwegian University of Life Sciences, Ås, Norway, **5** Center for International Forestry Research (CIFOR), Bogor, Indonesia, **6** Department of Ecology and Natural Resource Management, School of Environment, Science and Engineering, Southern Cross University, Lismore, New South Wales, Australia, **7** Institute of Tropical Forest Conservation (ITFC), Mbarara University of Science and Technology (MUST), Kabale, Uganda, **8** Texas Tech University, Lubbock, Texas, United States of America

Abstract

Uganda's forests are globally important for their conservation values but are under pressure from increasing human population and consumption. In this study, we examine how conversion of natural forest affects soil bacterial and fungal communities. Comparisons in paired natural forest and human-converted sites among four locations indicated that natural forest soils consistently had higher pH, organic carbon, nitrogen, and calcium, although variation among sites was large. Despite these differences, no effect on the diversity of dominant taxa for either bacterial or fungal communities was detected, using polymerase chain reaction-denaturing gradient gel electrophoresis (PCR-DGGE). Composition of fungal communities did generally appear different in converted sites, but surprisingly, we did not observe a consistent pattern among sites. The spatial distribution of some taxa and community composition was associated with soil pH, organic carbon, phosphorus and sodium, suggesting that changes in soil communities were nuanced and require more robust metagenomic methods to understand the various components of the community. Given the close geographic proximity of the paired sampling sites, the similarity between natural and converted sites might be due to continued dispersal between treatments. Fungal communities showed greater environmental differentiation than bacterial communities, particularly according to soil pH. We detected biotic homogenization in converted ecosystems and substantial contribution of β-diversity to total diversity, indicating considerable geographic structure in soil biota in these forest communities. Overall, our results suggest that soil microbial communities are relatively resilient to forest conversion and despite a substantial and consistent change in the soil environment, the effects of conversion differed widely among sites. The substantial difference in soil chemistry, with generally lower nutrient quantity in converted sites, does bring into question, how long this resilience will last.

Editor: Morag McDonald, Bangor University, United Kingdom

Funding: This work was funded by Xishuangbanna Tropical Botanical Garden (XTBG) and the Chinese Academy of Sciences. The funders had no role in study design, data collection and analysis, decision to publish, or preparation of the manuscript.

Competing Interests: The authors have declared that no competing interests exist.

* Email: alelepeter@gmail.com

Introduction

Tropical rainforests (TRF) possess most of the world's terrestrial biodiversity and deforestation is the leading cause of biodiversity loss [1,2]. Due to their high biodiversity and endemism, the tropical rainforests in Uganda's Nile river watershed are among the world's most important for their conservation values. But these areas are under pressure. The United Nations Population Division [3] predicts that the population of the Nile Basin states will increase by 57% from 2010 to 2030, reaching 647 million people. This rapid population growth, high levels of poverty and prevalent civil insecurity continue to exert severe pressure on natural resources in the region. Uganda in particular has one of the world's highest population growth rates (3.2% per year) [4]. Most of this growing population (nearly 80%) is dependent on

agriculture leading to large scale and continuing conversion of natural habitats [5].

Soil communities form the foundation of any ecosystem, in terms of nutrient cycling and availability, so understanding how land conversion affects these communities is an important first step. The effect of land use change on soil microbial communities has been studied in South American and Southeast Asian forests [6,7], but not in the biodiversity hotspots of the Nile river watershed. There is considerable global concern about the loss of biodiversity and the consequences for human well-being [8]. Microorganisms in particular play a vital role in many ecological processes and environmental services [9]: these roles are not always apparent or well characterized but if all microbes died the world would rapidly become buried in undecomposed dead material. Due to their significance in maintaining ecosystem function and productivity [9,10], our study offers a vital exploratory appraisal of microbial community dynamics in natural TRF and human-converted sites. We don't know if there are reasons to be concerned unless we look. Developing such knowledge is critical at this point, because populations in the Nile river watershed are highly dependent on forests for basic requirements such as food and fuel wood, with the environment contributing between 40–60% of the gross domestic product (GDP) of the Nile riparian states [11].

Because of widespread loss of biodiversity, focus from species conservation within particular habitats has been shifted to conservation of communities [12,13]. It is therefore important to explore and understand how composition and diversity changes across spatial scales in a given context [14–16]. Changes in ecosystems caused by conversion to intensive management can lead to biotic homogenization, the increase in community similarity over time and/or space and an implied loss of rare and vulnerable taxa when examined at larger scales [17–19]. Because microorganisms are the most diverse organisms on earth with most taxa and respective functions and behaviors as yet unknown, determining their sensitivities and biogeography remains a major challenge. But in the longer term such knowledge will help us better understand the sustainability of land-use systems and associated environmental values.

This study was therefore necessary as a first step in exploring these relationships, and to enhance understanding so as to contribute to the informed and appropriate stewardship of the region's natural resources. Our objective was to establish how forest conversion and soil factors affect soil bacterial and fungal diversity and community composition in the tropical rain-forests in the Nile river watershed of Uganda. We chose four forest sites found within protected areas, with paired treatments within each forest; (1) natural and (2) converted ecosystem sites. The natural forest ecosystem at each site had suffered minimal human disturbance, while converted areas had been transformed to cropland. These matched sites found in different locations and environmental conditions each experienced different land use histories, conservation circumstances and individual challenges for management.

In each matched set of natural and converted sites, we compared soil physical and chemical properties and microbial community diversity and composition using standard PCR-based genotyping techniques. We then calculated community similarity indices between sites. This approach would allow us to examine both environmental and biotic changes in the soil community associated with conversion. Disturbances of sufficient magnitude or duration may alter an ecosystem and force a different regime of predominant processes and structures that favor some populations over others [20].

We tested the null hypotheses that there was no difference in soil properties, band-types, and diversity between treatments. The influence of soil properties on microbial community diversity was measured by discriminant analysis and canonical correspondence analysis (CCA), [21,22]. Because additive partitioning of diversity provides a useful framework for quantifying the spatial patterns of diversity across hierarchical spatial scales [23,24], we partitioned total diversity (γ) in each ecosystem type (natural and converted) into additive components representing within-community diversity (α) and between-community diversity (β). Our objective was to identify the most important sources of total diversity so as to propose conservation measures for microbial communities in the TRF ecosystems of the Nile river watershed of Uganda.

Methods

Site description

We selected four tropical rainforest (TRF) sites because of their relative size, biodiversity, socio-economic and scientific importance (Fig 1). Mabira forest is located between the highly populated and urbanized Kampala city on the western side; the extensive and mechanized Lugazi sugar and tea plantations on the Eastern; and Lake Victoria on the southern side. Budongo forest is located next to the extensive Kinyala sugar plantations on one side and a densely populated mainly subsistence population scattered around it. Maramagambo and Kaniyo Pabidi are located within Queen Elizabeth and Murchison Falls national parks (NP) respectively. These two NP forests had perhaps the best protection due to presence of Uganda Wildlife Authority (UWA) personnel. However, Maramagambo's location starting on the steep slopes of the rift valley subjected it to frequent storms with strong runoff flow that swept away most of its top soil (Table 1).

Soil sampling design

We collected 400 core soil samples within 40 plots (1000 m^2 each) in four TRF sites (Fig 1). We sampled five plots from each site of the natural TRF and five plots from the converted TRF. We established the plots at least 100 m from the ecosystem edge and 500 m apart and collected 10 evenly placed core subsamples of top soil (0–15 cm) from each plot and homogenized them into one sample per plot. We then derived a 500 g composite sample from the mixture, sieved and packed it for physical and chemical analyses and DNA extraction.

Sample preparation

We sieved 100 g of the soil on-site through a 4 mm mesh, transported it to the laboratory on ice, and stored in a freezer at −40°C prior to nucleic acid extraction and analysis. We kept the rest of the soil for drying and physical and chemical analysis. We performed DNA extractions from 1 g of soil using the Ultra Clean soil DNA kit (Mo Bio Labs, Solana Beach, CA, USA) following the manufacturer's protocol. The purified DNA was detected by agarose gel electrophoresis, and the DNA was amplified by polymerase chain reaction (PCR).

Soil property analyses

We measured the soil pH in 2.5:1 water to soil suspension using a pH meter (10 g soil+ 25 ml of distilled water, shaken for 30 min and read on a calibrated pH meter). We then used the Walkley and Black method [25] to analyze soil organic carbon (SOC) and the Kjeldahl method [26] to determine soil nitrogen. We measured the soil phosphorus by the Bray and Kurtz no. 1 method [27]. The photoelectric flame photometer was used to determine the soil potassium, sodium and calcium after extraction with neutral ammonium acetate. We used the atomic absorption spectrometer to measure the soil magnesium after extraction with neutral ammonium acetate. The Bouyoucos hydrometer method adopted from Gee and Bauder [28] was used to determine soil texture. The soil copper and iron were then determined using the atomic absorption spectrometer after extraction with EDTA.

PCR amplification and DGGE analysis

Polymerase chain reaction-denaturing gradient gel electrophoresis (PCR-DGGE) method has been used extensively in microbial ecology and is a robust and cost effective method for exploratory classification of microbial communities [29]. Following soil DNA extraction, we performed a PCR for each DNA extraction to

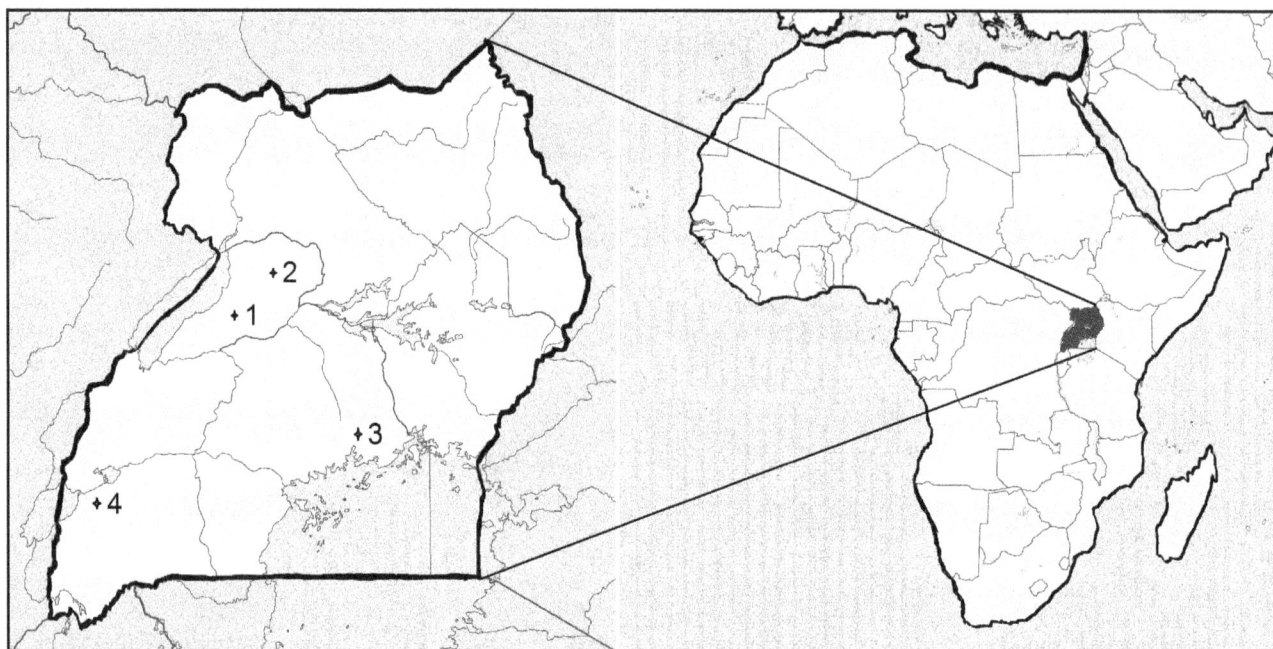

Figure 1. Map of Uganda showing the distribution of sampling sites; Budongo forest (1), Kaniyo Pabidi (2), Mabira forest (3), and Maramagambo forest (4).

amplify the 16S rRNA genes for bacteria and 18S rRNA genes for fungi using universal primers (Table 2).

PCR reactions had a final volume of 25 µl containing a final concentration of 1× TaKaRa ExTaq PCR buffer with $MgCl_2$, 300 pM of primers for bacteria. We then added 200 µM dNTPs, 2.5 U ExTaq DNA polymerase (TaKaRa Bio, Otsu, Japan) and milliQ H_2O to complete the volume, BSA was also added for the fungal community analysis. We performed PCR cycles with an initial denaturing temperature of 95 °C for 5 min, followed by 35 cycles of 95 °C for 30 sec, annealing temperature of 50 °C for 30 sec, extension of 72 °C for 1 min; and a final extension of 72 °C for 10 min. We checked the product of the PCR-rounds and quantified by agarose gel-electrophoresis.

We then performed 16S rRNA and 18S rRNA-DGGE analysis using a universal mutation detection system (Dcode Bio-Rad, Richmond, CA, USA) with a 6% and 8% acrylamide gel for bacteria and fungi respectively containing a gradient of 40–60% denaturant (100% denaturant contains 7 mol urea and 40% formamide). We applied 100 ng of PCR samples to the DGGE gel. DGGE was performed in 1 × TAE Buffer (40 mol Tris/acetate, pH 8; 1 mol ethylene diaminetetra acetic acid) at 60 °C and a constant voltage of 60 V for 16 hours. After staining with SYBR Green1, we recorded the DGGE gels as digital images and analyzed the DNA band numbers using image-processing software after subtracting background noise.

Data analysis

We used the Rolling disk method with Quantity One (Bio-Rad laboratories Inc.), which normalizes the band pattern from electrophoresis for identification of each band. We then converted the band patterns into binary data based on the presence or absence of each band for part of our analysis. The DGGE fingerprints were interpreted in terms of band richness (number of predominant DGGE bands/population). The pixel intensity of each band was detected by Quantity One software and is

expressed as relative abundance (P_i) [30]. Shannon Index (H') and Simpson index (D), the most widely used diversity indices were then calculated using the richness and relative abundance data following the equations:

$$H' = -\sum_{i=1}^{R} P_i In P_i \qquad (1)$$

$$D = \sum_{i=1}^{R} (P_i)^2 \qquad (2)$$

Where R, the richness, is number of different bands each data set contains, $P_i = \dfrac{n_i}{N}$ and n_i is the abundance of the ith band and N the total abundance of all bands in the sample.

Band-type data of the DGGE fingerprints was then used to derive the alpha diversity (bands per sample and ecosystem type), beta diversity (total bands per site) [31]. Jaccard's similarity indices [32] between converted and natural TRF sites were determined using the equation:

Jaccard's Similarity Index $= {}^A/_{(A+B+C)}$

Where,

A = Total number of bands present in both converted (C) and natural (N) ecosystem samples (plots) (also β-diversity)

B = Number of bands present in C but not in N

C = Number of bands present in N but not in C

We determined the influence of site factors as revealed by soil physicochemical properties on the variation of soil microbial communities by applying discriminant analysis using Statistical Package for the Social Sciences (SPSS). This was done to assess the relative importance of each predictor variable (pH, SOC, N, P, K, Na, Ca, Mg, and soil texture). We also used the Mann-Whitney

Table 1. Summary of study site description.

Forest Site	Location	Size (km²)	Altitude (masl)	Geology	Forest type	Habitat type	Ecosystem description
Budongo	31°N 35° E 1′S 45° N	793	700–1,270	Weathered pre-cambium with ferrallitic sandy clay loams	Ironwood forest (Cynometra alexandri); Mixed forest (Maesopis), and colonizing forest (Entandrophragma)	Primary forest	Consists of a medium altitude moist semi-deciduous forest with areas of savanna and woodland. Converted areas consisted of deforested agricultural land being cultivated and planted with maize, beans, sweet potatoes and cassava. This land has existed as agricultural land for at least 15years.
Mabira	33° 0.00′ E 0° 30.00′ N	300	1,070–1,340	Ferrallitic soils with mainly sandy clay loams	Mixed forest	Secondary forest, heavily influenced by humans	The forest is surrounded by a densely populated area. Converted areas were actively cultivated and used to grow maize, groundnuts, beans, yams, cassava, sweet potatoes and a few scattered plants of coffee and sugarcane. The converted land had existed as agricultural land for at least 10 years.
Maramagambo	00° 33′ 00″ S and 29° 53′ 00″ E	1,978 (QENP)	910–1390	Ferrallitic soils with undifferentiated dark horizons	Medium altitude, moist, semi-deciduous forest	Secondary forest influenced by wildlife and humans	Forms part of the Queen Elizabeth NP (QENP) which is 1,978 km2. Converted areas consisted of cultivated and grazed areas with gardens of sweet potatoes, beans, maize, and sorghum with areas commonly grazed by cattle and goats.
Kaniyo Pabidi	Lat 1.916667 Long 31.666667	N/A	700–1,270	Freely drained ferruginous tropical soils	Mixed forest	Primary forest	Located north of Budongo forest and part of Murchison Falls N.P. Converted areas consisted of cultivated areas with crops like maize, beans, cassava and sweet potatoes.

[5,61,70,71].

Table 2. Sequences of primers used in study.

Microorganism	Primer	Sequence (5'–3')	Reference
Bacteria	F357	CGC CCG CCG CGC GCG GCG GGC GGG GCG GGG GCA CGG GGG GCC TAC GGG AGG CAG CAG	[72]
	907R	CCG TCA ATT CMT TTG AGT TT	
Fungi	FF390	CGA TAA CGA ACG AGA CCT	[73]
	FR1GC	AIC CAT TCA ATC GGT AIT	

test to examine differences between soil properties in natural and converted ecosystems, and microbial communities in natural and converted ecosystems.

We tested the null hypothesis that diversity is uniform at all spatial scales by additive partitioning of total diversity (γ diversity). To determine contributions of α and β diversity to overall diversity across a range of spatial scales [14,23,33], an additive relationship between diversity components (i.e., $\beta = \gamma - \alpha$) was derived (Fig 2.). The scale at which diversity is maximized was therefore identified [23,34], to facilitate planning processes and management strategies to conserve natural levels of diversity accordingly [35–38].

We used PARTITION 3.0 software [39] to calculate average diversity at each scale and diversity was measured as band richness. Individual-based randomization procedure in the software was used to test whether the observed partitions of diversity within the ecosystem could have been obtained by a random allocation of lower-level samples nested among higher-level samples [34]. Null values of β_i obtained from 1,000 randomizations were used to obtain a p-value for the observed β_i at each hierarchical scale. Deviations of the observed diversity from the null expectation indicated a nonrandom spatial distribution of fungi or bacteria at a given scale.

Results

Soil property variations

Soil pH comparisons using a Mann-Whitney U test of significance, between five plots of natural and five plots of converted TRF ecosystems in each of the four forest sites, found significantly higher (less acidic) pH in three of the four sites at Budongo ($p = 0.0107$), Kaniyo Pabidi ($p = 0.0112$), and Mabira ($p = 0.0269$); and non-significant difference at Maramagambo ($p = 0.1706$). Percentage soil organic carbon (SOC) was signifi-

cantly higher in natural than converted ecosystems in all four sites i.e. Budongo ($p = 0.0119$), Kaniyo Pabidi ($p = 0.0212$), Mabira ($p = 0.0122$) and Maramagambo ($p = 0.0119$) with combined %SOC in natural sites more than double of that in converted sites; whereas %soil nitrogen was only significantly higher in natural forests at Budongo ($p = 0.0112$) and Kaniyo Pabidi ($p = 0.0119$), and non-significant at Mabira ($p = 0.6015$) and Maramagambo ($p = 0.0947$) (Table 3).

Ecosystem and site comparisons of microbial community diversity

Bacterial (B) communities were significantly richer ($p = 0.0304$; Mann-Whitney U) in detectable bands than fungal (F) communities in both converted (C) and natural (N) ecosystems (converted: medians; F = 36, B = 61.5; natural: medians; F = 39.5, B = 60.5). While total band richness (B+F) did not differ between natural and converted forests we observed greater fungal richness in natural than in converted forests (medians: C = 36, N = 39.5; test stat = 18.5) and more bacterial bands in converted than in natural ecosystems (medians: C = 61.5, N = 60.5; test stat = 18.5). Kaniyo Pabidi was the most diverse site overall with the highest number of bacterial and fungal bands, while Maramagambo had the least band richness (Fig 3).

Natural sites harbored more bands unique to one site than converted sites for bacteria at Kaniyo Pabidi and Maramagambo and for fungi at Maramagambo and Budongo. Mabira and Kaniyo Pabidi had higher numbers of unique bacterial bands than at Maramagambo and Budongo. There were also more unique fungal bands at Mabira and Budongo than at Maramagambo and Kaniyo Pabidi (Fig 4).

We also found that Mabira and Maramagambo had the lowest bacterial Jaccard's community similarity indices [32] between natural and converted ecosystems, whereas Budongo and Mabira had the lowest fungal community similarity between natural and converted ecosystems (Table 4). Dissimilarity between natural and converted ecosystems was nonetheless non-significant in all sites for both fungal and bacterial communities. Also, there was generally greater dissimilarity between sites of fungal communities than in bacterial communities suggesting a higher susceptibility to habitat change among fungi than bacteria (Table 4).

Ecosystem classification and importance of predictor variables

The CCA showed that, despite the relatively small amount of difference between sites, soil pH, average phosphorus, and texture (%sand) had strong influence on bacterial diversity in the TRF ecosystem (Fig 5); whereas organic carbon, sodium, pH and average phosphorus were strongly associated with fungal community variation in both natural and converted TRF ecosystems (Fig 5). The CCA also showed that bacterial communities in both Kaniyo Pabidi and Mabira were unique to bacterial communities

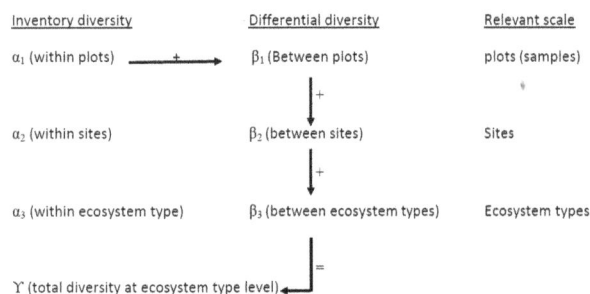

Figure 2. Illustration of hierarchical spatial scales in our additive partitioning model. The α scale is the within-level, and β scale, the between-level components. Because a diversity at a given scale is the sum of the α and β diversity at the next lower scale, the total diversity (γ) can be described by the following formula: $\alpha1+\beta1+\beta2+\beta3$ [14,22].

Table 3. Mean (Standard deviation) for diversity indices of Bacterial (B) and Fungal (F) communities and soil properties in natural (N) and converted (C) ecosystems.

	Budongo		Kaniyo Pabidi		Mabira		Maramagambo	
	Natural	Converted	Natural	Converted	Natural	Converted	Natural	Converted
Shannon (B)	2.49(0.68)	2.73(0.10)	3.25(0.21)	3.12(0.25)	3.06(0.18)	3.19(0.18)	3.04(0.16)	2.93(0.07)
Simpson (B)	0.89(0.10)	0.93(0.01)	0.95(0.13)	0.95(0.02)	0.95(0.01)	0.95(0.01)	0.95(0.01)	0.94(0.01)
Shannon (F)	2.44 (0.30)	2.49 (0.17)	2.83(0.19)	2.77(0.20)	2.57(0.62)	2.64(0.30)	2.10(0.50)	1.93(0.67)
Simpson (F)	0.90(0.29)	0.90(0.03)	0.93(0.01)	0.93(0.02)	0.90(0.06)	0.92(0.03)	0.84(0.08)	0.82(0.12)
pH	5.88(0.11)*	5.08(0.20)*	6.24(0.23)*	5.38(0.27)*	6.46(0.54)*	6.18(0.42)*	6.18(0.45)	5.80(0.24)
OC (%)	6.14(0.67)*	1.59(0.17)*	5.69(1.20)*	3.65(0.54)*	5.87(2.03)*	3.53(1.19)*	9.08(0.79)*	3.98(1.09)*
N (%)	0.43(0.05)*	0.11(0.01)*	0.43(0.08)*	0.25(0.03)*	0.21(0.10)	0.22(0.06)	0.28(0.04)	0.19(0.08)
Ca (Cmoles/kg)	8.00(2.56)*	4.00(0.58)*	13.54(3.36)*	6.74(0.98)*	8.14(1.19)*	6.20(0.76)*	12.12(2.13)*	7.34(1.16)*

*significant differences (p<0.05) between natural and converted ecosystems.

in the other sites and there was high contrast between bacterial communities of converted and natural ecosystems at Kaniyo Pabidi. Fungal communities at Maramagambo and Mabira were also unique to those in other sites and there was high contrast between fungal communities at Mabira's natural and converted ecosystems. Furthermore, the CCA showed that fungal communities responded more to soil pH levels than bacterial communities (Fig 5), with site-specific patterns showing that bacteria and fungi were grouping according to sites.

A discriminant analysis to predict whether bacterial or fungal communities were from natural or converted ecosystems found that only OC, Ca, N, and pH for bacterial communities; and OC, N, Ca, and pH for fungal communities (all ranked from most important to least important) were found to be significant predictors of soil physicochemical properties. All other variables were poor predictors in this context (Table 5).

Hierarchical scaling

We found 58 and 56 fungal bands in natural and converted forests respectively, from 17 plots of natural ecosystems and 20 plots of converted forests. There were also 92 and 88 bacterial bands in natural and converted ecosystems respectively found in 20 plots of converted ecosystems and 17 plots of natural ecosystem. All these were within four sites. β-diversity varied more than α-diversity between natural and converted ecosystems for both bacteria and fungi. We found higher bacterial and fungal β-diversity in converted ecosystems than in natural ecosystems at lower hierarchical scales (β_1); higher β-diversity in natural than converted at between-site scale (β_2), and higher β-diversity in converted than in natural ecosystems at the between-ecosystem type scale (β_3) (Fig 6).

We also found substantial contribution of observed β-diversity (β_1, β_2, and β_3) to total band richness (γ-diversity), while α-diversity of both bacteria and fungi in converted and natural ecosystems were similar. Spatial partitioning of total diversity also consistently showed that the beta components (β_1 and β_2) were always greater than expected by chance, whereas the alpha component (α_1) was always lower than expected. For both fungal and bacterial communities in natural and converted ecosystems, observed within plot diversity were substantially less than values expected from individual-based randomizations (Fig 7).

Discussion

Soil property variations and site differences

Studies in both tropical and temperate zones show that soils in converted or cropped areas normally have reduced soil aggregation, structural stability and organic matter, and an increase in bulk density when compared to forests [40,41]. Habitat conversion may also alter soil properties such as nutrient levels, and abiotic conditions and may affect associations between organisms. In our study there are some local details that may influence our results.

Both Maramagambo and Kaniyo Pabidi are located within Queen Elizabeth NP and Murchison Falls NP respectively and are protected by Uganda Wildlife Authority (UWA) personnel. They are well protected and there is little evidence of recent encroachment. There is significant wildlife populations including elephants, buffaloes, zebras and the areas are frequented by tourists. Protection by UWA and presence of dangerous animals (such as buffalos and lions) reduce damaging human activity at Kaniyo Pabidi and Maramagambo which should enhance the difference between natural and converted ecosystems. Maramagambo's location, in contrast, means the

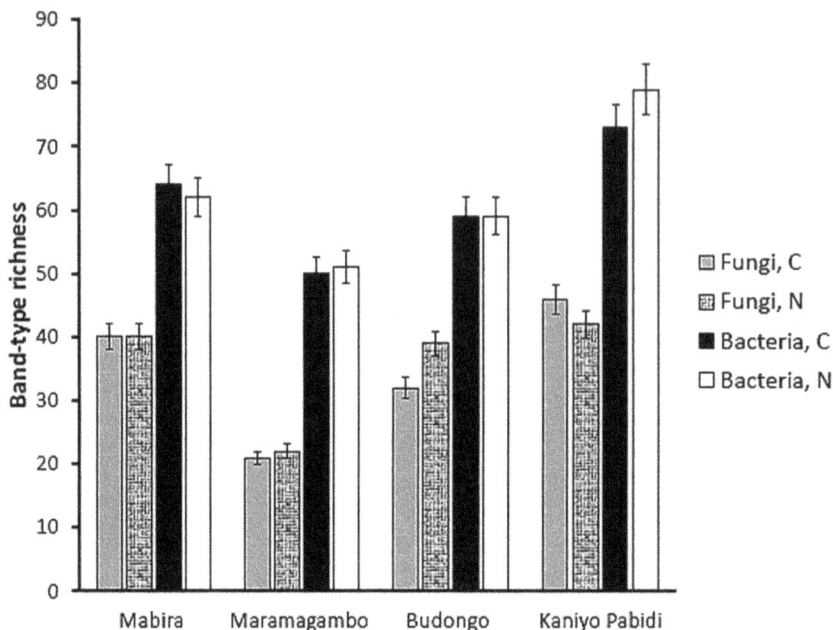

Figure 3. Band richness for fungal and bacterial communities in converted and natural ecosystems. All richness values are total bands present in five samples of each ecosystem treatment (error bars are 5% confidence interval).

forest is subjected to substantial natural disturbance from frequent storms and strong erosive runoff even within the natural forest, whereas tourist activity at Kaniyo Pabidi seemed to have little impact on soil properties. Converted areas at Kaniyo Pabidi were also sparsely populated with limited human impacts on the environment. Its sites were old and might have been cultivated for at least 20years.

In our study, Budongo is located next to a high, mainly subsistence population and resultant population activity. But even though encroachment, illegal hunting and logging in natural habitats in Budongo are not uncommon, there seems to be minimal impact of conversion on soil properties in our sample locations; whereas proximity of Mabira's natural forest to densely populated urban areas exposes it to increased human activities, likely reducing its difference with converted sites.

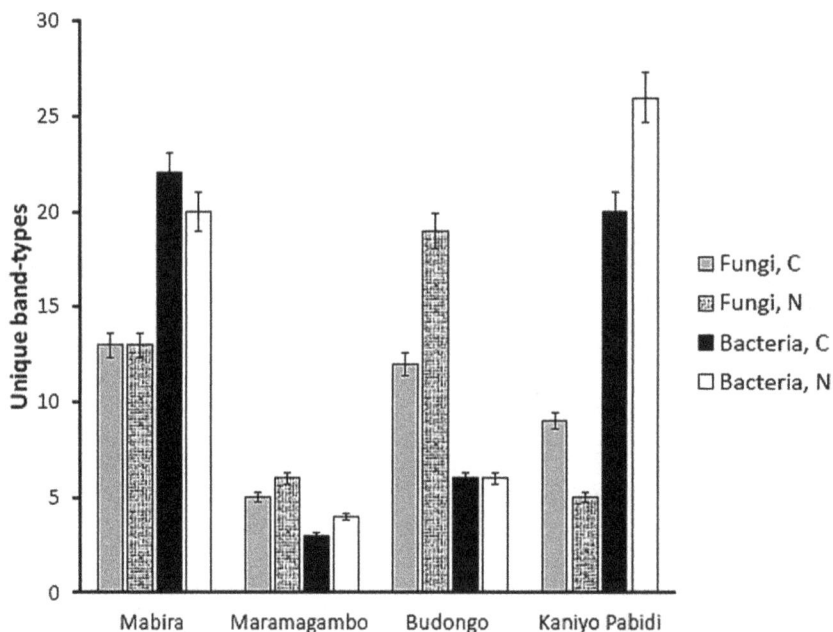

Figure 4. Bacterial and fungal bands unique to converted (C) and natural (N) ecosystems at each site (error bars are 5% confidence interval).

Table 4. Jaccard's similarity indices between bacterial and fungal communities in natural (N) and converted (C) sites of Mabira (Mb), Maramagambo (Mg), Budongo (Bd) and Kaniyo Pabidi (Kp).

			F	U	N	G	I			
			Mb	Mg	Bd	Kp	Mb	Mg	Bd	Kp
			N	N	N	N	C	C	C	C
B	Mb	N	1	0.713	0.769	0.732	0.671	0.685	0.696	0.768
A	Mg	N	0.841	1	0.733	0.666	0.666	0.711	0.646	0.666
C	Bd	N	0.803	0.840	1	0.804	0.808	0.693	0.622	0.827
T	Kp	N	0.731	0.765	0.698	1	0.774	0.670	0.794	0.785
E	Mb	C	0.667	0.788	0.739	0.741	1	0.663	0.710	0.761
R	Mg	C	0.866	0.885	0.846	0.755	0.809	1	0.684	0.698
I	Bd	C	0.825	0.857	0.844	0.696	0.731	0.856	1	0.786
A	Kp	C	0.805	0.869	0.716	0.683	0.759	0.776	0.733	1

Microbial community variations

Soil properties determine many aspects of soil microbial community structure [42–44]. Carbon availability [45–47], nitrogen availability [45,48,49] and soil pH [44,50,51] can all influence microbial community composition and diversity. In addition, correlation studies have shown that plant species [52–54] and soil type [45,54,55] are associated with variation in microbial communities. It has also been shown that land use indirectly affects bacterial community structure by modification of soil properties [56] but similarity between converted and natural ecosystem bacterial communities may also suggest a high number of generalists.

Nacke et al., [56] found that bacterial community composition in forests and grassland was largely determined by tree species and soil pH. Jesus et al. [6] also showed that bacterial community structure is influenced by changes linked to soil acidity and nutrient concentration. Other studies also suggest that soil pH is a major factor influencing microbial community composition [50,57–59]. This influence of soil pH has been recognized at different taxonomic levels [45,60] with most microorganisms thriving within a limited pH range. This is because acids can denature proteins and large pH changes may inhibit growth in microorganisms. Fierer and Jackson, 2006 [44] found, in contrast, that net carbon mineralization rate (an index of C availability) was the best predictor of phylum-level abundances of dominant bacterial groups, and Bisset et al., [42] found that soil microbial communities were consistent with disturbance gradients within different agricultural treatments and relatively undisturbed non-agricultural sites.

Because of widespread forest conversion in Uganda as a result of increasing population pressure, estimated at between 1.1% and 3.15% per year [61]; natural ecosystems and their inhabitant biodiversity are at risk [62]. Loss of diversity increases the likelihood of losing important functional roles and associated ecosystem processes. At landscape scale, spatial and temporal variations of microbial communities in forest soils are influenced by numerous biotic and abiotic factors. These factors may include climate, soil types, and vegetation associations [50,63,64]. Owing to this study design, many of these factors were assumed to be similar between natural and converted ecosystems. For instance, the proximity of natural and converted ecosystem sites meant that climate and geology were, we assume, similar in the two treatments. Even though there could still be a number of underlying causes of community differences, two likely influences were assumed to be soil properties [45,54,55] and vegetation types [52–54]. Clearly both of these sets of factors change when forest is converted for agriculture or range lands.

Despite substantial reductions in SOC, N, Ca and pH in converted sites in this study, differences in microbial communities were small meaning that converted sites still had sufficient SOC, N and Ca to sustain the same microbial populations. The close proximity of the matched pairs could also lead to a source-sink relationship between the natural and converted forests, with the presence of unique taxonomic groups a likely indication of habitat preference (endemism) for some taxonomic groups. It may also be an indicator of relative habitat dissimilarity. The high numbers of unique bacterial bands at Mabira and Kaniyo Pabidi and unique fungal bands at Mabira and Budongo (Fig 4) thus suggests that ecosystem alteration at these sites was sufficient to force a different regime of processes and structures enabling a new set of taxonomic groups to predominate. Mabira had high numbers of both unique bacterial and unique fungal bands that can be attributed to the extent of disturbance at its sites (Mabira

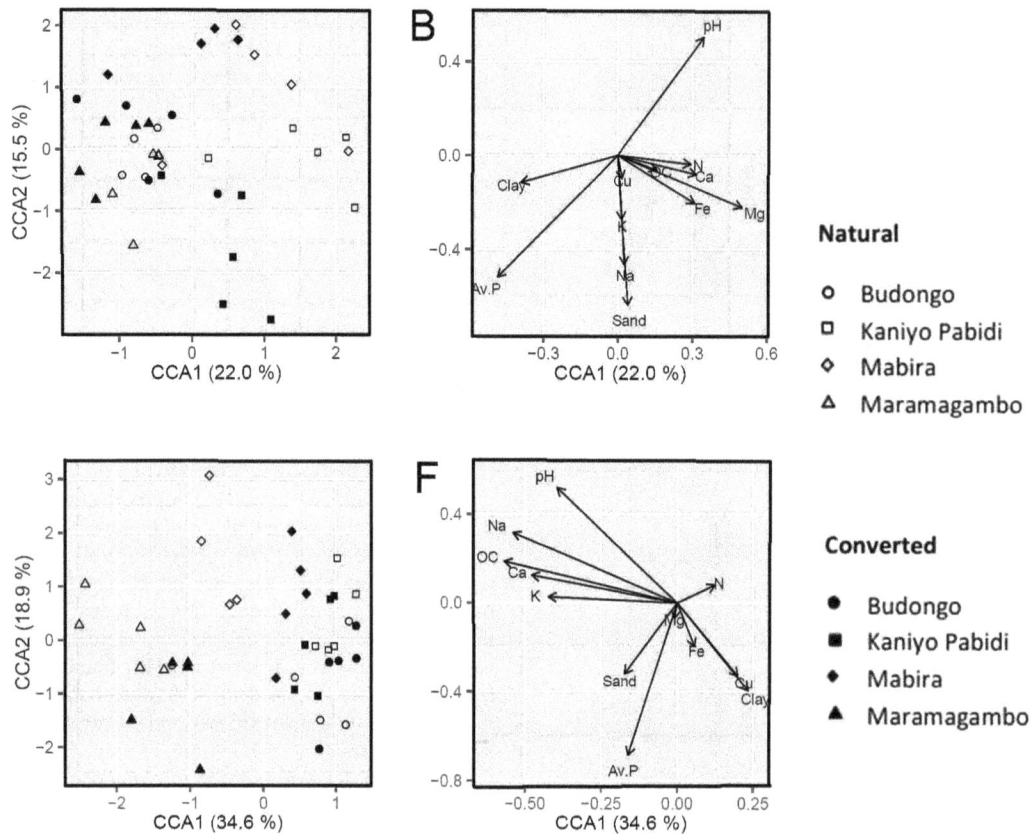

Figure 5. CCA for bacterial (B) and fungal (F) relationships using relative abundance of bands and soil physicochemical properties in natural and converted ecosystems. The symbols (left graphs) represent the similarity between each sample (plot) as defined by their diversity, and the vectors (right graphs) represent the structural matrix for soil properties and their influence on relative abundance of each band. The length of the vectors represents the relative strength of influence of the particular aspect of soil physicochemical property.

is the only peri-urban tropical rainforest site among the four selected sites).

The low numbers of unique bacterial and fungal bands at Maramagambo can be attributed to the high erosion at natural sites that reduced the contrast between the natural and converted sites. For the other sites, bacteria and fungi had different responses to ecosystem alterations. This could indicate separate influences on microbial distribution that exist when alteration is moderate. Similarity indices suggested that bacterial and fungal communities were determined by separate forces leading to distinct responses across the study locations.

Hierarchical scaling

Many studies have shown that specialist species are more negatively affected by current global changes than generalists [7,65]. The process of biotic homogenization can involve the replacement of native biota with non-natives or the introduction of generalist species [66]. In this study, the net decrease in β-diversity from natural to converted TRF ecosystem at the between-site scale (β_2) for both fungi and bacteria was an indication of biotic homogenization [18,66]. This can result from ecosystem alterations which can in-turn alter ecosystem function and reduce ecosystem resilience to disturbance [65,67].

We also showed that the β components of diversity (β_1 and β_2; the average diversity between the plots and sites, respectively) were consistently higher than those expected by chance, whereas the local α_1 diversity component (α_1, the average diversity within the plots) was consistently lower than that expected (Fig 7). Such scale-dependent deviations of the observed diversity from the expected can be generally explained by aggregation at a relatively small "local" scale and, spatial differentiation of diversity at a larger "landscape" scale [33,34,68,69].

Relatively lower diversity within converted ecosystems suggests that conversion of natural TRF ecosystems results in reduced diversity for both bacteria and fungi. This is compatible with recent studies that show that conversion of TRF ecosystems threatens microbial diversity [7] and because microorganisms, like all other organisms, have habitat preferences and may be affected by land-use changes [6,64]. While we cannot be certain that such decline in diversity has led to a decline in any particular ecosystem functions or services, this is a possibility that deserves further evaluation, and we speculate that such loss of diversity will at the very least cause a reduction in functional redundancy and associated resilience.

Higher β-diversity of both bacterial and fungal communities at the between-plot scale (β_1) in converted ecosystems indicates

Table 5. Structure matrix rank showing absolute size of correlation between discriminant analysis function from most important to least important predictor variable of site factors (soil physiochemical properties) and their influence on the variation of soil microbial communities.

Bacteria		Fungi	
Predictor Variables	**Function 1**	**Predictor variables**	**Function 1**
OC	0.625*	OC	0.625*
Ca	0.471*	N	0.493*
N	0.421*	Ca	0.473*
pH	0.355*	pH	0.340*
Mg	0.298	Mg	0.281
Cu	0.197	Cu	0.221
Na	0.181	Na	0.183
K	0.169	K	0.160
Av.P	−0.077	Sand	0.064
Sand	0.070	Av.P	−0.060
Simpson	−0.065	Fe	0.024
Fe	0.029	Shannon	0.012
Shannon	−0.023	Simpson	0.011

(* = important predictor variable, with 0.30 used as the threshold).

differentiation (reduced community similarity) in converted ecosystems at this hierarchical scale. Considering the multiple land-uses and cropping systems of converted areas, this was expected. There was also substantial contribution of β-diversity to total diversity (γ). This suggests the importance of nonrandom ecological processes at the between-plot and between-site scale in

determining total richness and community composition [14,34]. Differences between the observed and expected diversity components could be due to ecological processes that lead to a non-random dispersion of individuals. These processes could include intra-specific aggregation, habitat selection, and limited dispersal capacity [33].

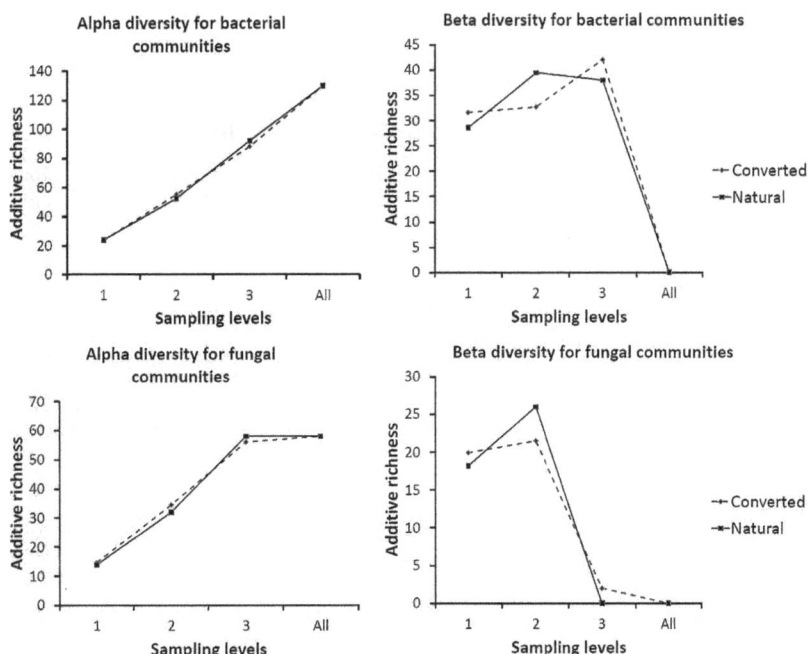

Figure 6. Additive partitioning of bacterial and fungal diversity (expressed as additive richness) across alpha and beta hierarchical spatial scales at three sampling levels (plot, site and ecosystem type) in natural and converted TRF ecosystems.

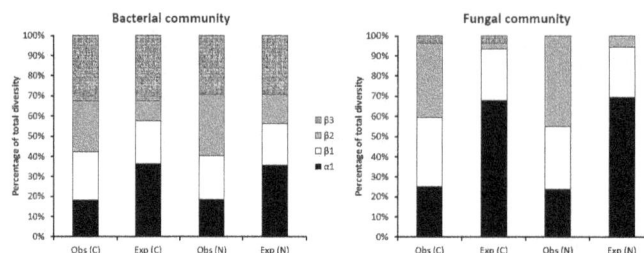

Figure 7. The additive partitioning of total bacterial and fungal community, γ -diversity into α and β components at three nested spatial scales, with each component expressing their relative contributions to γ -diversity; where γ -diversity is equal to $\alpha1+\beta1+\beta2+\beta3$. The observed (obs) partitions are compared with the expected (exp) values, as predicted by the null model based on 1000 iterations using individual-based randomization.

Conclusion

There is international concern about the threat to natural habitats in the Nile river watershed and the consequential loss of important biodiversity. Whereas aspects of microbial biogeography and influence of forest conversion in Uganda's Nile river watershed is largely unknown, this study offers an important first glimpse into indicators of spatial diversity patterns of soil fungal and bacterial communities in the Uganda's Nile river watershed. Our observations of reduced soil microbial diversity, both bacterial and fungal, in converted ecosystems though unsurprising in itself causes us some concern and would justify further work to determine the significance of the diversity lost and the wider implications.

By focusing on diversity patterns across multiple hierarchical spatial scales, we were able to identify the scale at which regional microbial diversity is maximized. We showed that there was substantial contribution of β-diversity to total ecosystem diversity (γ) which includes taxa at the between-plot, site and ecosystem scales and unique taxa, highlighting the necessity to conserve marginal habitats and ecotones. Soil microbial communities in Uganda's Nile river watershed exhibit considerable resilience to forest conversion even though SOC, N, Ca and pH were all significantly altered. This result is surprising given that these physical and chemical properties typically strongly influence microbial diversity. Additionally, the variation among sites was quite large, indicating that soil communities in this region vary considerably on a regional spatial scale. Our results do not explain this variation. Most studies suggest that biogeographic barriers play little role in the geographic structure of soil communities. Rather than a consistent general pattern of microbial community change following forest conversion we find that responses are largely site-specific and widely variable.

Acknowledgments

We confirm that our field work did not involve any endangered or protected species and that we were granted access to protected areas by Uganda Wildlife Authority (UWA) and National Forestry Authority (NFA). Our sample site coordinates included: (1) Mabira; Lat, N00°24.495'; Long, E033°02.464' (Required permission obtained from NFA); (2) Maramagambo; Lat, S00°04.421'; Long, E030°02.305' (Required permission obtained from UWA); (3) Budongo; Lat, N01°41.852'; Long, E031°29.363' (Required permission obtained from UWA/NFA); (4) Kaniyo Pabidi; Lat, N01°55.128'; Long, E031°43.116' (Required permission obtained from UWA).

We would like to thank the Institute of Tropical Forest Conservation (ITFC) of Mbarara University of Science and Technology for supporting and co-advising this research. We also thank the Genetics Lab at Faculty of Science and the soil department lab at Faculty of Agriculture both at Makerere University, Kampala, for their lab facilitation during extractions and analyses. And finally we acknowledge the ideas and insights of Masatoshi Katabuchi, Dossa Gbadamassi, and our field and lab technicians and assistants Solomon Echel, Francis Alele Ozirit and Boniface Balikuddembe.

Supporting Information

Data S1   Excel spreadsheet from Quantity One analysis of PCR-DGGE profiles of fungal and bacterial communities in natural and converted TRF ecosystems.
(XLSX)

Author Contributions

Conceived and designed the experiments: POA CHC. Performed the experiments: POA SL. Analyzed the data: POA DS CHC. Contributed reagents/materials/analysis tools: DS YSG CHC. Contributed to the writing of the manuscript: POA DS YSG CHC.

References

1. Turner IM (1996) Species loss in fragments of tropical rain forest: a review of the evidence. J Appl Ecol 33: 200–209.
2. Joseph Wright S, Muller-Landau HC (2006) The Future of Tropical Forest Species. Biotropica 38: 287–301.
3. United Nations (2013) World Population Prospects: The 2012 Revision. Highlights and Advance Tables. New York.
4. United Nations D of E and SAPD (2007) World Population Prospects: The 2006 Revision.
5. Kayanja FIB, Byarugaba D (2001) Disappearing forests of Uganda: The way forward. Curr Sci 81: 936–947.
6. Jesus da CE, Marsh TL, Tiedje, M J, de S Moreira FM (2009) Changes in land use alter the structure of bacterial communities in Western Amazon soils. ISME J 3: 1004–1011. Available: http://www.ncbi.nlm.nih.gov/pubmed/19440233. Accessed 15 November 2012.
7. Rodrigues JLM, Pellizari VH, Mueller R, Baek K, Jesus EDC, et al. (2012) Conversion of the Amazon rainforest to agriculture results in biotic homogenization of soil bacterial communities. Proc Natl Acad Sci U S A. Available: http://www.ncbi.nlm.nih.gov/pubmed/23271810. Accessed 29 December 2012.

8. Cardinale BJ, Duffy JE, Gonzalez A, Hooper DU, Perrings C, et al. (2012) Biodiversity loss and its impact on humanity. Nature 486: 59–67. Available: http://www.ncbi.nlm.nih.gov/pubmed/22678280. Accessed 25 May 2014.

9. Venail PA, Vives MJ (2013) Positive effects of bacterial diversity on ecosystem functioning driven by complementarity effects in a bioremediation context. PLoS One 8: e72561. Available: http://www.pubmedcentral.nih.gov/articlerender.fcgi?artid = 3762786&tool = pmcentrez&rendertype = abstract. Accessed 24 June 2014.

10. Torsvik V, Øvreås L (2002) Microbial diversity and function in soil: from genes to ecosystems. Curr Opin Microbiol 5: 240–245. Available: http://www.ncbi.nlm.nih.gov/pubmed/12057676.

11. Nile Basin Initiative (2012) State of the River Nile Basin. Entebbe (Uganda): Nile Basin Initiative Secretariat.

12. Olson DM, Dinerstein E, Powell GVN, Wikramanayake ED (2002) Conservation Biology for the Biodiversity Crisis. Conserv Biol 16: 1–3.

13. Ricklefs RE (2004) A comprehensive framework for global patterns in biodiversity. Ecol Lett 7: 1–15. Available: http://doi.wiley.com/10.1046/j.1461-0248.2003.00554.x. Accessed 21 January 2014.

14. Gering JC, Crist TO, Veech J a. (2003) Additive Partitioning of Species Diversity across Multiple Spatial Scales: Implications for Regional Conservation of Biodiversity. Conserv Biol 17: 488–499. Available: http://doi.wiley.com/10.1046/j.1523-1739.2003.01465.x.

15. Whittaker RJ, Willis KJ, Field R (2001) Scale and species richness: towards a general, hierarchical theory of species diversity. J Biogeogr 28: 453–470. Available: http://doi.wiley.com/10.1046/j.1365-2699.2001.00563.x.

16. Demeny P (2003) Population Policy: A Concise Summary. Internatio. Demeny P, McNicoll G, editors New York: MacMillan Reference. doi:173.

17. Olden JD (2006) Biotic homogenization: a new research agenda for conservation biogeography. J Biogeogr 33: 2027–2039. Available: http://doi.wiley.com/10.1111/j.1365-2699.2006.01572.x. Accessed 5 May 2014.

18. Olden JD, Poff NL (2004) Ecological processes driving biotic homogenization: testing a mechanistic model using fish faunas. Ecology 85: 1867–1875.

19. Olden JD, Poff NL (2003) Toward a mechanistic understanding and prediction of biotic homogenization. Am Nat 162: 442–460. Available: http://www.ncbi.nlm.nih.gov/pubmed/14582007.

20. Folke C, Carpenter S, Walker B, Scheffer M, Elmqvist T, et al. (2004) Regime Shifts, Resilience, and Biodiversity in Ecosystem Management. Annu Rev Ecol Evol Syst 35: 557–581. Available: http://www.annualreviews.org/doi/abs/10.1146/annurev.ecolsys.35.021103.105711. Accessed 26 October 2012.

21. Ter Braak CJ F. (1986) Canonical Correspondence Analysis: A New Eigenvector Technique for Multivariate Direct Gradient Analysis. Ecol Soc Am 67: 1167–1179.

22. Legendre P, Legendre L (1998) Numerical Ecology. 2nd ed. Amsterdam: Elsevier.

23. Lande R (1996) Statistics and partitioning of species diversity, and similarity among multiple communities. Oikos, Nord Soc 76: 5–13.

24. Godfray HCJ, Lawton JH (2001) Scale and species numbers. Trends Ecol Evol 16: 400–404. Available: http://www.ncbi.nlm.nih.gov/pubmed/11403873.

25. Walkley A., Black I (1934) An examination of the Degtjareff method for determining organic carbon in soils: effect of variations in digestion conditions and of inorganic soil constituents. Soil Sci: 251–263.

26. Kjeldahl J (1883) A new method for the determination of nitrogen in organic matter. Zeitschreft fur Anal Chemie 22.

27. Bray RH, Kurtz LT (1945) Determination of total, organic, and available forms of phosphorus in soils. Soil Sci 59: 39–45.

28. Gee G., Bauder J. (1986) Particle-size analysis. p. 383–411. In A Klute (ed.) Methods of Soil Analysis, Part 1. Physical and Mineralogical Methods. Agronomy Monograph No. 9 (2ed). Am Soc Agron Sci Soc Am Madison, WI.

29. Cleary DFR, Smalla K, Mendonça-Hagler LCS, Gomes NCM (2012) Assessment of variation in bacterial composition among microhabitats in a mangrove environment using DGGE fingerprints and barcoded pyrosequencing. PLoS One 7: e29380. Available: http://www.pubmedcentral.nih.gov/articlerender.fcgi?artid = 3256149&tool = pmcentrez&rendertype = abstract. Accessed 27 October 2012.

30. Reche I, Pulido-Villena E, Morales-Baquero R, Casamayor EO (2005) Does ecosystem size determine aquatic bacterial richness? Ecology 86: 1715–1722. Available: http://www.ncbi.nlm.nih.gov/pubmed/17489473.

31. Whittaker RH (1972) Evolution and Measurement of Species Diversity. Int Assoc Plant Taxon (IAPT) 21: 213–251.

32. Jaccard P (1908) Nouvelles recherches sur la distribution florale. Bull Soc Vaud Sci Nat 44: 223–270.

33. Veech JA, Summerville KS, Crist TO, Gering JC (2002) The additive partitioning of species diversity: recent revival of an old idea. Nord Soc Oikos 99: 3–9.

34. Crist TO, Veech J a, Gering JC, Summerville KS (2003) Partitioning species diversity across landscapes and regions: a hierarchical analysis of alpha, beta, and gamma diversity. Am Nat 162: 734–743. Available: http://www.ncbi.nlm.nih.gov/pubmed/14737711.

35. Chandy S, Gibson DJ, Robertson P a. (2006) Additive partitioning of diversity across hierarchical spatial scales in a forested landscape. J Appl Ecol 43: 792–801. Available: http://doi.wiley.com/10.1111/j.1365-2664.2006.01178.x. Accessed 22 January 2014.

36. Ribeiro DB, Prado PI, Brown Jr KS, Freitas AVL (2008) Additive partitioning of butterfly diversity in a fragmented landscape: importance of scale and implications for conservation. Divers Distrib 14: 961–968. Available: http://doi.wiley.com/10.1111/j.1472-4642.2008.00505.x. Accessed 3 February 2014.

37. Wu F, Yang XJ, Yang JX (2010) Additive diversity partitioning as a guide to regional montane reserve design in Asia: an example from Yunnan Province, China. Divers Distrib 16: 1022–1033. Available: http://doi.wiley.com/10.1111/j.1472-4642.2010.00710.x. Accessed 6 February 2014.

38. Sasaki T, Katabuchi M, Kamiyama C, Shimazaki M, Nakashizuka T, et al. (2012) Diversity partitioning of moorland plant communities across hierarchical spatial scales. Biodivers Conserv. doi:10.1007/s10531-012-0265-7.

39. Veech JA, Crist TO (2009) Partition: software for hierarchical partitioning of species diversity, version 3.0. Available: http://www.users.muohio.edu/cristto/partition.htm.

40. Monkiedje A, Spiteller M, Fotio D, Sukul P (2006) The effect of land use on soil health indicators in peri-urban agriculture in the humid forest zone of southern cameroon. J Environ Qual 35: 2402–2409. Available: http://www.ncbi.nlm.nih.gov/pubmed/17071911. Accessed 9 November 2012.

41. Neris J, Jiménez C, Fuentes J, Morillas G, Tejedor M (2012) Vegetation and land-use effects on soil properties and water infiltration of Andisols in Tenerife (Canary Islands, Spain). Catena 98: 55–62. Available: http://linkinghub.elsevier.com/retrieve/pii/S0341816212001270. Accessed 28 December 2012.

42. Bissett A, Richardson AE, Baker G, Thrall PH (2011) Long-term land use effects on soil microbial community structure and function. Appl Soil Ecol 51: 66–78. Available: http://linkinghub.elsevier.com/retrieve/pii/S0929139311001922. Accessed 9 November 2012.

43. Garbisu C, Alkorta I, Epelde L (2011) Assessment of soil quality using microbial properties and attributes of ecological relevance. Appl Soil Ecol 49: 1–4. Available: http://linkinghub.elsevier.com/retrieve/pii/S0929139311001089. Accessed 9 November 2012.

44. Fierer N, Jackson RB (2006) The diversity and biogeography of soil bacterial communities. Proc Natl Acad Sci U S A 103: 626–631. Available: http://www.pubmedcentral.nih.gov/articlerender.fcgi?artid = 1334650&tool = pmcentrez&rendertype = abstract.

45. Shange RS, Ankumah RO, Ibekwe AM, Zabawa R, Dowd SE (2012) Distinct soil bacterial communities revealed under a diversely managed agroecosystem. PLoS One 7: e40338. Available: http://www.pubmedcentral.nih.gov/articlerender.fcgi?artid = 3402512&tool = pmcentrez&rendertype = abstract. Accessed 1 November 2012.

46. Wang Y, Boyd E, Crane S, Lu-Irving P, Krabbenhoft D, et al. (2011) Environmental conditions constrain the distribution and diversity of archaeal merA in Yellowstone National Park, Wyoming, U.S.A. Microb Ecol62: 739–752. Available: http://www.ncbi.nlm.nih.gov/pubmed/21713435. Accessed 9 November 2012.

47. Fierer N, Bradford MA, Jackson RB (2007) Toward an ecological classification of soil bacteria. Ecology 88: 1354–1364. Available: http://www.ncbi.nlm.nih.gov/pubmed/17601128.

48. Saiya-Cork K., Sinsabaugh R., Zak D. (2002) The effects of long term nitrogen deposition on extracellular enzyme activity in an Acer saccharum forest soil. Soil Biol Biochem 34: 1309–1315. Available: http://linkinghub.elsevier.com/retrieve/pii/S0038071702000743.

49. Steenwerth K, Jackson L, Calderon F, Scow K, Rolston D (2005) Response of microbial community composition and activity in agricultural and grassland soils after a simulated rainfall. Soil Biol Biochem 37: 2249–2262. Available: http://linkinghub.elsevier.com/retrieve/pii/S0038071705001549. Accessed 9 November 2012.

50. De Vries FT, Manning P, Tallowin JRB, Mortimer SR, Pilgrim ES, et al. (2012) Abiotic drivers and plant traits explain landscape-scale patterns in soil microbial communities. Ecol Lett 15: 1230–1239. Available: http://www.ncbi.nlm.nih.gov/pubmed/22882451. Accessed 2 November 2012.

51. Hartman WH, Richardson CJ, Vilgalys R, Bruland GL (2008) Environmental and anthropogenic controls over bacterial communities in wetland soils. Proc Natl Acad Sci U S A 105: 17842–17847. Available: http://www.pubmedcentral.nih.gov/articlerender.fcgi?artid = 2584698&tool = pmcentrez&rendertype = abstract.

52. Cadotte MW, Cardinale BJ, Oakley TH (2008) Evolutionary history and the effect of biodiversity on plant productivity. Proc Natl Acad Sci U S A 105: 17012–17017. Available: http://www.pubmedcentral.nih.gov/articlerender.fcgi?artid = 2579369&tool = pmcentrez&rendertype = abstract.

53. Garbeva P, Veen JA Van, Elsas JD Van (2004) Microbial Diversity in Soil: Selection of Microbial Populations by Plant and Soil Type and Implications for Disease Suppressiveness. Annu Rev Phytopathol 42: 243–270. doi:10.1146/annurev.phyto.42.012604.135455.

54. Schulz S, Giebler J, Chatzinotas A, Wick LY, Fetzer I, et al. (2012) Plant litter and soil type drive abundance, activity and community structure of alkB harbouring microbes in different soil compartments. ISME J doi:10.103: 1763–1774.

55. Kuramae EE, Yergeau E, Wong LC, Pijl AS, van Veen J a, et al. (2012) Soil characteristics more strongly influence soil bacterial communities than land-use type. FEMS Microbiol Ecol 79: 12–24. Available: http://www.ncbi.nlm.nih.gov/pubmed/22066695. Accessed 7 November 2012.

56. Nacke H, Thürmer A, Wollherr A, Will C, Hodac L, et al. (2011) Pyrosequencing-based assessment of bacterial community structure along different management types in German forest and grassland soils. PLoS One 6: e17000. Available: http://www.pubmedcentral.nih.gov/articlerender.

fcgi?artid = 3040199&tool = pmcentrez&rendertype = abstract. Accessed 5 November 2012.

57. Waldrop MP, Balser TC, Firestone MK (2001) Linking microbial community composition to function in a tropical soil. Soil Biol Biochem 32: 1837–1846.

58. Dinsdale E a, Edwards R a, Hall D, Angly F, Breitbart M, et al. (2008) Functional metagenomic profiling of nine biomes. Nature 452: 629–632. Available: http://www.ncbi.nlm.nih.gov/pubmed/18337718. Accessed 26 October 2012.

59. Lauber CL, Hamady M, Knight R, Fierer N (2009) Pyrosequencing-based assessment of soil pH as a predictor of soil bacterial community structure at the continental scale. Appl Environ Microbiol 75: 5111–5120. Available: http://www. pubmedcentral.nih.gov/articlerender.fcgi?artid = 2725504&tool = pmcentrez& rendertype = abstract. Accessed 19 March 2014.

60. Russo SE, Legge R, Weber K a., Brodie EL, Goldfarb KC, et al. (2012) Bacterial community structure of contrasting soils underlying Bornean rain forests: Inferences from microarray and next-generation sequencing methods. Soil Biol Biochem 55: 48–59. Available: http://linkinghub.elsevier.com/retrieve/pii/ S0038071712002362. Accessed 8 November 2012.

61. Winterbottom B, Eilu G (2006) Uganda Biodiversity and Tropical Forest Assessment. Washington DC.

62. Laurance WF (1999) Reflections on the tropical deforestation crisis. Biol Conserv 91: 109–117.

63. Scheckenbach F, Hausmann K, Wylezich C, Weitere M, Arndt H (2010) Large-scale patterns in biodiversity of microbial eukaryotes from the abyssal sea floor. Proc Natl Acad Sci U S A 107: 115–120. Available: http://www. pubmedcentral.nih.gov/articlerender.fcgi?artid = 2806785&tool = pmcentrez& rendertype = abstract. Accessed 1 November 2012.

64. Martiny JBH, Bohannan BJM, Brown JH, Colwell RK, Fuhrman J a, et al. (2006) Microbial biogeography: putting microorganisms on the map. Nat Rev Microbiol 4: 102–112. Available: http://www.ncbi.nlm.nih.gov/pubmed/ 16415926. Accessed 27 October 2012.

65. McKinney M, Lockwood J (1999) Biotic homogenization: a few winners replacing many losers in the next mass extinction. Trends Ecol Evol 14: 450–453. Available: http://www.ncbi.nlm.nih.gov/pubmed/10511724.

66. Olden JD, Rooney TP (2006) On defining and quantifying biotic homogenization. Glob Ecol Biogeogr 15: 113–120. doi:10.1111/j.1466-822x.2006.00214.x.

67. Devictor V, Julliard R, Clavel J, Jiguet F, Lee A, et al. (2008) Functional biotic homogenization of bird communities in disturbed landscapes. Glob Ecol Biogeogr 17: 252–261. Available: http://doi.wiley.com/10.1111/j.1466-8238. 2007.00364.x. Accessed 21 January 2014.

68. Summerville KS, Boulware MJ, Veech J a., Crist TO (2003) Spatial Variation in Species Diversity and Composition of Forest Lepidoptera in Eastern Deciduous Forests of North America. Conserv Biol 17: 1045–1057. Available: http://doi. wiley.com/10.1046/j.1523-1739.2003.02059.x.

69. Weiher E, Howe A (2003) Scale-dependence of environmental effects on species richness in oak savannas. J Veg Sci 14: 917–920.

70. Obua J, Agea JG, Ogwal JJ (2010) Status of forests in Uganda. Afr J Ecol 48: 853–859.

71. National Environment Management Authority (NEMA) (2008) State of the Environment Report for Uganda.

72. Amann RI, Ludwig W, Schleifer K-H (1995) Phylogenetic identification and in situ detection of individual microbial cells without cultivation. Microb Rev 59: 143–169.

73. Vainio EJ, Hantula J (2000) Direct analysis of wood-inhabiting fungi using denaturing gradient gel electrophoresis of amplified ribosomal DNA. Mycol Res 104: 927–936.

Can Plants Grow on Mars and the Moon: A Growth Experiment on Mars and Moon Soil Simulants

G. W. Wieger Wamelink[1]*, Joep Y. Frissel[1], Wilfred H. J. Krijnen[2], M. Rinie Verwoert[2], Paul W. Goedhart[3]

1 Alterra, Wageningen UR, Wageningen, the Netherlands, **2** Unifarm, Wageningen UR, Wageningen, the Netherlands, **3** Biometris, Wageningen UR, Wageningen, the Netherlands

Abstract

When humans will settle on the moon or Mars they will have to eat there. Food may be flown in. An alternative could be to cultivate plants at the site itself, preferably in native soils. We report on the first large-scale controlled experiment to investigate the possibility of growing plants in Mars and moon soil simulants. The results show that plants are able to germinate and grow on both Martian and moon soil simulant for a period of 50 days without any addition of nutrients. Growth and flowering on Mars regolith simulant was much better than on moon regolith simulant and even slightly better than on our control nutrient poor river soil. Reflexed stonecrop (a wild plant); the crops tomato, wheat, and cress; and the green manure species field mustard performed particularly well. The latter three flowered, and cress and field mustard also produced seeds. Our results show that in principle it is possible to grow crops and other plant species in Martian and Lunar soil simulants. However, many questions remain about the simulants' water carrying capacity and other physical characteristics and also whether the simulants are representative of the real soils.

Editor: Alberto de la Fuente, Leibniz-Institute for Farm Animal Biology (FBN), Germany

Funding: This research was supported by the Dutch Ministery of Economic Affairs. The funders had no role in study design, data collection and analysis, decision to publish, or preparation of the manuscript.

Competing Interests: The authors have declared that no competing interests exist.

* Email: Wieger.wamelink@wur.nl

Introduction

Lunar and Mars explorations have provided information about the mineral composition of the soils of these solar objects. In addition to rocks they contain large amounts of sand-like soils or regoliths. All essential minerals for the growth of plants appear to be present in sufficient quantities in both soils probably with the exception of reactive nitrogen. Nitrogen in reactive form (NO_3, NH_4) is one of the essential minerals necessary for almost all plant growth [1]. The major source of reactive nitrogen on Earth is the mineralisation of organic matter [1]. However organic matter is absent on both Mars and moon although they do contain carbon [2–6]. Nitrogen in reactive form (NO_3, NH_4) is one of the essential minerals necessary for almost all plant growth [1]. Reactive nitrogen is part of the material in our solar system and is part of solar wind, a source of reactive nitrogen on the moon and Mars [3,7]. Reactive nitrogen may also arise as an effect of lightning or volcanic activity [8,9] and both processes may occur on Mars. This indicates that in principle reactive nitrogen could be present [7,10]. However, the Mars Pathfinder was not able to detect reactive nitrogen [11]. Thus the actual presence of major quantities of reactive nitrogen remains uncertain. The major source of reactive nitrogen on Earth is the mineralisation of organic matter [1], which is absent on both Mars and moon. The absence of sufficient reactive nitrogen may be solved by using nitrogen fixing species. In symbioses with bacteria [12,13] these nitrogen fixers are able to bind nitrogen from the air and transform it into nitrates, a process which requires nitrogen in the atmosphere. However, there is no atmosphere on the moon, and on Mars it is only minimally present and contains traces of nitrogen. Metals like aluminium and chromium are also present in the extra-terrestrial soils. Aluminium is known to disturb plant growth and even lead to plant death [14]. Another essential for plant growth is liquid water. Liquid water is not (moon) or possibly very limited present (Mars). Ice is present on both Mars and moon, and could be used after harvest [15–17]. Many plant species may be grown on water cultures, e.g. tomatoes or paprika, but not all. Therefore, local soils could be used to grow crops, at least partly.

During the Apollo project there has been no experiment with plant growth on the moon. However experiments on earth have been carried out with the brought back moon material. These experiments did not include growth of plants on moon soil. Instead plants were exposed to moon stones by rubbing them and even small amounts were added to growth medium. These experiments indicated that there were no toxic effects of moon soil on short term plant growth [18], for an overview see Ferl and Paul [19]. Ferl and Paul [19] also provide pictures of the model plant *Arabidopsis thaliana* grown on a moon regolith simulant (JSC1a). Studies with moon rock simulant (anorthosite) were carried out with the model plant *Tagetes patula* [20,21]. These studies revealed that these plants were able to grow with and without the addition of bacteria [20,21], and that plants were able to blossom [20]. There have been plant growth experiments with Mars regolith simulant as well. Experiments with bacteria on Mars soil simulant revealed that growth is possible, including nitrogen fixing bacteria [22].

Our goal was to investigate whether or not species of the three groups wild plants, crops and nitrogen fixers (Table 1), would germinate and live long enough to go through the first stages of plant development on artificial Mars and moon regoliths. If this would be the case it is conceivable that plant growth is possible within an artificial surrounding on Mars and moon surface, although our experiment was conducted on Earth with its deviating gravity. Moreover, we assumed that plant cultivation will be carried out in closed surroundings with Earth like light and atmospheric conditions.

Materials and Methods

Regoliths

Mars and moon regolith simulant were purchased from Orbitec (http://www.orbitec.com). Both regoliths were manufactured by NASA (for Mars we used JSC-1A Mars regolith simulant, for Moon we used the JSC1-1A lunar regolith simulant) [23,24]. Since the Mars and moon regolith simulants are comparable to Earth soils, at least in mineral composition [23–28], they can be mimicked by using volcanic Earth soils, as has been done by NASA [23,24].

As a control we used coarse river Rhine soil from 10 m deep layers which is nutrient poor, and free from organic matter and

seeds. Since the moon and Mars simulants had only been analysed for mineral content and particle size, we also analysed them for nutrients that are available for plant species. All three soil types were analysed for soil pH water, Organic matter content, Total N and P content (both destructive), NH_4, NO_2+, NO_3, PO_4, Al, Fe, K and Cr (all seven in $CaCl_2$ extract). All analyses were repeated two times according to standard protocol (RvA-accreditation for test laboratories; registration number scope: 342). These soil parameters are typically used to explain species occurrence on Earth [29].

The analysis revealed that the moon regolith simulant is truly nutrient poor, though it contains a small amount of nitrates and ammonium. The Mars regolith simulant also contains traces of nitrates of ammonium, and also a significant amount of carbon (Table 2). The pH of all three soils is high. The pH of the moon regolith is that high that it may be problematic for many plant species, especially for crops [30]. We applied the regoliths and the control earth sand as supplied, the sands were not sterilised, since sterilisation may alter its properties.

Table 1. Species used in the experiment, the species group it belongs to and information about the species trait partly based on Wamelink et al. [29,30,36].

Latin	English	Group	Abbreviation	description
Arnica montana	Leopards bane	Occurring naturally	ARM	Species of nutrient poor dry soil conditions with a light acidic pH.
Sinapsis arvensis	Field mustard	Occurring naturally	SIA	Species of nutrient rich soil conditions. Often used as green manure in winter.
Urtica dioica	Stinging nettle	Occurring naturally	URD	Ruderal species, can become dominant under nutrient rich soil conditions, mostly on soils with a light acidic till basic pH.
Cirsium palustre	Marsh thistle	Occurring naturally	CIP	Ruderal species, can become dominant under nutrient rich soil conditions, mostly on soils with a light acidic till basic pH.
Sedum reflexum	Reflexed stonecrop	Occurring naturally	SER	Species of (extreme) nutrient poor (extreme) dry soil conditions, mostly on soils with a light acidic till basic pH.
Festuca rubra	Red fescue	Occurring naturally	FER	Gras species that can withstand many circumstances from nutrient poor acidic dry till nutrient rich basic moist conditions.
Vicia sativa sativa	Common vetch	Nitrogen fixer	VIS	Species used as green manure or livestock fodder and eatable for humans. Cattle feed faster on vetch than on most grasses.
Lupinus angustifolius	Lupin	Nitrogen fixer	LUA	Known of soil improvement and is used as green manure or as a grain legume for human consumption or animal feed.
Melilotus officinalis	Yellow sweet clover	Nitrogen fixer	MEO	Biannual species that likes basic soils and is drought resistant. It does not like shaded places.
Lotus pedunculatus	Greater birds'-foot trefoil	Nitrogen fixer	LOP	Moist loving species of light acidic till neutral modest nutrient rich soils
Solanum lycopersicum	Tomato	Crop	SOL	The Tomato can be grown as an annual or perennial. It likes light acidic till basic soils that can be dry till wet.
Secale cereale	Rye	Crop	SEC	The seeds of Rye can be used for many eatable products. It is able to grow at relative low temperatures (winter hardy) and can grow in nutrient poor light acidic till basic dry soils.
Daucus carota s. sativus	Carrot	Crop	DAC	Biannual species, that likes sunny places and moist light acidic till basic not to nutrient rich soils.
Lepidium sativum	Garden cress	Crop	LES	Fast growing species that likes moist circumstances, but is known to grow almost anywhere.

Table 2. Analyses of the soil samples, in red the detection limits of the analysis.

Method		SFA-Nt/Pt destruction with H_2SO_4-H_2O_2-Se		ICP-AES extraction in 0.01 M $CaCl_2$			ICP-MS	SFA extraction with 0,01M $CaCl_2$			pH-H_2O	LECO-CHN		$CaCO_3$
Element		Nt	Pt	Al	Fe	K	Cr	N-NH4	N-(NO_3+NO_2)	P-PO4	at 20±1°C	C-elementary	N-elementary	Scheibler
Unit		[g/kg]	[mg/kg]	[mg/kg]	[mg/kg]	[mg/kg]	[µg/kg]	[mg/kg]	[mg/kg]	[mg/kg]	-	[g/kg]	[g/kg]	[%]
detection limit		0.3	100.0	0.5	3.0	3.0	5.0	1.0	0.5	0.4	-	3.0	0.3	
average	Earth	0.0	57.3	0.0	0.0	4.7	2.0	0.5	4.2	0.0	8.3	3.2	0.0	
	moon	0.0	1003.0	0.5	0.0	27.0	0.0	0.3	4.2	0.2	9.6	3.0	0.0	
	Mars	2.6	2487.7	0.0	0.0	138.0	0.0	3.9	2.1	0.0	7.3	30.1	2.5	0.2
s.e	Earth	0.2	1.5	0.2	0.0	0.6	3.5	0.1	0.2	0.0	0.0	0.1	0.1	
	moon	0.2	11.0	0.1	0.0	0.0	0.0	0.1	0.1	0.0	0.0	0.1	0.1	
	Mars	0.1	28.4	0.2	0.0	1.0	0.6	0.1	0.0	0.0	0.0	0.5	0.1	0.0

Species selection

Species were selected from three groups: four different crops, four nitrogen fixers and six wild plants which occur naturally in the Netherlands (Table 1). Only species with relatively small seeds were chosen so that the nutrient stock in the seeds would be quickly depleted and the plant becomes totally dependent on what is available in the soils for its growth. For the wild plants we chose species that are able to grow either under nutrient poor circumstances or under a wide range of circumstances (see Table 1) based on the responses of the species to abiotic conditions [29,30]. Note that although species may have limits for growth conditions in the field they are often able to grow in monocultures under different circumstances, e.g. more nutrient rich or nutrient poor conditions, because of lack of more competitive species. To be able to monitor the first growth stages we used seeds of the species. The crop and nitrogen fixer seed were bought at the local shop (Welkoop, Wageningen), and the wild plant seeds at Cruydt Hoeck (Nijeberkoop). The latter seeds were collected in the field. Externally present bacteria on the seeds, if any, were not killed.

Experimental design and observations

Small pots were filled with 100 g moon soil simulant, 100 g Earth soil or 50 g Mars soil simulant and 25 g demineralized water was added to each pot. The mass of the simulants added was different since we wanted to fill the pots with approximately the same volume to have the same column height. A filter was placed on the bottom of each pot to prevent soil from leaking. For each soil type and plant species twenty replica pots were used. This resulted in 840 pots (3 soils×14 species×20 replicas). In each pot we positioned five seeds, giving 100 seeds per species - soil combination. The pots were placed in a glasshouse in a completely randomized block design where each block constitutes a replicate (Fig. 1). Each pot was placed in a petri dish (without cap) to hold excessive water and to prevent roots growing into other pots. The pots were placed on a large table in the glasshouse (Fig. 2).

The experiment started of April 8th 2013. Temperature in the glasshouse was maintained at around 20°C. During the experimental period average temperature was 21.1±3.02°C and air humidity was 65.0±15.5% both based on 24 hour recording with a 5 minutes interval. Mean day time lasted for 16 hours. If the sunlight intensity was below 150 watt/m^2 lamps yielding 80 µmol (HS2000 from Hortilux Schréder) were switched on. The pots were watered once or twice a day depending on the evaporation rate by spraying with demineralised water (about 10 litres for the whole experiment for each occasion). We used demineralized water to mimic water from Mars and moon and to prevent pollution with (for example) nutrients that are present in tap water. Ambient air was used.

Seeds were scored on germination, first leaf production, bud forming, flowering and seed setting. At the end of the experiment, 50 days after April 8th, total biomass was harvested and, after cleaning, dried in a stove for 24 hours at 70°C; After cooling down above and below ground biomass were weighed separately. For 25 experimental units the total biomass was smaller than the weighting limit. For those units a value of 0.5 mg (for plants that germinated, but could not be recovered at the end of the experiment) or 0.1 mg (for plants that died before the end of the experiment directly after germination) was assigned to the total biomass. Above and below ground biomass was set to half this value. For 21 units the above ground biomass was smaller than the weighting limit and this was also true for the below ground biomass of 25 units. In these cases the corresponding biomass was set to 0.1 mg.

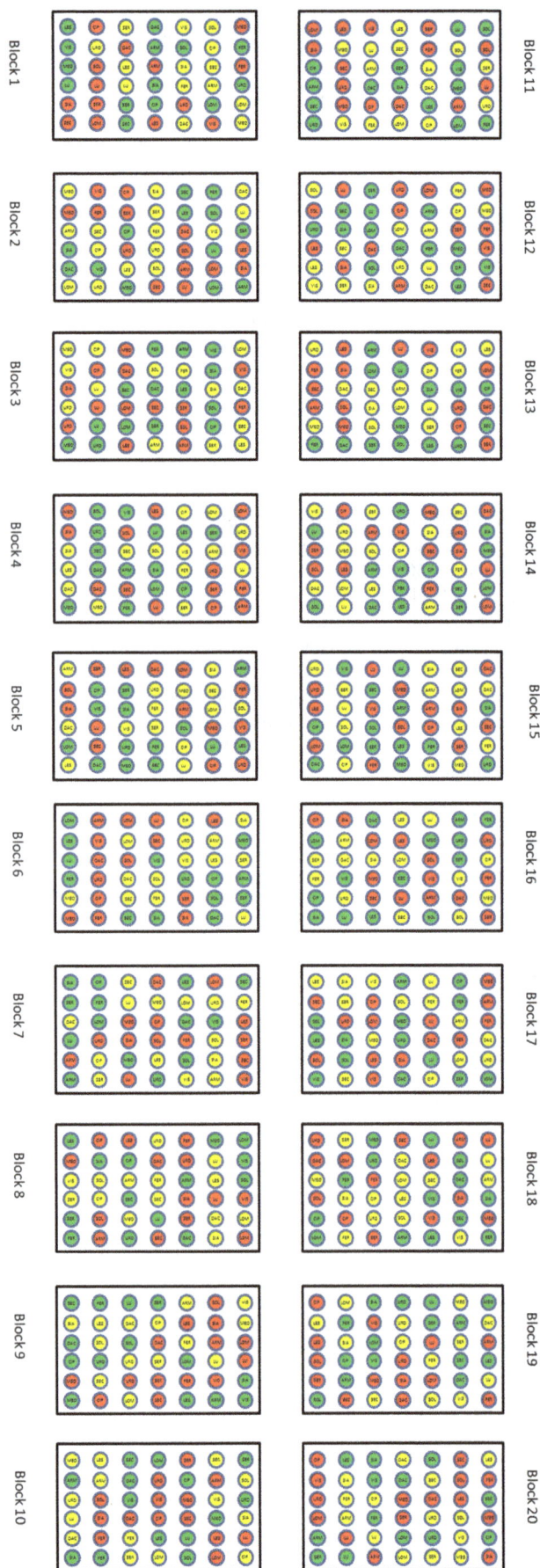

Figure 1. Design of the experiment with the first ten blocks the west oriented part of the experiment and the second ten blocks the east oriented part of the experiment. For abbreviations of the species see Table 1.

Statistical Analysis

Logistic regression was used to statistically analyse the number of germinated seeds in each pot, as well as the number of seeds which developed leaves, which developed flowers (including buds), and the numbers of plants which were still alive after 50 days. A pairwise likelihood ratio test, separately for each species and accounting for differences between blocks, was employed to test whether Earth, moon and Mars soil simulants give different results. When necessary, overdispersion was accounted for by inflating the binomial variance by an unknown factor and then using quasi likelihood rather than maximum likelihood [31].

An analysis of variance, again separately for each species and accounting for block effects, was performed on the logarithm of the total, above and below ground biomass, as well as on the ratio of the above and below ground biomass. The log transform was employed because this stabilizes the variance. Pairwise difference t-test between the soil types were carried out. Note that this is a conditional analysis since units with no biomass are excluded. This implies that no biomass is given for *V. sativa sativa* on the moon because none of these seeds germinated.

Results

Common vetch, a nitrogen fixer, did not germinate on moon soil. All other plant species did germinate with different proportions on all soils (Fig. 3; background information can be found in Table S1 and S2). In general the germination percentage on Martian soil simulant is highest and lowest on the moon soil simulant (Fig. 3). On average the four crop species have the highest germination percentages, although some species (Reflexed stonecrop, Red fescue, Yellow sweet clover and Greater birds'-foot trefoil) from the other two groups have similar germination percentages. Differences in germination percentages are most likely due to seed quality. The seeds of the crops Carrot, Cress and Tomato are controlled and have a high quality. The seeds of the other species are harvested from the field and except Rye have not been improved by plant breeding. These seed lots may therefore contain less or non-viable seeds. The percentages of plants that form leaves are sometimes considerably lower than the percentages for germination, indicating that some plants stop developing or even die. Leaf forming occurred most on Martian soil simulant and least on moon soil simulant. This trend is also present for species that form flowers or seeds. Only three species reach these stages, Field mustard, Rye and Cress (the last two being crops). Field mustard (only on Mars) and Cress (on Mars and Earth) also formed seeds. For examples see photo 1–10 (File S1). Also for the percentage plants still alive after 50 days, Martian soil simulant performed best and moon soil simulant worst. Martian soil simulant also performed better than Earth soil for most species. Leopards bane, Field mustard and Common vetch had no living plants left after 50 days on moon soil.

The biomass at the end of the experiment was significantly higher for eleven out of the fourteen species on Martian soil simulant as compared to both other soils. The biomass for earth and moon soil simulant is often quite similar (Fig. 4), although for nine species the biomass increment on Earth soil was significantly higher than on moon soil simulant. Apparently, in general, plants were able to develop at the same rate on Martian and Earth soil

Figure 2. Block 2 of the experiment, with randomly placed pots, 14 days after the start of the experiment. Each block contains 42 pots. Block 12 is visible in the background. The labels in the pots show the pot number, the species (from left to right on the first row Yellow sweet clover (twice), Leopards bane, Field Mustard, Carrot and Red fescue) and the soil type (L for moon or Lunar, M for Mars and E for Earth) combined with the block number (2).

simulants, but biomass increment was much higher on Mars simulant. This is reflected in both below and aboveground biomass, although there are differences at the species level.

Discussion

We found germination and plant growth for both moon and Mars soil simulants. Our results are in line with earlier research on *Arabidopsis thaliana* and *Tagetes patula* [19–21] on moon regolith simulant and moon rock simulant, though our results appear to be less promising. Kozyrovska et al. [20] had blossoming plants of *T. patula*, where we had only one plant of *Sinapsis arvensis* that formed a flower butt, but died before flowering.

On average species in Martian soil simulant performed significantly better than plants in Earth soil with respect to biomass increment. Although the Earth soil used, which was coarse and very nutrient poor, is not the best soil to grow crops on, we expected it to perform at least as well as the other two soils. However, in the warmer periods it was difficult to keep the water content in the pots high enough, despite spraying twice a day. The Mars soil simulant resembles loess-like soils from Europe and holds water better than the other two soils. Moon soil simulant dried out

fastest. It therefore is essential that further research on the physical characteristics of the extra-terrestrial soils is conducted, as well as the way they could be irrigated. The larger water holding capacity of Martian soil simulant may explain its better performance and, partly, the underperformance of moon soil simulant. The high pH may also explain the lagging growth on the moon soil simulant and also on the Earth soil. Important for plant growth is not only the presence of nutrients, but also the balance between them. Both soils are rather imbalanced for nutrients; where the artificial moon soil lacks nitrates, the artificial soil lacks of phosphate. If nutrients are added in future experiments this imbalance has to be corrected as well, besides the addition of nutrients itself. The presence of a high C-elementary content in the Mars soil simulant is surprising. We also chemically analysed organic matter content in the simulant, but that resulted in obviously wrong results. The standard procedure includes backing the soil at 550°C. The problem is that part of the oxides, especially the iron oxides, evaporates as well, clearly yielding wrong results. Nevertheless a part of the origin of the Carbon content may be from organic matter. It may be a result of the way the soil is 'harvested' on Hawaii, leaving traces of organic carbon in the soil. Kral et al. [22] found traces of organic material in the JSC-1 simulant. It may also

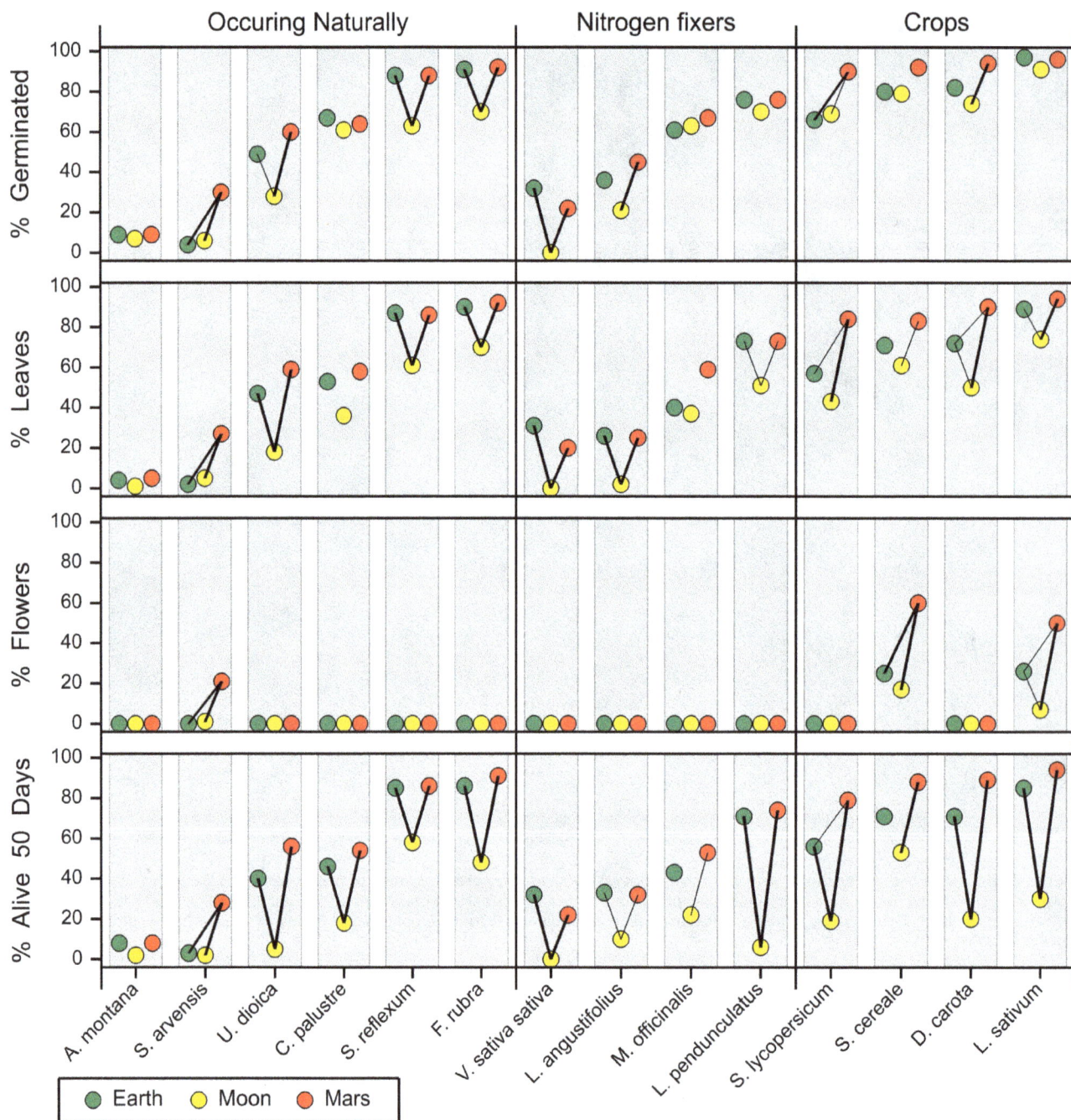

Figure 3. Percentage germination, leave formers, plants forming flowers and plants still alive after 50 days per species. All results are after 50 days and percentages are based on all 100 seeds per plant species-soil type combination Pairwise differences are displayed by a line which joins soil types which are significantly different at the 1% (thin line) and 0.1% (thick line) significance level. Background information can be found in Table S1 and S2.

partly explain why the Mars soil was able to hold water best, as organic matter is more capable of holding water than bare sand. There is no organic matter on Mars [2–6], as far as we now, so this would make the Mars simulant we used less suitable for experiments to investigate the potential of Mars soil, unless the experiment has as goal to test the potential of the soil after adding organic matter. In our experiment this was the reason to test the legumes. They can be used as green fertilizer and after growth mixed with the soil. Visual inspection of the Mars soil simulant did

not reveal large quantities of organic matter. However, further test on the simulant is advisable.

This experiment was carried out in pots. Some of the crops on Mars or moon may be cultivated in pots, but part of the crops may possibly be cultivated in full soil (in growth chambers or under domes). Moist conditions will then be different and may give rise to different results between pots and full soil. It is therefore of interest to conduct future experiments in full soil cultivation as well.

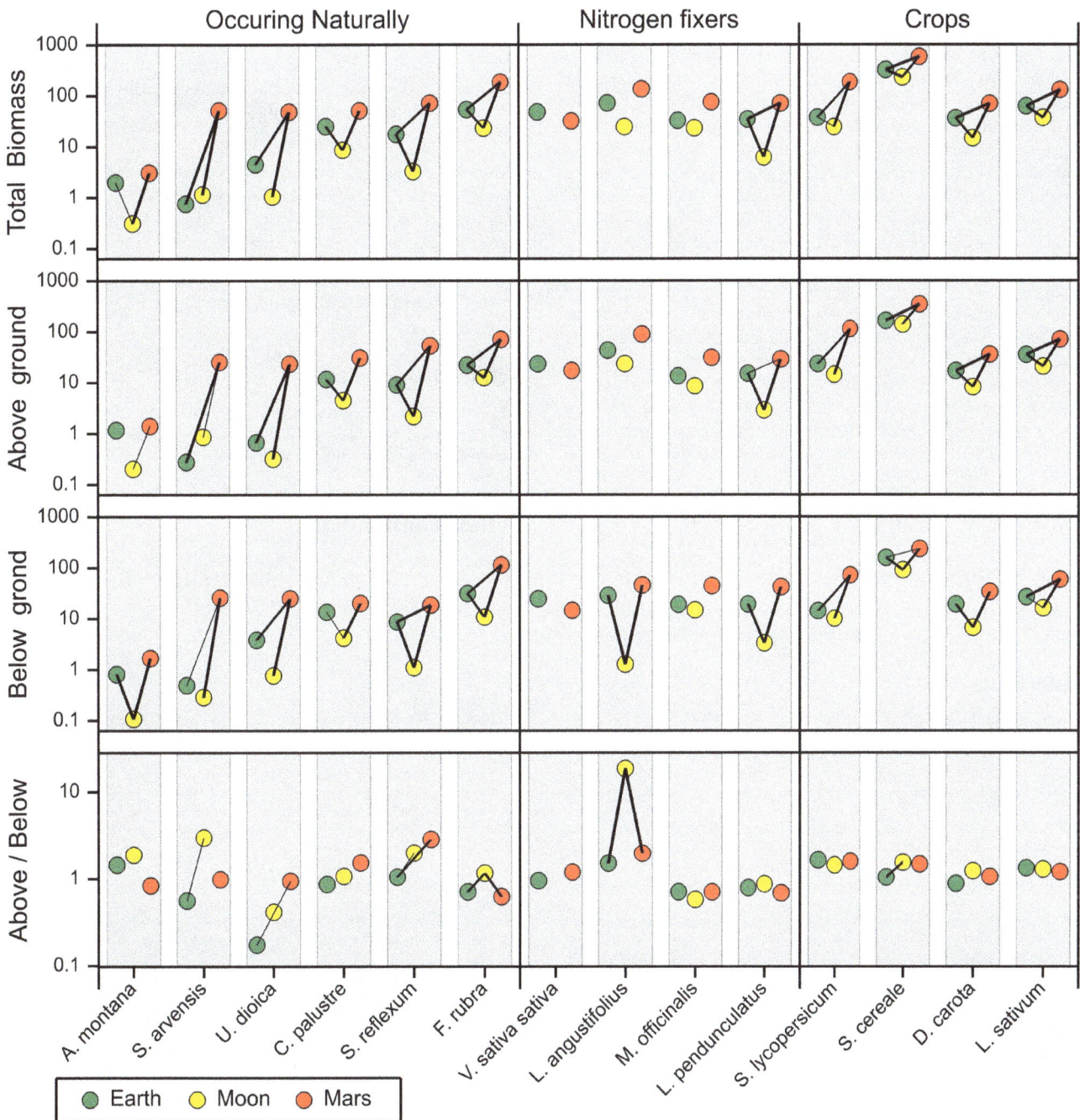

Figure 4. Average biomass results per species at the end of the 50 day experiment and the resulting aboveground belowground biomass ratio. Biomasses are given in mg dry weight on 10 log scale. The triangle indicates an outlier for Lupine (above/below 19.7). For Common vetch there is no ratio given because both above- and belowground biomass are zero. Pairwise differences are displayed by a line which joins soil types which are significantly different at the 1% (thin line) and 0.1% (thick line) significance level. Background information can be found in Table S1 and S2.

The reason for using nitrogen fixers in our experiment is that they may possibly compensate for the lack of sufficient reactive nitrogen in artificial Martian and moon soil. At the first stage of colonisation, these species can be used to enrich the soils with nitrogen, essential for all other plants, by mixing them with the soil after their growth as is commonly done in the Netherlands in winter [32–34]. This may be done in addition to manure brought from Earth or from human faeces. All chosen nitrogen fixers may perform this function; however Common vetch did not perform

very well on Martian soil simulant, which may indicate that inoculation with nitrogen fixing bacteria may be necessary. We did not inoculate the soil simulants with nitrogen fixing bacteria in this experiment, although we did not sterilise the simulants nor the seeds. The bacteria could thus be present, but we did not test that in our experiment. In future experiments we will inoculate the soils with these bacteria. The nitrogen fixers may also play a role in detoxifying soils polluted with metals [35].

Conclusions

Except for Common vetch all other plants germinated in some proportion on all three tested soils; the Mars soil simulant, the moon soil simulant and the River Rhine soil (control). Rye, cress and field mustard flowered, the latter two also formed seeds. Germination and biomass forming differed between species and soil types. The Mars soil simulant gave the highest biomass production, the moon soil simulant the lowest. On the moon soil simulant many germinated plants died or stayed very small. This may be due to the high soil pH, the moist holding capacity and or the free aluminium in the simulant. Our results show that it is in principle possible to grow plants in Martian and Lunar soil simulants although there was only one plant that formed a flower butt on moon soil simulant. Whether this extends to growing plants on Mars or the moon in full soils themselves remains an open question. More research is needed about the representativeness of the simulants, water holding capacity and other physical characteristics of the soils, whether our results extend to growing plants in full soil, the availability of reactive nitrogen on Mars and moon combined with the addition of nutrients and creating a balanced nutrient availability, and the influence of gravity, light and other conditions.

Supporting Information

Table S1 Percentages seeds which germinated, produced leaves, were flowering and were alive after 50 days. P values of pairwise difference tests, separately for each species, are given in the last three columns. P-values smaller than 0.01 are given in bold. All species soil type combinations had 20 replicas and five seeds were positioned in every pot. Note that due to the many replicas small differences are statistically significant.
(DOCX)

Table S2 Number of seeds that Germinated, formed green leaves, flowered, set seeds, number of plants alive after 50 days, total biomass per pot, below ground biomass per pot and above ground biomass per pot. (see Excel file).
(XLSX)

File S1 **Photos of the experiment.**
(DOCX)

Acknowledgments

R.M.A. Wegman, T. Busser and M. van Adrichem helped with start of the experiment and the harvest. We thank F. van der Helm and one anonymous reviewer for their helpful comments on a previous version of this manuscript.

Author Contributions

Conceived and designed the experiments: GWWW PWG. Performed the experiments: JYF WHJK MRV GWWW. Analyzed the data: GWWW PWG JYF. Wrote the paper: GWWW PWG JYF.

References

1. Stevens CJ, Manning P, van den Berg LJL, de Graaf MCC, Wamelink GWW, et al. (2011) Ecosystem responses to reduced and oxidised nitrogen inputs in European terrestrial habitats. Environmental Pollution 159: 665–676.
2. Palomba E, Zinzi A, Cloutis E-A, D'Amore M, Grassi D, et al. (2009) Evidence for Mg-rich carbonates on Mars from a 3.9 lm absorption feature. Icarus 203: 58–65.
3. Muller O (1974) Solar wind nitrogen and indigenous nitrogen in Apollo 17 lunar samples. Proc. Lunar Sci. Conf. 5th, Vol. 2, 1907–1918. Pergamon Press.
4. Parnell J (2005) Extraction of organic signatures from carbonates and evaporites: from mineral deposits to Mars. Proceedings of the Geologists Association 116: 281–291.
5. Brack A, Pillinger CT (1998). Life on Mars: chemical arguments and clues from Martian meteorites. Extremophilest 2, 313–319.
6. Leshin LA, Mahaffy PR, Webster CR, Cabane M, Coll P, et al. (2013) Volatile, isotope, and organic analysis of Martian fines with the Mars Curiosity Rover. Science 341: 1238937-1–1238937-9.
7. Mancinelli RL (1996) The search for nitrogen compounds on the surface. Adv Space Res 18: 241–248.
8. Golubyatnikov LL, Mokhov II, Eliseev AV (2013) Nitrogen cycle in the Earth climatic system and its modeling. Atmospheric and Oceanic Physics 49: 229–243.
9. Vitousek PM, Menge DNL, Reed SC, Cleveland CC (2013) Biological nitrogen fixation: rates, patterns and ecological controls in terrestrial ecosystems. Phil Trans R Soc B 368 no. 1621 20130119.
10. Mancinelli RL, Banin A (2003) Where is the nitrogen on Mars? International Journal of Astrobiology 2: 217–225.
11. Foley CN, Economou T, Clayton RN (2003) Final chemical results from the Mars Pathfinder alpha proton X-ray spectrometer. Journal of Geophysical Research 108: 37-1–37-21.
12. Mylona P, Pawlowski K, Bisseling T (1995) Symbiotic Nitrogen Fixation. Plant Cell 7, 869–885.
13. Foy CD (1984) Physiological Effects of Hydrogen, Aluminum, and Manganese Toxicities in Acid Soil. In: Adams F, editor. Soil Acidity and Liming. pp. 57–97. American Society of Agronomy, Crop Science Society of America, Soil Science Society of America.
14. Garg N (2007) Symbiotic nitrogen fixation in legume nodules: process and signalling. Agron Sustain Dev 27: 59–68.
15. Bullock MA, Moore JM, Mellon MT (2004) Laboratory simulations of Mars aqueous geochemistry. Icarus 170: 404–423.
16. Möhlmann DTF (2004) Water in the upper Martian surface at mid- and low-latitudes: presence, state, and consequences. Icarus 168: 318–323.
17. Hui H, Peslier AH, Zhang Y, Neal CR (2013) Water in lunar anorthosites and evidence for a wet early moon. Nature Geoscience 6: 177–180.
18. Baur PS, Clark RS, Walkinshaw CH, Scholes VE (1974) VE Uptake and translocation of elements from Apollo 11 lunar material by lettuce seedlings. Phyton 32: 133–142.
19. Ferl RJ, Paul AL (2010) Lunar Plant Biology—A Review of the Apollo Era. Astrobiology 10: 261–274.
20. Kozyrovska NO, Lutvynenko TL, Korniichuk OS, Kovalchuk MV, Voznyuk TM, et al. (2006) Growing pioneer plants for a lunar base. Advances in Space Research 37: 93–99.
21. Zaets I, Burlak O, Rogutskyy I, Vasilenkoa A, Mytrokhyn O, et al. (2011) Bioaugmentation in growing plants for lunar bases. Advances in Space Research 47: 1071–1078.
22. Kral TA, Bekkum CR, McKay CP (2004) Growth of methanogens on a mars soil simulant. Origins of Life and Evolution of the Biosphere 34: 615–626.
23. Carlton CA, Morris RV, Lindstrom DJ, Lindstrom MM, Lockwood JP (2014) JSC MARS-1: Martian regolith simulant. Orbitec website. Available: http://www.orbitec.com/store/JSC_Mars_1_Characterization.pdf. Accessed 2014 July 28.
24. Rickman D, McLemore CA, Fikes J (2007) Characterization summary of JSC-1a bulk lunar mare regolith simulant. Orbitec website. Available: http://www.orbitec.com/store/JSC-1A_Bulk_Data_Characterization.pdf; http://www.orbitec.com/store/JSC-1AF_Characterization.pdf. Accessed 2014 July 28.
25. Gibson EK (1977) Volatile elements, carbon, nitrogen, sulfur, sodium, potassium and rubidium in the lunar regolith. Phys Chem Earth Vol. X: 57–62.
26. Clark BC, Van Hart DC (1981) The Salts of Mars. Icarus 45: 370–378.
27. Clark BC (1993) Geochemical components in Martian soil. Geochimica et Cosmochimica acta 57: 4575–4581.
28. Chevriera V, Mathe PE (2007) Mineralogy and evolution of the surface of Mars: A review. Planetary and Space Science 55: 289–314.
29. Wamelink GWW, van Adrichem MHC, van Dobben HF, Frissel JY, Held M, et al. (2012) Vegetation relevés and soil measurements in the Netherlands; a database. Biodiversity and Ecology 2012: 125–132.
30. Wamelink GWW, Goedhart PW, van Dobben HF, Berendse F (2005) Plant species as predictors of soil pH: replacing expert judgement by measurements. Journal of vegetation Science 16: 461–470.
31. McCullagh P, Nelder JA (1989) Generalized linear models. Second edition. Chapman & Hall. London.
32. Wua SC, Caob ZH, Lib ZG, Cheunga KC, Wong MH (2005) Effects of biofertilizer containing N-fixer, P and K solubilizers and AM fungi on maize growth: a greenhouse trial. Geoderma 125: 155–166.
33. Houlton BZ, Wang Y, Vitousek PM, Field CB (2008) A unifying framework for dinitrogen fixation in the terrestrial biosphere. Nature 454: 327–330.

34. Cadotte MW, Cavender-Bares J, Tilman D, Oakley TH (2009) Using Phylogenetic, Functional and Trait Diversity to Understand Patterns of Plant Community Productivity. PLOS One 4: e5695.

35. Foy CD, Chaney RL, White MC (1978) The physiology of metal toxicity in plants. Ann Rev Plant Physiol 29:511–66.

36. Wamelink GWW, Goedhart PW, Malinowska AH, Frissel JY, Wegman RJM, et al. (2011) Ecological ranges for the pH and NO3 of syntaxa: a new basis for the estimation of critical loads for acid and nitrogen deposition. Journal of vegetation science 22: 741–749.

15

In Situ Measurement of Some Soil Properties in Paddy Soil Using Visible and Near-Infrared Spectroscopy

Ji Wenjun[1], Shi Zhou[1,3]*, Huang Jingyi[2], Li Shuo[1]

1 Institute of Agricultural Remote Sensing and Information Technology, College of Environmental and Resource Sciences, Zhejiang University, Hangzhou, China, **2** School of Biological, Earth and Environmental Science, The University of New South Wales, Kensington, Australia, **3** Zhejiang Provincial Key Laboratory of Subtropical Soil and Plant Nutrition, Zhejiang University, Hangzhou, China

Abstract

In situ measurements with visible and near-infrared spectroscopy (vis-NIR) provide an efficient way for acquiring soil information of paddy soils in the short time gap between the harvest and following rotation. The aim of this study was to evaluate its feasibility to predict a series of soil properties including organic matter (OM), organic carbon (OC), total nitrogen (TN), available nitrogen (AN), available phosphorus (AP), available potassium (AK) and pH of paddy soils in Zhejiang province, China. Firstly, the linear partial least squares regression (PLSR) was performed on the in situ spectra and the predictions were compared to those with laboratory-based recorded spectra. Then, the non-linear least-square support vector machine (LS-SVM) algorithm was carried out aiming to extract more useful information from the in situ spectra and improve predictions. Results show that in terms of OC, OM, TN, AN and pH, (i) the predictions were worse using in situ spectra compared to laboratory-based spectra with PLSR algorithm (ii) the prediction accuracy using LS-SVM ($R^2>0.75$, RPD>1.90) was obviously improved with in situ vis-NIR spectra compared to PLSR algorithm, and comparable or even better than results generated using laboratory-based spectra with PLSR; (iii) in terms of AP and AK, poor predictions were obtained with in situ spectra ($R^2<0.5$, RPD<1.50) either using PLSR or LS-SVM. The results highlight the use of LS-SVM for in situ vis-NIR spectroscopic estimation of soil properties of paddy soils.

Editor: Andrea Motta, National Research Council of Italy, Italy

Funding: This research was supported by National High Technology Research and Development Program of China (No. 2013AA102301); National Natural Science Foundation of China (No. 41271234); Zhejiang Provincial Science and Technology Project of China (No. 2011C13010). The funders had no role in study design, data collection and analysis, decision to publish, or preparation of the manuscript.

* Email: shizhou@zju.edu.cn

Introduction

Paddy soil is one of the most important soil resources for humans because more than half of the world's population takes rice, the typical farming product of paddy soils, as staple food. As one of the major rice producers, China has a large area of paddy fields of more than 25 million hectares, accounting for 29% of the cultivated lands of China and 23% of the world [1]. In the past 30 years, due to over-fertilization, significantly declined soil pH has been found in major crop production areas and enhanced nitrogen deposition has been identified in terrestrial and aquatic ecosystems as well as in rice [2,3]. As a result, characterizing the properties of paddy soils in an efficient way is of great importance for management of crop growth and yield.

Over the past decades, various agricultural sensors have been used to determine the soil properties as well as their spatial variabilities [4]. Among the agricultural sensors, visible and near-infrared (vis-NIR) spectroscopy has received popularity because it is fast, less labor-intensive and cost-effective compared to conventional chemistry experiments and enables rapid measurements of various soil physical and chemical properties. However, the flooded soil condition in paddy fields makes it difficult to perform soil sampling and analysis. The best time for soil

measurement is the short time gap between the harvest and following rotation, when irrigation water has been drained away. Despite the success of predicting various soil properties using laboratory-based measurement with vis-NIR spectra, the pre-treatment of samples (e.g. air-drying, grinding and sieving) is still tedious and time-consuming. With its faster and more effective characteristics compared to the laboratory-based spectroscopic measurement, in situ vis-NIR is a promising method in measuring and mapping soil properties of paddy fields [5].

Researchers have reported successful application of in situ vis-NIR spectroscopy to prediction of several soil properties. In terms of predicting clay content, Waiser *et al.* (2007) [6] found that in situ vis-NIR sensing can obtain similar results compared with laboratory-based sensing. With regard to soil organic and inorganic carbon, Morgan *et al.* (2009) [7] got slightly larger prediction errors using field-based vis-NIR measurements than using laboratory-based sensing method. When predicting soil color and mineral composition, Viscarra Rossel (2009) [8] concluded that results from in situ vis-NIR measurements were in good agreement with Munsell Book and X-ray diffraction methods. Furthermore, Mouazen *et al.* (2009) [9] improved prediction accuracy of available P by optimizing the field-based vis-NIR sensing system. In addition, a few other soil properties have been

Table 1. Statistics of paddy soil samples in this study.

Soil property	Unit	Dataset	NO. samples	Mean	St.Dev	Medium	Max	Min
OC	g/kg	All	183	15.87	6.65	14.53	36.29	4.12
		Training	138	15.89	6.73	14.51	36.29	4.12
		test	45	15.81	6.48	14.53	35.26	4.78
OM	g/kg	All	104	29.1	13.7	25.9	60.5	6.9
		Training	78	28.9	13.6	25.8	60.5	6.9
		test	26	29.8	14.0	26.6	60.5	7.4
TN	%	All	104	0.17	0.08	0.15	0.37	0.03
		Training	78	0.16	0.08	0.15	0.37	0.03
		test	26	0.17	0.09	0.15	0.36	0.04
AN	mg/kg	All	104	128.43	60.44	132.00	295.00	15.80
		Training	78	127.59	60.37	132.00	295.00	15.80
		test	26	130.96	61.77	132.50	264.00	18.00
AP	mg/kg	All	104	22.86	18.37	18.90	108.00	0.70
		Training	78	22.65	18.52	18.55	108.00	0.70
		test	26	23.48	18.26	19.65	75.60	2.00
AK	mg/kg	All	104	56.75	15.56	55.10	105.00	32.50
		Training	78	55.48	15.51	55.05	105.00	32.50
		test	26	57.55	15.99	55.35	97.30	34.60
pH	-	All	104	5.74	1.17	5.20	8.43	4.60
		Training	78	5.73	1.17	5.19	8.43	4.60
		test	26	5.79	1.21	5.22	8.32	4.62

predicted with acceptable accuracy, including soil organic matter [10], nitrogen [11,12], pH [13] and water content [12,14]. However, most of the studies on predicting soil properties using in situ vis-NIR spectroscopy were conducted on dry farming land.

Although a couple of studies conducted on determining properties of paddy soils based on laboratory-based vis-NIR spectroscopy [15,16], to the best of our knowledge, there are few papers published describing the systematic use of in situ vis-NIR measurements to predict soil properties in paddy fields.

The aims of this study were to evaluate the feasibility of in situ vis-NIR sensing for prediction of soil properties in paddy soils by (i) predicting various soil properties of paddy soils (i.e. organic carbon (OC), organic matter (OM), total nitrogen (TN), available nitrogen (AN), available phosphorus (AP), available potassium (AK) and pH using in situ vis-NIR spectroscopy; (ii) comparing the prediction accuracy between in situ vis-NIR spectra and laboratory-based spectra for paddy soils; (iii) evaluating the prediction accuracy of in situ vis-NIR measurements of soil properties by implementing a multivariate calibration algorithm, i.e., linear partial least square regression (PLSR), and a data-mining algorithm, i.e., least-square support vector machine (LS-SVM).

Materials and Methods

Ethics Statement

We randomly chose 11 paddy fields from close vicinity to 6 cities in Zhejiang province and got permission from Agricultural Bureaus from these six cities, i.e. Tonglu (2 fields, 16 samples), Jiande (2 fields, 11 samples), Pujiang (1 field, 8 samples), Zhuji (1 field, 8 samples), Yiwu (1 field, 24 samples) and Fuyang(4 fields, 117 samples). Three of the four fields we chose in Fuyang were the experimental fields in the China National Rice Research Institute. There is no endangered or protected species involved.

Soil sampling and spectroscopic measurements

In this study, the spectra of the soil samples were recorded by proximal in situ stationary vis-NIR sensing and by laboratory-based vis-NIR measurements. A total of 184 sampling sites were randomly selected in eleven paddy fields in Zhejiang Province, China, with latitudes ranging from 29°03′N to 30°10′N, and

longitudes from 119°10′E to 122°48′E. The water in the paddy fields was drained and left to dry for 10 days prior to sampling and vis-NIR measurement.

vis-NIR measurements at 104 sampling sites were taken in November 2011, while the remaining 80 sites were surveyed in August 2013. At each site, the water content of the surface soil (i.e. 0–20 cm) was firstly measured using a TDR-300 (Spectrum Technologies Inc., USA) with a 20-cm guide. Then, a soil sample was collected using a cube soil sampler to a depth of 20 cm. The surface of the sample profile was flattened and evened, without smearing the soil. Spectra were recorded at three randomly selected locations at different depths within A horizon. If there were stones, roots or voids within the soil sample, spectroscopic measurements were made on the adjacent area. For each of the three sensing locations, 10 spectra were recorded, and the mean value of the whole 30 spectra was used to represent the spectra of the soil at that site. In total, 184 spectra were recorded under the field condition with one spectrum per site.

After in situ vis-NIR measurements, the samples were packed into plastic bags, labeled and transported to laboratory. The soil samples were air-dried, ground and sieved to less than 2 mm. The vis-NIR spectra of these 184 samples were then measured again under laboratory condition. The chemical analyses of soil properties were also conducted using these samples, which would be described later.

A Fieldspec ProFR vis–NIR spectrometer (Analytical Spectral Devices, Boulder, CO, USA) was used for in situ and laboratory-based measurements. The instrument measures the spectra between 350 and 2500 nm, with a resolution of 3 nm at 700 nm and 10 nm at 1400 nm and 2100 nm. The sampling resolution of the spectra is 1 nm. To implement in situ sensing, a high intensity contact probe (Analytical Spectral Devices) was used to prevent the interference from stray light during measurement. The probe has its own light source and a viewing window of 2 cm in diameter through which the measurements are made. To keep the measurement consistent, the contact probe was also used in the laboratory-based measurement. A Spectralon panel with 99% reflectance was used to calibrate the spectrometer before each measurement.

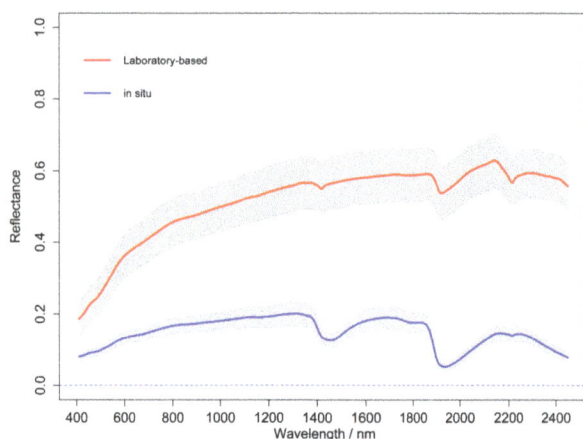

Figure 1. The average reflectance spectra measured in laboratory (red) and in situ (blue) and their corresponding standard deviation values (shaded regions).

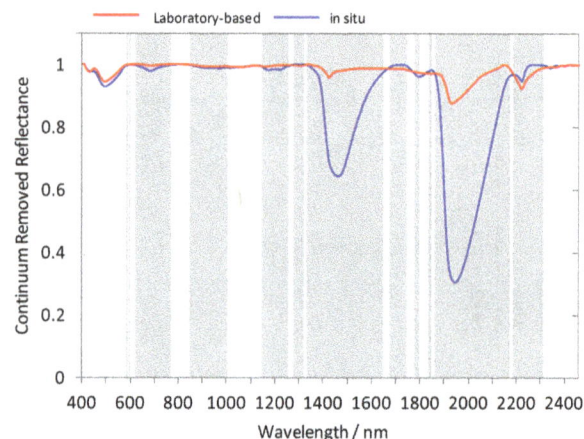

Figure 2. Wavelength specific t-tests between continuum removed laboratory-based and in situ spectra. Note: The shaded regions show where significant differences occur between the spectra at $\alpha=0.01$ significance level.

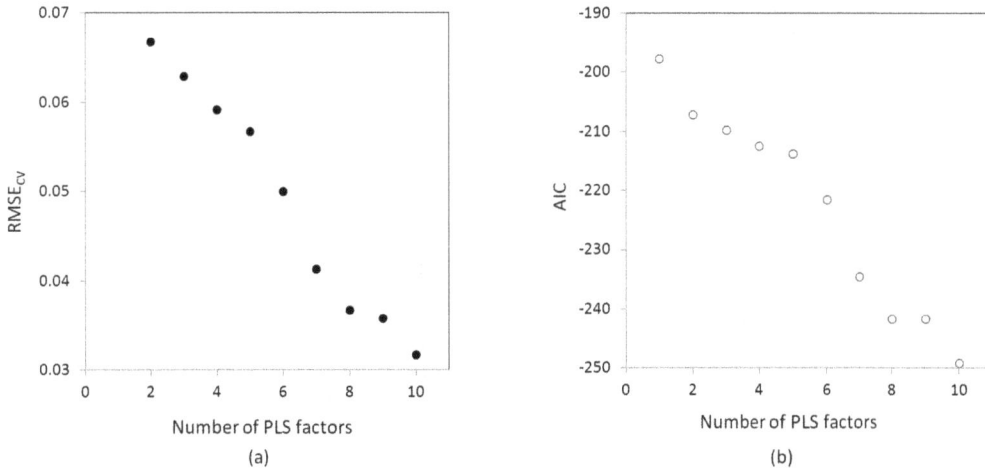

Figure 3. Number of factors (NF) used in partial least square regression versus (a) cross-validated root mean square error (RMSEcv) and (b) Akaike Information Criterion (AIC).

Chemical analysis

Soil OM was measured using the H_2SO_4-$K_2Cr_2O_7$ oxidation method at $180°C$ for 5 minutes [17]. Soil TC and OC content were determined by dry combustion at $1100°C$ with a multi N/C 3100 (Analytik Jena AG, Germany). Before the determination of soil OC, soil samples were acidized by hydrochloric acid to remove the inorganic carbon in the soil. Soil TN was measured using Semi-micro Kjeldahl Method and soil AN was measured by the alkaline hydrolysis diffusion method [18]. Soil AP was measured by the NH_4F-HCl method [18]. Soil AK was measured using the NH_4OAC extraction method and analyzed using a flame photometer [18]. Soil pH was measured in a 1:1 soil: water suspension [18]. The statistics of measured soil properties are listed in Table 1.

Data pre-processing

The spectral regions for 350–399 nm and 2451–2500 nm were deleted because of noise. The reflectance spectra were transformed to apparent absorbance (log1/R) and then mean centered. The smoothing process of the spectra was made using the Savitzky-Golay algorithm with a window size of 11 and polynomial of order 2 [19]. One sample was regarded as outlier and removed from the dataset because its spectra were strange. For each soil property, corresponding values were sorted from small to large, and then every forth one was selected into test dataset, leaving the rest in training dataset.

Partial least square regression (PLSR)

Among the multiple linear calibration algorithms, partial least square regression (PLSR) [20] is one of the most popular algorithms used for spectral calibration and prediction. It is closely related to principal component regression (PCR) yet with a slight difference. Both of them compress the data before prediction while PLSR avoids the dilemma encountered by PCR of choosing components for the regression [21].

We assume the spectral data matrix used as independent variable into PLSR is X, where $X = [x_1, x_2, \cdots, x_i]$, and soil properties as dependent variable is y, with both mean-centered. The first step to perform PLSR is to extract a few linear combinations (called components or factors), T, of the original spectral matrix X:

$$T = \omega^T X \qquad (1)$$

where ω are the scaled weights and can be calculated as the eigenvectors of the matrix $X'yy'X$. Then both X and y can be regressed onto T as follows:

$$X = TP^T + E \qquad (2)$$

$$y = Tq + f \qquad (3)$$

where P are spectral loadings and q are chemical loadings, describing how the variables in T relate to X and y. E and f are residuals and represent noise or irrelevant variability in X and y. After the model parameters are estimated, they can be combined into the final prediction model as

$$\hat{y} = b_0 + x_i \hat{b}_i \qquad (4)$$

where b_0 is the intercept and $\hat{b}_i$ are regression vectors. The detailed description of $\hat{b}_i$ can be found in the book of [21].

To avoid over-fitting or under-fitting, leave-one-out cross validation was used to determine the number of factors to retain in the calibration models [22]. Root mean square error of cross validation (RMSE) and Akaike information criterion (AIC) [23] were used to decide the number of factors.

$$RMSE_{CV} = \sqrt{\sum_{i=1}^{n} (\hat{y}_i - y_i)^2 / n} \qquad (5)$$

Where $\hat{y}_i$ is the predicted value and y_i is the observed value, n is the number of calibration samples.

$$AIC = nln RMSE_{CV} + 2p \qquad (6)$$

Table 2. Comparison of prediction accuracy for in situ PLSR, laboratory-based PLSR and in situ LS-SVM.

Soil property	Unit	in situ spectra + PLSR					Laboratory-based spectra + PLSR					in situ spectra + LS-SVM					
		No. factors	R^2	RMSE	RPD	Grade	No. factors	R^2	RMSE	RPD	Grade	γ	σ^2	R^2	RMSE	RPD	Grade
OC	g/kg	10	0.75	3.33	1.95	D	9	0.81	2.94	2.20	C	26	6346	0.79	2.95	2.20	C
OM	g/kg	10	0.75	7.66	1.83	D	8	0.81	6.11	2.30	C	559	19647	0.81	6.41	2.18	C
TN	%	8	0.86	0.03	2.68	B	8	0.87	0.03	2.81	B	32	1064	0.88	0.03	3.05	A
AN	mg/kg	8	0.76	32.41	1.91	D	8	0.86	24.76	2.49	B	171	1326	0.76	32.27	1.91	D
AP	mg/kg	4	0.43	13.71	1.33	E	8	0.29	19.33	1.17	E	2	714	0.36	14.33	1.27	E
AK	mg/kg	6	0.03	17.92	0.89	E	10	0.07	20.82	0.77	E	23	236	0.14	17.66	0.91	E
pH	Unit	9	0.77	0.58	2.11	C	8	0.82	0.51	2.42	B-C	2548	813	0.80	0.54	2.23	C

Table 3. Upper triangular correlation matrix among six soil properties.

Correlations	OC	OM	TN	AN	AP	AK	pH
OC	1	0.96	0.96	0.93	0.00	0.42	−0.21
OM		1	0.98	0.93	0.06	0.45	−0.29
TN			1	0.95	0.09	0.471	−0.27
AN				1	0.12	0.41	−0.39
AP					1	−0.04	−0.46
AK						1	0.13
pH							1

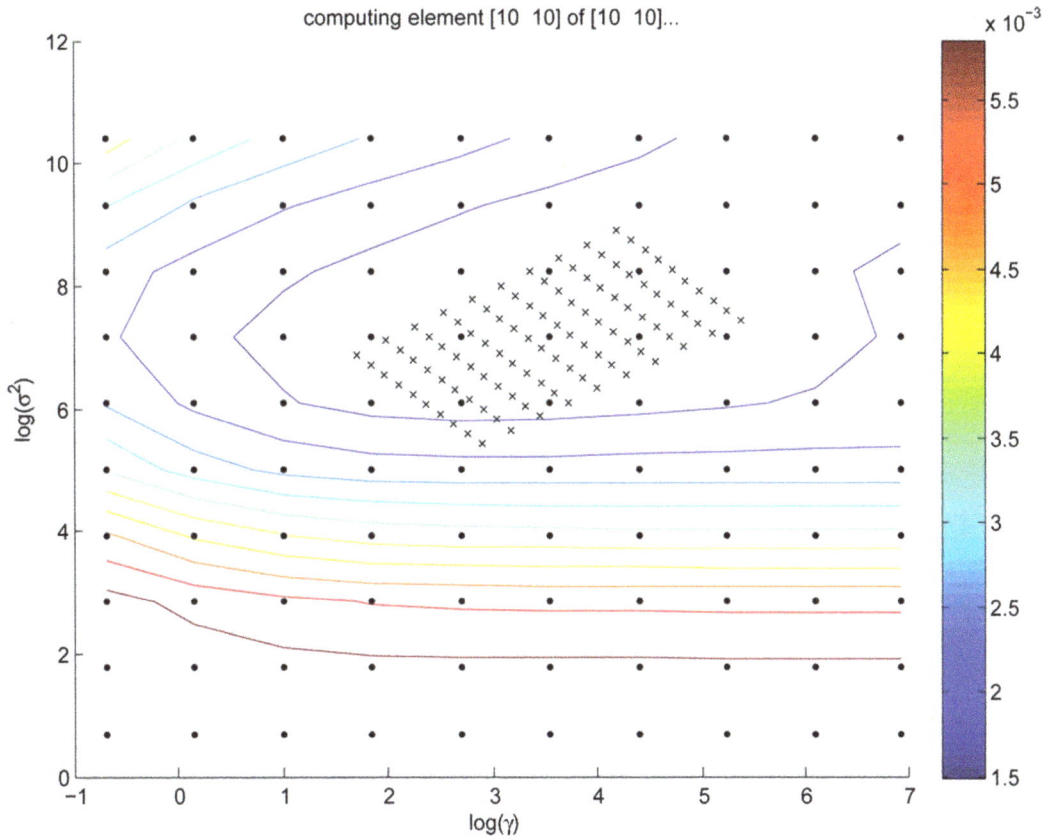

Figure 4. Grid search on γ and σ^2 using least square support vector machine (LS-SVM).

Where n is the number of samples and p is the number of features used in the prediction. The best model has the smallest $RMSE_{CV}$ and AIC.

Least square support vector machine (LS-SVM)

Support vector machine (SVM) is a kernel-based learning algorithm [24] and has been widely used in the pattern classification and regression. The kernel-based learning methods use an implicit mapping of the input data in a high dimensional feature space, a special type of hyperplane defined by a kernel function, in which a regression model is built. As an optimized algorithm based on standard SVM, the least-squares support vector machine (LS-SVM) [25] uses a squared loss function instead of the e-insensitive loss function, from which equality constraints rather than inequality constraints follow. Compared to SVM, complex calculations are avoided in LS-SVM and the multivariate calibration problem can be solved in a relatively fast way. The theory of LS-SVM has been introduced by Suykens *et al.* (2002) [25].

Similarly, the spectral data matrix used as independent variable is $\boldsymbol{X}$, where $\boldsymbol{X} = [\boldsymbol{x}_1, \boldsymbol{x}_2, \cdots, \boldsymbol{x}_i]$, and soil properties as dependent variable is y. The LS-SVM uses nonlinear regression function:

$$y(x) = \sum_{i=1}^{n} \alpha_i K(x, x_i) + b_0 \qquad (7)$$

where $\boldsymbol{b}_0$ is the bias; n is the number of samples; $\boldsymbol{x}_i$ is the measured vis-NIR spectra of different samples; $\boldsymbol{K}(\boldsymbol{x}, \boldsymbol{x}_i)$ is defined by the kernel function. We used radial basis function kernel (RBF), which is the typical general-purpose kernel:

$$K(x, x_i) = e^{-(\|x - x_i\|^2)/2\sigma^2} \qquad (8)$$

where σ^2 is the RBF kernel function parameter, determining the width of the kernel.

α_i is Lagrange multipliers (i.e. support value), which is used by solving the linear Karush-Kuhn-Tucker (KKT) system:

$$\begin{bmatrix} 0 & \boldsymbol{I}_n^T \\ \boldsymbol{I}_n & \boldsymbol{K} + \gamma^{-1}\boldsymbol{I} \end{bmatrix} \begin{bmatrix} \boldsymbol{b}_0 \\ \boldsymbol{\alpha} \end{bmatrix} = \begin{bmatrix} \boldsymbol{0} \\ \boldsymbol{y} \end{bmatrix} \qquad (9)$$

where $\boldsymbol{I}$ refers to an (n×n) identity matrix; γ is the regularization parameter which balances the model's complexity and the training errors; $\boldsymbol{I}_n$ is a (n×1) vector, with all elements ones; $\boldsymbol{y}$ is an (n×1) vector of observed properties values and K denotes elements in kernel matrix.

As we can see from these formulas, in order to make an LS-SVM model, two additional parameters (i.e. γ and σ^2) need to be determined by users. The regularization parameter γ determines the trade-off between the fitting error minimization and smoothness of the estimated function, and is important to improve the generalization performance of the LS-SVM model. An increase in

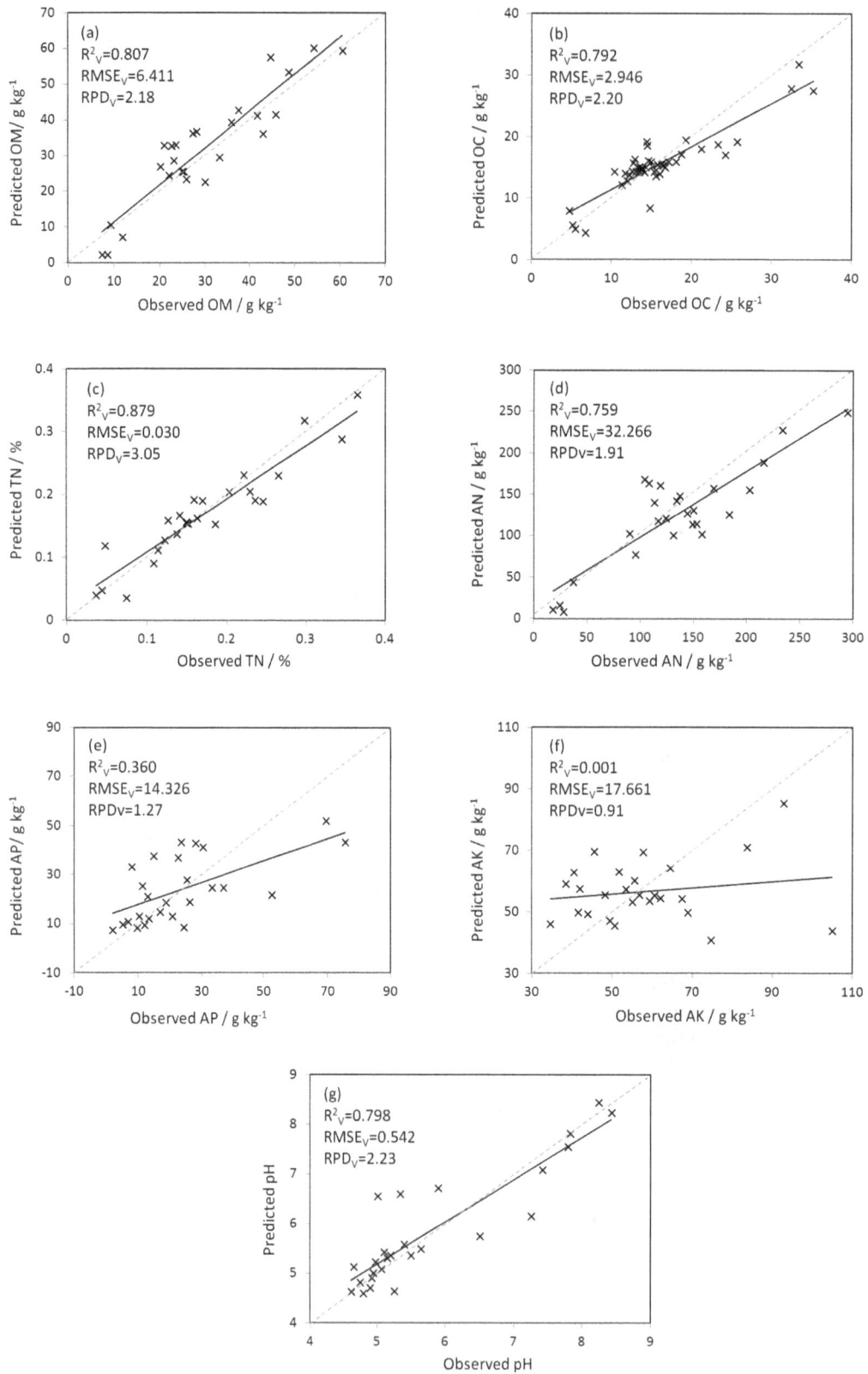

Figure 5. Predicted versus observed values of soil (a) OM, (b) OC, (c) TN, (d) AN, (e) AP, (f) AK, and (g) pH using least square support vector machines (LS-SVM) with in situ vis-NIR spectra.

γ is analogous to an increase in the number of latent variables in a PLS model [26]. The RBF kernel function parameter σ^2 changes the width of the kernel, and thus the degree of the non-linearity can be modeled. When σ^2 increases, the kernel becomes confined, forcing the model towards a linear regression, and its accuracy decreases as well. By contrast, decreased σ^2 and increased γ may lead to over-fit and thus should be treated cautiously [26].

Assessment of statistics

Coefficients of determination (R^2), root mean square error (RMSE) and the ratio of prediction derivation (RPD) were used to compare the prediction accuracies.

$$R^2 = \frac{\left[\sum_{i=1}^{n}(\hat{y}_i - \bar{\hat{y}}_i)(y_i - \bar{y}_i)\right]^2}{\sum_{i=1}^{n}(\hat{y}_i - \bar{\hat{y}}_i)^2 \sum_{i=1}^{n}(y_i - \bar{y}_i)^2} \quad (10)$$

$$RMSE = \sqrt{\frac{1}{n}\sum_{i=1}^{n}(\hat{y}_i - y_i)^2} \quad (11)$$

$$RPD = SD/RMSE \quad (12)$$

Where $\hat{y}_i$ is the predicted value and y_i is the observed value; $\bar{y}_i$ is the mean of observed value; $\bar{\hat{y}}_i$ is the mean of predicted value; SD is standard deviation of observed values; n is the number samples.

Williams (2003) [27] and Saeys et al. (2005) [28] proposed a criterion for the classification of R^2 and RPD: an R^2 value below 0.50 or an RPD value below 1.5 indicates very poor model predictions and such a value could not be useful; an R^2 value between 0.50 and 0.65 an RPD value between 1.5 and 2.0 indicates a possibility of distinguishing between large and small values, while an R^2 value between 0.66 and 0.81 or an RPD value between 2.0 and 2.5 makes approximate quantitative predictions possible. For an R^2 value between 0.82 and 0.90 or an RPD value between 2.5 and 3.0 and above 3.0, the prediction is classified as good. If R^2 value is larger than 0.91 and RPD value is larger than 3.0, the prediction is considered excellent. Generally, a good model prediction would have large values of R^2 and RPD, and a small value of RMSE. In order to simplify the classification, Grade A to E was assigned to the accuracy classes from excellent to not useful.

The LS-SVM toolbox (LS-SVM v.1.5, Suykens, Leuven, Belgium) was applied with Matlab R2009 (MathWorks, Inc., Natick, MA) to perform the LS-SVM models. And other data analysis was conducted in R 2.15.0 [29].

Results and Discussions

Comparison of in situ spectra and laboratory-based spectra

The average reflectance (R) of in situ and laboratory-based measurements of 183 samples and their respective standard deviations are given in Fig. 1. In brief, in situ spectra have smaller reflectance values compared with the laboratory-based spectra. This is because the presence of soil moisture, replacing the air within the soil gaps, increases forward scattering of light and thus the whole absorption of soil moisture at each wavelength increases [30].

Near Infrared (NIR) spectra are dominated by weak overtones and combinations of fundamental vibration which occurs in the MIR region, while visible spectra mainly comprise of electronic transitions [22]. The absorption features of the raw reflectance spectra are usually broad and weak and some of them are difficult to distinguish with the naked eye. As such, continuum removal was applied to all spectra to emphasize absorption features in the spectra. The averaged continuum removed reflectance (CR) is given in Fig. 2, and wavelength specific t-tests were performed between the continuum removed laboratory-based and in situ spectra. In Fig. 2, shaded regions show where there were significant differences between the spectra at $\alpha = 0.01$ significance level. The absorption features due to soil iron oxides near 430 nm and 480 nm [31] have similar size and shape in both in situ and laboratory-based spectra. However, the absorption feature near 650 nm probably correlated with haematite (Fe_2O_3) [32,33] of in situ spectra is greater than that of laboratory-based measurements. The shallow absorption near 1000 nm may be due to amidogen group present in both in situ and laboratory-based spectra, and they are significantly different. The most obvious differences between the two types of spectra are located in two primary water absorption regions within the NIR spectrum, i.e. one around 1450 nm and the other near 1950 nm. It can be explained by the permanently waterlogged conditions of the paddy soil samples. The absorptions caused by soil moisture increases when soil is wet and their features broaden and deepen compared to laboratory collected spectra. However, the strong water absorption near 1950 nm of in situ field collected spectra partly masks the absorptions of clay minerals near 2200 nm which can be identified in the dry laboratory-based spectra. It might affect the prediction accuracies of the spectroscopic models [30].

Prediction of soil properties using PLSR

PLSR algorithms were performed on the training dataset with the optimal number of factors decided by leave-one-out cross validation, and the test dataset was used to validate the PLSR model independently. Taking TN for example, the cross-validated RMSEcv and AIC were plotted against the number of factors (Fig. 3). The optimal number of factors was selected based on the minimum $RMSE_{CV}$ and AIC. Meanwhile, a small number of factors should be included in the model to reduce its complexity when comparable predictions can be obtained. As a result, 8 factors were selected to be used in PLSR with laboratory vis-NIR spectra.

Prediction accuracy of seven soil properties with laboratory-based soil spectra using PLSR method and their accuracy classes are presented in Table 2. Of all the measured soil properties, TN was best predicted with R^2 of 0.87 and RPD of 2.81 (Grade B). OM and OC were approximately quantitatively predicted (Grade C), with R^2 of 0.81, RPD of 2.30 and R^2 of 0.81, RPD of 2.20 for OM and OC, respectively. The predictions of TN, OM and OC is comparable to previous studies [34,35] The successful predictions of these properties are mainly because carbon and nitrogen have direct spectral responses due to the overtones and combinations of N-H, C-H+C-H and C-H+C-C in the vis-NIR spectra [36,37].

However, the prediction accuracy often varies with the forms of carbon and nitrogen present in the soils [36,38]. The phenomenon also occurs in our results. For example, prediction of AN shows a lower accuracy than that of TN with R^2 of 0.86 and RPD of 2.49 (Grade B). This is because most of AN in soil is inorganic, which have no characteristic absorption in vis-NIR region, and the amount of AN is usually small, generally less than 5% of TN, which have a slighter effect on soil spectra.

Table 4. Comparison of prediction accuracy of soil properties with in situ vis-NIR for paddy soils and irrigated soils (dry-farming).

Soil property	Paddy soils (PLSR)	Paddy soils (LS_SVM)	irrigated soils [36]
OC	D	C	B–C
OM	D	C	N.A.
TN	B	A	B
AN	D	D	N.A.
AP	E	E	C
AK	E	E	D
pH	C	C	C

Although some researchers have reported successful prediction of AP and AK using vis-NIR [39–42], it is not the case in this study. AP was not well predicted in consideration of R^2 of 0.29 and RPD of 1.17 (Grade E); prediction of AK was even worse with R^2 of 0.07 and RPD of 0.77 (Grade E). It is because there is no direct spectral absorption features in the vis-NIR region for AP and AK. The occasionally successful prediction of these soil properties may be due to the covariation with other soil properties which have directly spectral responses in the vis-NIR range [37]. However, in the present study, poor correlations of AP or AK with carbon and nitrogen have been found (see Table 3).

Additionally, pH can be predicted with approximately quantitative accuracy with R^2 of 0.82 and RPD of 2.42 (Grade B–C). Although without direct spectral responses in the vis-NIR region, measurements of pH were always reported to be more successful compared to P and K [43,44]. It might be because pH is related to wavelengths of minerals [33]. However, further investigation was needed.

Spectroscopic prediction using PLSR: in situ vs. laboratory-based

Prediction accuracies with in situ collected spectra using PLSR are given in Table 2. Compared to laboratory-based spectroscopic measurements, predictions of soil properties, such as OC, OM, TN, AN and pH, with in situ measured spectra were worse. For example, predictions of soil OM and OC using laboratory-based spectra were considered to be approximately quantitatively accuracy (Grade C) while those using in situ measurements were only able to be distinguished between high and low values (Grade D). Besides, the prediction accuracy of AN decreases to Grade D using in situ spectra ($R^2 = 0.76$ and RPD = 1.91) from Grade B using laboratory-based spectra ($R^2 = 0.86$ and RPD = 2.49). It may be caused by the environmental factors existing during the in situ measurement, such as soil moisture, ambient light, temperature and condition of the soil surface, which would partly mask the absorption features of some soil properties.

As prediction of soil properties with in situ vis-NIR spectra is less accurate than with laboratory-based measurement when linear calibration algorithm was used, a non-linear data mining (i.e. LS-SVM) algorithm was carried out aiming to extract more useful information from the in situ spectra and improve predictions.

Spectroscopic prediction of soil properties: PLSR vs. LS-SVM

In attempt to improve the prediction accuracy using in situ soil spectra, LS-SVM was used to build the models. In order to determine the parameters of γ and σ^2 for LS-SVM models, γ ranging from 2^{-1} to 2^{10} and σ^2 ranging from 2 to 2^{15} were tested.

The ranges were based on previous studies. For each combination of γ and σ^2, the root mean square error of cross-validation ($RMSE_{cv}$) was calculated and the optimal parameters were determined when smaller $RMSE_{cv}$ occurred. The optimizing process of predicting TN is shown in Fig. 4. The grid search and leave-one-out cross validation were employed to find the optimal combination of γ and σ^2. Grid search is a two-dimensional minimization procedure based on exhaustive search in a limited range. The grids of "." in the first step was 10×10, and the searching step at this stage was relatively large. The grids of "×" in the second step was 10×10, and the searching step in the second stage was relatively small. The optimal search area was determined using the contour lines of $RMSE_{cv}$ plotted in Fig. 4.

Predictions with in situ spectra using LS-SVM can be found in Table 2. Firstly, comparison between PLSR and LS-SVM was made with in situ spectra. Soil OM and OC can only be distinguished by high and low values (i.e. Grade D) when PLSR method was performed (OM: $R^2 = 0.75$ and RPD = 1.83; OC: $R^2 = 0.75$ and RPD = 1.95). However, using LS-SVM method, both OM and OC can be approximately quantitatively estimated (i.e. Grade C), with the prediction accuracies of $R^2 = 0.81$ and RPD = 2.18 for OM, and $R^2 = 0.79$ and RPD = 2.20 for OC. Prediction of TN was even more accurate using LS-SVM with $R^2 = 0.88$ and RPD = 3.05 (i.e. Grade A) compared to PLSR with $R^2 = 0.86$ and RPD = 2.68 (i.e. Grade B). Besides, comparable prediction accuracies of AN were obtained between LS-SVM and PLSR, both with $R^2 = 0.76$ and RPD = 1.91 (Grade D). In terms of pH, LS-SVM only slightly improved the prediction compared to PLSR. However, AK and AP remained unpredictable (Grade E) using two methods. The use of the data-mining algorithm (i.e. LS-SVM here) improved the prediction accuracy of most of soil properties compared with the linear PLSR algorithm with in situ vis-NIR spectra. Fig. 5 shows the predicted values of seven soil properties against the observed ones using LS-SVM with in situ vis-NIR spectra.

Surprisingly, the predictions of OM, OC and pH with in situ spectra using LS-SVM were comparable to those using PLSR with laboratory-based spectra; prediction of TN using in situ spectra with LS-SVM was one grade better than using laboratory-based measurement with PLSR. The prediction accuracy of TN is comparable to the result from Kleinebecker et al. (2013) [45] with air dried samples. However, in term of AN, laboratory-based model with PLSR still offers better prediction. Given the improved prediction results of OM, OC, TN and pH using LS-SVM, in situ vis-NIR spectroscopy would become an effective tool for rapid and reliable measurement of soil properties in the field.

In *situ* prediction: paddy soils vs. irrigated soils

The prediction of paddy soil properties with in situ vis-NIR spectra were compared to a recent review of in situ vis-NIR measurements [36] of irrigated (arable) soils, i.e. dry-farming soils (Table 4). Prediction accuracy of TN and pH of paddy soils is similar to that of irrigated soils. However, due to the presence of considerable amount of soil water in paddy soils, which affects the in situ measured soil vis-NIR spectra, OC, AP and AK are better predicted in irrigated soils compared to paddy soils.

Conclusions

Compared with laboratory-based vis-NIR spectroscopic measurement, field-based measurement is more efficient by measuring soil spectra directly in situ. It thus offers a promising way to analysis soil properties quickly in paddy fields when water is drained away before and after harvest. In our study, systematic research on paddy soil properties using in situ vis-NIR spectra and laboratory-based vis-NIR spectroscopy were carried out, including soil organic matter (OM), total organic carbon (OC), total nitrogen (TN), available nitrogen (AN) available phosphorus (AP), available potassium (AK) and pH.

Using the PLSR algorithm with laboratory-based vis-NIR spectra, soil OM, OC, TN, AN and pH can be quantitatively estimated with various accuracies while AP and AK can be poorly predicted. However, the prediction accuracy of soil properties decreased to some extent when in situ spectra were used for modeling. It happened especially for the prediction of soil OM, OC, AN and pH, with one grade decreasing. It might be due to the existence of soil moisture and ambient light, as well as the environment temperature and soil surface condition, which might mask or partly mask the absorption information on spectra, and influence their prediction accuracies.

By performing the non-linear LS-SVM algorithm, prediction of soil OM, OC, TN and pH with in situ vis-NIR spectra was obviously improved. Their predictions were comparable or even better than laboratory-based spectroscopic measurement using PLSR algorithm. Prediction of AN was not improved and AP and AK remained unpredictable. Thus, we propose the use of LS-SVM algorithm for in situ vis-NIR spectroscopic estimation of soil properties of paddy soils.

Owing to the permanently waterlogged conditions of paddy soils, in situ prediction of several soil properties of paddy fields is less accurate compared with irrigated soils. Other data mining methods are expected to be tested on the in situ paddy soil spectra. Besides, further research on the chemometic algorithms for removing the effects of water and other environmental factors from the spectra might fundamentally improve the prediction of soil properties with in situ spectra.

Supporting Information

File S1   In situ measured vis-NIR spectra of 184 samples. To every tenth wavelength was retained to reduce the size of the file. (XLSX)

Acknowledgments

The authors thank Tian Yanfeng, Liu Xiang, Li Chengxue and Ma Ziqiang for their assistance in the field-based measurements of paddy soil samples, and Chen Songchao and Wang Qianlong for the help with the laboratory-based vis-NIR measurements.

Author Contributions

Conceived and designed the experiments: SZ JW. Performed the experiments: JW SZ LS. Analyzed the data: JW. Contributed reagents/materials/analysis tools: SZ JW LS. Contributed to the writing of the manuscript: JW HJ.

References

1. Li Q (1992) Paddy soils of China. Beijing: Science Press. (In [Chinese])
2. Guo J, Liu X, Zhang Y, Shen J, Han W, et al. (2010) Significant acidification in major Chinese croplands. Science, 327: 1008–1010.
3. Liu X, Zhang Y, Han W, Tang A, Shen J, et al. (2013) Enhanced nitrogen deposition over China. Nature 494: 459–462.
4. Gebbers R, Adamchuk VI (2010) Precision agriculture and food security. Science 327: 828–831.
5. Guo Y, Ji W, Wu H, Shi Z (2012) Estimation and mapping of soil organic matter based on vis-NIR reflectance spectroscopy. Spectroscopy and spectral analysis. 33: 1135–1140.
6. Waiser TH, Morgan CL, Brown DJ, Hallmark CT (2007) In situ characterization of soil clay content with visible near-infrared diffuse reflectance spectroscopy. Soil Sci Soc Am J 71: 389–396.
7. Morgan RPC (2009) Soil erosion and conservation. John Wiley & Sons.
8. Viscarra Rossel RA, Cattle SR, Ortega A, Fouad Y (2009) In situ measurements of soil colour, mineral composition and clay content by vis–NIR spectroscopy. Geoderma 150: 253–266.
9. Mouazen AM, Maleki M, Cockx L, Van Meirvenne M, Van Holm L, et al. (2009) Optimum three-point linkage set up for improving the quality of soil spectra and the accuracy of soil phosphorus measured using an on-line visible and near infrared sensor. Soil and Tillage Research 103: 144–152.
10. Christy CD (2008) Real-time measurement of soil attributes using on-the-go near infrared reflectance spectroscopy. Comput Electron Agr 61: 10–19.
11. Kusumo BH, Hedley MJ, Hedley CB, Arnold GC, Tuohy MP (2010). Predicting pasture root density from soil spectral reflectance: field measurement. Eur J Soil Sci 61: 1–13.
12. Kuang B, Mouazen AM (2013) Effect of spiking strategy and ratio on calibration of on-line visible and near infrared soil sensor for measurement in European farms. Soil and Tillage Research 128: 125–136.
13. Tekin Y, Kuang B, Mouazen AM (2013) Potential of On-Line Visible and Near Infrared Spectroscopy for Measurement of pH for Deriving Variable Rate Lime Recommendations. Sensors 13(8): 10177–10190.
14. Mouazen AM, Maleki MR, De Baerdemaeker J, Ramon H (2007). On-line measurement of some selected soil properties using a VIS–NIR sensor. Soil and Tillage Research 93(1): 13–27.
15. Gholizade A, Soom MAM, Saberioon MM, BorůvkaP L (2013) Visible and near infrared reflectance spectroscopy to determine chemical properties of paddy soils. Journal of Food, Agriculture & Environment 11: 859–866.
16. Kim YJ, Choi CH (2013) The analysis of paddy soils in Korea using visible-near infrared spectroscopy for development of real-time soil measurement system. Journal of the Korean Society for Applied Biological Chemistry 56: 559–565.
17. Soil Science Society of China (2000) Soil and Agricultural Chemistry Analysis. Beijing: China Agricultural Science and Technology. (In [Chinese])
18. Bao SD (1981) Soil and agricultural chemistry analysis. Beijing: China Agricultural Press.
19. Savitzky A, Golay MJ (1964) Smoothing and differentiation of data by simplified least squares procedures. Anal Chem 36: 1627–1639.
20. Wold S, Martens H, Wold H (1983) The multivariate calibration problem in chemistry solved by the PLS method. Matrix Pencils: Springer. pp. 286–293.
21. Naes T, Isaksson T, Fearn T, Davies T (2002) A User-Friendly Guide to Multivariate Calibration and Classification. Chichester: NIR Publications.
22. Viscarra Rossel RA, Walvoort DJJ, McBratney AB, Janik LJ, Skjemstad JO (2006) Visible, near infrared, mid infrared or combined diffuse reflectance spectroscopy for simultaneous assessment of various soil properties. Geoderma 131: 59–75.
23. Li B, Morris J, Martin EB (2002) Model selection for partial least squares regression. Chemometr Intell Lab 64: 79–89.
24. Vapnik V (1995) The Nature of Statistical Learning Theory. New York: Springer.
25. Suykens JA, Van Gestel T, De Brabanter J, De Moor B, Vandewalle J, et al. (2002) Least squares support vector machines. Singapore: World Scientific.
26. Cogdill RP, Dardenne P (2004) Least-squares support vector machines for chemometrics: an introduction and evaluation. J Near Infrared Spec 12 93–100.
27. Williams P (2003) Near-infrared technology — Getting the best out of light. PDK Grain, Nanaimo, Canada.
28. Saeys W, Mouazen AM, Ramon H (2005) Potential for onsite and online analysis of pig manure using visible and near infrared reflectance spectroscopy. Biosyst Eng 91: 393–402.

29. R Development Core Team (2012) R: A language and environment for statistical computing. R Foundation for Statistical Computing, Vienna. ISBN 3-900051-07-0, URL http://www.R-project.org [accessed 1 March, 2012].

30. Lobell DB, Asner GP (2002) Moisture effects on soil reflectance. Soil Sci Soc Am J 66: 722–727.

31. Clark RN (1999) Spectroscopy of rocks and minerals, and principles of spectroscopy. Manual of remote sensing 3: 3–58.

32. Sherman DM, Waite TD (1985) Electronic spectra of Fe (super 3+) oxides and oxide hydroxides in the near IR to near UV. Am Mineral 70: 1262–1269.

33. Viscarra Rossel RA, Behrens T (2010) Using data mining to model and interpret soil diffuse reflectance spectra. Geoderma 158: 46–54.

34. Fystro G (2002) The prediction of C and N content and their potential mineralization in heterogeneous soil samples using Vis-NIR spectroscopy and comparative methods. Plant Soil 246:139–149.

35. Ji W, Li X, Li C, Zhou Y, Shi Z (2012) Using different data mining algorithms to predict soil organic matter based on visible-near infrared spectroscopy. Spectroscopy and spectral analysis. 32: 2392–2398.

36. Kuang B, Mahmood HS, Quraishi MZ, Hoogmoed WB, Mouazen AM, et al. (2012) Chapter four Sensing Soil Properties in the Laboratory, In Situ, and On-Line: A Review. Adv Agron 114: 155–223.

37. Stenberg B, Viscarra Rossel RA, Mouazen AM, Wetterlind J (2010) Chapter Five-Visible and Near Infrared Spectroscopy in Soil Science. Adv Agron 107: 163–215.

38. Weyer L, Lo S-C (2002) Spectra–structure correlations in the near-infrared. In: Chalmers JM, Griffiths PR, editors. Handbook of Vibrational Spectroscopy. Chichester: John Wiley and Sons, Ltd. pp.1817–1837.

39. Udelhoven T, Emmerling C, Jarmer T (2003) Quantitative analysis of soil chemical properties with diffuse reflectance spectrometry and partial least-square regression: A feasibility study. Plant Soil 251: 319–329.

40. Bogrekci I, Lee WS (2005a) Spectral phosphorus mapping using diffuse reflectance of soils and grass. Biosyst Eng 91: 305–312.

41. Bogrekci I, Lee WS (2005b) Spectral soil signatures and sensing phosphorus. Biosyst Eng 92: 527–533.

42. Shao Y, He Y (2011) Nitrogen, phosphorus, and potassium prediction in soils, using infrared spectroscopy. Soil Research 49: 166–172.

43. Shepherd KD, Walsh MG (2002) Development of reflectance spectral libraries for characterization of soil properties. Soil Sci Soc Am J 66: 988–998.

44. Cohen MJ, Prenger JP, DeBusk WF (2005) Visible-near infrared reflectance spectroscopy for rapid, nondestructive assessment of wetland soil quality. J Environ Qual 34: 1422–1434.

45. Kleinebecker T, Poelen MDM, Smolders AJP, Lamers LPM, Hölzel N (2013) Fast and inexpensive detection of total and extractable element concentrations in aquatic sediments using near-Infrared reflectance spectroscopy(NIRS). PLoS ONE 8(7): e70517. doi:10.1371/journal.pone.0070517

16

Soil Type Dependent Rhizosphere Competence and Biocontrol of Two Bacterial Inoculant Strains and Their Effects on the Rhizosphere Microbial Community of Field-Grown Lettuce

Susanne Schreiter[1,2], Martin Sandmann[2], Kornelia Smalla[1], Rita Grosch[2*]

1 Julius Kühn-Institut – Federal Research Centre for Cultivated Plants (JKI), Institute for Epidemiology and Pathogen Diagnostics, Braunschweig, Germany, **2** Leibniz Institute of Vegetable and Ornamental Crops Großbeeren/Erfurt e.V., Department Plant Health, Großbeeren, Germany

Abstract

Rhizosphere competence of bacterial inoculants is assumed to be important for successful biocontrol. Knowledge of factors influencing rhizosphere competence under field conditions is largely lacking. The present study is aimed to unravel the effects of soil types on the rhizosphere competence and biocontrol activity of the two inoculant strains *Pseudomonas jessenii* RU47 and *Serratia plymuthica* 3Re4-18 in field-grown lettuce in soils inoculated with *Rhizoctonia solani* AG1-IB or not. Two independent experiments were carried out in 2011 on an experimental plot system with three soil types sharing the same cropping history and weather conditions for more than 10 years. Rifampicin resistant mutants of the inoculants were used to evaluate their colonization in the rhizosphere of lettuce. The rhizosphere bacterial community structure was analyzed by denaturing gradient gel electrophoresis of 16S rRNA gene fragments amplified from total community DNA to get insights into the effects of the inoculants and *R. solani* on the indigenous rhizosphere bacterial communities. Both inoculants showed a good colonization ability of the rhizosphere of lettuce with more than 10^6 colony forming units per g root dry mass two weeks after planting. An effect of the soil type on rhizosphere competence was observed for 3Re4-18 but not for RU47. In both experiments a comparable rhizosphere competence was observed and in the presence of the inoculants disease symptoms were either significantly reduced, or at least a non-significant trend was shown. Disease severity was highest in diluvial sand followed by alluvial loam and loess loam suggesting that the soil types differed in their conduciveness for bottom rot disease. Compared to effect of the soil type of the rhizosphere bacterial communities, the effects of the pathogen and the inoculants were less pronounced. The soil types had a surprisingly low influence on rhizosphere competence and biocontrol activity while they significantly affected the bottom rot disease severity.

Editor: Martha E. Trujillo, Universidad de Salamanca, Spain

Funding: The study was funded by 'German Research Foundation' (DFG), SM 59/11-1; GR 1729/8-1: KS RG. The funders had no role in study design, data collection and analysis, decision to publish, or preparation of the manuscript.

Competing Interests: The authors have declared that no competing interests exist.

* Email: grosch@igzev.de

Introduction

Plant pathogens are a limiting factor in crop productivity worldwide and responsible for yield losses [1]. Crop rotation, use of resistant cultivars and application of chemicals are strategies to minimize disease incidence and severity in Integrated Pest Management (IPM). However, resistant cultivars and effective fungicides for the control of diseases caused by soil-borne pathogens such as *Rhizoctonia solani* (Kühn) are often not available. Moreover, adverse eco-toxicological effects of chemical fungicides urge the development of alternative strategies to combat fungal diseases on crops [2–4]. In terms of disease control, it is well-documented that microbial inoculants as part of IPM can contribute to the reduction of adverse environmental effects caused by the exclusive reliance on fungicides [5-7] and thus represent a promising strategy for more sustainable agriculture [8]. Currently, the worldwide bio-pesticide market offers products

including 60 bacterial and 60 fungal species [9]. Nonetheless, the exploitation of microbial inoculants as biocontrol agents in agriculture has been hampered by inconsistent results at the field scale [10,11]. The inconsistency observed in biocontrol effects limits the attraction of microbial inoculants for growers but the reasons for this variability remain largely unexplored. Variation in the colonization ability of bacterial inoculants in the rhizosphere (rhizosphere competence) is assumed to be one of the factors contributing to this inconsistency. Several studies showed that the expression of genes, responsible for the capability of a biocontrol strain to suppress a disease, is often regulated in a cell density dependent manner [12,13]. Therefore, the ability of inoculants to colonize the rhizosphere at sufficiently high numbers for an extended period was identified as a prerequisite for their beneficial effect on plants [11,14,15]. Thanks to the application of advanced genomics and microscopy methods the understanding of factors contributing to the biocontrol activity of bacterial strains has clearly improved [11,15–18], especially for *Pseudomonas* strains such as *P. fluorescens* CHA0 or Pf-5. The complex regulation of genes involved in biocontrol and plant-microbe interaction has been studied in more detail [19–23]. Although the mode of action differs from strain to strain, numerous studies supported the assumption that biocontrol activity most likely results from multi-factorial processes such as antibiosis, production of cell wall degrading enzymes, surfactants, volatile substances or sidero-phores, competition for nutrients and space and/or the enhance-ment of plant innate defense responses [24,25]. Several properties of bacterial inoculants such as motility [26], attachment [27], growth [16], production of antifungal metabolites or siderophores [24] and uptake and catabolism of root exudates [17,28] have been shown to be linked to rhizosphere competence.

However, knowledge of factors influencing rhizosphere compe-tence of bacterial inoculants under field conditions is largely lacking. Only in a few studies efforts were made to quantify the inoculant densities in the rhizosphere of field-grown crops or to evaluate the influence of inoculants and/or pathogens on the indigenous rhizosphere microbial community [7,29,30]. In agri-culture, crops are cultivated under various ecological conditions. Therefore, a better understanding of the complex relationships among inoculant, pathogen, plant, and ecological factors such as the soil type is a prerequisite for improved and reliable biocontrol effects. So the goal of the present study was to investigate the influence of soil types on the rhizosphere competence and biocontrol activity of bacterial inoculants and their effects on the indigenous soil bacterial community.

The strains *Pseudomonas jessenii* RU47 [31] and *Serratia plymuthica* 3Re4-18 [32], which revealed remarkably good control effects against bottom rot in previous experiments [7,32], were selected for this study. The causal agent of bottom rot on lettuce, the soil-borne fungus *R. solani* AG1-IB whose genome was recently sequenced [33] was used as model pathogen and lettuce as model host plant. The experimental plot system with three soil types under the same cropping history at the same field site enabled us to study the effects of different soil types on the rhizosphere competence and the biocontrol activity of the bacterial inoculants for the first time. We hypothesized that the soil types influence the rhizosphere competence and the biocontrol activity of the inoculant strains applied. Furthermore, we hypothesized that both the inoculants and the presence of the pathogen (*R. solani* AG1-IB) also influences the structural diversity of microbial communities in the rhizosphere of lettuce, and that the extent of this effect depends on the soil type.

Materials and Methods

Bacterial inoculants

The bacterial inoculant *P. jessenii* RU47 was isolated from a disease-suppressive soil [31,34] and the strain *S. plymuthica* 3Re4-18 originated from the endorhiza of potato [35]. To monitor the survival of inoculants in the rhizosphere spontaneous rifampicin resistant mutants were used [34]. Both strains were stored at −80°C in Luria-Bertani broth (Carl Roth GmbH & Co. KG, Karlsruhe, Germany) with 20% glycerol.

Design of field experiments

To evaluate the effect of soil types on the rhizosphere competence and disease suppression of the bacterial inoculants *P. jessenii* RU47 and *S. plymuthica* 3Re4-18 without and with *R. solani* inoculation, two independent field experiments were carried out in a unique experimental plot system at the Leibniz Institute of

Vegetable and Ornamental Crops (Großbeeren, Germany, 52° 33′ N, 13° 22′ E). The first experiment was performed in unit 5, with planting on 8 June, harvest on 18 July 2011, the second experiment in unit 6, with planting on 27 July and harvest on 5 September 2011. Each unit was comprised out of three blocks (one block per soil type). The three soil types were characterized as Arenic-Luvisol (diluvial sand, DS), Gleyic-Fluvisol (alluvial loam, AL) and Luvic-Phaeozem (loess loam, LL) [36,37]. Each block consisted of 24 plots of 2 m×2 m in size and a depth of 75 cm. In unit 5, the following crops were cultivated from 2000 to 2010: pumpkin, nasturtium, nasturtium, phacelia, amaranth, wheat, pumpkin, nasturtium, wheat, broccoli, wheat, Teltow turnip and lettuce, and in unit 6 pumpkin, nasturtium, pumpkin, amaranth, wheat, wheat, pumpkin, nasturtium, wheat, wheat and lettuce.

Lettuce seeds (cv. Tizian, Syngenta, Bad Salzuflen, Germany) were sown in seedling trays filled with the respective soil type and incubated at 12°C for 48 h and then grown in the greenhouse at approximately 20/15°C (day/night). To maintain the substrate moisture all seedling trays were watered daily and fertilized weekly (0.2% Wuxal TOP N, Wilhelm Haug GmbH & Co. KG, Düsseldorf, Germany). Lettuce seedlings were transplanted at the 3–4-leaf stage (BBCH 14) in the experimental plot system. Plants were placed in a within-row and intra-row distance of 30 cm (36 plants per plot), and lettuce plantlets were overhead irrigated based on the computer program 'BEREST' [38]. The daily soil water content in the rooted soil layer using the water holding capacity of the soil under consideration of the plant growth stage and the potential evapotranspiration were the input variables for the irrigation program. Irrigation decisions were made on the basis of the calculated soil water content and the expected evapotrans-piration and precipitation of the next five days. The soil temperature (reflectometer PT100b1/3 DIN, Messtechnik Gera-berg GmbH, Martinroda, Germany) and the matric potential (CS616-L water content reflectometer, Campbell Scientific, North Logan, Utah, USA) were recorded by data logger (4MbSRAm data logger, Campbell Scientific). Both reflectometers determine an average value of a 20 cm top soil layer. The fertilizer was added to each plot based on a chemical analysis of soils before planting, done according to the certified protocols of Agricultural Tests and Research Institutions Association (VdLUFA, Germany). Each soil type was adjusted to the same amount of nitrogen (162 mg/100 g) by fertilization with Kalkamon (27% N, TDG mbh Lommatzsch, Germany). Lettuce was harvested six weeks after planting (6WAP, BBCH 49) in both experiments. The lettuce shoot dry mass (SDM) of each plant and the disease severity of bottom rot were scored at harvest. For assessment of SDM each lettuce head was cut in four portions and dried at 80°C until a constant dry mass was achieved. The disease severity was rated in four categories: 1–without bottom rot symptoms; 3–symptoms only on first lower leaves and small brown spots on the underside of leaf midribs; 5–brown spots on leaf midribs on lower and next upper leaf layer and 7–severe disease symptoms on upper leaf layers and beginning of head rot to total head rot according to Grosch et al. [39].

The following treatments of lettuce were investigated for each soil type: no treatment with inoculants without (control) and with *R. solani* (*Rs*) inoculation (control+*Rs*), plants treated with inoculants without (RU47; 3Re4-18) and with *R. solani* inocula-tion (RU47+*Rs*; 3Re4-18+*Rs*). Each treatment included four replicates with 36 plants per replicate.

Preparation of pathogen inoculum and inoculation

The *R. solani* AG1-IB isolate 7/3 from the strain collection of the Leibniz Institute of Vegetable and Ornamental Crops (Großbeeren) was used in the present study. The inoculum was

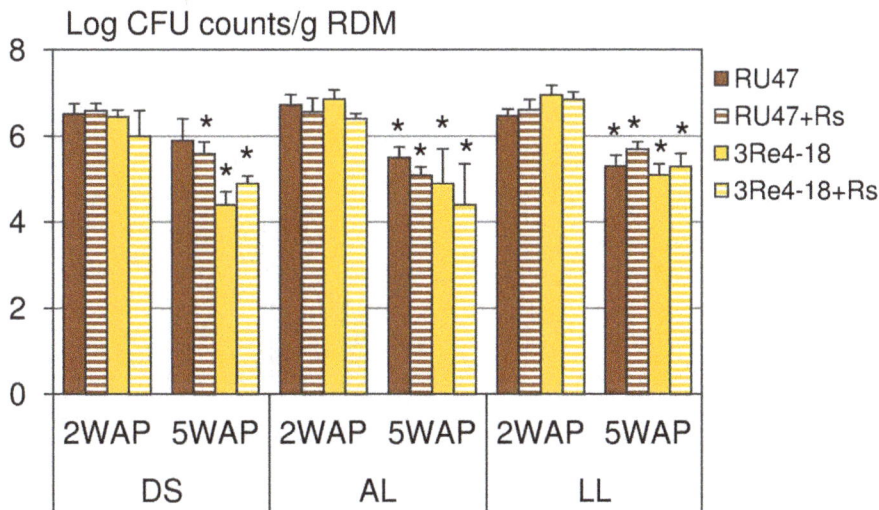

A)

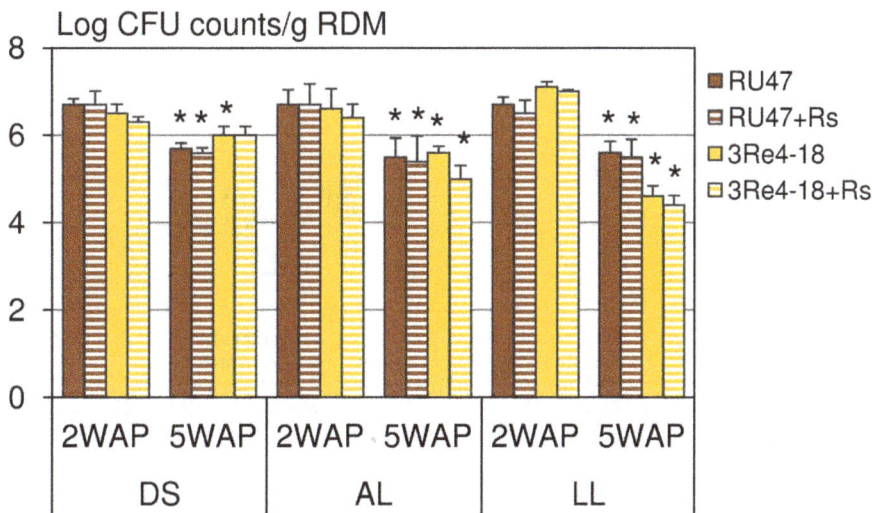

B)

Figure 1. CFU counts of *Pseudomonas jessenii* **RU47 and** *Serratia plymuthica* **3Re4-18 per gram of root dry mass (RDM) without (RU47, 3Re4-18) and with** *Rhizoctonia solani* **inoculation (RU47+***Rs***; 3Re4-18+***Rs***) in two experiments, A) and B), two and five weeks after planting (2WAP, 5WAP) in the 2011-season.** Plants were grown in three soil types (DS, AL, LL) at the same field site. An asterisk indicates significant differences in CFU counts of RU47 or 3Re4-18 between 2WAP and 5WAP for each soil type (Tukey post-hoc test, $P<0.05$). The bars show the standard deviation.

Table 1. ANOVA results: Factor [soil type, plant growth development stage (PGDS), pathogen] dependent *P*-values for CFU counts of *Pseudomonas jessenii* RU47 and *Serratia plymuthica* 3Re4-18 ($P<0.05$).

	Experiment 1		Experiment 2	
Factor	**RU47**	**3Re4-18**	**RU47**	**3Re4-18**
Soil type	0.079	0.0007	0.544	<0.0001
PGDS	<0.0001	<0.0001	<0.0001	<0.0001
Pathogen	0.623	0.2609	0.3872	0.006

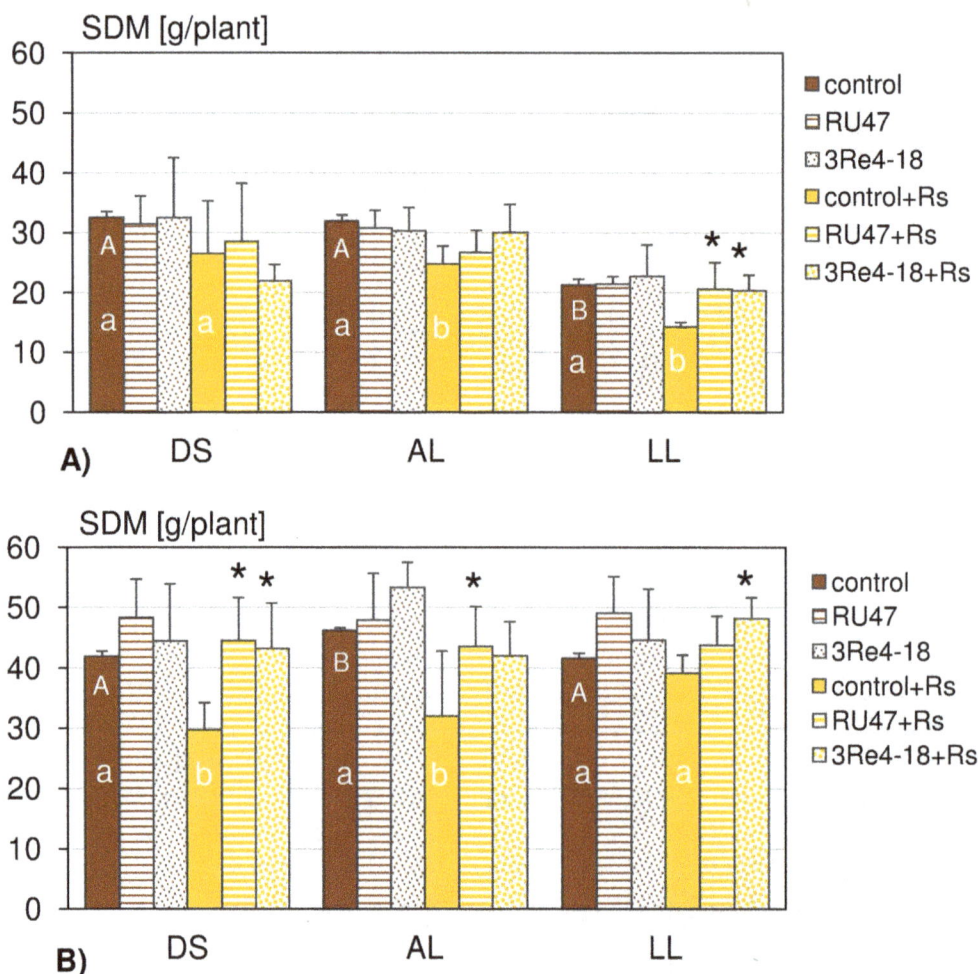

Figure 2. Shoot dry mass (SDM) of lettuce (cv. Tizian) determined for the following treatments: control, RU47, 3Re4-18, control+*Rs*, RU47+*Rs*, 3Re4-18+*Rs*) of two experiments, A) and B), in the 2011-season. Plants were grown in three soil types (DS, AL, LL) for six weeks each, at the same field site. Different capital letters denote significant differences in SDM of lettuce in controls or control+*Rs* between soil types (ANOVA; *P*<0.1). Different lower-case letters indicate significant differences in SDM between control and control+*Rs* within each soil type (Tukey post-hoc test, *P*<0.1). An asterisk denotes significant effects of the inoculants RU47 and 3Re4-18 on SDM to the corresponding control or control+*Rs* in each soil type. The bars show the standard deviation.

prepared as described by Schneider et al. [40] on barley kernels. To ensure a higher pathogen pressure the following inoculation procedure was applied: 36 lettuce plants were planted in each of the 24 plots as described above; after a cultivation time of three weeks they were evenly incorporated into the top soil (10 cm) by means of a rotary hoe, together with 40 g of barley kernels without or with *R. solani* infestation. The experiment started two weeks later assuming a decomposition of incorporated infested or non-infested lettuce plant material.

Preparation of bacterial inocula and application mode

For seed treatment King's B agar (Merck KGaA, Darmstadt, Germany) supplemented with rifampicin (75 µg/ml) were inoculated with the *P. jessenii* RU47 of *S. plymuthica* 3Re4-18 and incubated overnight at 29°C. The bacterial cells were harvested from the Petri dishes by resuspension in 15 ml sterile 0.3% NaCl and the concentration was adjusted in a spectrophotometer to a density of 10^8 colony forming units (CFU)/ml. A total of 200 lettuce seeds (cv. Tizian) were coated with 500 µl of a bacterial cell suspension dripping on the seed during vortexing in a 50 ml Falcon tube.

To prepare the inoculum for the treatment of young plants the inoculant strains were grown in nutrient broth (NB II, SIFIN GmbH, Berlin, Germany) amended with rifampicin (75 µg/ml) on a rotary shaker (90 rpm) at 29°C. After a cultivation time of 16 h the overnight culture was centrifuged at 13,000 *g* for 5 min, the supernatant discarded and the pellet was resuspended in sterile 0.3% NaCl solution. The cell density was adjusted to 10^7 CFU/ml or 10^8 CFU/ml for the drenching before and after planting, respectively. Lettuce plants were treated by drenching with 20 ml bacterial cell suspension per plant at the 3-leaf stage one week before planting in the field. A second treatment of young plants with 30 ml bacterial cell suspension 10^8 CFU/ml per plant was carried out at the 4-leaf stage two days after planting. The control plants were drenched with 20 ml or 30 ml of 0.3% NaCl solution, respectively, instead of bacterial suspension.

Sampling and sample processing

Rhizosphere samples were collected two and five weeks after planting (2WAP and 5WAP; BBCH 19 and BBCH 49) the lettuce in the experimental plot system. For each treatment and sampling time the roots of three plants per replicate (plot) were combined as

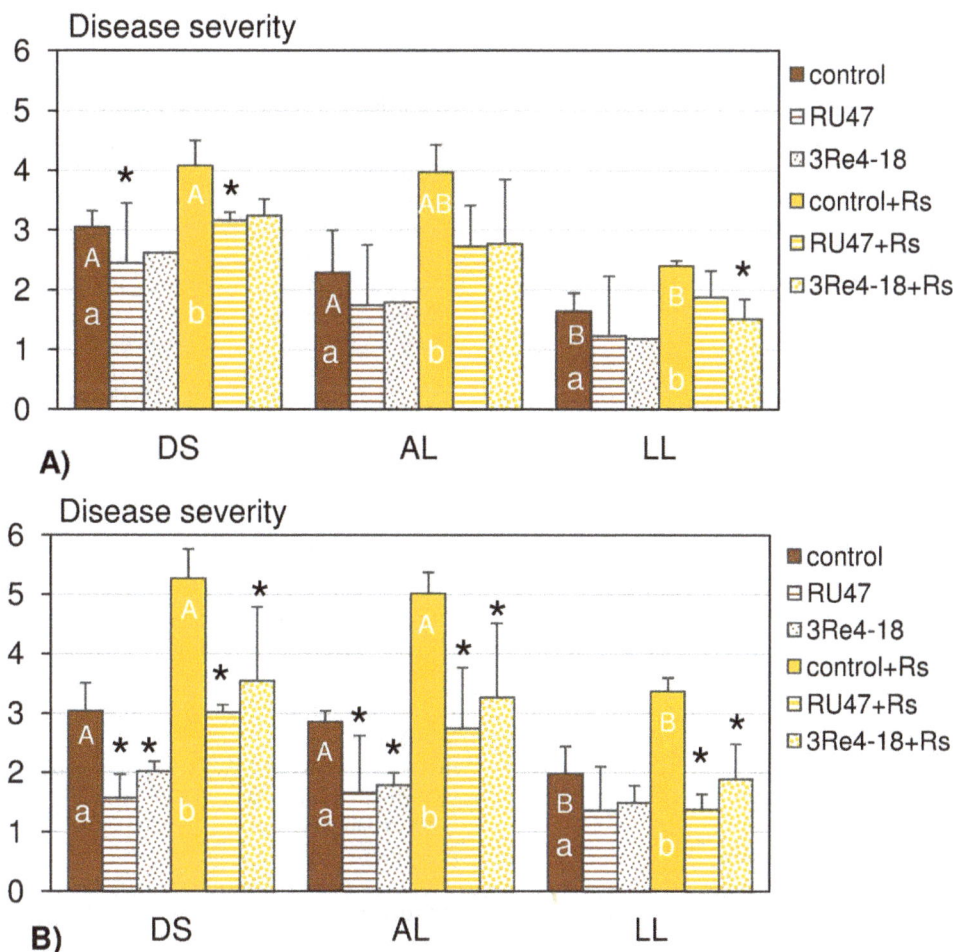

Figure 3. Disease severity of bottom rot on lettuce (cv. Tizian) determined for the following treatments: control, RU47, 3Re4-18, control+Rs, RU47+Rs and 3Re4-18+Rs the in two experiments, A) and B) in the 2011-season. Plants were grown in three soil types (DS, AL, LL) for six weeks each, at the same field site. Different capital letters indicate significant differences in disease severity of bottom rot in controls or control+Rs between soil types and different lower-case letters indicate significant differences between control and control+Rs within each soil type (Mann-Whitney U-test; $P<0.1$). An asterisk denotes significant differences in disease severity of the RU47 or 3Re4-18 treatment compared to the control and of RU47+Rs or 3Re4-18+Rs treatments compared to the control+Rs in each soil type. The bars show the standard deviation.

a composite sample and considered as one replicate; four replicates were used per treatment. Adhering soil was removed by washing the roots with sterile tap water before microbial cells were extracted as follows: the roots were cut into pieces of approximately 1 cm length and carefully mixed. Five gram of roots were placed in sterile Stomacher bags and treated by a Stomacher 400 Circulator (Seward Ltd, Worthing, UK) for 30 s at high speed after adding 15 ml of sterile 0.3% NaCl. The Stomacher blending step was repeated three times and followed by centrifugation steps as described by Schreiter et al. [37].

Analysis of rhizosphere competence of the bacterial inoculants

The ability of *P. jessenii* RU47 and *S. plymuthica* 3Re4-18 to colonize the rhizosphere of lettuce grown in the three soil types was determined 2WAP and 5WAP. Aliquots of the rhizosphere microbial cell suspension resulting from the combined supernatants of three Stomacher blending steps were immediately processed to determine the inoculant CFU counts by plating serial dilutions onto King's B agar supplemented with rifampicin (75 µg/ml) and cycloheximide (100 µg/ml) and incubated at 29°C

for 48 h. The CFU counts were calculated per gram of root dry mass (RDM). For all soil types Stomacher supernatants obtained from the control plots were plated as well to determine the background of the rifampicin resistant indigenous bacteria.

Data analysis

Data of SDM, disease severity and inoculant plate counts were analyzed with the STATISTICA program (StatSoft Inc., Tulsa, OK, USA). The impact of the soil type, the pathogen and the inoculants on SDM was determined using three-way ANOVA ($P<0.1$) combined with Tukey post-hoc test ($P<0.1$). The data of disease severity was evaluated using the nonparametric Kruskal Wallis test followed by Mann-Whitney U-test ($P<0.1$). The determined inoculants density (CFU counts/g RDM) was calculated and logarithmically (Log_{10}) converted before the impact of the soil type, plant growth development stage, and the presence of the pathogen *R. solani* on the plate counts of each inoculant strain was analyzed using three-way ANOVA ($P<0.05$) combined with Tukey post-hoc test ($P<0.05$). Average values for soil temperature of each soil type were analyzed by Tukey post-hoc test using

Table 2. Treatment-dependent differences (d-values) of bacterial communities in the rhizosphere of lettuce, grown in three soil types (DS, AL, LL), in the 2011-season.

Soil type	Experiment	Figure	Differences between control and				
			control+Rs	RU47	3Re4-18	RU47+Rs	3Re4-18+Rs
DS	1	S1	6.1*	2.6*	6.2*	3.6*	10.3*
AL	1	S2	1.9*	13.6*	20.6*	12.3*	13.2*
LL	1	S3	1.7*	6.8*	10.1*	7.8*	10.8*
DS	2	S4	4.9*	4.4*	8.8*	5.7*	7.6*
AL	2	4a	2.9*	2.8	10.0*	9.0*	17.3*
LL	2	S5	3.5	14.6*	16.1*	16.2*	21.8*

The asterisks indicate the significant differences (P<0.05) between the untreated control and the respective treatment.

standard errors of difference values and calculation of variance (P<0.05).

Analysis of 16S rRNA gene fragments PCR amplified from total community DNA by denaturing gradient gel electrophoresis (DGGE)

Total community DNA (TC-DNA) was extracted from the microbial pellets using the FastDNA SPIN Kit (MP Biomedicals, Heidelberg, Germany) as described by the manufacturer after a harsh lysis step with the FastPrep-24 Instrument (MP Biomedicals, Heidelberg, Germany). The TC-DNA was purified with GENE-CLEAN SPIN Kit (MP Biomedicals, Heidelberg, Germany) according to the manufacturer. The purified TC-DNA was diluted 1:10 with 10 mM Tris HCl before use.

For amplification of 16S rRNA gene fragments, PCR reactions were performed with TC-DNA obtained from rhizosphere samples with the primers F984-GC and R1378 as described by Heuer [41] using Taq DNA polymerase (Stoffel fragment, ABI, Darmstadt, Germany). The PCR products were analyzed by DGGE approach as described by Weinert et al. [42].

Bacterial fingerprints were evaluated with GELCOMPAR II version 6.5 (Applied Maths, Sint-Martens-Latern, Belgium) as described by Schreiter et al. [37]. The obtained Pearson similarity matrices were used for construction of a dendrogram by an Unweighted Pair-Group Method with Arithmetic mean (UPGMA) as well as of statistical analysis by the permutation test, calculating the d-value from the average overall correlation coefficients within the groups minus the average overall correlation coefficients between samples from treatments compared as suggested by Kropf et al. [43].

Results

Soil parameters of both field experiments

The concentration of N, P and K measured for each soil type before planting revealed only minor variations between both experiments (Table S1). In the second experiment, the temperature measured for all three soils in the 20 cm top layer was approximately 0.6°C below that of the first experiment (Table S2).

In contrast to the temperature the volumetric soil water content (VWC) in the top soils varied significantly among the soil types and experiments (except for AL and LL soil in the first experiment; Table S3). The lowest percentage of VWC was recorded in DS, and the highest in AL soil in both experiments. The calculated percentages of VWC in the first experiment were 18.9%, 39.6% and 39.1% in DS, AL and LL soil, and 17.0%, 35.7% and 27.4% in DS, AL and LL soil, in the second experiment. In the first experiment the VWC of AL and LL soil did not differ significantly. In both experiments VWC values for DS soil were significantly lower than those for AL and LL (Tukey post-hoc tests and confidence limits in Table S3).

Effect of the soil types on rhizosphere competence of the bacterial inoculants

In both experiments the inoculants were able to colonize the rhizosphere of lettuce grown in the different soil types at comparable CFU counts (Figure 1). Three-way ANOVA (soil type, plant growth development stage, pathogen) revealed that in both experiments the soil type had no significant effect on the CFU of RU47 in the rhizosphere of lettuce (P>0.05) but on the CFU of 3Re4-18 (P<0.05) (Table 1). In the first and second experiment 2WAP the CFU counts of 3Re4-18 were significantly higher in LL soil than in DS and AL soil. With increasing plant

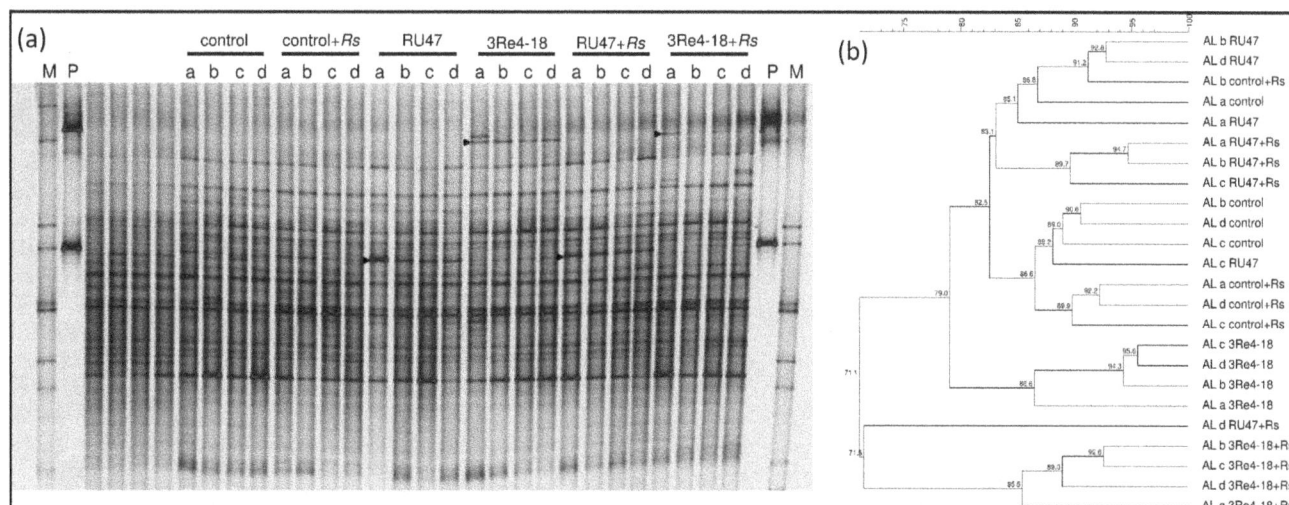

Figure 4. DGGE fingerprints (a) of bacterial 16Sr-RNA gene fragments from community DNA extracts were obtained from lettuce rhizosphere 2WAP of lettuce into the AL soil (second experiment). The corresponding UPGMA dendrogram (b) of these DGGE fingerprints based on the Pearson similarity matrix. The four replicates of the treatments: control, control+Rs, RU47, 3Re4-18, RU47+Rs, 3Re4-18+Rs were indicated by a–d. M: marker [57].

age the CFU counts of both inoculants decreased in the rhizosphere of lettuce grown in all three soils and in both experiments ($P<0.05$) except for the treatment with RU47 in DS soil in the first experiment and the treatment with 3Re4-18+Rs in DS soil in the second experiment (Figure 1). No significant effect of *R. solani* on CFU counts of both inoculants was revealed ($P>0.05$) in both experiments. The natural background of the rifampicin mutation was very low in all soil types.

Effects of the soil type, the bacterial inoculants and the pathogen on the lettuce growth

The effects of the soil type, the inoculants and *R. solani* on lettuce growth assessed by comparing the SDM of lettuce plants harvested 6WAP revealed lower lettuce growth in the first experiment compared with the second experiment in particular for LL soil (Figure 2). A significant effect of the soil type on lettuce growth was observed in both experiments. However, the results differed between the experiments. In the first experiment plants grown in LL soil showed the lowest SDM (21.3 g/plant) compared to SDM of plants from DS (32.5 g/plant) and AL soil (31.9 g/plant). In the second experiment the SDM was highest for plants grown in AL soil (46.2 g/plant) while the SDM of plants from DS (41.9 g/plant) and LL soil (41.6 g/plant) were comparable. In contrast to the first experiment an improved lettuce growth was observed in the treatments with the inoculants RU47 and 3Re4-18 in all three soils in the second experiment. However, the inoculant effects on lettuce growth were not significant (Figure 2).

In general the pathogen *R. solani* had a negative effect on lettuce growth in all three soils and in both experiments. The effect was significant in AL soil in both experiments, in LL soil only in the first experiment, and in DS soil in the second experiment (Figure 2). The treatment of lettuce with the inoculants resulted in an improved lettuce growth in RU47+Rs and 3Re4-18+Rs in all three soils and in both experiments but significant effects were only recorded in the first experiment for the treatments RU47+Rs in DS soil and 3Re4-18+Rs in LL soil, and in the second experiment for RU47+Rs and 3Re4-18+Rs in DS soil, for RU47+Rs in AL soil and for 3Re4-18+Rs in LL soil (Figure 2).

Effect of the soil type on the biocontrol activity of the bacterial inoculants

Bottom rot symptoms were also recorded in the untreated controls in each soil and in both experiments (Figure 3). The inoculation of *R. solani* (control+Rs) resulted in a significantly increased disease severity in all three soils and in both experiments. More severe bottom rot symptoms were scored in the second compared to the first experiment based on the disease severity in the pathogen controls in all three soils. The soil type influenced the disease severity of bottom rot in both experiments (Figure 3). The lowest disease severity was always recorded on lettuce grown in LL soil (control and control+Rs) compared to disease severity observed for lettuce grown in DS and AL soil.

RU47 and 3Re4-18 were able to reduce severity of bottom rot on lettuce in the treatments without and with *R. solani* inoculation in all three soils and in both experiments (Figure 3). Significant biocontrol effects were observed by the inoculant RU47 in DS soil (RU47; RU47+Rs) in both experiments and in AL (RU47; RU47+Rs) and LL (RU47+Rs) soil in the second experiment only. The inoculant 3Re4-18 showed a significant control effect in LL (3Re4-18+Rs) in the first experiment and in all soils in the second experiment (3Re4-18; 3Re4-18+Rs) except for 3Re4-18 in LL soil (Figure 3).

Effects of the pathogen and the bacterial inoculants on the bacterial community composition in the rhizosphere of lettuce

The effect of *R. solani* on the bacterial community composition were assessed by comparing the bacterial DGGE fingerprints of the control treatments (control) with those of the pathogen control (control+Rs) for each of the soils. In both experiments, the inoculation of *R. solani* had low but an significant effect in all soils except LL in the second experiment, the highest d-values were observed in DS soil (Table 2).

The effects of the inoculants on the bacterial community composition in the rhizosphere of lettuce analyzed by DGGE varied between the two experiments. The bacterial 16S rRNA gene fragments, amplified from TC-DNA of all treatments, were analyzed for each soil type separately and the DGGE gel for AL soil of the second experiment that had high d-values (Figure 4a). The DGGE fingerprint (Figure 4a) revealed that 2WAP the inoculant strain 3Re4-18 belonged to the dominant members of the bacterial community in the rhizosphere as bands with the electrophoretic mobility of 3Re4-18 were detected in the corresponding treatments. However, bands with electrophoretic mobility of the inoculant RU47 could not be detected as it co-migrated with bands of the indigenous bacterial community. Thus RU47 could only be identified in the second experiment in DS and AL soil, while 3Re4-18 was detected in AL and LL soil in the first, and in DS and AL soils in the second experiment (Figures S1–S5). Furthermore, the analysis of DGGE fingerprints by UPGMA revealed that the bacterial community composition of the different treatments shared approximately 70% similarity for all soils (Figure 4b). Although treatment dependent clusters were not always observed, the treatment effects were significant based on the permutation test ($P<0.05$) for all treatments with the inoculants RU47 or 3Re4-18 except for RU47 in AL soil in the second field experiment. However, the differences between the fingerprints of the controls and the treatments given by the calculated d-values varied for both experiments (Table 2). In the first experiment highest d-values were observed for AL soil for both inoculants while in the second experiment the highest d-values were observed for LL soil (Table 2). In both experiments and in all soils higher d-values for the treatments with 3Re4-18 indicated a stronger effect of this inoculant on the indigenous rhizosphere bacterial community compared to RU47 (Table 2). The effects of both inoculants on the bacterial community composition were typically increased for the treatments with *R. solani* inoculation in both experiments (except for RU47 and 3Re4-18 in AL soil in the first experiment).

Discussion

The availability of three soil types differing in their properties but sharing the same cropping history and weather conditions at the same field site enabled us to investigate the effect of the soil type on the rhizosphere competence of the two bacterial inoculants *P. jessenii* RU47 and *S. plymuthica* 3Re4-18 without and with *R. solani* inoculation for the first time. We hypothesized that the soil types with their distinct microbial community composition and physico-chemical properties but also the soil type dependent root exudates of the model plant lettuce [44] might affect the rhizosphere competence of the inoculants and in consequence their biocontrol efficacy. The composition of root exudates collected from lettuce grown in DS, AL and LL soil was recently shown to differ only quantitatively as similar compounds were detected as independent of the soil type [44]. Several studies showed that root exudates are important drivers of the rhizosphere microbial community composition [45,46] and the metabolic activity of the bacterial community including inoculant strains [47]. Recently, we could show by means of DGGE and pyrosequence analysis of 16S rRNA gene fragments, amplified from TC-DNA of bulk soil samples taken from the experimental unit 6 in 2010, that the three soil types indeed harbored a distinct bacterial community structure indicating the importance of the mineral composition and the soil organic matter for shaping the bacterial community composition [37]. Interestingly, in the rhizosphere of lettuce grown in the different soils numerous similar genera were increased in relative abundance, and a high proportion of dominant operational taxonomic units were shared among the rhizosphere samples from different soils [37]. Among the genera, significantly increased in the rhizosphere, was the genus *Pseudomonas* which might explain the negligible effect of

soil types on the ability of RU47 to colonize the lettuce rhizosphere in both experiments performed in 2011. Remarkably, similar CFU counts of RU47 were observed 2WAP in all soil types and in both experiments. The decrease of RU47 CFU counts in the rhizosphere observed 5WAP in both experiments might well be a decrease in its relative abundance per gram of RDM as the rhizosphere pellet was obtained from the complete root system due to changes in the rhizoplane surface/RDM ratio or/and possibly newly emerging roots were not colonized by RU47 as previously reported for *gfp*-tagged inoculants [48]. A decrease in inoculant CFU densities with increasing plant age of lettuce confirmed previous observations for the inoculants used in this study [34,49] and other biocontrol strains [24]. Nonetheless, our results showed an influence of the soil types on the rhizosphere competence of 3Re4-18. While higher CFU counts were observed in LL soil in both experiments 2WAP in comparison to DS and AL soil, a pronounced decline of 3Re4-18 counts was recorded in LL soil in the second experiment 5WAP. Indeed, in the second experiment the SDM was almost doubled for lettuce grown in LL soil compared to the first experiment which might have resulted in a decreased relative abundance of 3Re4-18. Neumann et al. [44] reported that the lettuce growth and the root morphology were influenced by soil type which might have contributed to the soil type dependent differences in rhizosphere competence of 3Re4-18. The differences in lettuce growth between the first and the second experiment in particular in LL soil were not unusual for agricultural practice. Factors that might have contributed to these differences range from slightly lower VWC, temperatures in the field directly after planting to differences in sun light intensity. In addition, the slightly different cropping history in both experimental units could have affected the soil microbiome with potential effects on plant growth. Plant growth in turn is assumed to influence the quantity of root exudates [50]. Several studies have underlined a clear relationship between the level of disease suppression and inoculant densities in the rhizosphere of an introduced biocontrol strain [17,51,52]. Inconsistent results in biological control were often assumed to be associated with inefficient root colonization at the field scale [11,53]. However, despite the good colonization of the lettuce rhizosphere by both inoculants in the present study, the biocontrol effects varied in both experiments. Although in the presence of the inoculant strains the disease severity was reduced. Compared to the control and the pathogen control (control+*Rs*), the biocontrol effects were not always significant. In fact, in the first experiment significant biocontrol effects were only observed for RU47 and RU47+*Rs* in DS soil and 3Re4-18+*Rs* in LL soil while biocontrol effects were found to be much more pronounced and significant in almost all treatments in the second experiment. The improved biocontrol effect seems to be correlated with an increased lettuce growth in the second experiment. It is tempting to speculate that an improved plant growth might be linked to an increased amount of photosynthates released via the roots. The increased root exudation in turn might have not only affected the metabolic activity of the inoculants but also facilitated the interaction of *R. solani* with the lettuce plant. Therefore, more severe bottom rot symptoms in the control and the control+*Rs* treatments were recorded in the second experiment. Furthermore, the disease severity in the control (natural background of *R. solani*) and in the pathogen control (control+*Rs*) was highest in DS soil and lowest in LL soil in both experiments. The higher conduciveness of the DS soil may be caused by higher oxygen availability because of the bigger pore sizes in sandy soils but might also be due to differences in the microbial community structure. Also, Steinberg et al. [54] reported that the capacity of a soil to antagonize the action of a plant pathogen is related to the structure and function of the microbial communities in soil.

Molecular fingerprinting techniques confirmed that the inoculants were indeed dominant members of the bacterial communities in the rhizosphere of lettuce. The additional DGGE bands of the inoculants strains 3Re4-18 and the increased intensity of a band co-migrating with RU47 which did not occur in the control treatments likely contributed to the effects of the inoculants on the rhizosphere bacterial community composition determined by the permutation test (Table 2). An approach to prevent the amplification of the inoculant strains is the use of taxon-specific primers which would exclude the taxonomic group to which the inoculant belongs, as proposed by Gomes et al. [55]. The actinobacterial fingerprint, done for the samples from 2WAP of the second experiment still revealed a significant effect except for RU47 in DS soil (Schreiter et al. unpublished data). The low effects of *R. solani* on the bacterial community composition can most likely be explained by the pathogenesis of *R. solani* AG1-IB which is primarily not a root pathogen but attacks the lower leaves of lettuce [56]. In conclusion, in the present study which was based on two independent field experiments performed in the 2011-season we could not confirm our hypothesis that the soil type influences the rhizosphere competence of biocontrol strains RU47 and 3Re4-18. We are aware that this finding cannot be generalized and can be different for other plants–inoculant combinations. The soil type independent enrichment of *Gammaproteobacteria* such as *Pseudomonas*, as observed by Schreiter et al. [37], in the rhizosphere of lettuce may be an explanation for the fact that the rhizosphere competence of RU47 was not influenced by soil type. Only minor changes in the bacterial rhizosphere composition were observed due to the inoculation of RU47, 3Re4-18 and *R. solani*, and these did not depend on the soil type. However, the plant growth and the disease severity of bottom rot were influenced by the soil type which in turn also influenced the biocontrol effects. The present study is unique as the rhizosphere competence, biocontrol effects, plant growth and treatment effects on the indigenous rhizosphere bacterial community were assessed in three soil types at the same site in two independent field experiments. The present study showed that the often reported inconsistency of biocontrol is likely more due to plants which in turn are influenced by weather conditions. Thus we conclude that the multitrophic interaction between plant, inoculant, pathogen and indigenous microbial community deserves far more attention in the future.

Supporting Information

Figure S1 *Bacteria* **DGGE fingerprint of rhizosphere-samples were obtained from DS soil 2WAP of the lettuce (first experiment).** Lanes a–d replicates of each of the treatments: untreated control (control), control inoculated with *Rhizoctonia solani* (control+*Rs*), inoculation with *Pseudomonas jessenii* RU47 (RU47), inoculation with *Serratia plymuthica* 3Re4-18 (3Re4-18), inoculation with RU47 and *R. solani* (RU47+*Rs*), inoculation with 3Re4-18 and *R. solani* (3Re4-18+*Rs*); M: marker [57].
(TIF)

Figure S2 *Bacteria* **DGGE fingerprint of rhizosphere-samples were obtained from AL soil 2WAP of the lettuce (first experiment).** Lanes a–d replicates of each of the treatments: untreated control (control), control inoculated with *Rhizoctonia solani* (control+*Rs*), inoculation with *Pseudomonas jessenii* RU47 (RU47), inoculation with *Serratia plymuthica* 3Re4-18 (3Re4-18), inoculation with RU47 and *R. solani* (RU47+*Rs*),

inoculation with 3Re4-18 and *R. solani* (3Re4-18+*Rs*); P: inoculants (upper band 3Re4-18; lower band RU47); M: marker [57].
(TIF)

Figure S3 *Bacteria* **DGGE fingerprint of rhizosphere-samples were obtained from LL soil 2WAP of the lettuce (first experiment).** Lanes a–d replicates of each of the treatments: untreated control (control), control inoculated with *Rhizoctonia solani* (control+*Rs*), inoculation with *Pseudomonas jessenii* RU47 (RU47), inoculation with *Serratia plymuthica* 3Re4-18 (3Re4-18), inoculation with RU47 and *R. solani* (RU47+*Rs*), inoculation with 3Re4-18 and *R. solani* (3Re4-18+*Rs*); P: inoculants (upper band 3Re4-18; lower band RU47); M: marker [57].
(TIF)

Figure S4 *Bacteria* **DGGE fingerprint of rhizosphere-samples were obtained from DS soil 2WAP of the lettuce (second experiment).** Lanes a–d replicates of each of the treatments: untreated control (control), control inoculated with *Rhizoctonia solani* (control+*Rs*), inoculation with *Pseudomonas jessenii* RU47 (RU47), inoculation with *Serratia plymuthica* 3Re4-18 (3Re4-18), inoculation with RU47 and *R. solani* (RU47+*Rs*), inoculation with 3Re4-18 and *R. solani* (3Re4-18+*Rs*); P: inoculants (upper band 3Re4-18; lower band RU47); M: marker [57].
(TIF)

Figure S5 *Bacteria* **DGGE fingerprint of rhizosphere-samples were obtained from LL soil 2WAP of the lettuce (second experiment).** Lanes a–d replicates of each of the treatments: untreated control (control), control inoculated with *Rhizoctonia solani* (control+*Rs*), inoculation with *Pseudomonas jessenii* RU47 (RU47), inoculation with *Serratia plymuthica* 3Re4-18 (3Re4-18), inoculation with RU47 and *R. solani* (RU47+*Rs*), inoculation with 3Re4-18 and *R. solani* (3Re4-18+*Rs*); P: inoculants (upper band 3Re4-18; lower band RU47); M: marker [57].
(TIF)

Table S1 **Concentration of nitrogen (N), phosphorus (P) and potassium (K) in soil measured before planting lettuce in three soil types (diluvial sand, DS; alluvial loam, AL; loess loam LL) in the season 2011 of the experimental plot system at the same field site.**
(DOCX)

Table S2 **Soil temperature recorded on average of the upper 20 cm top soil in three soil types (diluvial sand, DS; alluvial loam, AL; loess loam, LL) in the season 2011 at the same field site.**
(DOCX)

Table S3 **Comparison of the volumetric soil water content (VWC) of three soil types (diluvial sand, DS; alluvial loam, AL; loess loam, LL) in the season 2011 at the same field site.**
(DOCX)

Acknowledgments

We would also like to thank Petra Zocher, Ute Zimmerling, Sabine Breitkopf and Angelika Fandrey for their skilled technical assistance. We also thank Christin Zachow for providing the *Serratia plymuthica* 3Re4-18 strain and Ilse-Marie Jungkurth for helpful comments on the manuscript.

Author Contributions

Conceived and designed the experiments: KS RG. Performed the experiments: SS RG KS. Analyzed the data: SS RG MS. Contributed reagents/materials/analysis tools: KS RG SS MS. Contributed to the writing of the manuscript: SS RG KS. Designed software used in analysis of volumetric soil water content: MS.

References

1. Oerke EC (2006) Crop losses to pests. J Agr Sci 144: 31–43.
2. Alabouvette C, Olivain C, Steinberg C (2006) Biological control of plant diseases: the European situation. Eur J Plant Pathol 114: 329–341.
3. Leistra M, Matser AM (2004) Adsorption, transformation, and bioavailability of the fungicides carbendazim and iprodione in soil, alone and in combination. J Environ Sci Heal B 39: 1–17.
4. Wang YS, Wen CY, Chiu TC, Yen JH (2004) Effect of fungicide iprodione on soil bacterial community. Ecotox Environ Safe 59: 127–132.
5. Weller DM, Raaijmakers JM, Gardener BBM, Thomashow LS (2002) Microbial populations responsible for specific soil suppressiveness to plant pathogens. Annu Rev Phytopathol 40: 309–348.
6. Kazempour MN (2004) Biological control of *Rhizoctonia solani*, the causal agent of rice sheath blight by antagonistics bacteria in greenhouse and field conditions. Plant Pathol J 3: 88–96.
7. Scherwinski K, Grosch R, Berg G (2008) Effect of bacterial antagonists on lettuce: active biocontrol of *Rhizoctonia solani* and negligible, short-term effects on nontarget microorganisms. FEMS Microbiol Ecol 64: 106–116.
8. Tikhonovich IA, Provorov NA (2011) Microbiology is the basis of sustainable agriculture: an opinion. Ann Appl Biol 159: 155–168.
9. Quinlan RJ, Lisansky SG (2010) North America: Biopesticides Market. CPL Business Consultants, Oxfordshire, UK.
10. Mark GL, Morrissey JP, Higgins P, O'Gara F (2006) Molecular-based strategies to exploit *Pseudomonas* biocontrol strains for environmental biotechnology applications. FEMS Microbiol Ecol 56: 167–177.
11. Barret M, Morrissey JP, O'Gara F (2011) Functional genomics analysis of plant growth-promoting rhizobacterial traits involved in rhizosphere competence. Biol Fert Soils 47: 729–743.
12. Pierson III LS, Pierson EA (1996) Phenazine antibiotic production in *Pseudomonas aureofaciens*: role in rhizosphere ecology and pathogen suppression. FEMS Microbiol Lett 136: 101–108.
13. Steidle A, Allesen-Holm M, Riedel K, Berg G, Givskov M, et al. (2002) Identification and characterization of an N-acylhomoserine lactone-dependent quorum-sensing system in *Pseudomonas putida* strain IsoF. Appl Environ Microbiol 68: 6371–6382.
14. De Bellis P, Ercolani GL (2001) Growth interactions during bacterial colonization of seedling rootlets. Appl Environ Microbiol 67: 1945–1948.
15. Ghirardi S, Dessaint F, Mazurier S, Corberand T, Raaijmakers JM, et al. (2012) Identification of traits shared by rhizosphere-competent strains of fluorescent Pseudomonads. Microb Ecol 64: 725–737.
16. Bloemberg GV, Lugtenberg BJJ (2001) Molecular basis of plant growth promotion and biocontrol by rhizobacteria. Curr Opin Plant Biol 4: 343–350.
17. Lugtenberg BJJ, Dekkers L, Bloemberg GV (2001) Molecular determinants of rhizosphere colonization by *Pseudomonas*. Annu Rev Phytopathol 39: 461–493.
18. Persello-Cartieaux F, Nussaume L, Robaglia C (2003) Tales from the underground: molecular plant-rhizobacteria interactions. Plant Cell Enviro 26: 189–199.
19. Schnider U, Keel C, Blumer C, Troxler J, Defago G, et al. (1995) Amplification of the housekeeping sigma-factor in *Pseudomonas fluorescens* CHA0 enhances antibiotic production and improves biocontrol abilities. J Bacteriol 177: 5387–5392.
20. Dekkers LC, Mulders IHM, Phoelich CC, Chin-A-Woeng TFC, Wijfjes AHM, et al. (2000) The sss colonization gene of the tomato-*Fusarium oxysporum* f. sp *radicis-lycopersici* biocontrol strain *Pseudomonas fluorescens* WCS365 can improve root colonization of other wild-type *Pseudomonas* spp. bacteria. Mol Plant Microbe In 13: 1177–1183.
21. Patten CL, Glick BR (2002) Role of *Pseudomonas putida* indoleacetic acid in development of the host plant root system. Appl Environ Microbiol 68: 3795–3801.
22. Loper JE, Hassan KA, Mavrodi DV, Davis EW, II, Lim C, et al. (2012) Comparative genomics of plant-associated *Pseudomonas* spp.: insights into diversity and inheritance of traits involved in multitrophic interactions. PLoS Genet 8: e1002784–e1002784.
23. Raaijmakers JM, Mazzola M (2012) Diversity and natural functions of antibiotics produced by beneficial and plant pathogenic bacteria. Annu Rev Phytopathol 50: 403–424.
24. Haas D, Defago G (2005) Biological control of soil-borne pathogens by fluorescent Pseudomonads. Nat Rev Microbiol 3: 307–319.
25. Raaijmakers JM, Paulitz TC, Steinberg C, Alabouvette C, Moënne-Loccoz Y (2009) The rhizosphere: a playground and battlefield for soilborne pathogens and beneficial microorganisms. Plant Soil 321: 341–361.
26. Capdevila S, Martinez-Granero FM, Sanchez-Contreras M, Rivilla R, Martin M (2004) Analysis of *Pseudomonas fluorescens* F113 genes implicated in flagellar filament synthesis and their role in competitive root colonization. Microbiol-Uk 150: 3889–3897.
27. Rodriguez-Navarro DN, Dardanelli MS, Ruiz-Sainz JE (2007) Attachment of bacteria to the roots of higher plants. FEMS Microbiol Lett 272: 127–136.
28. Mavrodi DV, Joe A, Mavrodi OV, Hassan KA, Weller DM, et al. (2011) Structural and functional analysis of the type III secretion system from *Pseudomonas fluorescens* q8r1-96. J Bacteriol 193: 177–189.
29. Lottmann J, Heuer H, de Vries J, Mahn A, Düring K, et al. (2000) Establishment of introduced antagonistic bacteria in the rhizosphere of transgenic potatoes and their effect on the bacterial community. FEMS Microbiol Ecol 33: 41–49.
30. Chowdhury SP, Dietel K, Randler M, Schmid M, Junge H, et al. (2013) Effects of *Bacillus amyloliquefaciens* FZB42 on lettuce growth and health under pathogen pressure and its impact on the rhizosphere bacterial community. PLoS ONE 8: e68818–e68818. doi:10.1371/journal.pone.0068818.
31. Adesina MF, Lembke A, Costa R, Speksnijder A, Smalla K (2007) Screening of bacterial isolates from various European soils for in vitro antagonistic activity towards *Rhizoctonia solani* and *Fusarium oxysporum*: Site-dependent composition and diversity revealed. Soil Biol Biochem 39: 2818–2828.
32. Grosch R, Faltin F, Lottmann J, Kofoet A, Berg G (2005) Effectiveness of 3 antagonistic bacterial isolates to control *Rhizoctonia solani* Kühn on lettuce and potato. Can J Microbiol 51: 345–353.
33. Wibberg D, Jelonek L, Rupp O, Hennig M, Eikmeyer F, et al. (2013) Establishment and interpretation of the genome sequence of the phytopathogenic fungus *Rhizoctonia solani* AG1-IB isolate 7/3/14. J Biotechnol 167: 142–155.
34. Adesina MF, Grosch R, Lembke A, Vatchev TD, Smalla K (2009) In vitro antagonists of *Rhizoctonia solani* tested on lettuce: rhizosphere competence, biocontrol efficiency and rhizosphere microbial community response. FEMS Microbiol Ecol 69: 62–74.
35. Berg G, Krechel A, Ditz M, Sikora RA, Ulrich A, et al. (2005) Endophytic and ectophytic potato-associated bacterial communities differ in structure and antagonistic function against plant pathogenic fungi. FEMS Microbiol Ecol 51: 215–229.
36. Rühlmann J, Ruppel S (2005) Effects of organic amendments on soil carbon content and microbial biomass-results of the long-term box plot experiment in Grossbeeren. Arch Agron Soil Sci 51: 163–170.
37. Schreiter S, Ding G-C, Heuer H, Neumann G, Sandmann M, et al. (2014) Effect of the soil type on the microbiome in the rhizosphere of field-grown lettuce. Front Microbiol 5: 144.
38. Gutezeit B, Herzog FN, Wenkel KO (1993) Das Beregnungsbedarfssystem für Freilandgemüse. Gemüse 29: 106–108.
39. Grosch R, Schneider JHM, Kofoet A, Feller C (2011) Impact of continuous cropping of lettuce on the disease dynamics of bottom rot and genotypic diversity of *Rhizoctonia solani* AG 1-IB. J Phytopathol 159: 35–44.
40. Schneider JHM, Schilder MT, Dijst G (1997) Characterization of *Rhizoctonia solani* AG 2 isolates causing bare patch in field grown tulips in the Netherlands. Eur J Plant Pathol 103: 265–279.
41. Heuer H, Krsek M, Baker P, Smalla K, Wellington EMH (1997) Analysis of actinomycete communities by specific amplification of genes encoding 16S rRNA and gel-electrophoretic separation in denaturing gradients. Appl Environ Microbiol 63: 3233–3241.
42. Weinert N, Meincke R, Gottwald C, Heuer H, Gomes NCM, et al. (2009) Rhizosphere communities of genetically modified zeaxanthin-accumulating potato plants and their parent cultivar differ less than those of different potato cultivars. Appl Environ Microbiol 75: 3859–3865.
43. Kropf S, Heuer H, Grüning M, Smalla K (2004) Significance test for comparing complex microbial community fingerprints using pairwise similarity measures. J Microbiol Meth 57: 187–195.
44. Neumann G, Bott S, Ohler M, Mock H, Lippman R, et al. (2014) Root exudation and root development of lettuce (*Lactuca sativa* L. cv. Tizian) as affected by different soils. Front Microbiol 5: 2.
45. Paterson E, Gebbing T, Abel C, Sim A, Telfer G (2007) Rhizodeposition shapes rhizosphere microbial community structure in organic soil. New Phytol 173: 600–610.
46. Henry S, Texier S, Hallet S, Bru D, Dambreville C, et al. (2008) Disentangling the rhizosphere effect on nitrate reducers and denitrifiers: insight into the role of root exudates. Environ Microbiol 10: 3082–3092.
47. Benizri E, Nguyen C, Piutti S, Slezack-Deschaumes S, Philippot L (2007) Additions of maize root mucilage to soil changed the structure of the bacterial community. Soil Biol Biochem 39: 1230–1233.
48. Götz M, Gomes NCM, Dratwinski A, Costa R, Berg G, et al. (2006) Survival of *gfp*-tagged antagonistic bacteria in the rhizosphere of tomato plants and their effects on the indigenous bacterial community. FEMS Microbiol Ecol 56: 207–218.
49. Grosch R, Dealtry S, Schreiter S, Berg G, Mendonça-Hagler L, et al. (2012) Biocontrol of *Rhizoctonia solani*: complex interaction of biocontrol strains, pathogen and indigenous microbial community in the rhizosphere of lettuce shown by molecular methods. Plant Soil 361: 343–357.

Soil Type Dependent Rhizosphere Competence and Biocontrol of Two Bacterial Inoculant Strains...

187

50. Baudoin E, Benizri E, Guckert A (2002) Impact of growth stage on the bacterial community structure along maize roots, as determined by metabolic and genetic fingerprinting. Appl Soil Ecol 19: 135–145.

51. Bull CT, Weller DM, Thomashow LS (1991) Relationship between root colonization and suppression of *Gaeumannomyces graminis* var. tritici by *Pseudomonas fluorescens* strain 2–79. Phytopathology 81: 954–959.

52. Raaijmakers JM, Weller DM (2001) Exploiting genotypic diversity of 2,4-diacetylphloroglucinol-producing *Pseudomonas* spp.: Characterization of superior root-colonizing *P. fluorescens* strain Q8r1-96. Appl Environ Microbiol 67: 2545–2554.

53. Lemanceau P, Alabouvette C (1993) Suppression of fusarium-wilt by fluorescent pseudomonads: Mechanisms and applications. Biocontrol Sci Techn 3: 219–234.

54. Steinberg C, Edel-Hermann V, Alabouvette C, Lemanceau P (2007) Soil suppressiveness to plant diseases. Modern Soil Microbiology: 455–478.

55. Gomes NCM, Kosheleva IA, Abraham WR, Smalla K (2005) Effects of the inoculant strain *Pseudomonas putida* KT2442 (pNF142) and of naphthalene contamination on the soil bacterial community. FEMS Microbiol Ecol 54: 21–33.

56. Davis R, Subbarao K, Raid R, Kurtz E (1997) Compendium of Lettuce Diseases. APS Press: 15–16.

57. Heuer H, Wieland G, Schönfeld J, Schönwalder A, Gomes NCM, et al. (2001) Bacterial community profiling using DGGE or TGGE analysis; Rochelle PA, editor. 177–190 p.

17

Soil Carbon and Nitrogen Fractions and Crop Yields Affected by Residue Placement and Crop Types

Jun Wang[1], Upendra M. Sainju[2]*

1 College of Urban and Environmental Sciences, Northwest University, Xian, Shaanxi Province, China, **2** U.S. Department of Agriculture, Agricultural Research Service, Northern Plains Agricultural Research Laboratory, Sidney, Montana, United States of America

Abstract

Soil labile C and N fractions can change rapidly in response to management practices compared to non-labile fractions. High variability in soil properties in the field, however, results in nonresponse to management practices on these parameters. We evaluated the effects of residue placement (surface application [or simulated no-tillage] and incorporation into the soil [or simulated conventional tillage]) and crop types (spring wheat [*Triticum aestivum* L.], pea [*Pisum sativum* L.], and fallow) on crop yields and soil C and N fractions at the 0–20 cm depth within a crop growing season in the greenhouse and the field. Soil C and N fractions were soil organic C (SOC), total N (STN), particulate organic C and N (POC and PON), microbial biomass C and N (MBC and MBN), potential C and N mineralization (PCM and PNM), NH_4-N, and NO_3-N concentrations. Yields of both wheat and pea varied with residue placement in the greenhouse as well as in the field. In the greenhouse, SOC, PCM, STN, MBN, and NH_4-N concentrations were greater in surface placement than incorporation of residue and greater under wheat than pea or fallow. In the field, MBN and NH_4-N concentrations were greater in no-tillage than conventional tillage, but the trend reversed for NO_3-N. The PNM was greater under pea or fallow than wheat in the greenhouse and the field. Average SOC, POC, MBC, PON, PNM, MBN, and NO_3-N concentrations across treatments were higher, but STN, PCM and NH_4-N concentrations were lower in the greenhouse than the field. The coefficient of variation for soil parameters ranged from 2.6 to 15.9% in the greenhouse and 8.0 to 36.7% in the field. Although crop yields varied, most soil C and N fractions were greater in surface placement than incorporation of residue and greater under wheat than pea or fallow in the greenhouse than the field within a crop growing season. Short-term management effect on soil C and N fractions were readily obtained with reduced variability under controlled soil and environmental conditions in the greenhouse compared to the field. Changes occurred more in soil labile than non-labile C and N fractions in the greenhouse than the field.

Editor: Raffaella Balestrini, Institute for Sustainable Plant Protection, C.N.R., Italy

Funding: Funding came from the U.S. Department of Agriculture-Agricultural Research Service, Sidney, MT, USA and the National Natural Science Foundation of China (No. 31270484). The funders had no role in study design, data collection and analysis, decision to publish, or preparation of the manuscript.

* Email: upendra.sainju@ars.usda.gov

Introduction

Soil organic matter, as indicated by C and N levels, is an important component of soil quality and productivity. Increasing soil organic matter through enhanced C and N sequestration can also reduce the potentials for global warming by mitigating greenhouse gas emissions and N leaching by increasing N storage in the soil [1,2]. Carbon and N sequestration usually occur when non-harvested crop residues, such as stems, leaves, and roots, are placed at the soil surface due to no-tillage [3,4,5]. Carbon and N sequestration rates, however, depend on the balance between the amounts of plant residue C and N inputs and rates of C and N mineralized in the nonmanured soil [6,7]. Other benefits of increasing C and N storage include enhancement of soil structure and soil water-nutrient-crop productivity relationships [8].

Soil and crop management practices can alter the quantity, quality, and placement of crop residues in the soil, thereby influencing soil C and N storage, microbial biomass and activity, and N mineralization–immobilization [9,10]. Residue placement in the soil under different tillage systems can influence C and N levels by affecting soil aggregation, aeration, and C and N

mineralization [9,11]. Crop types can affect the quantity and quality (C/N ratio) of crop residue returned to the soil and therefore on soil C and N levels [9,12]. Legumes, such as pea, because of its higher N concentration and lower C/N ratio, decompose more rapidly in the soil and supply greater amount of N to succeeding crops than nonlegumes [12,13]. As a result, N fertilization rates to crops following pea can be reduced to sustain yields [14,15].

Because of large pool sizes and inherent spatial variability, soil organic C (SOC) and total N (STN) (slow or non-labile fractions) change slowly with management practices [16]. Therefore, measurements of SOC and STN alone may not adequately reflect changes in soil quality and nutrient status [16,17]. Active (or labile) C and N fractions, such as potential C and N mineralization (PCM and PNM) that indicate microbial activity and N mineralization, and microbial biomass C and N (MBC and MBN) that refer to microbial biomass and N immobilization, change seasonally [16,18]. Similarly, particulate organic C and N (POC and PON) that represent coarse organic matter and considered as intermediate C and N levels between slow and active fractions, provide substrates for microbes and influence soil aggregation [19,20].

Available N fractions that influence plant growth and N losses due to leaching, denitrification, or volatilization are NH_4-N and NO_3-N [10,12].

Although active C and N fractions in the soil can change more rapidly than the other fractions, these fractions sometime may not be readily changed within a crop growing season due to high variability in soil properties within a short distance in the field or in regions with limited precipitation, cold weather, and a short growing season [10,12,15]. Under controlled soil and environmental conditions, such as in the greenhouse, it may be possible to detect changes in these fractions more rapidly as affected by management practices than in the field. We hypothesized that surface placement of crop residue (a simulation of no-tillage in the field) under spring wheat can increase soil labile and non-labile C and N fractions and sustain crop yields compared to residue incorporation into the soil (a simulation of conventional tillage) under pea or fallow more in the greenhouse than in the field. Our objectives were to: (1) evaluate the effects of residue placement and crop types on crop yields, residue C and N losses, and soil labile and non-labile C and N fractions within a growing season in the greenhouse and the field and (2) determine if soil C and N fractions change more readily in the greenhouse than the field within a growing season.

Materials and Methods

Greenhouse experiment

The experiment was conducted under controlled soil and environmental conditions in the greenhouse with air temperatures of 25°C in the day and 15°C in the night. Soil samples were collected manually from an area of 5 m^2 using a shovel to a depth of 20 cm under a mixture of crested wheatgrass [*Agropyron cristatum* (L.) Gaertn] and western wheatgrass [*Pascopyrum smithii* (Rydb.) A. Love] from a dryland farm site, 11 km east of Sidney, Montana, USA. The research farm site where soil samples were collected is under the management of USDA, Agricultural Research Service, Sidney, Montana and no endangered or protected species were involved or was negatively impacted by this research. The soil was a Williams loam (fine-loamy, mixed, frigid, Typic Argiborolls [International classification: Luvisols]) with 350 g kg^{-1} sand, 325 g kg^{-1} silt, 325 g kg^{-1} clay, 1.42 Mg m^{-3} bulk density, and 7.2 pH at the 0–20 cm depth. Soil C and N fractions in the sample before the initiation of the experiment are shown in Table 1. Soil was air-dried and sieved to 4.75 mm after discarding coarse organic materials and rock fragments. Eight kilograms of soil was placed in a plastic pot, 25 cm high by 25 cm diameter, above 3 cm of gravel at the bottom.

Treatments consisted of two residue placements (surface placement vs. incorporation into the soil) and three crop types (spring wheat, pea, and fallow [or no crop]) arranged in a completely randomized design with three replications. In order to match the residue and crop type, spring wheat residue was placed under spring wheat and fallow and pea residue under pea. Residues included nine-week old spring wheat and pea plants collected from the field without grains, chopped to 2 cm, and oven-dried at 60°C for 3 d. Fifteen grams of residues per pot (corresponding to 2.6 Mg ha^{-1} of residue found in the field) were either placed uniformly at the soil surface or incorporated into the soil by mixing the residue with the soil by hand. The surface placement of residue corresponded to the simulated no-tillage system in the field, although the soil was disturbed during collection, and incorporated residue to the simulated conventional tillage system. Spring wheat received 0.96 g N pot^{-1} as urea, similar to the recommended N fertilization rate (80 kg N ha^{-1}) in the field, while pea received 0.11 g N pot^{-1} (or 9 kg N ha^{-1}) while applying monoammonium phosphate as the P fertilizer. Half of 0.96 g N pot^{-1} was applied at planting and other half at four weeks later. Both spring wheat and pea also received P fertilizer (monoammonium phosphate) at 0.25 g P pot^{-1} (or 27 kg P ha^{-1}) and K fertilizer (muriate of potash) at 0.50 g K pot^{-1} (or 29 kg K ha^{-1}). No fertilizers were applied to the fallow treatment.

In July 2012, five spring wheat (cultivar Reeder) and pea (cultivar Majoret) seeds were planted per pot, except in the fallow treatment. At a height of 3 cm, seedlings were thinned to two plants per pot. In order to compensate for the water received as rainfall in the field, water was applied to all treatments in the greenhouse experiment to field capacity (0.25 m^3 m^{-3}) [21] at 300 to 500 mL pot^{-1}. Water was applied at planting and at 3 to 7 d intervals thereafter, depending on soil water content (as determined by a soil water probe [TDR 300, Spectrum Technologies Inc., Aurora, IL] installed to a depth of 15 cm). Since measured amount of water was applied according to soil water content and crop demand, only a negligible amount of water was leached below the pot that was not determined. Herbicides and pesticides were applied to plants as needed. At 105 d after planting, shoot biomass including grains was harvested from the pot, washed with water, oven-dried at 60°C for 3 to 7 d, and dry matter yield was determined. Because of the small amount of grain production, grains were also included in the shoot biomass. After crop harvest, soil from the entire pot was sieved to 2 mm to separate coarse residue and root fragments, which were picked by hand, washed with water, and oven-dried at 60°C for 3 to 7 d to determine dry matter yields. A portion (100 g) of residue and root-free soil sample visible to the naked eye was collected from each pot, air-dried, and used for determinations of C and N fractions. The remaining soil samples were further washed in a nest of 1.0 and 0.5 mm sieves under a continuous stream of water to separate fine roots. Roots left in the sieves were picked using a tweezers, oven-dried at 60°C for 3 to 7 d, and dry matter yield was determined. Total root biomass was determined by adding biomass of coarse and fine roots.

Shoot and root biomass and crop residues added to the soil at the initiation of the experiment and those (>2.00 mm) recovered from the soil at the end were ground to 1 mm and C and N concentrations (g kg^{-1}) were determined with a high induction furnace C and N analyzer (LECO, St. Joseph, MI). Amounts of C and N in the residue added and recovered from the soil were determined by multiplying C and N concentrations by the weight of the soil in the pot. Carbon and N losses from the residue were determined as: Residue C and N losses (g kg^{-1}) = (Residue C and N added − Residue C and N recovered) × 1000/Residue C and N added. While determining the amount of C and N recovered in the residue, it was assumed that fine residue (<2.00 mm) was a part of soil organic matter.

Field experiment

The field experiment was conducted using identical treatments, design, and replications as in the greenhouse from April to August 2012 near the place where soil samples were collected for the greenhouse experiment. As a result, soils were similar in both field and greenhouse experiments. The field site has mean monthly air temperature ranging from −8°C in January to 23°C in July and August. The mean annual precipitation (105-yr average) is 340 mm, 80% of which occurs during the crop growing season (April-October). Equivalent amounts of crop residues and fertilizers using the same treatments as in the greenhouse were applied to spring wheat, pea, and fallow in the field. Because the amount of residue applied was similar, the amounts of C and N

Table 1. Average soil organic C (SOC), total N (STN), particulate organic C and N (POC and PON), potential C and N mineralization (PCM and PNM), microbial biomass C and N (MBC and MBN), and NH_4-N and NO_3-N concentrations at the start of the experiment (n = 4).

Parameter	Concentration
SOC (g C kg^{-1})	11.80
POC (g C kg^{-1})	3.18
PCM (mg C kg^{-1})	9.25
MBC (mg C kg^{-1})	117.6
STN (g N kg^{-1})	1.29
PON (g N kg^{-1})	0.34
PNM (mg N kg^{-1})	8.95
MBN (mg N kg^{-1})	69.0
NH_4-N (mg N kg^{-1})	2.86
NO_3-N (mg N kg^{-1})	5.04

added in residue to the soil were also identical in the greenhouse and field. Residues and fertilizers were placed at the soil surface in the no-till system and incorporated to a depth of 10 cm using tillage with a field cultivator in the conventional tillage system. Plot size was 12.2×6.1 m.

Spring wheat and pea were planted in April with a no-till drill at a spacing of 20.3 cm. Growing season weeds were controlled with selective post emergence herbicides appropriate for each crop. Contact herbicides were applied at postharvest and preplanting. Crops were grown under dryland condition receiving only precipitation without irrigation. In August, biomass yield of spring wheat and pea was determined from two 0.5 m^2 areas outside yield rows within each plot and grain yield was determined by harvesting grains from a swath of 1.5 m×12.0 m using a combine harvester. Carbon and nitrogen concentrations in the grain and biomass were determined after oven drying subsamples at 55°C and using the C and N analyzer as above. Carbon and N contents (Mg C or N ha^{-1}) in grain and biomass were determined by multiplying C and N concentrations by grain and biomass yields, respectively. Total aboveground biomass and C and N contents were determined by adding yields and C and N contents of grain and biomass.

Soil samples were collected from five random locations in central rows of the plot to a depth of 20 cm using a truck-mounted hydraulic probe (3.5 cm inside diameter). Samples were composited within a plot, air-dried, ground, and sieved to 2 mm for determining C and N concentrations. No attempts were made to collect the surface residue at soil sampling because of residue loss and contamination with soil and residue from one plot to another due to actions of wind and water. Therefore, residue C and N losses were not determined in the field.

Soil carbon and nitrogen fractions measurements

The SOC concentration in the greenhouse and field soils were determined with a high induction furnace C and N analyzer as above after pretreating the soil with 5% H_2SO_3 to remove inorganic C [22]. The STN concentration was determined by using the analyzer without pretreating the soil with the acid. For determining POC and PON concentrations, 10 g soil sample was dispersed with 30 mL of 5 g L^{-1} sodium hexametaphosphate by shaking for 16 h and the solution was poured through a 0.053 mm sieve [19]. The solution and particles that passed through the sieve and contained mineral-associated and water-soluble C and N were

dried at 50°C for 3 to 4 d and SOC and STN concentrations were determined by using the analyzer as above. The POC and PON concentrations were determined by the difference between SOC and STN in the whole-soil and that in the particles that passed through the sieve after correcting for the sand content.

The PCM and PNM concentrations in air-dried soils were determined by the modified method of Haney et al. [23]. Two 10 g soil subsamples were moistened with water at 50% field capacity [21] and placed in a 1 L jar containing beakers with 4 mL of 0.5 mol L^{-1} NaOH to trap evolved CO_2 and 20 mL of water to maintain high humidity. Soils were incubated in the jar at 21°C for 10 d. At 10 d, the beaker containing NaOH was removed from the jar and PCM was determined by measuring CO_2 absorbed in NaOH, which was back-titrated with 1.5 mol L^{-1} BaCl$_2$ and 0.1 mol L^{-1} HCl. One beaker containing soil was removed from the jar and extracted with 100 mL of 2 mol L^{-1} KCl for 1 h. The NH_4-N and NO_3-N concentrations in the extract were determined by using the autoanalyzer (Lachat Instrument, Loveland, CO). The PNM was calculated as the difference between the sum of NH_4-N and NO_3-N concentrations in the soil before and after incubation.

The other beaker containing moist soil and incubated for 10 d (used for PCM determination above) was used for determining MBC and MBN concentrations by the modified fumigation–incubation method for air-dried soils [24]. The moist soil was fumigated with ethanol-free chloroform for 24 h and placed in a 1 L jar containing beakers with 2 mL of 0.5 mol L^{-1} NaOH and 20 mL water. As with PCM, fumigated moist soil was incubated for 10 d and CO_2 absorbed in NaOH was back-titrated with BaCl$_2$ and HCl. The MBC was calculated by dividing the amount of CO_2–C absorbed in NaOH by a factor of 0.41 [25] without subtracting the values from the nonfumigated control [24]. For MBN, the fumigated–incubated sample at 10 d was extracted with 100 mL of 2 mol L^{-1} KCl for 1 h and NH_4-N and NO_3-N concentrations were determined by using the autoanalyzer as above. The MBN was calculated by the difference between the sum of NH_4-N and NO_3-N concentrations in the sample before and after fumigation–incubation and divided by a factor of 0.41 [25,26]. The NH_4-N and NO_3-N concentrations determined in the nonfumigated–nonincubated samples were used as available fractions of N.

Data analysis

Data for C and N contents in crop biomass and residue and soil C and N fractions were analyzed by using the MIXED model of SAS [27]. Treatment was considered as the fixed effect and replication as the random effect. Means were separated by using the least square means test when treatments and interactions were significant [27]. Statistical significance was evaluated at $P \leq 0.05$, unless otherwise stated.

Results

Greenhouse experiment

Shoot and root biomass yields and carbon and nitrogen contents. Shoot and root biomass yields and C and N contents varied among residue placements and crop types (Table 2). Interaction between residue placement and crop types on these parameters was not significant. Shoot and root biomass yields and C and N contents were greater in surface placement than incorporation of residue into the soil. Shoot biomass yield and C and N contents were also greater in wheat than in pea. Because of the negligible amount of roots, root biomass yield and C and N contents in pea were not determined. Absence of plants in the fallow also resulted in non-existence of crop data in this treatment. The coefficient of variation (CV) for crop parameters ranged from 38.2 to 62.5%.

Residue carbon and nitrogen losses. Total amounts of C and N added through residue application and leaf fall and those recovered in coarse fractions (>2 mm) after crop harvest varied with residue placements and crop types, with the significant residue placement × crop type interaction for C and N recovered in the residue (Table 3). Although the amount of residue applied was similar in all treatments (15 g of wheat or pea residue pot^{-1}), differences in C and N concentrations between residues and those added through leaf fall during crop growth varied residue C and N additions among treatments. Averaged across crop types, residue C addition was greater in surface placement than incorporation of residue into the soil. Averaged across residue placements, residue C addition was greater under wheat than pea or fallow, but residue N addition was greater under pea than wheat or fallow. Residue C recovery was greater in surface placement under wheat and fallow than surface placement under pea and incorporation under fallow. Residue N recovery was also greater in surface placement under wheat and fallow than surface placement under pea and incorporation under fallow and wheat. Averaged across crop types, residue N recovery was greater in surface placement than incorporation of residue into the soil. Averaged across residue placements, residue C recovery was greater under wheat than pea. The coefficient of variation for residue C and N addition and recovery varied from 7.2 to 17.8%.

Residue C and N losses also varied with residue placements and crop species, with the significant residue placement × crop species interaction (Table 3). Residue C loss was greater in surface placement under pea and incorporation under fallow than surface placement under fallow. Residue N loss was in the order: surface placement and incorporation under pea > incorporation under wheat and fallow > surface placement under wheat > surface placement under fallow. Averaged across crop types, residue N loss was greater in residue incorporation than surface placement. Averaged across residue placements, residue N loss was greater under pea than under fallow and wheat. The coefficient of variation for residue C and N losses varied from 14.3 to 31.6%.

Soil carbon and nitrogen fractions. The SOC, POC, and PCM concentrations varied among residue placements and crop types (Table 4). Averaged across crop types, SOC and PCM were greater in surface placement than incorporation of the residue into the soil. Averaged across residue placements, SOC was greater under wheat than pea and fallow and POC was greater under wheat than pea. The MBC was not influenced by treatments. The coefficient of variation for soil C fractions ranged from 2.6 to 14.3%.

The STN, PNM, MBN, NH_4-N, and NO_3-N concentrations also varied among residue placements and crop types (Table 4). Averaged across crop types, PNM and NH_4-N were greater in surface placement than incorporation of residue into the soil. Averaged across residue placements, STN was greater under wheat than pea and MBN was greater under wheat than fallow. In contrast, PNM was greater under pea and fallow than wheat and NO_3-N was greater under fallow than wheat. The PON was not influenced by treatments. The coefficient of variation for soil N fractions ranged from 4.6 to 15.9%.

Field experiment

Aboveground total crop biomass yield and C and N contents varied with crop types (Table 5). Averaged across tillage practices, crop biomass yield and C content were greater in wheat than pea, but the trend reversed for N content. Tillage and its interaction with crop type were not significant for crop biomass yield and C and N contents. The coefficient of variation for crop biomass yield and C and N contents ranged from 28.1 to 41.9%.

Soil MBN, NH_4-N, and NO_3-N concentrations varied with tillage practices and MBC and PNM varied with crop types (Table 5). Averaged across crop types, MBN and NH_4-N were greater in no-tillage than conventional tillage, but NO_3-N was greater in conventional tillage than no-tillage. Averaged across tillage practices, MBC was greater under wheat than fallow and PNM was greater under pea than wheat and fallow. Tillage, crop type, and their interaction were not significant for SOC, POC, PCM, STN, and PON. The coefficient of variations for soil C and N fractions ranged from 8.0 to 36.7%.

Discussion

Enhanced soil water conservation due to mulch action of the residue at the soil surface [28] may have increased shoot and root biomass yields and C and N contents in surface placement compared to incorporation of residue into the soil in the greenhouse (Table 2). It has been reported that surface placement of residue in the no-till system increased spring wheat yield compared to residue incorporation in the conventional till system [3,15]. In our field experiment, crop biomass yield and C and N contents, however, were not influenced by tillage (Table 5). It may be possible that wheat and pea residues applied by hand at the soil surface were more uniformly distributed in the greenhouse than in the field where residues were distributed by a machine sprayer. As a result, soil water was probably conserved more, resulting in increased crop yield and C and N contents with the surface placement than incorporation of residue in the greenhouse compared to the field.

Differences in the amount of N fertilizer applied and N fixation capacity may have resulted in variation in crop biomass yields and C contents among crop species in the greenhouse and the field (Tables 2 and 5). Higher amount of N fertilizer application may have increased biomass yield and C and N contents in wheat than pea in the greenhouse. Higher amount of N fertilizer application also may have increased biomass yield and C content in wheat and pea, but greater N fixation may have increased N content in pea than wheat in the field [14,28]. Grain and biomass yields are usually greater in wheat which receives N fertilizer than pea which

Table 2. Effects of residue placement and crop type on crop shoot (grains+leaves+stems) and root biomass C and N contents in the greenhouse.

Residue placement	Crop type	Shoot biomass	Root biomass	Shoot biomass C	Root biomass C	Shoot biomass N	Root biomass N	Total biomass C	Total biomass N
		—— g pot⁻¹ ——		—— g C pot⁻¹ ——		—— g Npot⁻¹ ——		g C pot⁻¹	g N pot⁻¹
Incorporated		4.51b[a]	2.37b	1.83b	0.74b	0.14b	0.05b	2.91b	0.25b
Surface		7.41a	6.14a	3.07a	1.74a	0.24a	0.11a	5.55a	0.44a
	Pea	4.55b	—[b]	1.91b	—	0.12b	—	1.91b	0.12b
	Wheat	7.36a	4.25	3.00a	1.24	0.26a	0.08	4.23a	0.34a
CV (%)[c]		42.4	61.9	42.4	56.4	55.0	62.5	40.4	38.2
Significance									
Residue placement (R)		*	—	*	*	**	*	*	*
Crop species (C)		*	*	*	—	**	—	*	—
R×C		NS[d]	NS	NS	—	NS	—	—	—

*Significant at $P=0.05$.
**Significant at $P=0.01$.
[a]Numbers followed by different letters within a column in a set are significantly different at $P \leq 0.05$ by the least square means test.
[b]Non-measurable values due to negligible amount of root biomass.
[c]Coefficient of variation.
[d]Not significant.

Table 3. Effects of residue placement and crop type on residue C and N addition, recovered in coarse fragments (>2 mm), and losses during the crop growing period in the greenhouse.

Residue placement	Crop type	Crop residue					
		C added[a]	N added[a]	C recovered	N recovered	C loss	N loss
		g C pot⁻¹	g N pot⁻¹	g C pot⁻¹	g N pot⁻¹	g kg⁻¹	g kg⁻¹
Incorporated	Fallow	7.80ab[b]	0.40a	3.84b	0.23b	508a	438b
	Pea	7.80a	0.64a	4.24ab	0.26ab	457ab	588a
	Wheat	8.76a	0.44a	4.89ab	0.24b	430ab	456b
Surface	Fallow	7.80a	0.40a	5.16a	0.31a	339b	213d
	Pea	7.80a	0.64a	3.82b	0.23b	511a	640a
	Wheat	8.92a	0.46a	4.95a	0.31a	446ab	325c
CV (%)[c]		7.2	17.8	15.4	16.5	143	316
Means							
Incorporated		8.07b	0.49a	4.32a	0.24b	465a	494a
Surface		8.17a	0.50a	4.64a	0.28a	432a	393b
	Fallow	7.80b	0.40b	4.50ab	0.27a	423a	326b
	Pea	7.80b	0.64a	4.02b	0.25a	484a	614a
	Wheat	8.76a	0.45b	4.92a	0.28a	438a	390b
Significance							
Residue placement (R)		*	NS[d]	NS	*	NS	**
Crop species (C)		***	***	*	NS	*	***
R×C		NS	NS	*	*	*	**

*Significant at $P = 0.05$.
**Significant at $P = 0.01$.
*** Significant at $P = 0.001$.
[a]Includes C and N added from the residue application and leaf fall.
[b]Numbers followed by different letters within a column in a set are significantly different at $P \leq 0.05$ by the least square means test.
[c]Coefficient of variation.
[d]Not significant.

Table 4. Effects of residue placement and crop type on soil organic C (SOC), total N (STN), particulate organic C and N (POC and PON), potential C and N mineralization (PCM and PNM), microbial biomass C and N (MBC and MBN), and NH$_4$-N and NO$_3$-N concentrations in the greenhouse.

Residue placement	Crop type	SOC	POC	PCM	MBC	STN	PON	PNM	MBN	NH$_4$-N	NO$_3$-N
		g C kg⁻¹		mg C kg⁻¹		g N kg⁻¹		mg N kg⁻¹			
Incorporated		12.0b[a]	3.28a	11.6b	128.0a	1.29a	0.33a	5.80b	50.8a	1.14b	11.8a
Surface		12.3a	3.28a	14.4a	143.8a	1.31a	0.31a	9.56a	57.5a	1.65a	16.5a
	Fallow	12.0b	3.26ab	11.4a	118.9a	1.30ab	0.32a	9.36a	46.1b	1.44a	23.7a
	Pea	12.0b	3.18b	13.1a	144.0a	1.26b	0.29a	9.68a	56.3ab	1.38a	12.8ab
	Wheat	12.4a	3.40a	14.5a	145.1a	1.34a	0.34a	5.98b	97.6a	1.36a	6.1b
CV (%)[b]		2.6	11.8	14.3	13.3	4.6	15.9	14.2	14.3	14.8	13.6
Significance											
Residue placement (R)		**	NS[c]	*	NS	NS	NS	*	NS	*	NS
Crop species (C)		**	*	NS	NS	*	NS	*	*	NS	*
R×C		NS	NS	NS	NS	NS	NS	NS	NS	NS	NS

Soil samples were collected at the 0–20 cm depth in the field and used for the greenhouse experiment.
*Significant at $P = 0.05$.
**Significant at $P = 0.01$.
[a]Numbers followed by different letters within a column in a set are significantly different at $P \leq 0.05$ by the least square means test.
[b]Coefficient of variation.
[c]Not significant.

Table 5. Effects of residue placement and crop type on crop aboveground biomass (grains+stems+leaves) yield, C and N contents, and soil organic C (SOC), total N (STN), particulate organic C and N (POC and PON), potential C and N mineralization (PCM and PNM), microbial biomass C and N (MBC and MBN), and NH_4-N and NO_3-N concentrations at the 0–20 cm depth in the field.

Tillage[a]	Crop type	Crop biomass yield	Crop C content	Crop N content	SOC	POC	PCM	MBC	STN	PON	PNM	MBN	NH_4-N	NO_3-N
		Mg ha^{-1}	Mg C ha^{-1}	kg N ha^{-1}	g C kg^{-1}		mg C kg^{-1}		g N kg^{-1}		mg N kg^{-1}		mg N kg^{-1}	
CT		4.91b	2.06a	59.6a	11.0a	2.56a	45.1a	114.4a	1.33a	0.24a	3.21a	12.6b	3.05b	4.54a
NT		5.08a	2.11a	65.3a	110.a	2.56a	56.8a	122.9a	1.37a	0.28a	4.28a	19.6a	3.82a	2.08b
	Fallow	---[c]	---	---	10.6a	2.55a	45.61a	111.4b	1.30a	0.25a	2.85b	14.2a	2.93a	3.36a
	Pea	4.68b	1.87b	72.3a	11.4a	2.56a	51.0a	118.6ab	1.42a	0.27a	5.56a	15.5a	3.43a	2.87a
	Wheat	5.31a	2.18a	52.6b	11.0a	2.57a	56.0a	126.0a	1.34a	0.26a	2.83b	18.6a	3.93a	3.71a
CV (%)		28.1	30.0	41.9	8.0	18.7	27.1	28.9	9.3	18.4	36.4	36.7	27.9	25.0
Significance														
Tillage (T)		NS[d]	NS	NS	NS	NS	NS	NS	NS	NS	NS	*	*	*
Crop species (C)		*	*	**	NS	NS	NS	*	NS	NS	*	NS	NS	NS
T×C		NS	NS	NS	NS	NS	NS	NS	NS	NS	NS	NS	NS	NS

*Significant at $P=0.05$.
**Significant at $P=0.01$.
[a]Tillage are CT, conventional tillage; and NT, no-tillage.
[b]Numbers followed by different letters within a column in a set are significantly different at $P \leq 0.05$ by the least square means test.
[c]Crop absent in the fallow.
[d]Not significant.

receives no N fertilizer due to increased water-use efficiency, but higher N concentration due to increased atmospheric N fixation can increase N content in pea than wheat [10,15,28]. The fact that different trends in N content in pea vs. wheat occurred in the field and the greenhouse was probably related to root growing soil volume. It may be possible that roots exploited greater soil volume that resulted in increased N fixation by pea and therefore increased its N content in the field compared to the greenhouse where plants were grown in a limited soil volume in the pot.

Greater residue input due to higher biomass yield may have increased residue C addition in surface placement than incorporation of residue into the soil or increased under wheat than pea or fallow in the greenhouse (Tables 2 and 3). In contrast, higher N concentration may have increased residue N addition under pea than wheat or fallow. Greater C and N recovered in the residue placed at the soil surface under wheat and fallow were probably due to reduced mineralization of wheat residue as a result of its higher C/N ratio than pea residue. While surface placement of residue reduces its contact with soil microorganisms that result in reduced mineralization [29,30], increased mineralization of pea residue due to its lower C/N ratio may have resulted in reduced C and N recovery in the residue placed at the soil surface under pea. Residues of legumes, such as pea with lower C/N ratio, decompose more rapidly than those of nonlegumes, such as wheat with higher C/N ratio [12]. When incorporated into the soil, residue C and N recovery were lower under fallow and wheat. As a result, C and N losses were higher in surface placement of residue under pea or residue incorporation under pea and fallow than the other treatments. It may be possible that some of C and N lost from the residue converted into soil C and N fractions, as discussed below.

Reduced mineralization of residue may have increased SOC, PCM, PNM, and NH_4-N concentrations in surface placement than incorporation of residue into the soil in the greenhouse (Table 4). Similar increases in MBN and NH_4-N concentrations in no-tillage compared to conventional tillage were found in the field (Table 5). Several researchers [5,16,31,32,33] have reported greater SOC, POC, MBC, PCM, PNM, and MBN in surface residue placement in the no-tillage system than residue incorporation into the soil in the conventional tillage system. Increased N mineralization due to residue incorporation, however, may have increased NO_3-N concentration in conventional tillage than no-tillage in the field.

Higher C and N substrate availability due to increased yield probably increased SOC, POC, STN, and MBN under wheat than under pea or fallow in the greenhouse (Table 4) or increased MBC under wheat than fallow in the field (Table 5). Root biomass C, residue C addition (Tables 2 and 3), and amount of applied N fertilizer were greater in wheat than pea or fallow. Similar results probably occurred in the field, since treatments were identical in the greenhouse and the field and crop biomass C was higher in wheat than pea in the field (Table 5). Rhizodeposit C released by roots can increase microbial biomass and activity and soil C storage [34]. Liebig et al. [35] also found higher MBC under spring wheat than under fallow. In contrast, greater PNM and NO_3-N under pea and fallow than wheat in the greenhouse were probably either due to increased mineralization of pea residue as a result of its lower C/N ratio than wheat residue [12] or to greater mineralization of soil and wheat residue as a result of enhanced microbial activity from higher soil temperature and water content and absence of plants to uptake N under fallow [11,13,36]. Since residue N loss was greater under pea than wheat and fallow (Table 3), part of N from pea residue may have contributed to increased PNM and NO_3-N concentrations under pea. Similar

result of increased PNM under pea than wheat and fallow was also found in the field, since crop biomass N was greater in pea than wheat (Table 5).

Comparison of soil C and N fractions at the beginning and end of the experiment due to residue placement (Tables 1 and 4) showed that SOC increased by 4.2%, PCM by 55.7%, and PNM by 6.1% with surface residue placement in the greenhouse. Corresponding values in SOC, PCM, and PNM with residue incorporation were 1.7, 25.4, and −35.1%, respectively. In the field, MBN reduced by 71.5% in no-tillage and 81.7% in conventional tillage from the beginning to the end of the experiment. This shows that residue placement at the surface either increased soil C and N fractions in the greenhouse or reduced their losses in the field within a crop growing season compared to residue incorporation. Since soil NH_4-N and NO_3-N concentrations vary seasonally due to N mineralization from crop residue and soil, N fertilization, crop N uptake, and N losses due to leaching, volatilization, and denitrification [10,15], variations in their levels from the beginning to the end of the experiment were not taken into account.

Among crop types, SOC increased by 5.1%, POC by 6.9%, STN by 3.9%, and MBN by 41.4%, but PNM decreased by 33.1% under wheat from the beginning to the end of the experiment in the greenhouse. In the field, MBC increased by 7.1%, but PNM decreased by 68.3% under wheat during this period. The corresponding increases in SOC, POC, STN, and MBN or decrease in PNM during this period were lower under pea and fallow. This suggests that wheat increased more soil C and N fractions, except PNM, than pea or fallow due to increased substrate availability from root and rhizodeposition and/or to slow decomposition of wheat than pea residue due to differences in residue quality (e.g. C/N ratio). The greater PNM under pea than wheat or fallow was due to increased N contribution from its residue (Table 5).

When the greenhouse and field experiments were compared, trends in changes in soil C and N fractions due to treatments within a crop growing season were similar. However, greater changes in labile than nonlabile C and N fractions occurred more in the greenhouse than in the field. Furthermore, the coefficient of variations in soil C and N fractions were lower in the greenhouse (2.6 to 15.9%) than in the field (8.0 to 36.7%) (Tables 4 and 5). This indicates that soil C and N fractions changed more readily but with lower variability with management practices within a crop growing season when soil and environmental conditions are controlled in the greenhouse than in field where soil heterogeneity often results in non-significant differences among treatments in these fractions [16,18,30]. Use of disturbed soil in the greenhouse vs. undisturbed (especially in the no-till system) in the field also may have an influence on differences in changes in soil C and N fractions between the two experiments. The greater changes in labile than nonlabile C and N fractions as influenced by management practices within a short period in the greenhouse and the field suggests that labile C and N fractions are better indicators of changes in soil organic matter quality than nonlabile fractions, a case similar to that reported by various researchers [10,11,13,16,30]. The fact that more changes in labile than nonlabile C and N fractions occurred in the greenhouse than in the field suggests that better measurements of changes in soil organic matter due to management practices within a short period can be observed when soil and environmental conditions are controlled. Greater levels of most soil C and N fractions in the greenhouse than in field was probably a result of increased turnover rate plant C and N into soil C and N, because disturbed soil was used in the greenhouse and environmental condition for microbial transformation was more favorable in the greenhouse than the field.

Greenhouse study provided more information on plant and residue parameters, such as measurement of root biomass and C and N contents and residue C and N losses, which cannot be measured easily in the field. This resulted in the measurement of turnover rate of plant C and N into soil C and N in the greenhouse, a fact that was absent in the field. Because of greater changes in soil C and N fractions, greenhouse study provided a more robust method of evaluating C and N cycling and soil quality within a short period of time as affected by management practices than the field experiment. Such changes can also be measured in the field but it may take longer time. While all results from the greenhouse study may not be readily applied in the field, some information, such as root biomass and residue C and N losses, measured in the greenhouse can be extrapolated to the field condition. The effects of short-term study in the greenhouse can be useful to predict the long-term impact of management practices on soil C and N fractions in the field.

Conclusions

Crop yields, residue C and N losses, and soil C and N fractions varied with residue placement and crop types in the greenhouse and the field. Surface placement of residue increased crop yields, residue C and N losses, and enhanced SOC, PCM, MBN, and NH$_4$-N concentrations, but residue incorporation increased PNM and NO$_3$-N concentrations. Similarly, spring wheat had higher yield and increased SOC, POC, MBC, STN, and MBN than pea or fallow, but pea had higher N content and increased PNM than wheat or fallow. Placing nonlegume residue at the soil surface

using no-tillage can increase soil C and N sequestration and microbial biomass and activity that can improve soil health and quality. Using this practice, producers can claim for C credit. Incorporation of legume and nonlegume residues into the soil using conventional tillage can increase N mineralization and availability which can reduce N fertilization rate to succeeding crops, but can degrade soil quality due to reduced organic matter and increased erosion. Although soil labile C and N fractions changed more readily than nonlabile fractions within a crop growing season both in the greenhouse and field, greater changes in labile than nonlabile fractions occurred with reduced variability more in the greenhouse than in the field. Results suggest that greenhouse study provided a more robust measurement of crop growth and changes in soil C and N fractions within a short period as influenced by management practices than the field experiment. Longer time will be probably needed in the field to obtain results similar to those in the greenhouse. Additional information, such as root growth, residue C and N losses, turnover of plant C and N to soil C and N, and results of short-term study on soil C and N fractions as influenced by management practices in the greenhouse can be used to predict the long-term impact in the field.

Acknowledgments

We appreciate the excellent support of Joy Barsotti and Thecan Caesar-TonThat for analyzing soil and plant samples in the laboratory.

Author Contributions

Conceived and designed the experiments: UMS. Performed the experiments: JW. Analyzed the data: UMS. Contributed reagents/materials/analysis tools: UMS. Wrote the paper: UMS.

References

1. Lal R, Kimble JM, Stewart BA (1995) World soils as a source or sink for radiatively-active gases. In: Lal R, editor, Soil management and greenhouse effect. Advances in soil science. CRC Press, Boca Raton, FL, pp. 1–8
2. Paustian K, Robertson GP, Elliott ET (1995) Management impacts on carbon storage and gas fluxes in mid-latitudes cropland. In: Lal R, editor, Soils and global climate change. Advances in soil science. CRC Press, Boca Raton, FL, USA, pp. 69–83.
3. Halvorson AD, Peterson GA, Reule CA (2002a) Tillage system and crop rotation effects on dryland crop yields and soil carbon in the central Great Plains. Agron J 94:1429–1436.
4. Sherrod LA, Peterson GA, Westfall DG, Ahuja LR (2003) Cropping intensity enhances soil organic carbon and nitrogen in a no-till agroecosystem. Soil Sci Soc Am J 67:1533–1543.
5. Sainju UM, Caesar-TonThat T, Lenssen AW, Evans RG, Kolberg R (2007) Long-term tillage and cropping sequence effects on dryland residue and soil carbon fractions. Soil Sci Soc Am J 71:1730–1739.
6. Rasmussen PE, Allmaras RR, Rhoade CR, Roager NC Jr (1980) Crop residue influences on soil carbon and nitrogen in a wheat-fallow system. Soil Sci Soc Am J 44:596–600.
7. Peterson GA, Halvorson AD, Havlin JL, Jones OR, Lyon DG, Tanaka DL (1998) Reduced tillage and increasing cropping intensity in the Great Plains conserve soil carbon. Soil Tillage Res 47:207–218.
8. Bauer A, Black AL (1994) Quantification of the effect of soil organic matter content on soil productivity. Soil Sci Soc Am J 58:185–193.
9. Ghidey F, Alberts EE (1993) Residue type and placement effects on decomposition: Field study and model evaluation. Trans ASAE 36:1611–1617.
10. Sainju UM, Lenssen AW, Caesar-Tonthat T, Waddell J (2006b) Tillage and crop rotation effects on dryland soil and residue carbon and nitrogen. Soil Sci Soc Am J 70:668–678.
11. Halvorson AD, Wienhold BJ, Black AL (2002b) Tillage, nitrogen, and cropping system effects on soil carbon sequestration. Soil Sci Soc Am J 66:906–912.
12. Kuo S, Sainju UM, Jellum EJ (1997) Winter cover cropping influence on nitrogen in soil. Soil Sci Soc Am J 61:1392–1399.
13. Sainju UM, Lenssen AW, Caesar-TonThat T, Waddell J (2006a) Carbon sequestration in dryland soils and plant residue as influenced by tillage and crop rotation. J Environ Qual 35:1341–1349.
14. Miller PR, McConkey B, Clayton GW, Brandt SA, Staricka JA, Johnston AM, Lafond GP, Schatz BG, Baltensperger DD, Neill KE (2002) Pulse crop adaptation in the northern Great Plains. Agron J 94:261–272.
15. Sainju UM, Lenssen AW, Caesar-TonThat T, Evans RG (2009) Dryland crop yields and soil organic matter as influenced by long-term tillage and cropping sequence. Agron J 101:243–251.
16. Franzluebbers AJ, Hons FM, Zuberer DA (1995) Soil organic carbon, microbial biomass, and mineralizable carbon and nitrogen in sorghum. Soil Sci Soc Am J 59:460–466.
17. Bezdicek DF, Papendick DF, Lal R (1996) Introduction: Importance of soil quality to health and sustainable land management. In: Doran JW, Jones AJ, editors, Methods of assessing soil quality, Spec. Publ. 49, Soil Science Society of America, Madison, USA, pp. 1–18.
18. Franzluebbers AJ, Arshad MA (1997) Soil microbial biomass and mineralizable carbon of water-stable aggregates. Soil Sci Soc Am J 67:1090–1097.
19. Cambardella CA, Elliott ET (1992) Particulate soil organic matter changes across a grassland cultivation sequence. Soil Sci Soc Am J 56:777–783.
20. Six J, Elliott ET, Paustian K (1999) Aggregate and soil organic matter dynamics under conventional and no-tillage systems. Soil Sci Soc Am J 63:1350–1358.
21. Pikul JL Jr, Aase JK (2003) Water infiltration and storage affected by subsoiling and subsequent tillage. Soil Sci Soc Am J 67:859–866.
22. Nelson DW, Sommers LE (1996) Total carbon, organic carbon, and organic matter. In: Sparks DL, editor, Methods of soil analysis. Part 3. Chemical method. SSSA Book Ser. 5. Soil Science Society of America, Madison, pp. 961–1010.
23. Haney RL, Franzluebbers AJ, Porter EB, Hons FM, Zuberer DA (2004) Soil carbon and nitrogen mineralization: Influence of drying temperature. Soil Sci Soc Am J 68:489–492.
24. Franzluebbers AJ, Haney RL, Hons FM, Zuberer DA (1996) Determination of microbial biomass and nitrogen mineralization following rewetting of dried soil. Soil Sci Soc Am J 60:1133–1139.
25. Voroney RP, Paul EA (1984) Determination of k$_C$ and k$_N$ in situ for calibration of the chloroform fumigation-incubation method. Soil Biol Biochem 16:9–14.
26. Brookes PC, Landman A, Pruden G, Jenkinson DJ (1985) Chloroform fumigation and the release of soil nitrogen: A rapid direct-extraction method to measure microbial biomass nitrogen in soil. Soil Biol Biochem 17:937–942.
27. Littell RC, Milliken GA, Stroup WW, Wolfinger RR (1996) SAS system for mixed models. SAS Institute Inc., Cary, NC, USA.
28. Lenssen AW, Johnson GD, Carlson GR (2007) Cropping sequence and tillage system influences annual crop production and water use in semiarid Montana. Field Crops Res 100:32–43.

29. Coppens F, Garnier P, de Gryze P, Merckx R, Recous S (2006) Soil moisture, carbon and nitrogen dynamics following incorporation and surface application of labelled crop residues in soil columns. Europ J Soil Sci 57:894–905.

30. Giacomini SJ, Recous S, Mary B, Aita C (2007) Simulating the effects of nitrogen availability, straw particle size and location in soil on carbon and nitrogen mineralization. Plant Soil 301: 289–301.

31. Malhi SS, Lemke R (2007) Tillage, crop residue and nitrogen fertilizer effects on crop yield, nutrient uptake, soil quality and greenhouse gas emissions in the second 4-yr rotation cycle. Soil Tillage Res 96:269–283.

32. Wright AL, Hons FM, Lemon RG, MacFarland ML, Nichols RL (2008) Microbial activity and soil carbon sequestration for reduced and conventional tillage cotton. Appl Soil Ecol 38:168–173.

33. Lupwayi NZ, Lafond GP, Ziadi N, Grant CA (2012) Soil microbial response to nitrogen fertilizer and tillage in barley and corn. Soil Tillage Res 118:139–146.

34. Lu Y, Watanabe A, Kimura M (2002) Contribution of plant-derived carbon to soil microbial biomass dynamics in a paddy rice microcosm. Biol Fertil Soils 36:136–142.

35. Liebig MA, Tanaka DL, Wienhold BJ (2004) Tillage and cropping effects on soil quality indicators in the northern Great Plains. Soil Tillage Res 78:131–141.

36. Kuzyakov Y, Domanski G (2000) Carbon input by plants into the soil: Review. J. Plant Nutri Soil Sci 163:421–431.

18

Influence of Residue and Nitrogen Fertilizer Additions on Carbon Mineralization in Soils with Different Texture and Cropping Histories

Xianni Chen[1], Xudong Wang[1]*, Matt Liebman[2], Michel Cavigelli[3], Michelle Wander[4]*

1 College of Resources and Environment, Northwest A&F University, Yangling, Shaanxi, PR China, **2** Department of Agronomy, Agronomy Hall, Iowa State University, Ames, Iowa, United States of America, **3** Sustainable Agricultural Systems Lab, Agricultural Research Center, Beltsville, Maryland, United States of America, **4** Department of NRES, University of Illinois, Urbana-Champaign, Illinois, United States of America

Abstract

To improve our ability to predict SOC mineralization response to residue and N additions in soils with different inherent and dynamic organic matter properties, a 330-day incubation was conducted using samples from two long-term experiments (clay loam Mollisols in Iowa [IAsoil] and silt loam Ultisols in Maryland [MDsoil]) comparing conventional grain systems (Conv) amended with inorganic fertilizers with 3 yr (Med) and longer (Long), more diverse cropping systems amended with manure. A double exponential model was used to estimate the size (C_a, C_s) and decay rates (k_a, k_s) of active and slow C pools which we compared with total particulate organic matter (POM) and occluded-POM (OPOM). The high-SOC IAsoil containing highly active smectite clays maintained smaller labile pools and higher decay rates than the low-SOC MDsoil containing semi-active kaolinitic clays. Net SOC loss was greater (2.6 g kg^{-1}; 8.6%) from the IAsoil than the MDsoil (0.9 g kg^{-1}, 6.3%); fractions and coefficients suggest losses were principally from IAsoil's resistant pool. Cropping history did not alter SOC pool size or decay rates in IAsoil where rotation-based differences in OPOM-C were small. In MDsoil, use of diversified rotations and manure increased k_a by 32% and k_s by 46% compared to Conv; differences mirrored in POM- and OPOM-C contents. Residue addition prompted greater increases in C_a (340% vs 230%) and C_s (38% vs 21%) and decreases in k_a (58% vs 9%) in IAsoil than MDsoil. Reduced losses of SOC from residue-amended MDsoil were associated with increased OPOM-C. Nitrogen addition dampened CO_2-C release. Clay type and C saturation dominated the IAsoil's response to external inputs and made labile and stable fractions more vulnerable to decay. Trends in OPOM suggest aggregate protection influences C turnover in the low active MDsoil. Clay charge and OPOM-C contents were better predictors of soil C dynamics than clay or POM-C contents.

Editor: Upendra M. Sainju, Agricultural Research Service, United States of America

Data Availability: The authors confirm that all data underlying the findings are fully available without restriction. All relevant data are within the paper.

Funding: Financial support was given by NIFA (Hatch) LLU-875-320, and grants from the Leopold Center for Sustainable Agriculture (Project 2010-E02) and the Iowa Soybean Association. This project was also partially funded by the National 15th Key Technology R&D Program of the Ministry of Science and Technology-Technology Integration and Demonstration of Agriculture-Fruit-Livestock Industry Recycling in Loss Plantae, China (2012BAD14B11). The funders had no role in study design, data collection and analysis, decision to publish, or preparation of the manuscript.

Competing Interests: The authors received funding from the professional association 'Iowa Soybean Association'. There are absolutely no restrictions on sharing of data and/or materials derived from the authors' work.

* Email: wangxudong01@126.com (XW); mwander@illinois.edu (MW)

Introduction

To manage soil sustainability, crop rotation, tillage, fertilization and other management practices must be combined in ways that improve or maintain soil carbon stocks and reduce net carbon loss [1–3]. Use of diversified farming systems that rely on crop rotation and return of crop residues and/or manures is advocated as one way to improve agricultural sustainability and increase soil organic carbon (SOC) [4–6]. Diversified crop rotations benefit the soil by varying the quantity, quality, and spatial and temporal placement of organic matter inputs, which therefore altered the physical and biochemical factors and influenced decay of SOC [7]. The ability of the management practice to alter SOC is likely to vary with inherent soil properties, such as clay content, mineralogy and pH [8–10]. Dynamic properties like particulate organic matter (POM) that are sensitive to management and change within relatively short time frames (years to decades) may help us predict whether or how SOM status might be improved. Failure to accumulate SOC with increased C additions can occur when soil is already C saturated [11], or in instances where decay rates are high due to stoichiometric imbalance usually caused by high levels of available N [12,13]. Several studies suggest that manure application can accelerate SOC decay rates [14,15]. Both the frequency of manure addition and crop rotation length (diversity) have been linked to increased levels of available soil N [16]. This suggests that longer rotations with less frequent manure application might be better able to sequester C in soil. In order to optimize soil C cycling to maintain soil productivity and environmental function, we need to be able to predict how soils C dynamics will respond to management.

Table 1. Summary of the management, inherent and dynamic soil properties at Marsden and the Farming Systems Project.

	Iowa (Marsden)				Maryland (Farming Systems Project)			
	Site mean	Conv	Med	Long	Site mean	Conv	Med	Long
Crop sequence[†]		C-S	C-S-oat/rc	C-S-oat/A-A		C-r/S-W/S	hv/C-r/S-W	C-r/S-W/A-A-A
Tillage[‡]		Ch	Ch, MB	Ch, MB		Ch	D, Ch or MB	D, Ch or MB
Fertilizer sources[†]		N	GM, AM, N	GM, AM, N		N, P, K	GM, AM, K	GM, AM, K
Sand (%)	35.2±3.6 A	35.4±4.0 a	35.3±4.4 a	34.8±3.7 a	20.6±6.7 B	23.6±11.6 a	19.0±2.2 a	19.3±3.5 a
Silt (%)	38.5±2.7 B	38.5±4.1 a	38.5±2.3 a	38.4±2.4 a	59.3±5.3 A	55.9±7.4 a	59.5±4.7 a	62.4±1.4 a
Clay (%)	26.4±2.3 A	26.2±2.7 a	26.4±2.9 a	26.8±2.2 a	20.1±3.8 B	20.5±4.7 a	21.5±4.4 a	18.4±2.8 a
pH	7.2±0.4 A	6.9±0.4 a	7.3±0.4 a	7.4±0.3 a	6.4±0.3 B	6.3±0.7 a	6.4±0.1 a	6.4±0.1 a
SOC (g kg^{-1})	28.0±5.6 A	26.4±2.8 a	28.0±9.3 a	29.6±4.1 a	14.1±2.0 B	12.2±2.3 a	15.1±1.0 a	14.9±1.3 a
C:N	12.7±0.6 A	12.9±0.7 a	12.5±0.7 a	12.8±0.5 a	10.7±0.7 B	11.3±0.4 a	10.1±0.3 b	10.7±0.8 b
POM-C (g kg^{-1})	2.8±0.9 A	2.4±0.4 a	3.5±1.4 a	2.7±0.3 a	3.6±0.8 A	2.7±0.5 b	3.7±0.6 a	4.4±0.5 a
POM-C:N	15.7±2.1 A	16.3±1.9 a	15.0±1.9 a	15.9±2.8 a	17.7±3.9 A	21.9±4.3 a	15.0±1.5 b	16.3±1.3 b
OPOM-C (g kg^{-1})	2.2±0.5 B	2.0±0.1 a	2.2±0.2 a	2.3±0.8 a	3.2±0.6 A	2.4±0.4 b	3.2±0.2 a	3.8±0.6 a
OPOM-C:N	12.1±0.6 B	12.7±0.2 a	11.5±0.2 b	12.4±0.8 ab	13.7±0.9 A	14.6±0.5 a	12.9±0.3 b	13.6±1.1 ab

Variables include texture, pH and soil organic carbon(SOC), soil carbon to nitrogen ratio (C:N), particulate organic matter-carbon (POM-C) and POM-C:N ratio, occluded-POM carbon (OPOM-C) and POM-C:N ratio. Data in table are means ± standard deviation.

[†] C-corn, S-soybean, rc-red clover, A-alfalfa, r-rye cover crop, W-wheat, hv-hairy vetch, W/S-wheat followed by double-cropped soybean. Conv in Maryland followed a 2 yr C-W/S rotation from 1996–1999, Long in Maryland followed a 4 yr C-r/S-W/(r+ orchard grass hay) rotation from 1996–1999.

[‡] Ch-chisel plow, MB-moldboard plow, D-disk.

[§] N-urea ammonium nitrate, GM-green manure, AM-animal manures, P-triple super phosphate, K-potassium sulfate. For Iowa site, N fertilization rate was 100 kg N ha−1 with side dressing (0–100 kg N ha−1) as needed based on standard soil tests; green manure was red clover (15.7 Mg ha−1, fresh weight basis) for Med system and was second-year alfalfa for Long system; composted beef cattle manure (on average 128 kg N ha−1) was supplied to both Med and Long systems. For Maryland site, the Conv system received on average 160 kg N ha−1 each year; Med and Long systems received green manure (hairy vetch for Med system, and alfalfa for Long system) and cattle manure (on average 150 kg N ha−1) as N sources.

Values not followed by the same upper case letter differ between two sites (Iowa, Maryland), values not followed by the same lower case letter differ among cropping systems (Conv, Med, Long) within each site. Statistical significances were performed at p<0.05.

Soil incubations provide one method to study mechanisms controlling soil C turnover under variable amendment regimes. Incubation data are often divided into 'active' (C_a) and 'slow' (C_s) carbon pools by fitting CO_2 release data to a two-pool model [3,17–19]. Estimates of C_a and C_s are often compared against direct measures of SOC fractions, such as total POM and occluded-POM or microbial activity, in an attempt to validate/evaluate models and improve their structure [20,21]. Density or particle size fractionation may provide a useful way to quantify changes in physical protection and improve our ability to delineate active and slow C pools and understand their dynamics [22]. Incubations provide a way to carry out controlled evaluation of how inherent and dynamic soil properties influence soil response to residue and N additions.

Residue additions can increase C mineralization and positive priming (which is the acceleration of native SOC decomposition caused by residue inputs) especially in soils with high levels of SOC or high C:N ratios [23]. Increases in C mineralization caused by nitrogen fertilization have been positively related to labile C concentrations [24]. Nitrogen fertilization and its interaction with residue addition can prompt C mineralization or immobilization depending on soil type [25,26]. Greater understanding of how soil physicochemical properties influence soil response to C and N additions is needed to help us manage the soil C cycle to mitigate climate change [27,28].

The objective of this study was to improve our understanding of SOC mineralization response to residue and N additions in soils with different inherent and dynamic soil properties. Residue additions were expected to result in C accrual in soils that were more degraded or further away from C saturation. Nitrogen additions were expected to increase C mineralization in soils with higher levels of labile C and this effect was expected to be more pronounced in soils with longer rotations assuming that they are better able to increase labile C stocks..

Materials and Methods

Soil samples and sites description

Soil samples were collected from two long-term agricultural research sites, one located in central Iowa (IA) and the other in central Maryland (MD). Both studies included cropping system treatments that differ in C input diversity and N fertility sources. The Marsden Farm Cropping Systems Experiment was initiated in 2002 and is located at the Iowa State University Marsden Farm, in Boone County, IA (42°01′ N; 93°47′ W). Average annual precipitation since 1981 is 844 mm and temperature is 9.1°C. Before initiation of the experiment, the IA site had been managed for at least 20 yr with a corn-soybean rotation receiving conventional fertilizer and herbicide inputs. Soils vary across the experimental site and are predominantly Clarion loam (fine-loamy, mixed, superactive, mesic, Typic Hapludolls, 2%–5% slope), Nicollet loam (fine-loamy, mixed, superactive, mesic, Aquic Hapludolls, 1–3% slope) and Webster silty clay loam (fine-loamy, mixed, superactive, mesic, Typic Endoaquolls, 0–2% slope), with smaller areas of Harps loam (fine-loamy, mixed, superactive, mesic Typic Calciaquolls, 0–2% slope), and Canisteo silty clay loam (fine-loamy, mixed, superactive, calcareous, mesic Typic Endoaquolls, 0–2% slope). The experiment compared a conventionally managed 2-yr rotation with 3-yr and 4-yr rotations that are diversified farming systems. Details about rotations, tillage and fertilization are given by Liebman [29] and Davis [30] and summarized briefly in Table 1. The experiment was a randomized complete block design with each crop phase of each rotation

system present every year in four replicate blocks. Plot size was 18 m by 85 m.

The Farming Systems Project was located at the western edge of the Atlantic Coastal Plain at the United States Department of Agriculture-Agricultural Research Service (USDA-ARS) Beltsville Agricultural Research Center in Beltsville, MD (39°03′ N, 76°90′ W). The 30-yr average annual precipitation is 1110 mm; rainfall is distributed evenly through the year. Average annual temperature is 12.8°C. The site had been managed as a row crop production field with continuous no-till for at least 11 years before the study was initiated in 1996. The dominant soil types are Christiana (fine, kaolinitic, mesic Typic Paleudults), Matapeake (fine-silty, mixed, semiactive, mesic Typic Hapludults), Keyport (fine, mixed, semiactive, mesic Aquic Hapludults), and Mattapex (fine-silty, mixed, active, mesic Aquic Hapludults) silt loams. The MD site included five cropping systems [16]; we only sampled the conventional tilled treatment and two of the organic cropping systems (Table 1) to cover a gradient of crop and input diversity that echoed the series of comparisons made in the Marsden Farm plots. Farming systems are replicated four times with each crop phase of each rotation system present every year in a split-plot design with system assigned to whole plots and crop rotation entry point assigned to 111 m by 9.1 m subplots.

These two field studies did not involve endangered or protected species. No written permissions or permits were required to secure access to sample the two experimental trials. Both are publically funded research sites established with the purpose of being sampled.

Incubation set up

Soil samples were taken from plots or subplots before entering the corn planting phase on 9 May, 2011 in Iowa and 3 June, 2011 in Maryland using four 3–4 cm diameter soil cores to a depth of 20 cm. Composite samples representing the separate blocks were air-dried and sieved to 2 mm for use in the incubation study that compared four treatments for each cropping system soil: un-amended control (Control), N-fertilized treatment (N), residue-amended treatment (R) and treatment with both residue and N-fertilizer (RN). Microcosms were established in 800 mL Mason jars containing 40 g soil. Half of the jars containing soil from each cropping system treatment were amended with residues of wheat (hard red spring wheat, *Triticum aestivum* L.) that had been grown in a greenhouse. Both shoots and roots were cleaned and dried at 40°C, then ground to pass a 1 mm sieve. Dry residue, consisting of 38.6% C and 1.3% N, was added at rate of 0.153 g $(40 \text{ g})^{-1}$ soil to approximate a typical field return rate of 8580 kg ha^{-1}. The soil was wetted by pipetting an appropriate amount of H_2O (determined gravimetrically) to reach 50% soil water holding capacity (WHC) and pre-incubated at 4°C for 2 days to allow soil moisture to diffuse evenly and to stimulate microbial activity. Soils were then allowed to warm to room temperature and brought to 60% WHC using either H_2O or $(NH_4)_2SO_4$ to produce "no N fertilized" and "N fertilized" (170 kg N ha^{-1}) treatments, respectively. There are four replications of each treatment, and the study was a completely randomized block design. The incubation lasted for 330 days.

Soil analysis and statistics

Soil respiration was quantified by periodic sampling from the headspace of jars incubated in the dark at 24°C. Gas samples were collected from the Mason jars every day for the first 3 days, every other day until day 9, every 3 days until day 30, every 10 days for the next two months, and once per month for the last 8 months of the 11 month study. The CO_2 concentration in the headspace was measured using an LI-800 CO_2 Gas Hound Analyzer (Model LI-800, LI-COR). Ports (3 mm diameter) in the top of each jar were sealed with Butly rubber stoppers. Jars were re-aerated and moisture was adjusted to 60% WHC based on a weight estimate as needed after headspace samples were analyzed.

Soil texture, pH and particulate organic matter (POM) content of soils were measured prior to establishing the incubation. Soil texture was estimated by the hydrometer method [31] and pH was measured using an Orion pH electrode method after dispersing soil in distilled water at a soil:water ratio of 1:1. Soil POM was determined by dispersing soil in sodium metaphosphate solution and collecting material >0.53 μm on a sieve [32]. A subset of soil samples were destructively used for organic matter characterization at 6 and 11 months. The aggregate occluded particulate organic matter (OPOM) was determined using the procedure outlined by Yoo and Wander [33]. Briefly, the light residue of a 20 g soil sample was floated in a centrifuge tube using sodium polytungstate (1.6 g cm^{-3}; Geoliquids Inc., Chicago, IL), then removed using 1 μm polycarbonate membrane filters (Osmonics Inc., Minnetonka, MN) after being centrifuged at 5000 rpm for 30 min. The remaining material was shaken at 350 oscillations min^{-1} for 60 min with 50 ml deionized water, then OPOM was collected by 53 μm sieving using poly-carbonate mesh (Gilson Co., Columbus, OH). The POM and OPOM were dried at 80°C and their C and N contents were determined by dry combustion with an Elemental Analyzer (Costech 4010, Costech Analytical Technologies Inc. Valencia, CA).

In soil C mineralization studies, the commonly used models to simulate CO_2 release include first-order kinetics models, hybrid model (a simplified special case of the two-component model) and double exponential model [19,34,35]. Pre-analysis showed that highest r square values ($r^2 > 0.99$) were observed when CO_2 emission data were fit to a double exponential model. Based on this, and reports that the double exponential model can provide an accurate description of C mineralization for incubations of 200 days or longer [19,36], we used the double exponential model to describe results from our 330-day incubation. Data were fitted using the following model in PROC NLIN (nonlinear regression) in SAS 9.3:

$$C_t = C_a\left(1 - e^{at}\right) + C_s\left(1 - e^{st}\right)$$

where C_t is the cumulative amount of CO_2-C released (g C kg^{-1} soil) in time t; and, C_a and C_s represent the active and slow pools of mineralizable C (g C kg^{-1} soil) with decomposition rates of k_a and k_s (day^{-1}), respectively. Estimates of C_a, C_s and k_a, k_s derived from empirically fitting the equation represent predictions about the size and lability of mineralizable fractions. The double exponential model assumes that "resistant carbon pool" (C_r) does not contribute to CO_2 emissions in a relatively short period [36]. In this study, the C_r was estimated by subtracting the sum of C_a and C_s from total SOC.

The four parameters estimated by nonlinear regression were checked for normality in SAS 9.3 using "univariate normal plot". The "Mixed" procedure was then performed to quantify the overall effects of site, N fertilization and residue addition (Control, N, R and RN), and their interactions using data from both sites. There was a significant three way interaction between Site, R and N. Associated analysis produced similar information to that supplied by site-based analyses of data. Accordingly, for simplicity's sake, and to avoid being repetitive, we only report on analyses of sites performed separately. Analyses of our treatments (cropping systems [Conv, Med, Long] and their interactions with N fertilization and residue addition) on measured

fractions and estimated coefficients are reported for each site separately because the Conv, Med, and Long cropping systems were not identical at the two locations. Common labeling was used to determine whether generalizations about cropping system effects (eg: rotation length and manure application) might be made. Site-based comparisons were only reported for analyses performed on unamend controls. The multiple comparison of Least Squares Means differences was conducted only when ANOVA results for main factors or factor interactions were significant (Turkey adjusted p<0.05).

Results and Discussion

Key site-based differences

Table 1 summarizes crop management and soil properties measured at each site at the start of the incubation. The Iowa soil (IAsoil) had higher sand and clay contents and pH and a lower silt content than the Maryland soil (MDsoil). The IAsoil also had higher SOC content (28.0 g kg^{-1}) and SOM-C:N ratio (12.7) than the MDsoil (SOC = 14.1 g kg^{-1}, C:N = 10.7), indicating that organic matter was less decomposed in IAsoil than in MDsoil [22]. The amount of C mineralized from soils was similar at the two sites. Cumulative losses of CO_2-C ranged from 0.94–1.93 g C kg^{-1} soil in IAsoil, and from 1.02–2.07 g C kg^{-1} soil in MDsoil (Fig. 1). We assumed that temperature and moisture were maintained at or near optimal levels for mineralization [37–39] and noted that mineralization rates observed in this study were consistent with soils incubated under similar near-optimal laboratory conditions [40,41].

When considered as a percentage of SOC, CO_2-C losses from all treatments in the MDsoil (Control = 7.93, N = 7.26, R = 14.7, RN = 13.2% of SOC) were about twice those from the IAsoil (Control = 3.68, N = 3.33, R = 6.88, RN = 6.59% of SOC). The percentage loss rates from controls are consistent with introductory soils text books [42] that suggest 2%–3.5% of SOC will typically mineralize annually and that larger losses can be expected from soils with less clay. The clay contents of the MDsoil are however, only slightly lower than those of the IAsoil (20% vs 26%). Difference in soil texture, which is commonly used to regulate kinetics in organic matter models as it alters soil water availability, pore size distribution, nutrient availability and surface area [27], does not adequately explain our results. Soil physical protection of SOC is often assumed to be positively related to clay and silt contents [43] but such particle-size based classes poorly reflect critical differences in mineralogy that likely influenced C dynamics in these two soils. The IAsoils are dominated by highly-active smectite clays while the MDsoils are dominated by semi-active kaolinitic clays [44]. By multiplying clay content by typical cation exchange capacity (CEC, smectite = 150–115 cmol$_c$ kg^{-1}, kaolinite = 0.4–1.5 cmol$_c$ kg^{-1}) values, we noted that the IAsoil had about 50–100 times more surface activity than the MDsoil [45,46]. Use of CEC of different clay types to reflect the different types of mineral surfaces that influence surface binding with organic matter [47] rather than clay content to represent differences in inherent soil properties would allow modelers to better estimate soil's organic C protective capacity. Such an approach could have particular value for efforts seeking to use texture-organic matter associations to target efforts to increase soil C sequestration [48,49].

In addition to physical protection, modeling efforts assume organic matter quality and quantity control mineralization patterns [50]. Levels of POM, which are often used as an index of SOC status, help predict soil response to amendment. We found that the MDsoil contained a greater percent of SOC as POM-C

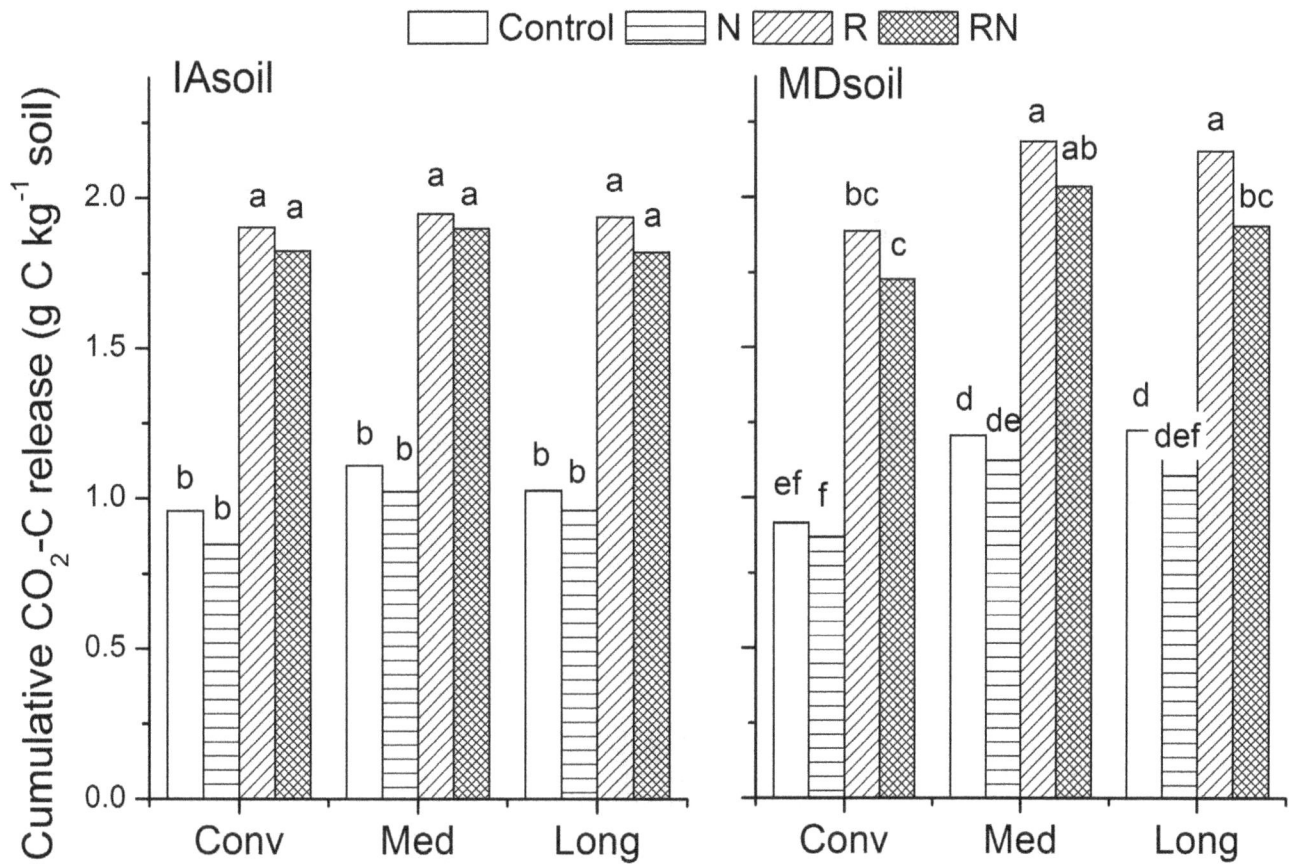

Figure 1. Cumulative CO_2-C release from the IAsoil and the MDsoil after 330 days incubation. Control: treatments with no residue or N fertilizer added. N: treatments with N fertilization. R: treatments with residue application. RN: treatments applied with both residue and N fertilizer. Different letters above bars suggest significant differences at p<0.05, comparisons were made for the interaction of cropping system (Conv, Med, Long) by treatments (Control, N, R, RN) within each site (IAsoil, MDsoil).

(25.5%) than did the IAsoil (10.0%) (Table 1), which is consistent with the IAsoil having smaller C_a and C_s pools that have slightly larger specific decay rates (k_a, k_s) than the MDsoil (Table 2). By subtracting the sum of C_a and C_s from SOC, we estimated that the IAsoil's protected pool (C_r) was about twice the size of that in the MDsoil. These differences in C pool dynamics are consistent with the notion that the IAsoil afford superior physical protection of SOC by reducing microbial access to SOC [8,51–53].

Those results are also consistent with the greater percentage of SOC loss as CO_2 from the MDsoil but at odds with directly measured losses in SOC, which were greater from the IAsoil (1.91–2.84 g kg^{-1}) than from the MDsoil (0.76–1.35 g kg^{-1}) (p< 0.01) (Fig. 2). Note, however, that the percentage of SOC loss relative to initial SOC contents was similar for the two sites (6.6%– 9.4% in the IAsoil vs 5.4%–9.7% in the MDsoil) (Fig. 2). Losses in SOC were greater than could be accounted for by CO_2-mineralization in the IAsoil and in the Control treatment of MDsoil; SOC must have been lost to dissolved organic carbon (DOC) or in volatile forms including CH_4 that were not measured during the incubation. Work by Bellamy [54] suggesting DOC losses are greater from high active soils like the IAsoil supports this notion. Observed losses of SOC are consistent with the faster decay rates (k_a and k_s) observed for IAsoil. More rapid decay in the IAsoil might be due to the porosity which allowed those soils to maintain more water at 60% WHC (moisture = 35.7% in IAsoil vs 29.6% in MDsoil).

Efforts to predict soils' response to management often focus more on labile fractions than on resistant fractions. Models of C dynamics commonly assume that about half of the C contained in residues or active fractions will be mineralized annually [55]. Direct measurement of particulate organic matter has been proposed as one way to estimate mineralization [10]. The observed CO_2 mineralization from un-amended controls (IA-soil = 0.99 g C kg^{-1} soil and MDsoil = 1.07 g C kg^{-1} soil) (Fig. 1) were about one third that contained in POM-C (IAsoil = 2.8 g C kg^{-1} soil and MDsoil = 3.6 g C kg^{-1} soil) (Table 1) at the start of the incubation. Differences in the size of estimated labile pools parallel differences in POM-C and OPOM-C observed in the two sites; this suggests that these direct measures can provide useful estimates of C vulnerability to decay [52]. Changes in OPOM-C contents observed over the course of the incubation were positively correlated to net SOC loss (IAsoil: $r^2 = 0.40$, n = 48, p<0.05; MDsoil: $r^2 = 0.38$, n = 48, p<0.05). Similarly, total POM-C contents were also correlated with SOC loss (IAsoil: $r^2 = 0.45$, n = 48, p<0.01; MDsoil: $r^2 = 0.45$, n = 48, p<0.01). However, no correlation was found between initial OPOM-C contents and C_s size, and OPOM-C net loss during incubation was negatively related to estimates of C_s in both soils (IAsoil: $r^2 = -0.35$, n = 35, p<0.05; MDsoil: $r^2 = -0.31$, n = 46, p<0.05), and to C_a in the MDsoil ($r^2 = -0.74$, n = 46, p<0.01). Negative correlations suggest organic matter fractions other than OPOM contribute to mineralization.

Table 2. Influence of site, management and treatment on active (C_a) and slow (C_s) carbon pool sizes and their decomposition rate constants (k_a, k_s respectively).

Factors		C_a	k_a	C_s	k_s	C_t
Site	Treatment	(g C kg^{-1} soil)	(day^{-1})	(g C kg^{-1} soil)	(day^{-1})	(g C kg^{-1} soil)
Iowa (IAsoil)	Control	0.16±0.03 B	0.16±0.02 A	1.07±0.18 B	0.0047±0.0010 A	25.2±3.9 A
	Conv	0.14±0.02 b	0.18±0.01 a	1.17±0.15 a	0.0036±0.0008 a	25.1±2.7 a
	Med	0.17±0.01 ab	0.15±0.01 a	1.02±0.14 a	0.0054±0.0005 a	22.2±3.4 a
	Long	0.19±0.02 a	0.16±0.02 a	1.01±0.23 a	0.0051±0.0009 a	28.3±4.2 a
All trts	Conv	0.43±0.30 a	0.15±0.07 a	1.27±0.26 a	0.0042±0.0007 a	25.0±2.4 a
	Med	0.48±0.32 a	0.13±0.05 a	1.29±0.25 a	0.0051±0.0007 a	22.1±2.6 a
	Long	0.47±0.29 a	0.13±0.05 a	1.19±0.28 a	0.0050±0.0008 a	27.9±4.0 a
	Control, N	0.17±0.03 b	0.19±0.03 a	1.05±0.16 b	0.0046±0.0009 a	25.2±3.7 a
	R, R+N	0.75±0.06 a	0.08±0.01 b	1.45±0.18 a	0.0049±0.0007 a	24.8±4.0 b
	Control, R	0.47±0.31 a	0.12±0.04 b	1.28±0.25 a	0.0049±0.0009 a	24.9±3.8 a
	N, R+N	0.45±0.29 a	0.15±0.06 a	1.22±0.27 a	0.0046±0.0007 a	25.0±3.9 a
Maryland (MDsoil)	Control	0.23±0.02 A	0.13±0.02 B	1.29±0.16 A	0.0038±0.0011 B	12.5±1.9 B
	Conv	0.22±0.02 a	0.11±0.02 a	1.29±0.22 a	0.0024±0.0006 b	10.6±2.0 b
	Med	0.23±0.03 a	0.14±0.02 a	1.35±0.14 a	0.0040±0.0002 a	13.5±0.8 a
	Long	0.23±0.03 a	0.15±0.03 a	1.21±0.09 a	0.0049±0.0007 a	13.4±1.1 a
All trts	Conv	0.51±0.29 a	0.11±0.01 b	1.33±0.29 a	0.0031±0.0008 c	10.2±1.6 b
	Med	0.53±0.28 a	0.13±0.01 a	1.38±0.23 a	0.0050±0.0010 a	13.3±0.8 a
	Long	0.51±0.28 a	0.12±0.02 ab	1.46±0.20 a	0.0041±0.0009 b	13.0±1.3 a
	Control, N	0.87±0.03 b	0.12±0.02 a	1.26±0.18 b	0.0034±0.0010 b	12.5±1.6 a
	R, RN	2.90±0.05 a	0.11±0.01 b	1.52±0.22 a	0.0047±0.0010 a	11.8±1.9 b
	Control, R	1.92±0.30 a	0.12±0.02 a	1.42±0.20 a	0.0044±0.0012 a	12.1±1.9 a
	N, RN	1.86±0.26 a	0.12±0.01 a	1.36±0.28 a	0.0037±0.0010 b	12.2±1.7 a

All coefficients were estimated by modeling Ct = Ca (1−e$^{-ka \cdot t}$)+ Cs (1−e$^{-ks \cdot t}$) using cumulative CO_2 emission data. Results are means ± standard deviation for unamend control soils (Control) and all treatments (All trts). Control: treatments with no residue or N fertilizer added. N: treatments with N fertilization. R: treatments with residue application. RN: treatments applied with both residue and N fertilizer. "Control, N" and "R, RN" represent the treatments without or with residue addition respectively; "Control, R" and "N, RN" represent the treatments without or with N fertilization respectively. Grand mean comparisons between sites were made with controls, values not followed by the same upper case letter differ at p<0.05. Means comparisons within each site were made within treatment groups, i.e. cropping system (Conv, Med, Long), residue addition ("Control, N" , "R, RN"), N fertilization ("Control, R" , "N, RN"), values not followed by the same lower case letter differ at p<0.05.

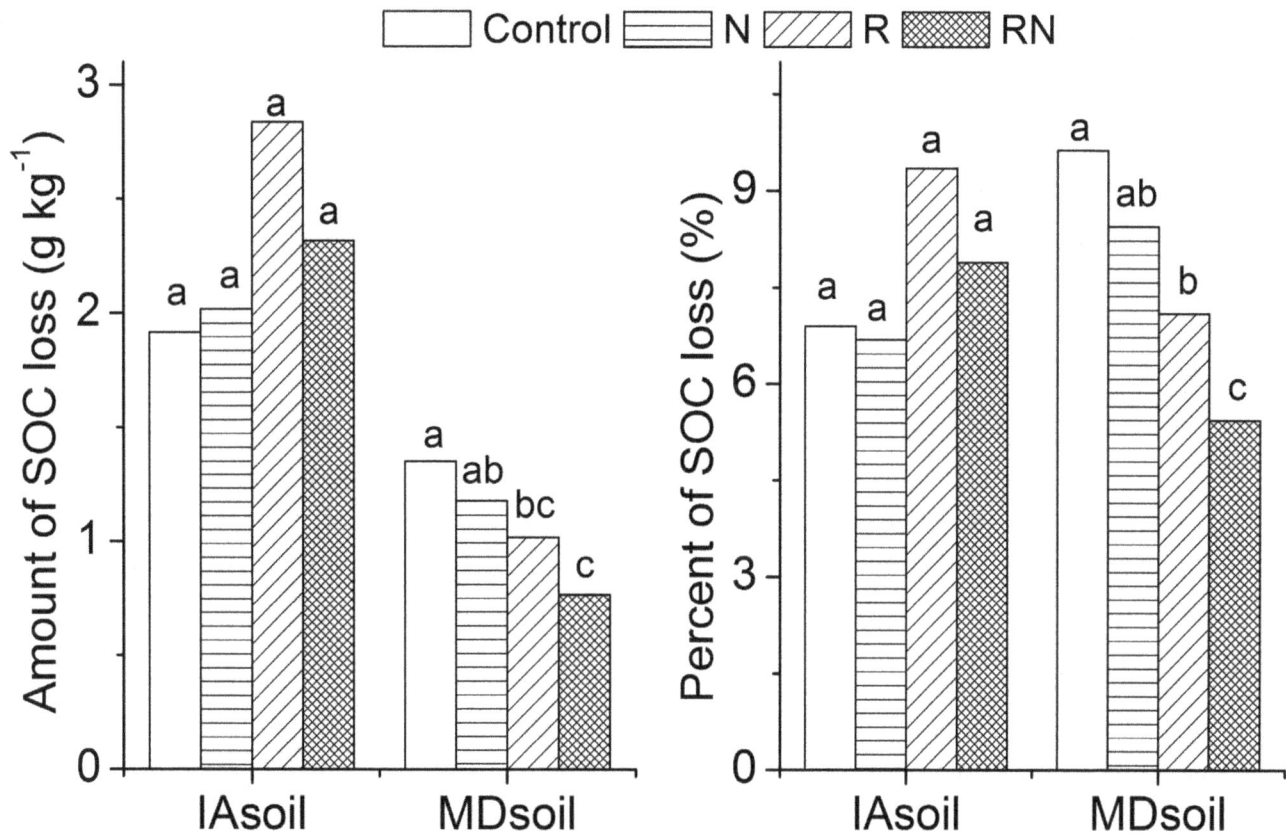

Figure 2. Amount and percentage of SOC loss comparing with initial SOC content. Control: treatments with no residue or N fertilizer added. N: treatments with N fertilization. R: treatments with residue application. RN: treatments applied with both residue and N fertilizer. Different letters above the bar suggest significant differences at p<0.05, comparisons were performed in different treatments within each site (IAsoil, MDsoil).

Cropping Systems

Use of diversified cropping along with organic fertilizers established greater differences in dynamic soil properties in the MDsoil than the IAsoil. Cropping system-based differences in IAsoil's dynamic properties were not statistically significant in this study (Table 1) due to high variability among a small number of samples. The POM-C patterns in IAsoil (Med≥Long≥Conv) agree with recent work [56] using more extensive sampling from two soil depth (0–10 cm, 10–20 cm). That recent work showed the POM-C contents in the IAsoil were significantly higher in the Med and Long systems than the Conv system, which were principally due to increases in POM-C and potentially mineralizable nitrogen in the 10–20 cm depth of soils in the diversified rotations. Our study considered the 0–20 cm depth as a whole, and sampled only plots about to be planted with corn. Variability among IAsoil from different blocks, which were kept separate in 4 replications, prevented us from finding significant differences of POM-C fractions among systems; but the OPOM-C:N ratio was lower in the Med than the Conv system (p<0.05). In the MDsoil, longer cropping systems with more diverse C input (poultry litter and legume cover crop inputs) had significantly increased POM-C and OPOM-C contents (Long≥Med>Conv), and decreased C:N ratios of SOM, POM (Med≤Long<Conv) and OPOM (Med< Conv) (Table 1). These findings are consistent with Spargo [16] who reported that organic cropping systems increased POM-C contents and decreased POM-C:N ratios in the MDsoil.

The more notable influence of cropping system on soil C pools observed in the MDsoil was expressed in cumulative CO_2 emissions; losses of CO_2-C (1.61 g C kg^{-1} soil) from Med and Long were greater than from the Conv treatment (1.30 g C·kg^{-1} soil) (p<0.05) (Fig. 1). Cumulative CO2 emissions did not vary among systems in the IAsoil. Use of diversified rotations in IAsoil did increase the C_a pool size in control soils (Long≥Med≥Conv) but only differences between the Long and Conv were significant; while for the MDsoil controls, both k_s and C_r were significantly greater in the Long and Med than in the Conv system (p<0.05) (Table 2). For both sites, greater OPOM-C values were observed in soils from Med than from Conv systems, and OPOM-C stocks declined significantly in the Med systems in both soils and Long system in the IAsoil during 0–180 days of the incubation (p<0.05). During the last 180–330 days, the OPOM stocks remained unchanged in all systems in the IAsoil and increased slightly in the Med system in the MDsoil (p = 0.0507) (Fig. 3).

By removing loose-POM, which contains organic residues and roots that are not well decomposed or bound to minerals, we resolved differences among cropping systems treatments that were obscured by the presence of the more labile residues, suggesting stronger predictive power of OPOM than total POM. We had expected differences of C_a and C_s among systems to parallel those observed for POM-C and OPOM-C but cropping system had little influence on these estimated coefficients (Table 2). Here both the directly measured fractions and coefficients derived from incubations suggest that C dynamics in the MDsoil are more strongly influenced by management and that aggregate protection plays a more important role. Recent work [56] asserted decay rates in the IAsoil must vary among systems to maintain similar levels of SOC despite differences in C input levels. The absence of

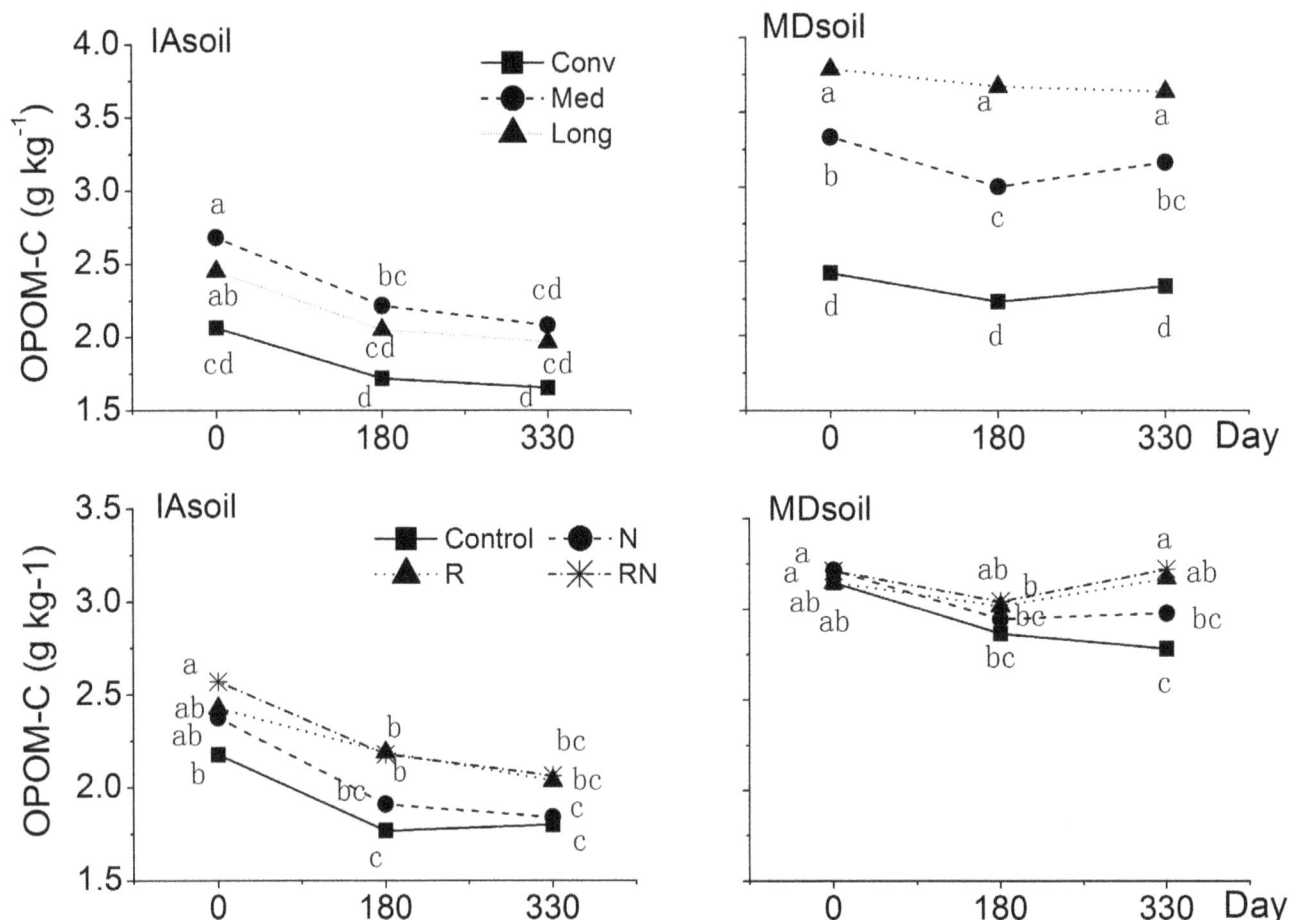

Figure 3. Effect of cropping system and treatment on occluded POM-C (OPOM-C) dynamics during 330 day incubation. Control: treatments with no residue nor N fertilizer added. N: treatments with N fertilized, R: treatments with residue applied. RN: treatments with both residue and N fertilizer added. Different letters suggest significant differences at p<0.05, comparisons were were made for the interaction of cropping system (Conv, Med, Long) by date (0, 180, 330 days), and treatments (Control, N, R, RN) by date within each site (IAsoil, MDsoil).

differences in C_r pool size and modest differences of labile fractions observed among cropping systems (Table 2) in the IAsoil are consistent with the characteristics [57] assigned to C saturated soils. The faster decay rates observed in IAsoil (Table 2) are also consistent with the notion that SOC stabilization capacity could be reduced in C saturated soils [11,58]. Carbon saturation of the IAsoil might help explain why net SOC losses might be derived from stocks attributed to the resistant pool.

Residue and Nitrogen Additions

Wheat residue addition doubled the amount of C mineralized to CO_2 at both sites (Fig. 1) presumably by increasing the activity, size and possibly the composition of the microbial community [59,60]. The CO_2 loss ranked R = RN>Control = N. Shifts in the microbial community resulting from changes in C availability likely occurred during the incubation, shifting from those specializing in the decay of fresh organic-materials (r-strategists), to K-strategists that can consume soil native SOM [59,61,62] but those changes would not be clearly revealed by our analyses. Factors regulating CO_2 loss from this type of 'curve fitting' study are typically assumed to include organic matter quality and quantity and soil's physical protection while assuming that microbial community acts as a catalyst [50]. Accordingly, soil

amendment will only change pool sizes and rate coefficients with changes in microbial activity being reflected in the rate coefficient.

Residue addition increased C_a by 340% and C_s by 38% in the IAsoil (p<0.05). For the MDsoil, residue addition increased C_a by 230%, C_s by 21% and k_s by 38% (p<0.05) (Table 2). Ladd et al. [63] found decay rates were reduced more by residue addition in soils with reactive minerals than in lower activity soils, particularly during the early phases of decay. This is consistent with our findings, where k_a declined by 58% in IAsoil compared to just 9.1% in MDsoil after residue addition (Table 2). In the IAsoil, residue additions altered k_a more than k_s, suggesting that the C_s pool were less affected by amendment. Even though residue addition increased the size of both C_a and C_s in both sites, reductions in C_r in the MDsoil suggest amendment primed losses of SOC from the resistant pool. Residue additions slightly but non-significantly increased the amount and percentage of SOC loss from IAsoil but significantly reduced both the amount and percentage of SOC loss compared to the Control and N treatments in MDsoil (p<0.05) (Fig. 2). These opposite responses to residue amendment might suggest that there was more labile C vulnerable to priming effects present in IAsoil than in MDsoil [23]. This notion is consistent with the higher C:N ratios observed in IAsoil but does not agree with estimates of labile reserves derived from curve fitting that suggest the IAsoil stocks (C_a+C_s) are smaller

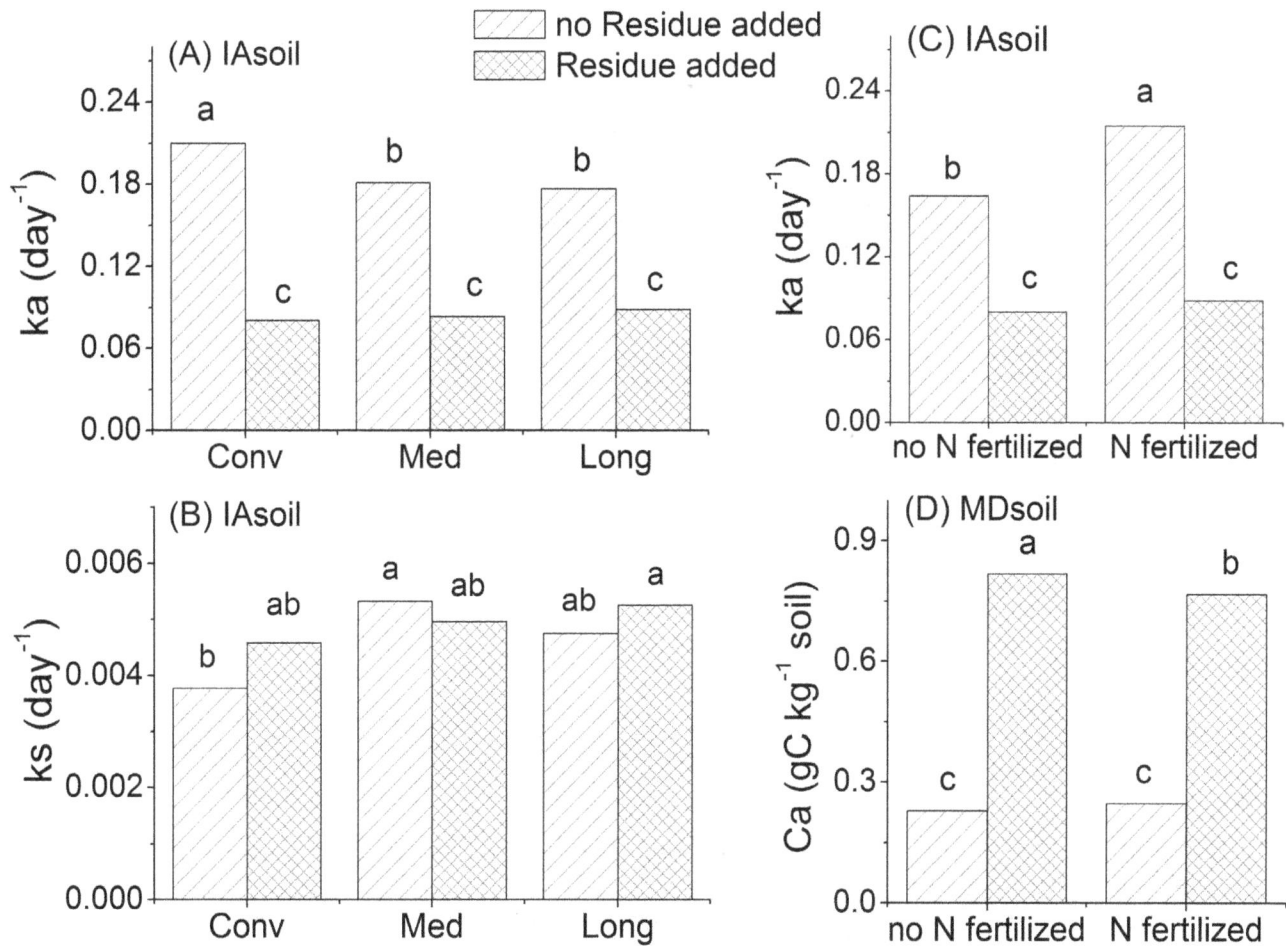

Figure 4. Interactions between residue addition and cropping system or N fertilization on C mineralization coefficients. Different letters above bars indicate statistically significant difference at p<0.05, comparisons were made within each sub-figure.

than those present in the MDsoil. The notion that C_r contributes to decay in the IAsoil due to C saturation could explain this discrepancy. Neither N fertilization nor residue application increased OPOM-C (p>0.05). However, trends in OPOM-C, which decreased throughout the incubation in IAsoil, reinforce the notion that that soil has limited capacity to absorb additional C. In the MDsoil, residue addition caused OPOM-C stocks to increase during the later phase of the incubation (Fig. 3), suggesting that the MDsoil was better able to stabilize additional C and that aggregate protection of labile C played an important role at that site.

Nitrogen fertilization consistently but non-significantly reduced the amount of CO_2-C loss from both soils (Fig. 1). Reduction of C mineralization caused by N fertilization is likely due to two reasons: 1) the increased microbial growth resulting from a reduction of soil C:N ratios from 10.1–13.7 (C:N ratio of residue used is 30) to values closer to that of soil microbes (4.1–12.1) [64] following fertilization, and 2) soil microbes with N addition have a lower requirement for organic N sources and thus the amount of organic matter mineralization needed to support microbial growth was reduced. Nitrogen fertilization increased k_a of IAsoil by 25% and decreased k_s of MDsoil by 16% (p<0.05) (Table 2). These differences again seem to be associated with differences in SOM saturation and protective capacity of the two soils.

Residue additions interacted with cropping system and/or with N additions in both soils (Fig. 4). In the IAsoil, residue addition reduced k_a by 58%, and removed differences in k_a observed among controls with different cropping histories where k_a was greater in Conv than Med and Long (Fig. 4A). Residue addition also alleviated differences in the k_s observed for controls (Med> Conv) from different cropping systems (Fig. 4B). When residue and N fertilizer were applied together, N addition increased k_a values of the IAsoil by 31% over the control when no residue was applied but k_a remained unchanged after N addition when residue was added (Fig. 4C). There was also a significant interaction between residue addition and N fertilization on C_a in the MDsoil, where residue addition prompted a smaller (2.1 fold) increase in C_a in N fertilized soils and a 2.6 fold increase in unfertilized soils (Fig. 4D).

Conclusion

This work showed that by using CEC or surface charge, rather than texture or clay content, to represent inherent differences in soil protective capacity, we were able to account for differences in C mineralization response to residue and nitrogen fertilization observed in different soil types. By using dynamic soil properties (POM, OPOM) we were able to anticipate important differences in C dynamics observed in the soils from different sites. More diverse cropping notably improved soil dynamic properties and accelerated C mineralization in the MDsoil. Use of diversified rotations produced more limited effects on dynamic properties in the IAsoil. Residue application increased the size of C_a and decreased k_a in both soils, but the effect was more pronounced in the IAsoil than the MDsoil. As predicted, residue addition reduced SOC loss and promoted C immobilization in OPOM in the MDsoil where SOC levels were low. The effect of residue addition on SOC loss from the IAsoil, which appears to be at or near C saturation levels, is consistent with the observations of Zhang [23] that linked high SOC to priming. The greater vulnerability of the IAsoil to C priming was not associated with POM-C levels, which were low compared to those observed in the MDsoil, but was associated with higher SOM-C:N ratios and loss from the resistant pool. Our results suggest that the proportion of SOC in POM-C does not reveal soil C saturation status in a way that can be generalized or be used on its own to predict the ability of soils to sequester or lose SOC. Use of OPOM, rather than POM or C mineralization fitted parameters, to evaluate protection by aggregates, helped resolve important differences in C cycling within the two soils. Nitrogen fertilization did not induce priming (net SOC loss or CO_2 loss) from either soil; in fact, additions suppressed net C mineralization from both soils. Nitrogen addition caused increased k_a in the IA soil and in C_s in the MDsoil when added with residues; this indicates N might have limited early stages of decay in the IAsoil and C assimilation in the MDsoil. Future research should explore potential for use of CEC and POM to initialize, or adjust coefficients used within carbon cycle process models.

Acknowledgments

We thank Patricia Lazicki, and Carmen Ugarte for the help with laboratory work and managers of the Marsden site for access to samples. We also thank Susanne Aref (statistician with a PhD in biometrics from Cornell, taught and led statistical consulting programs at two universities), Ho-young Kwon and Xubo Zhang for their assistance with statistical analysis.

Author Contributions

Conceived and designed the experiments: XC XW MW. Performed the experiments: XC. Analyzed the data: XC MW. Contributed reagents/materials/analysis tools: ML MC MW. Contributed to the writing of the manuscript: XC. Revised the paper: XW ML MC MW.

References

1. Yang X, Kay B (2001) Rotation and tillage effects on soil organic carbon sequestration in a typic Hapludalf in Southern Ontario. Soil Till Res 59: 107–114.

2. Peterson G, Halvorson A, Havlin J, Jones O, Lyon D, et al. (1998) Reduced tillage and increasing cropping intensity in the Great Plains conserves soil C. Soil Till Res 47: 207–218.

3. Li LJ, Han XZ, You MY, Yuan YR, Ding XL, et al. (2012) Carbon and nitrogen mineralization patterns of two contrasting crop residues in a Mollisol: Effects of residue type and placement in soils. Eur J Soil Biol 54: 1–6.

4. Marriott EE, Wander M (2006) Qualitative and quantitative differences in particulate organic matter fractions in organic and conventional farming systems. Soil Biol Biochem 38: 1527–1536.

5. Zentner RP, Wall DD, Nagy CN, Smith EG, Young DL, et al. (2002) Economics of crop diversification and soil tillage opportunities in the Canadian prairies. Agron J 94: 216–230.

6. Kremen C, Miles A (2012) Ecosystem services in biologically diversified versus conventional farming systems: Benefits, externalities, and trade-offs. Ecol Soc 17(4): 40.

7. Govaerts B, Verhulst N, Castellanos-Navarrete A, Sayre K, Dixon J, et al. (2009) Conservation agriculture and soil carbon sequestration: Between myth and farmer reality. Crit Rev Plant Sci 28: 97–122.

8. Franzluebbers A, Haney R, Hons F, Zuberer D (1996) Active fractions of organic matter in soils with different texture. Soil Biol Biochem 28: 1367–1372.

9. Jindaluang W, Kheoruenromne I, Suddhiprakarn A, Singh BP, Singh B (2013) Influence of soil texture and mineralogy on organic matter content and composition in physically separated fractions soils of Thailand. Geoderma 195: 207–219.

10. Franzluebbers A, Arshad M (1997) Particulate organic carbon content and potential mineralization as affected by tillage and texture. Soil Sci Soc Am J 61: 1382–1386.

11. Six J, Conant R, Paul E, Paustian K (2002) Stabilization mechanisms of soil organic matter: Implications for C-saturation of soils. Plant Soil 241: 155–176.

12. Hobbie SE (2000) Interactions between litter lignin and soil nitrogen availability during leaf litter decomposition in a Hawaiian montane forest. Ecosystems 3: 484–494.

13. Neff JC, Townsend AR, Gleixner G, Lehman SJ, Turnbull J, et al. (2002) Variable effects of nitrogen additions on the stability and turnover of soil carbon. Nature 419: 915–917.

14. Wander M, Yun W, Goldstein W, Aref S, Khan S (2007) Organic N and particulate organic matter fractions in organic and conventional farming systems with a history of manure application. Plant Soil 291: 311–321.

15. Wander M, Traina S, Stinner B, Peters S (1994) Organic and conventional management effects on biologically active soil organic matter pools. Soil Sci Soc Am J 58: 1130–1139.

16. Spargo JT, Cavigelli MA, Mirsky SB, Maul JE, Meisinger JJ (2011) Mineralizable soil nitrogen and labile soil organic matter in diverse long-term cropping systems. Nutr Cycl in Agroecosys 90: 253–266.

17. Adiku SGK, Narh S, Jones J, Laryea K, Dowuona G (2008) Short-term effects of crop rotation, residue management, and soil water on carbon mineralization in a tropical cropping system. Plant Soil 311: 29–38.

18. Zhang X, LI L, PAN G (2007) Topsoil organic carbon mineralization and CO_2 evolution of three paddy soils from South China and the temperature dependence. J Environ Sci-China 19: 319–326.

19. Martin JV, de Imperial RM, Calvo R, Garcia M, Leon-Cofreces C, et al. (2012) Carbon mineralisation kinetics for analysis of poultry manure in two soils. Aust J Soil Res 50: 222–228.

20. Paustian K, Elliott E, Collins H, Cole CV, Paul E (1995) Use of a network of long-term experiments for analysis of soil carbon dynamics and global change: The North American model. Anim Prod Sci 35: 929–939.

21. Cochran R, Collins H, Kennedy A, Bezdicek D (2007) Soil carbon pools and fluxes after land conversion in a semiarid shrub-steppe ecosystem. Biol Fert Soils 43: 479–489.

22. Wander M (2004) Soil organic matter fractions and their relevance to soil function. Soil organic matter in sustainable agriculture. CRC Press, Boca Raton, FL, USA. 67–102.

23. Zhang W, Wang X, Wang S (2013) Addition of external organic carbon and native soil organic carbon decomposition: A meta-analysis. PloS One 8: e54779.

24. Ding W, Yu H, Cai Z, Han F, Xu Z (2010) Responses of soil respiration to N fertilization in a loamy soil under maize cultivation. Geoderma 155: 381–389.

25. Green C, Blackmer A, Horton R (1995) Nitrogen effects on conservation of carbon during corn residue decomposition in soil. Soil Sci Soc Am J 59: 453–459.

26. Sakala WD, Cadisch G, Giller KE (2000) Interactions between residues of maize and pigeonpea and mineral N fertilizers during decomposition and N mineralization. Soil Biol Biochem 32: 679–688.

27. Scott NA, Cole CV, Elliott ET, Huffman SA (1996) Soil textural control on decomposition and soil organic matter dynamics. Soil Sci Soc Am J 60: 1102–1109.

28. Van Veen J, Ladd J, Amato M (1985) Turnover of carbon and nitrogen through the microbial biomass in a sandy loam and a clay soil incubated with $[^{14}C(U)]$ glucose and $[^{15}N](NH_4)_2SO_4$ under different moisture regimes. Soil Biol Biochem 17: 747–756.

29. Liebman M, Gibson LR, Sundberg DN, Heggenstaller AH, Westerman PR, et al. (2008) Agronomic and economic performance characteristics of conventional and low-external-input cropping systems in the central Corn Belt. Agron J 100: 600–610.

30. Davis AS, Hill JD, Chase CA, Johanns AM, Liebman M (2012) Increasing cropping system diversity balances productivity, profitability and environmental health. PloS One 7: e47149.

31. Gee GW, Bauder JW (1986) Particle-size analysis. In: Klute A, editor. Methods of soil analysis: Part 1 Physical and mineralogical methods. Soil Science Society of America, American Society of Agronomy. 383–411.

32. Marriott EE, Wander MM (2006) Total and labile soil organic matter in organic and conventional farming systems. Soil Sci Soc Am J 70: 950–959.

33. Yoo G, Wander MM (2008) Tillage effects on aggregate turnover and sequestration of particulate and humified soil organic carbon. Soil Sci Soc Am J 72: 670–676.

34. Saviozzi A, Levi-Minzi R, Riffaldi R, Vanni G (1997) Role of chemical constituents of wheat straw and pig slurry on their decomposition in soil. Biol Fert Soils 25: 401–406.

35. Bonde TA, Rosswall T (1987) Seasonal variation of potentially mineralizable nitrogen in four cropping systems. Soil Sci Soc Am J 51: 1508–1514.

36. Wang W, Baldock JA, Dalal R, Moody P (2004) Decomposition dynamics of plant materials in relation to nitrogen availability and biochemistry determined by NMR and wet-chemical analysis. Soil Biol Biochem 36: 2045–2058.

37. Rodrigo A, Recous S, Neel C, Mary B (1997) Modelling temperature and moisture effects on CN transformations in soils: Comparison of nine models. Ecol Model 102: 325–339.

38. Bauer J, Herbst M, Huisman J, Weihermueller L, Vereecken H (2008) Sensitivity of simulated soil heterotrophic respiration to temperature and moisture reduction functions. Geoderma 145: 17–27.

39. Qi Y, Xu M (2001) Separating the effects of moisture and temperature on soil CO_2 efflux in a coniferous forest in the Sierra Nevada mountains. Plant Soil 237: 15–23.

40. Townsend AR, Vitousek PM, Desmarais DJ, Tharpe A (1997) Soil carbon pool structure and temperature sensitivity inferred using CO_2 and $^{13}CO_2$ incubation fluxes from five Hawaiian soils. Biogeochemistry 38: 1–17.

41. Miyittah M, Inubushi K (2003) Decomposition and CO_2-C evolution of okara, sewage sludge, cow and poultry manure composts in soils. Soil Sci Plant nutr 49: 61–68.

42. Brady NC, Weil RR (2004) Soil organic matter. In: Elements of the nature and properties of soils. Pearson Prentice Hall Inc. USA. 386–408.

43. Hassink J, Whitmore AP (1997) A model of the physical protection of organic matter in soils. Soil Sci Soc Am J 61: 131–139.

44. Soil Survey Staff. Natural Resources Conservation Service, United States Department of Agriculture. Web Soil Survey. Available: http://websoilsurvey.nrcs.usda.gov/. Accessed 2013 May 27.

45. Dixon J (1989) Kaolin and serpentine group minerals. In: Dixon JB, Weed SB, editors. Minerals and soil environments, 2nd Edition. SSSA Book Series. Madison, WI. USA. 506 p.

46. Borchardt G (1989) Smectites. In: Dixon JB, Weed SB, editors. Minerals and soil environments, 2nd Edition. SSSA Book Series. Madison, WI. USA. 703 p.

47. Kleber M, Sollins P, Sutton R (2007) A conceptual model of organo-mineral interactions in soils: Self-assembly of organic molecular fragments into zonal structures on mineral surfaces. Biogeochemistry 85: 9–24.

48. Arrouays D, Saby N, Walter C, Lemercier B, Schvartz C (2006) Relationships between particle-size distribution and organic carbon in French arable topsoils. Soil Use Manage 22: 48–51.

49. Saby NPA, Arrouays D, Antoni V, Lemercier B, Follain S, et al. (2008) Changes in soil organic carbon in a mountainous French region, 1990–2004. Soil Use Manage 24: 254–262.

50. McGill WB (1996) Evaluation of soil organic matter models: Review and classification of ten soil organic matter (SOM) models. NATO ASI Series 38: 111–132.

51. Wang W, Dalal R, Moody P, Smith C (2003) Relationships of soil respiration to microbial biomass, substrate availability and clay content. Soil Biol Biochem 35: 273–284.

52. Setia R, Marschner P, Baldock J, Chittleborough D, Verma V (2011) Relationships between carbon dioxide emission and soil properties in salt-affected landscapes. Soil Biol Biochem 43: 667–674.

53. Chivenge P, Vanlauwe B, Gentile R, Six J (2011) Comparison of organic versus mineral resource effects on short-term aggregate carbon and nitrogen dynamics in a sandy soil versus a fine textured soil. Agr Ecosys Environ 140: 361–371.

54. Bellamy PH, Loveland PJ, Bradley RI, Lark RM, Kirk GJ (2005) Carbon losses from all soils across England and Wales 1978–2003. Nature 437: 245–248.

55. Parton WJ, Stewart JW, Cole CV (1988) Dynamics of C, N, P and S in grassland soils: A model. Biogeochemistry 5: 109–131.

56. Lazicki PA (2011) Effect of rotation, organic inputs and tillage on crop performance and soil quality in conventional and low-input rotations in central Iowa. M.Sc. Thesis. University of Illinois, Champaign-Urbana. Available: https://www.ideals.illinois.edu/handle/2142/26085 Accessed 2012 Oct 15.

57. Gulde S, Chung H, Amelung W, Chang C, Six J (2008) Soil carbon saturation controls labile and stable carbon pool dynamics. Soil Sci Soc Am J 72: 605–612.

58. Stewart CE, Paustian K, Conant RT, Plante AF, Six J (2008) Soil carbon saturation: Evaluation and corroboration by long-term incubations. Soil Biol Biochem 40: 1741–1750.

59. Waldrop MP, Firestone MK (2004) Microbial community utilization of recalcitrant and simple carbon compounds: impact of oak-woodland plant communities. Oecologia 138: 275–284.

60. Bailey VL, Smith JL, Bolton Jr H (2002) Fungal-to-bacterial ratios in soils investigated for enhanced C sequestration. Soil Biol Biochem 34: 997–1007.

61. Lipson DA, Schmidt SK, Monson RK (2000) Carbon availability and temperature control the post-snowmelt decline in alpine soil microbial biomass. Soil Biol Biochem 32: 441–448.

62. Fontaine S, Mariotti A, Abbadie L (2003) The priming effect of organic matter: a question of microbial competition? Soil Biol Biochem 35: 837–843.

63. Ladd J, Oades J, Amato M (1981) Microbial biomass formed from ^{14}C, ^{15}N-labelled plant material decomposing in soils in the field. Soil Biol Biochem 13: 119–126.

64. Kaye JP, Hart SC (1997) Competition for nitrogen between plants and soil microorganisms. Trends Ecol Evol 12: 139–143.

Soil Organic Carbon Loss and Selective Transportation under Field Simulated Rainfall Events

Xiaodong Nie[1,2], Zhongwu Li[1,2]*, Jinquan Huang[3], Bin Huang[1,2], Yan Zhang[1,2], Wenming Ma[1,2], Yanbiao Hu[1,2], Guangming Zeng[1,2]

1 College of Environmental Science and Engineering, Hunan University, Changsha, PR China, 2 Key Laboratory of Environmental Biology and Pollution Control (Hunan University), Ministry of Education, Changsha, PR China, 3 Department of Soil and Water Conservation, Yangtze River Scientific Research Institute, Wuhan, PR China

Abstract

The study on the lateral movement of soil organic carbon (SOC) during soil erosion can improve the understanding of global carbon budget. Simulated rainfall experiments on small field plots were conducted to investigate the SOC lateral movement under different rainfall intensities and tillage practices. Two rainfall intensities (High intensity (HI) and Low intensity (LI)) and two tillage practices (No tillage (NT) and Conventional tillage (CT)) were maintained on three plots (2 m width × 5 m length): HI-NT, LI-NT and LI-CT. The rainfall lasted 60 minutes after the runoff generated, the sediment yield and runoff volume were measured and sampled at 6-min intervals. SOC concentration of sediment and runoff as well as the sediment particle size distribution were measured. The results showed that most of the eroded organic carbon (OC) was lost in form of sediment-bound organic carbon in all events. The amount of lost SOC in LI-NT event was 12.76 times greater than that in LI-CT event, whereas this measure in HI-NT event was 3.25 times greater than that in LI-NT event. These results suggest that conventional tillage as well as lower rainfall intensity can reduce the amount of lost SOC during short-term soil erosion. Meanwhile, the eroded sediment in all events was enriched in OC, and higher enrichment ratio of OC (ERoc) in sediment was observed in LI events than that in HI event, whereas similar ERoc curves were found in LI-CT and LI-NT events. Furthermore, significant correlations between ERoc and different size sediment particles were only observed in HI-NT event. This indicates that the enrichment of OC is dependent on the erosion process, and the specific enrichment mechanisms with respect to different erosion processes should be studied in future.

Editor: Vanesa Magar, Centro de Investigacion Cientifica y Educacion Superior de Ensenada, Mexico

Funding: The study was funded by the National Natural Science Foundation of China (40971179, 41271294), the Program for New Century Excellent Talents in University (NCET-09-330), and the Natural Science Foundation of Hunan Province of China (11JJ3041). The funders had no role in study design, data collection and analysis, decision to publish, or preparation of the manuscript.

Competing Interests: The authors have declared that no competing interests exist.

* Email: lzw@hnu.edu.cn

Introduction

Soil erosion has attracted more and more attention from all over the world for its impact on carbon geochemical cycles between soils and the atmosphere [1,2]. However, it is still a controversial issue on the role of soil erosion on carbon cycles, with the most famous debate is the carbon source or sink [3–7]. The substance of the issue is the poor understanding of soil erosion process and the included carbon dynamics. The Intergovernmental Panel on Climate Change (IPCC) [8] suggested that lateral carbon movement was the source of the greatest uncertainty in the global carbon balance. Furthermore, Kuhn et al. [1] indicated that the movement of soil organic carbon (SOC), both its particulate and dissolved forms, through agricultural landscapes is not fully understood.

Loss of SOC from the ecosystem occurs as a result of three processes: (i) physical removal by water (erosion); (ii) release of carbon into the atmosphere; and (iii) leaching [9,10]. While in water erosion, most of the SOC is lost through the procedure (i). Researchers considered that the physical removal of SOC undergo four stages during the erosion processes [11]. Firstly, the macroaggregates are detached and dispersed into microaggregates by raindrop impact, and release organic carbon (OC) at the same time. Secondly, SOC is transported by runoff in form of either dissolved organic carbon (DOC) or sediment-bound organic carbon (SBOC). Thirdly, the coarse or heavy particles were deposited in micro-depression during the migration path. Eventually, the SOC with the transportable particles or runoff are transported to outlet and deposited in concave slopes and floodplains. However, these processes are related to a number of factors, namely, rainfall intensity and kinetic energy, infiltration and runoff rates, soil properties and soil surface conditions such as soil moisture, roughness, crop residues, slope length and steepness [12,13]. Among them, rainfall intensity and tillage practice have become the focus of the erosion study. Lots of experiments were conducted to study the impact of rainfall intensity and tillage practice on soil delivery and nutrient loss [13–15].

However, most of the researches, under different rainfall intensities and tillage practices, focused on the SOC dynamics during erosion, were conducted in watershed [16–18] or laboratory [10,13,19]. Different points of view were observed between them for the variety of research conditions. For example, Lal et al. [20] suggested that no-till would decrease silt in rivers and lakes, which would lower transport of SOC and pollutant-laden sediments to aquatic ecosystems and reduce hypoxia. Also, some researchers indicated that conservation tillage practice reduce losses in soil and SOC [21,22]. However, Cogle et al. [23] found that the lost carbon from 20 cm deep tillage was consistently less from zero tillage. In addition, the study scale is also considered to be an important factor impact on the movement of SOC [24]. Schiettecatte et al. [19] indicated that at a largerscale, due to the increased probability of sediment deposition by topography and vegetation, the sediment became more enriched in OC. The problem obtained within watershed scale is the representativeness of field conditions and the extent to which the data obtained with these microcosms can be extrapolated [25]. And for laboratory experiments, the experiment condition is too ideal to simulate the natural state. While simulated rainfalls at small plot scale had been applied to investigate the detachment and sediment transport capacity of runoff [26] and the effects of water erosion on soil properties and productivity [27], and important and meaningful results were obtained. So the study at plot scale in field condition is essential for improving the understanding of SOC dynamics under different erosion processes.

Therefore, field simulated rainfall events at small plot scale were performed in this study, and the objectives of present study were to: (i) examine the carbon lateral movement at plot scale in field runoff area, (ii) investigate the selective migration processes of SOC. Such information will be useful for the study of SOC transportation, also, provide basic data for SOC migration model.

Materials and Methods

2.1. Ethics Statement

In this study, soil sampling and sample determinations conducted were permitted by the local authorities (i.e. Soil and Water Conservation Monitoring Station). We also obtained a permission from the local authorities for reporting research results to the public. In addition, the field studies did not involve endangered or protected species.

2.2. Study site

The simulated rainfall experiments were conducted at the Soil and Water Conservation Monitoring Station (111°22′ E, 27°03′ N), Hunan province, China (Fig. 1). The study area is located in subtropics humid monsoon climate zone, with an annual precipitation of 1 218.5 ~1 473.5 mm and average annual temperature of 17.1°C. The record rainfall intensity of the last five years varied between 0.10 and 2.11 mm min^{-1}, with 90% between 0.50 and 1.44 mm min^{-1}. The area is characterized by hills and sloping lands with a gradient of 3 to 30%. The soil was developed from Quaternary red soil with sandy and clay loam texture. The area is a typical red soil hilly region. Due to the dense population and unreasonable land use, this region has suffered serious soil erosion, and more than 4 195.05 km^2 cropland suffered water erosion with different degree. This area is representative of the agricultural, socio-economic and environmental situation of many slope farming areas in the region. The study carried out on this area is typical and representative of general situation in red soil hilly region.

2.3. Plot set-up and rainfall simulation

A 7 m×5 m block was taken from a typical sloping land with a 17% gradient to conduct the rainfall simulation. The block was previously planted with slope cultivated *Polygonatum odoratum*

(Mill.) Druce. After harvesting the crops, this block was abandoned for almost one year, and it was almost bare before the experiments. The soil had a bulk density of 1.65±0.15 g cm^{-3}, meanwhile with a water content of 0.15±0.02%. The soil pH was 4.47±0.10 (acidic soil), and the soil carbon is considered to be SOC. The mean SOC concentration of the surface layer was 7.47±2.48 g kg^{-1} dry soil (the mean value of 45 replicate samples ± standard error), the mean DOC concentration was 29.26±0.21 mg kg^{-1}. The soil had a clay-loamy texture with 33.44±1.27% clay particle size distribution, 27.82±1.63% silt, and 38.73±1.74% sand. Before the experiments, the block was divided into three equal plots, and each plot was designed with 2 m (width) × 5 m (length), and named Plot I, II, III respectively. Two tillage practices were maintained on the three plots: plot I and II were applied no tillage (NT) and plot III was maintained conventional tillage (CT). The plot III was disk ploughed (10 cm), while others kept natural state. The three plots were separated with a 0.5 m wide space. Each plot was bound with a metal frame inserted into the ground 15 cm in order to prevent runoff from adjacent areas. To determine ERoc, plot soil was sampled in all plots at depth of 10 cm. The boreholes were later filled and carefully leveled in order to reduce the effects of soil sampling.

For the erosion experiments, a rainfall simulator with a SPRACO cone jet nozzle mounted on the top of fixed 4.57 m long stand pipes was built. The nozzles were placed on the boundary of the plots. The median drop size was 2.4 mm with a uniformity of 89.7%. According to the local rainfall intensity variation for the past five years, rainfall intensities of 0.4~0.6 and 1.3~1.5 mm min^{-1} were used, representing the low intensity (LI) and high intensity (HI) storms of this region. Plot I, II, III were treated with HI-NT, LI-NT and LI-CT, respectively. Calibrations of rainfall intensities were conducted prior to the experiments. Four simulators were used in HI event, and two were used in LI events. For the HI event, each simulator was located on the longer side and closed to the conner of the runoff plot, the two simulators in LI events were located on the similar positions and distributed in diagonal direction. Meanwhile, five rain gauges were used to measure the actual rainfall intensity, one was placed on the top of the plot, and four were distributed around the two sides. For each rainfall event, rainfall lasted 60 min after the overland surface runoff began. Once overland surface runoff began, random runoff samples were manually and intermittently collected at 6 min intervals using a 1000 mL kettle. Each collected sample were deposited, separated from the water, dried in a forced-air oven at 105°C until constant mass was achieved and weighed for the determination of sediment concentration. All other runoff and sediment samples were collected in a marked pail and the total runoff volume in 6 min was recorded. Another sample was taken from the thoroughly mixed pail, and this sample was splitted into two portions. One portion was dried in an oven at 105°C until constant mass and then weighed for the determination of physicochemical properties, the other portion was sieved with1, 0.5 and 0.25 mm pore openings for the separation of aggregates. The separated aggregates were dried and weighed separately. The actual rainfall intensity was determined after the simulated rainfall event through the rainfall gauges. The mean rainfall intensity was found to be 1.38 mm min^{-1} for the HI–NT event plot, 0.53 mm min^{-1} for the LI-NT and LI-CT event plot.

2.4. Sample treatment and data analysis

Soil bulk density was determined by cutting ring method. SOC concentrations of soil and sediment were determined with the dichromate oxidation method of Walkley and Black [28]. Soil particle sizes were analyzed using the pipette method [29]. Total

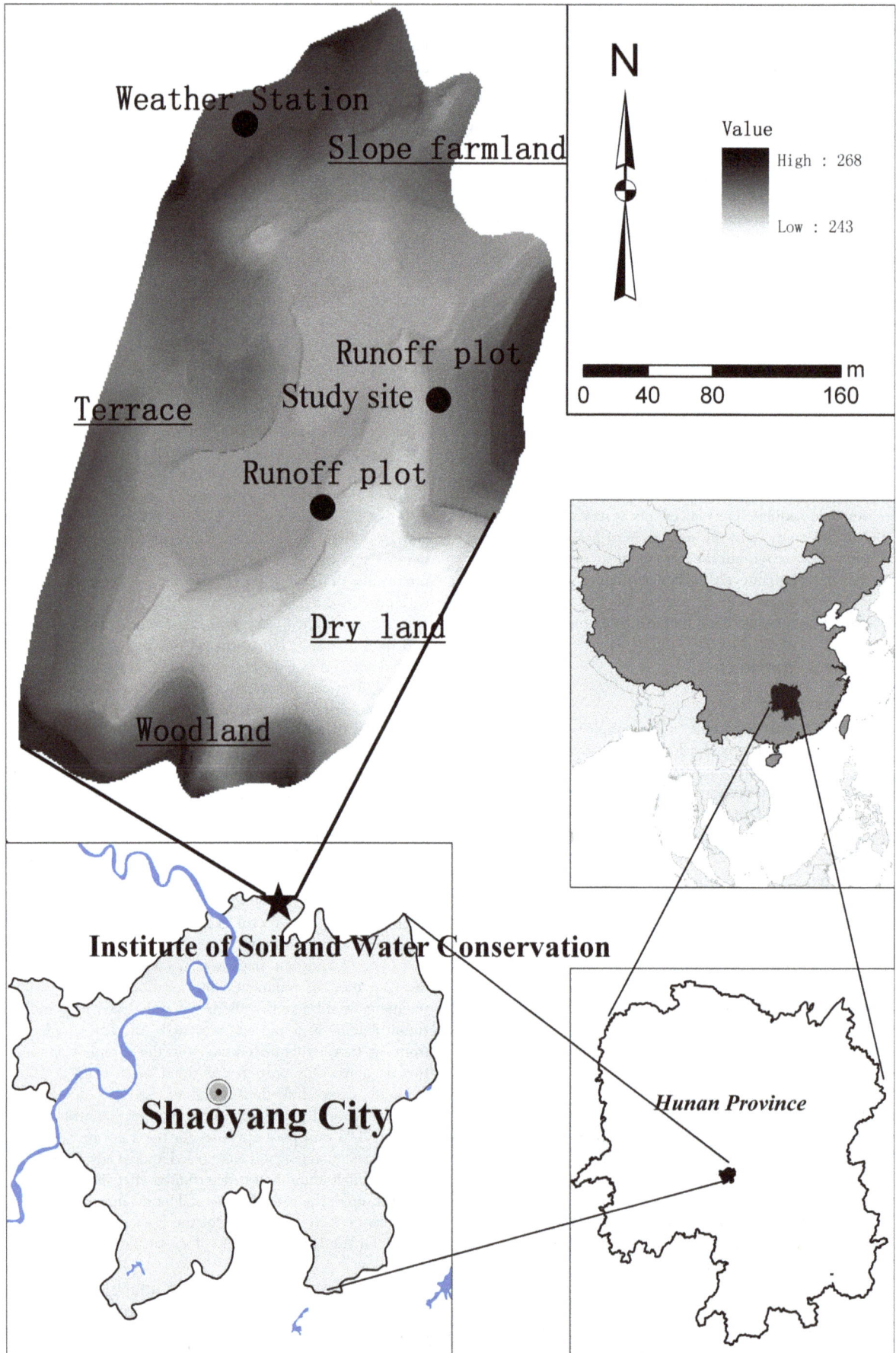

Figure 1. Location of the study area.

Table 1. The regular patterns of sediment and runoff transportation during water erosion.

Sample number	HI-NT (1.33 mm min⁻¹), TSR: 1'31"				LI-NT (0.53 mm min⁻¹), TSR: 2'31"				LI-CT (0.53 mm min⁻¹), TSR:48'0"			
	SYR ($g\,min^{-1}$)	CS (g)	RR ($L\,min^{-1}$)	CR (L)	SYR ($g\,min^{-1}$)	CS (g)	RR ($L\,min^{-1}$)	CR (L)	SYR ($g\,min^{-1}$)	CS (g)	RR ($L\,min^{-1}$)	CR (L)
1	248.93	1493.58	5.62	33.70	45.33	271.99	1.62	9.70	1.25	7.48	0.42	2.50
2	815.48	6386.479	12.25	107.20	111.35	940.08	4.77	38.30	2.74	23.91	0.55	5.80
3	659.31	10342.35	11.87	178.40	92.08	1492.56	4.78	67.00	4.37	50.12	0.80	10.60
4	514.51	13429.40	12.08	250.90	99.52	2089.68	5.18	98.10	4.93	79.70	1.00	16.60
5	680.28	17511.10	11.67	320.90	94.67	2657.68	4.75	126.60	4.88	108.95	1.08	23.10
6	524.83	20660.10	13.22	400.20	92.05	3209.98	5.00	156.60	5.80	143.77	1.22	30.40
7	586.71	24180.35	12.88	477.50	119.98	3929.86	4.93	186.20	11.53	212.95	1.57	39.80
8	508.14	27229.17	11.48	546.40	161.07	4896.29	4.95	215.90	15.77	307.60	1.65	49.70
9	564.65	30617.05	11.18	613.50	212.03	6168.45	5.08	246.40	16.69	407.76	1.70	59.90
10	665.98	34612.93	12.37	687.70	171.08	7194.95	6.03	282.60	15.95	503.46	2.18	73.00

The first sample was collected at the time of runoff began, and samples were collected at 6 minutes intervals.
TSR, time to start runoff; SYR, sediment yield rate; CS, cumulative sediment yield; RR, runoff rate; CR, cumulative runoff volume.

organic carbon concentrations for the runoff samples were measured with a Shimadzu TOC-TN analyzer.

The ERoc of sediment was calculated by dividing the SOC content of the sediment by its content in the original soil material. In this study, the ERoc of sediment was the ratio between the SOC concentration of sediment and the value of the source soil for each plot.

2.5. Statistical analysis

Statistical data analysis was performed using SPSS 20.0 for Windows. Pearson correlation was used to test the significance of correlations among ERoc and sediment size as well as the sediment particle size distribution. Differences in correlation analysis were detected using the least significant difference procedure for a multiple range test at the 0.05 significance level.

Results

3.1. Rainfall features

3.1.1. Sediment and runoff loss. Through the trials, different sediment and runoff yield rates were distinguished. For the events of HI-NT and LI-NT, two stages of sediment and runoff loss were found. In the first stage, plot soil underwent infiltration and runoff starting, the rates of sediment and runoff loss rapidly increased in the initial rainfall time (0~12 min). In the second stage, runoff loss rates reached steady state. The sediment loss rate in HI-NT event was consistent with the runoff loss rate. However, the values of sediment loss in LI-NT were first stable and then increased. In addition, different from these trends, the sediment and runoff yield rates in LI-CT increased consistently with the rainfall time.

The time to start runoff in LI-NT event was 1.66 times longer than that in HI-NT event, while, the figure in LI-CT event was 19.05 times longer than that in LI-NT event (Table 1). Despite longer time (additional 48 min) spent in runoff starting, the LI-CT event generated less runoff than LI-NT event did. The total runoff in HI-NT event was 2.43 times greater than that in LI-NT event which was 3.87 times higher than that in LI-CT event. Moreover, the sediments yield in HI-NT and LI-NT event was 68.34 and 14.38 times than that in LI-CT event, and which was correspond to the regular pattern of the lost runoff.

3.1.2. Sediment sorting. For the soil aggregates, microaggregates can be preferentially transported and macroaggregates deposit easily during water erosion. In this experiment, the sediment was principally composed of <0.25 mm aggregate which accounted for more than 58% of the sediment yield (Fig. 2). The average proportion of 0.25–1 mm aggregate was lower than 20%, and the proportion of >1 mm aggregate was lower than 10%. For all the events, the proportions of aggregates in sediment decreased with increasing size. Further, the composition of aggregates varied with rainfall duration. For LI-NT and LI-CT events, the proportions of microaggregate (<0.25 mm) first increased and then decreased, eventually reached steady state. However, for HI-NT event, the proportions of aggregates were in stable state except in 6–18 min. The dynamics of aggregates also depend with rainfall events.

3.2. SOC loss and selective migration

3.2.1. SOC loss. During water erosion, SOC loss in two forms: SBOC and DOC. Changes in loss rates of the SOC with respect to time were found to be very different among different rainfall events (Table 2). During the HI-NT event the loss rate of SBOC increased rapidly at the initial rainfall time, and then entered into a stable state. While For the LI-NT and LI-CT

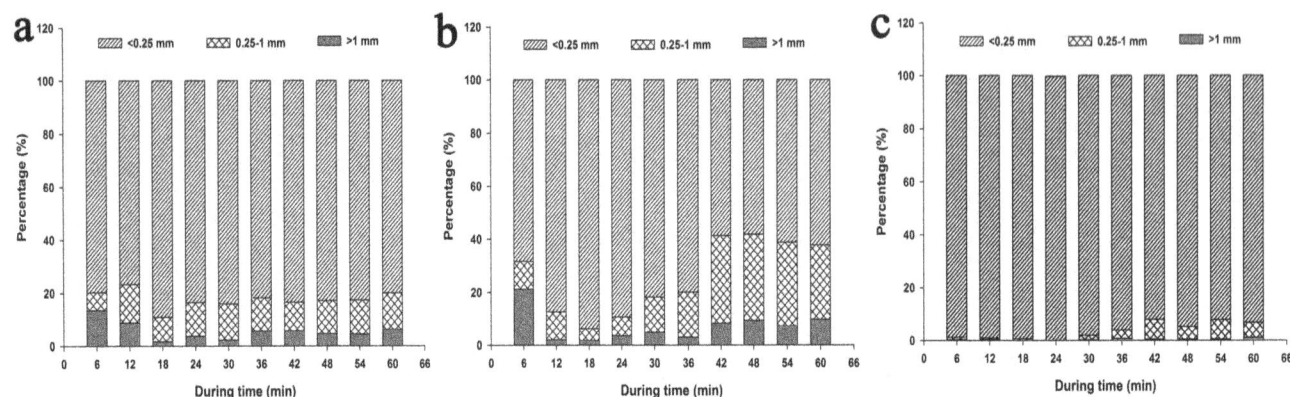

Figure 2. The distribution of different size aggregates in sediment in (a) HI-NT event, (b) LI-NT event, and (c) LI-CT event. HI-NT, high rainfall intensity-no tillage event; LI-NT, low rainfall intensity-no tillage event; LI-CT, low rainfall intensity– conventional tillage event.

events, the loss rates of SBOC increased with the whole rainfall time. In comparison to SBOC, the loss rates of DOC presented various trends. For the HI-NT event, the DOC loss rate was relatively stable during the first 30 minutes and then decreased to the end. Meanwhile, the trends of fluctuation and increasing in DOC loss rates were found in LI-NT and LI-CT events, respectively.

The lost SOC (SBOC+DOC), SBOC and DOC in different events decreased in the order: HI-NT>LI-NT>LI-CT (Table 2). For NT plots, HI rainfall event had 3.25 times higher lost SOC than the LI events did. While for the LI events, NT plot had 12.76 times higher lost SOC than CT plot did, despite the CT plot had longer rainfall duration. Further, compared to the lost SOC, it was little and approximate for the DOC loss in each event. However, a considerable of lost SBOC under different rainfall events was observed. The values of SBOC/SOC reached 94% in HI-NT and LI-NT plot, the least was 67.02% in LI-CT. SBOC was the main form of the lost SOC.

3.2.2. ERoc in sediment. In addition to the amount of the lost SOC, the selective migration processes were also studied. OC enrichment ratio (ERoc) in sediment for each event is presented in Fig. 3. ERoc curves for all the events were >1 except one value (0.96) in HI-NT event (48 min). For the events of LI-NT and LI-CT, ERoc curves had similar shapes, for example, the increasing stage (18–36 min), decline stage (36–54 min) and the peak value (36 min) were occurred at the same time. However, for HI-NT event, the ERoc curves decreased rapidly in fluctuation, and an exponential relationship (ER = 1.01+1.11 exp(−0.058t), R^2 = 0.73) between ERoc and duration time was found. Under LI events, the tillage practices (NT and CT) had a moderate influence on ERoc, while in NT plots, the rainfall intensity (HI and LI) had a great impact on ERoc.

3.3. Correlation analysis
3.3.1. The correlations of SOC and sediment and runoff. As the direct or indirect carrier of the lost SOC, runoff and sediment are important factors impact on the transportation of SOC. Fig. 4 displays the correlations between sediment yield, runoff volume and lost SOC for different rainfall events. Significant positive correlations ($P<0.05$) between sediment yield and lost SOC were observed in all rainfall events. First of all, the amount of the lost SOC increased with the increasing of the eroded sediment. Secondly, the linear correlations decreased with the increasing of the sediment yield (the correlation coefficient r decrease in the order: LI-CT>LI-NT>HI-NT). However, there

was not significant linear relationship ($P>0.05$) between lost SOC and runoff volume for all the rainfall events except in LI-CT event. In LI-CT event, the runoff volume was very low, the lost SOC increased with runoff volume. While in LI-NT and HI-NT events, runoff increased rapidly, and finally got into a stable state.

3.3.2. The correlations between ERoc and different size particles. Sediment particles transportation is often considered to be a cause of the selective migration of SOC. The correlations of sediment particles (different size aggregates (non-disperd particles) and sediment particle size distribution) and ERoc were analyzed (Table 3). The result shows that the correlations varied with rainfall events. In HI-NT event, ERoc had significant positive correlations with the content of clay ($P = 0.011$) and >1 mm aggregate ($P = 0.042$). Meanwhile, a significant negative correlation between ERoc and sand ($P = 0.041$) was observed. However, for the events of LI-NT and LI-CT, there were not significant correlations between ERoc and soil particles, neither sediment aggregates nor sediment particle size distribution.

Discussion

4.1. SOC loss
The eroded carbon in all events was found to be mainly in form of SBOC (more than 67%, and as high as 90% in NT events). Similar results were obtained by Lowrance and Williams [30], who found that up to 90% of SOC in runoff may be in particulate phase. This result could be mainly explained by the distribution of different forms carbon in plot soil. In this study, the content of the DOC in original soil was 0.4% of the SOC, this means that most of the SOC is insoluble in water. Thus the original source of SBOC would be greatly guaranteed. Furthermore, the erosion intensity was also an important factor impact on the composition of the lost SOC. As showed in table 2, both the lost DOC and SBOC increased with erosion intensity, but the SBOC growth rate was much higher than DOC. This indicates that the more SOC lost, the higher proportion of SBOC in lost SOC.

The correlation analysis indicated that the lost SOC was significantly correlated to the eroded sediment ($P<0.05$), while not always correlated to runoff. This result showed that the lost SOC was more close to sediment, as well as the lost SOC was mainly in form of SBOC (Table 2), which indicated that sediment was the direct and main carrier of the lost SOC. While, runoff, as the limiting condition of sediment transport and detachment [26], did not affect the SOC transportation directly. This is consistent with the study result that nutrient loss during soil erosion was not

Table 2. The regular patterns of SOC transportation during water erosion.

sample number	HI-NT (1.33 mm min⁻¹), TSR: 1'31"				LI-NT (0.53 mm min⁻¹), TSR: 2'31"				LI-CT (0.53 mm min⁻¹), TSR:48'0"			
	SBOCR (g min⁻¹)	CSBOC (g)	DOCR (g min⁻¹)	CDOC (g)	SBOCR (g min⁻¹)	CSBOC (g)	DOCR (g min⁻¹)	CDOC (g)	SBOCR (g min⁻¹)	CSBOC (g)	DOCR (g min⁻¹)	CDOC (g)
1	3.41	20.46	0.08	0.47	0.61	3.64	0.01	0.03	0.01	0.09	0.02	0.11
2	11.34	88.47	0.08	0.97	1.47	12.46	0.02	0.13	0.03	0.27	0.02	0.22
3	5.88	123.76	0.08	1.45	1.30	20.25	0.05	0.44	0.05	0.58	0.03	0.41
4	6.02	159.88	0.09	1.98	1.40	28.67	0.03	0.60	0.06	0.94	0.04	0.65
5	6.60	199.47	0.07	2.38	1.51	37.76	0.01	0.65	0.06	1.31	0.04	0.91
6	4.29	225.20	0.05	2.70	1.55	47.04	0.01	0.70	0.08	1.78	0.04	1.17
7	5.20	256.42	0.20	3.91	1.90	58.41	0.29	2.42	0.14	2.59	0.06	1.51
8	3.86	279.56	0.32	5.82	2.40	72.81	0.29	4.15	0.18	3.66	0.07	1.90
9	5.35	311.65	0.25	7.31	3.05	91.13	0.01	4.22	0.17	4.66	0.08	2.36
10	5.47	344.45	0.06	7.70	1.93	102.73	0.24	5.62	0.17	5.69	0.07	2.80

The first sample was collected at the time of runoff began, and samples were collected at 6 minutes intervals. SBOCR, soil-bound organic carbon loss rate; CSBOC, cumulative soil-bound organic carbon; DOCR, dissolved organic carbon loss rate; CDOC, cumulative dissolved organic carbon.

directly related to runoff volume [23]. Consequently, it is appropriate to study the lost SOC through the study of sediment and sediment bound organic carbon.

Arnaez et al. [31] found that runoff increased linearly with rainfall intensity resulting in soil losses that also increased with rainfall intensity. Further, Zhang et al. [32] suggested that the amounts of eroded SOC were found to be strongly influenced by rainfall intensity. This study showed that more SOC was lost in HI-NT event in comparison to LI-NT and the amount of the lost SOC was significantly associated with sediment ($P<0.05$) (Fig. 4). In fact, the high rainfall intensity made seal formation and ponding time become shorter (Table 2), therefore, the infiltrated rainfall decreased quickly, meanwhile, runoff volume and velocity increased. In this way, rill erosion could become more intense and more soil and nutrients will be lost under high rainfall intensity event [13]. Consequently, higher rainfall intensity leads to great amount sediment eroded which result in more SOC loss.

While the amount of the lost SOC in LI-CT event can be ignored in comparison to that in LI-NT event. There are big difference between this result with the view that CT degrades soil structure and loosens soil surface which can accelerate soil erosion and SOC loss [21,22,33,34]. Low amount of runoff and sediment yield were considered to be the reason of the negligible lost SOC in LI-CT event. Researchers suggested that under CT condition, the amount of macropores increases, infiltration is improved and the water storage capacity of soil becomes larger, also roughness of the field surface is reported to decrease runoff velocity [22,35–37]. As a result, most of the rain water infiltrated to ground, and erosive power became very low. In general, CT changed the underlying surface properties and prevented soil loss. This result is consistent with that in the study of Cogle et al. [23], in which SOC loss in tillage areas was consistently less than that in NT areas. However, they considered that this benefit is of limited temporal value and not persistent. It is considered that the benefit occurred immediately post tillage, and before the soil had crusted [38]. Therefore, through the study, we believe that CT can make the underlying surface becomes rough and then reduce the loss of SOC, but this is limited by rainfall conditions. And we tend to attribute this result to the short rainfall time and low rainfall intensity used in our experiments. Therefore, more studies which were conducted with higher rainfall intensity and long-term monitor on CT plot should be taken in future. Nevertheless, the result is important for soil and water conservation in the areas which suffer from frequent short rains, for example the central southern China.

4.2. The selective transportation of SOC

Massey and Jackson [39] suggested that the ERoc often reflect the selectivity of OC. The study result showed that the ERoc values were >1 and indicated that the SOC could be transported preferentially. As the main carrier of the lost SOC, the eroded sediment also showed selectivity in migration process during erosion. The study result, the proportion of aggregates decreased with the increasing of aggregate size, indicated that the finer particles were preferential transported. Red soils in subtropical China are low in exchangeable sodium potential [40], and the main mechanisms of aggregate breakdown were by slaking due to fast wetting, mechanical breakdown due to raindrop impact and by runoff shear stress [41,42]. With the impact of rainfall kinetic energy [43], stream power [42] on the aggregate breakdown and transportation, a sorting process in sediment (both sediment size and density) transport was found [42]. Due to nonhomogeneous distribution of nutrients among particles of various sizes and density [10], the selective migration of particles was considered to

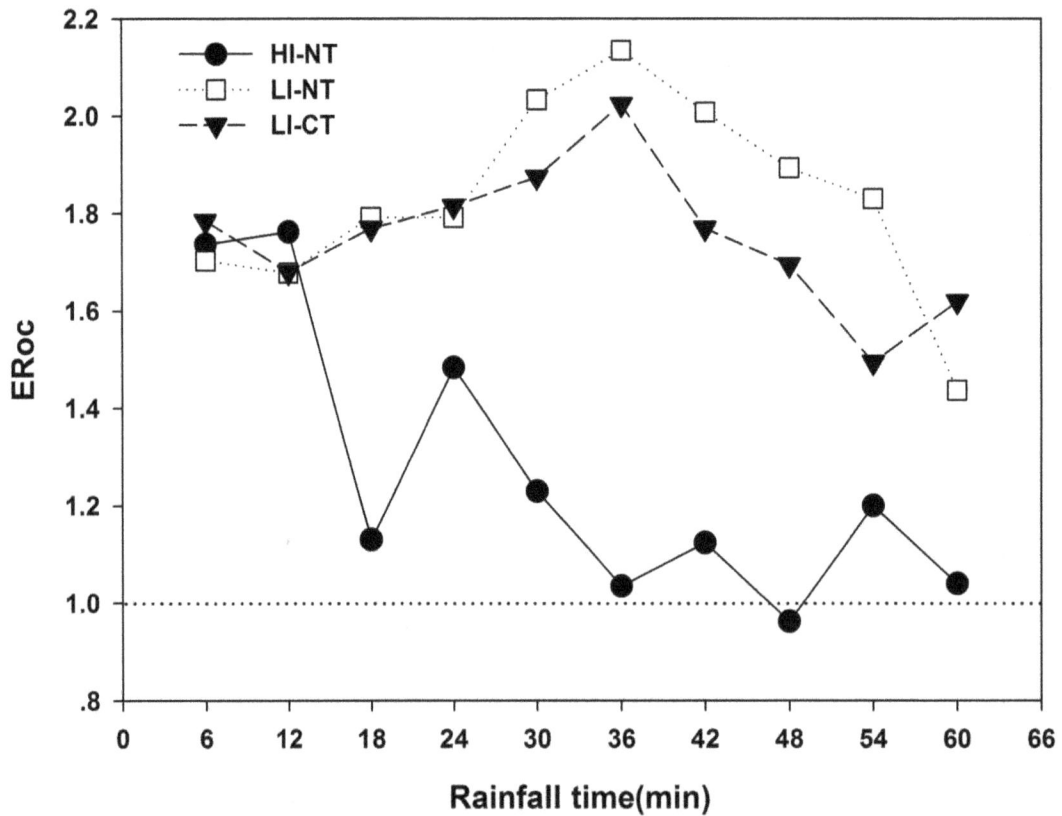

Figure 3. Dynamics of ERoc in sediment for different rainfall events. ERoc, enrichment ratio of organic carbon in sediment. HI-NT, high rainfall intensity-no tillage event; LI-NT, low rainfall intensity-no tillage event; LI-CT, low rainfall intensity-conventional tillage event.

be one of the reasons of the enrichment of OC in sediment [44]. Martinez-Mena et al. [25] suggested that the selectivity has been partly attributed to the transported of fine-sized sediments which are richer in silt and clay particles. However, this study found that the correlations between ERoc and different size particles depend on rainfall events. Significant correlations between ERoc and

sediment particles can be only found in HI-NT event. The large amount of the lost SOC (most of C was associate with the sediment particles) may help explain those significant correlations. While in LI-CT and LI-NT events, the ERoc had not any significant correlation with soil particles or sediment particle size distribution (Table 3). Jacinthe et al. [45] showed sediment collected during

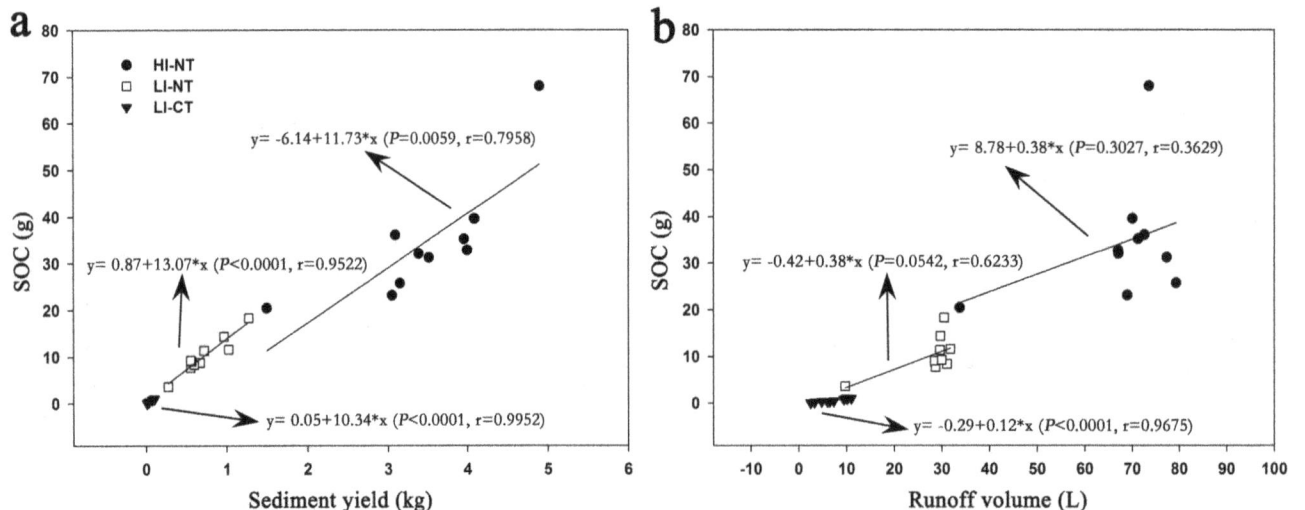

Figure 4. Correlations between the lost SOC and (a) sediment yield, (b) runoff volume. HI-NT, high rainfall intensity-no tillage event; LI-NT, low rainfall intensity-no tillage event; LI-CT, low rainfall intensity-conventional tillage event.

Table 3. Correlation analysis of ERoc and different size particles in sediment.

ERoc	Non-disperd particles			particle-size distribution		
	>1 mm	0.25–1 mm	<0.25 mm	sand	silt	clay
HI-NT	0.649*	−0.224	−0.523	−0.652*	0.236	0.760*
LI-NT	−0.343	0.105	−0.064	0.019	0.213	−0.088
LI-CT	−0.191	−0.468	0.462	–	–	–

Due to the small amount of sediment in LI-CT event, there were not enough samples to measure the particle-size distribution, and then default values were produced.
HI-NT, high rainfall intensity-no tillage event; LI-NT, low rainfall intensity-no tillage event; LI-CT, low rainfall intensity-conventional tillage event.
*Significant at 0.05 level.

the low-intensity storms contained more mineralized carbon (30–40% of sediment carbon) than materials displaced during the high-intensity summer storms. And the preferentially transportation of poorly decomposed non-cohesive plant fragments are often attributed to the higher ERoc in sediment in low rainfall events [46].

Researchers suggested that sediment will become less enriched in carbon as time passes during an event since the more carbon-rich fine aggregates are depleted early in the event [25]. Whereas the increasing transport capacity of runoff is also considered to be related to the decreasing of ERoc [47,48]. As a result, an exponential relationship in character (ERoc = 1.18+0.76exp(−0.046t), $R^2 = 0.81$) between ERoc and rainfall duration (t) was found by Polyakov and Lal [10]. The same result of ERoc dynamics was observed in HI-NT event. However, for LI events, despite with increasing erosive power and sediment yield, an increasing trend of the ERoc was found during 12~36 min. This suggests that the easiest transport substance were not the particles which have the highest OC concentration in LI events. Therefore, rainfall intensity had great impact on the ERoc in sediment and different enrichment mechanisms accounted for different ERoc dynamics. However, in contrast to rainfall intensity, tillage practices had a smaller impact on ERoc. For LI-NT and LI-CT events, ERoc curves had similar shapes. And the ERoc was almost the same in LI-CT and LI-NT plot when the erosive power of runoff was limited, while the difference could be observed when the erosive power growth at different degrees. Our results support the view that tillage affected the sediment yields but did not directly influence the ERoc [23]. Nevertheless, the impact of tillage on the ERoc mechanism is still unclear and further study is required.

Conclusions

The regular patterns of SOC transportation during soil erosion processes were studied through field simulated rainfall experiments. These experiments showed that the LI-NT event had 12.76 times higher lost SOC than LI-CT event did. Conventional tillage can increase rainwater infiltration and reduce soil erosion and SOC loss. However, these results were obtained under short-term and low rainfall intensity event, and more experiments at long-term and storm events are needed to confirm these results. It was also found that SOC as well as sediment particles presented selective in erosion processes, and these processes were affected by rainfall intensity but not tillage practice. The selective transportation of finer particles was not always the reason for the enrichment of OC in sediment. The specific enrichment mechanisms of OC in sediment in relation to different erosion processes have to be studied in future. In addition, due to the complexity and uncertainties of field rainfall experiments, more similar field experiments should be carried out in the future.

Acknowledgments

We would like to thank Shuguang Wang, Guiping Liu and Kunjun Li of Soil and Water Conservation Monitoring Station of Shaoyang for the help of providing the study area and facilities for field work.

Author Contributions

Conceived and designed the experiments: ZWL XDN GMZ. Performed the experiments: ZWL XDN JQH WMM BH. Analyzed the data: XDN YZ YBH. Contributed reagents/materials/analysis tools: XDN. Contributed to the writing of the manuscript: XDN ZWL.

References

1. Kuhn NJ, van Oost K, Cammeraat E (2012) Soil erosion, sedimentation and the carbon cycle Preface. Catena 94: 1–2.
2. Doetterl S, van Oost K, Six J (2012) Towards constraining the magnitude of global agricultural sediment and soil organic carbon fluxes. Earth Surf Proc Land 37: 642–655.
3. Van Oost K, Six J, Govers G, Quine T, De Gryze S (2008) Soil erosion: A carbon sink or source? Response. Science 319: 1042–1042.
4. Lal R, Pimentel D (2008) Soil erosion: A carbon sink or source? Science 319: 1040–1042.
5. Harden JW, Berhe AA, Torn M, Harte J, Liu S, et al. (2008) Soil erosion: Data say C sink. Science 320: 178–179.
6. Fullen MA, Booth CA (2006) Grass ley set-aside and soil organic matter dynamics on sandy soils in Shropshire, UK. Earth Surf Proc Land 31: 570–578.
7. Dymond JR (2010) Soil erosion in New Zealand is a net sink of CO(2). Earth Surf Proc Land 35: 1763–1772.
8. Denman KL, Brasseur G, Chidthaisong A, Ciais P, Cox PM, et al. (2007) Couplings between changes in the climate system and biogeochemistry. In: Solomon S, Qin D, Manning M, Chen Z, Marquis M, Averyt KB, Tignor M, Miller HL, editors, Climate Change 2007: The Physical Science Basis. Contribution of Working Group I to the Fourth Assessment Report of the Inter-governmental Panel on Climate Change. Cambridge, United Kingdom and New York, NY, USA.: Cambridge University Press.
9. Schreiber J (1999) Nutrient leaching from corn residues under simulated rainfall. J Environ Qual 28: 1864–1870.
10. Polyakov VO, Lal R (2004) Soil erosion and carbon dynamics under simulated rainfall. Soil Sci 169: 590–599.
11. Lal R (2005) Soil erosion and carbon dynamics. Soil Till Res 81: 137–142.
12. Chaplot VA, Le Bissonnais Y (2003) Runoff features for interrill erosion at different rainfall intensities, slope lengths, and gradients in an agricultural loessial hillslope. Soil Sci Soc Am J 67: 844–851.
13. Assouline S, Ben-Hur A (2006) Effects of rainfall intensity and slope gradient on the dynamics of interrill erosion during soil surface sealing. Catena 66: 211–220.
14. Jin K, Cornelis WM, Gabriels D, Baert M, Wu HJ, et al. (2009) Residue cover and rainfall intensity effects on runoff soil organic carbon losses. Catena 78: 81–86.
15. Girmay G, Singh B, Nyssen J, Borrosen T (2009) Runoff and sediment-associated nutrient losses under different land uses in Tigray, Northern Ethiopia. J Hydrol 376: 70–80.
16. Zhou H, Li BG, Lu YH (2009) Micromorphological analysis of soil structure under no tillage management in the black soil zone of Northeast China. J Mt Sci-Engl 6: 173–180.

17. Moebius-Clune BN, van Es HM, Idowu OJ, Schindelbeck RR, Moebius-Clune DJ, et al. (2008) Long-term effects of harvesting maize stover and tillage on soil quality. Soil Sci Soc Am J 72: 960–969.
18. Pinheiro EFM, Pereira MG, Anjos LHC (2004) Aggregate distribution and soil organic matter under different tillage systems for vegetable crops in a Red Latosol from Brazil. Soil Till Res 77: 79–84.
19. Schiettecatte W, Gabriels D, Cornelis WM, Hofman G (2008) Impact of deposition on the enrichment of organic carbon in eroded sediment. Catena 72: 340–347.
20. Lal R, Griffin M, Apt J, Lave L, Morgan MG (2004) Ecology - Managing soil carbon. Science 304: 393–393.
21. Kisic I, Basic F, Nestroy O, Mesic M, Butorac A (2002) Chemical properties of eroded soil material. J Agron Crop Sci 188: 323–334.
22. Puustinen M, Koskiaho J, Peltonen K (2005) Influence of cultivation methods on suspended solids and phosphorus concentrations in surface runoff on clayey sloped fields in boreal climate. Agr Ecosyst Environ 105: 565–579.
23. Cogle A, Rao K, Yule D, Smith G, George P, et al. (2002) Soil management for Alfisols in the semiarid tropics: erosion, enrichment ratios and runoff. Soil Use Manage 18: 10–17.
24. van Noordwijk M, Cerri C, Woomer PL, Nugroho K, Bernoux M (1997) Soil carbon dynamics in the humid tropical forest zone. Geoderma 79: 187–225.
25. Martinez-Mena M, Lopez J, Almagro M, Albaladejo J, Castillo V, et al. (2012) Organic carbon enrichment in sediments: Effects of rainfall characteristics under different land uses in a Mediterranean area. Catena 94: 36–42.
26. Schiettecatte W, Verbist K, Gabriels D (2008) Assessment of detachment and sediment transport capacity of runoff by field experiments on a silt loam soil. Earth Surf Proc Land 33: 1302–1314.
27. Li Z, Huang J, Zeng G, Nie X, Ma W, et al. (2013) Effect of Erosion on Productivity in Subtropical Red Soil Hilly Region: A Multi-Scale Spatio-Temporal Study by Simulated Rainfall. PloS one 8: e77838.
28. Walkley A, Black IA (1934) An examination of the Degtjareff method for determining soil organic matter, and a proposed modification of the chromic acid titration method. Soil Sci 37: 29–38.
29. Gee GW, Bauder JW (1986) Particle size analysis. In: Klute A, editor, Methods of Soil Analysis (2nd ed.). Am. Soc. Agron., Madison, 383–411.
30. Lowrance R, Williams RG (1988) Carbon movement in runoff and erosion under simulated rainfall conditions. Soil Sci Soc Am J 52: 1445–1448.
31. Arnaez J, Lasanta T, Ruiz-Flano P, Ortigosa L (2007) Factors affecting runoff and erosion under simulated rainfall in Mediterranean vineyards. Soil Till Res 93: 324–334.
32. Zhang X, Li Z, Tang Z, Zeng G, Huang J, et al. (2013) Effects of water erosion on the redistribution of soil organic carbon in the hilly red soil region of southern China. Geomorphology 197: 137–144.
33. West TO, Post WM (2002) Soil organic carbon sequestration rates by tillage and crop rotation: A global data analysis. Soil Sci Soc Am J 66: 1930–1946.
34. Sa JCD, Cerri CC, Dick WA, Lal R, Venske SP, et al. (2001) Organic matter dynamics and carbon sequestration rates for a tillage chronosequence in a Brazilian Oxisol. Soil Sci Soc Am J 65: 1486–1499.
35. Alakukku L (1998) Properties of compacted fine-textured soils as affected by crop rotation and reduced tillage. Soil Till Res 47: 83–89.
36. Schnug E, Haneklaus S (2002) Agricultural production technique and infiltration significance of organic farming for preventive flood protection. Landbauforsch Volk 52: 197–203.
37. Gowda P, Mulla D, Dalzell B (2003) Examining the targeting of conservation tillage practices to steep vs. flat landscapes in the Minnesota River Basin. J Soil Water Conserv 58: 53–57.
38. Yule D, Cogle A, Smith G, Rao K, George P (1991) Soil management of Alfisols for water conservation and utilization. J Indian Water Resour Soc 11: 10–13.
39. Massey H, Jackson M (1952) Selective erosion of soil fertility constituents. Soil Sci Soc Am J 16: 353–356.
40. Shi ZH, Yan FL, Li L, Li ZX, Cai CF (2010) Interrill erosion from disturbed and undisturbed samples in relation to topsoil aggregate stability in red soils from subtropical China. Catena 81: 240–248.
41. Li ZX, Cai CF, Shi ZH, Wang TW (2005) Aggregate stability and its relationship with some chemical properties of red soils in subtropical China. Pedosphere 15: 129–136.
42. Shi ZH, Fang NF, Wu FZ, Wang L, Yue BJ, et al. (2012) Soil erosion processes and sediment sorting associated with transport mechanisms on steep slopes. J Hydrol 454: 123–130.
43. Wang L, Shi ZH, Wang J, Fang NF, Wu GL, et al. (2014) Rainfall kinetic energy controlling erosion processes and sediment sorting on steep hillslopes: A case study of clay loam soil from the Loess Plateau, China. J Hydrol 512: 168–176.
44. Palis R, Ghandiri H, Rose C, Saffigna P (1997) Soil erosion and nutrient loss. III. Changes in the enrichment ratio of total nitrogen and organic carbon under rainfall detachment and entrainment. Aust J Soil Res 35: 891–905.
45. Jacinthe PA, Lal R, Owens LB, Hothem DL (2004) Transport of labile carbon in runoff as affected by land use and rainfall characteristics. Soil Till Res 77: 111–123.
46. Ghadiri H, Rose C (1991) Sorbed chemical transport in overland flow: II. Enrichment ratio variation with erosion processes. J Environ Qual 20: 634–641.
47. Sharpley A (1985) The Selection Erosion of Plant Nutrients in Runoff. Soil Sci Soc Am J 49: 1527–1534.
48. Weigand S, Schimmack W, Auerswald K (1998) The enrichment of 137Cs in the soil loss from small agricultural watersheds. Zeitschrift für Pflanzenernährung und Bodenkunde 161: 479–484.

20

Comparison of Soil Quality Index Using Three Methods

Atanu Mukherjee*, Rattan Lal

Carbon Management and Sequestration Center, School of Environment and Natural Resources, The Ohio State University, Columbus, Ohio, United States of America

Abstract

Assessment of management-induced changes in soil quality is important to sustaining high crop yield. A large diversity of cultivated soils necessitate identification development of an appropriate soil quality index (SQI) based on relative soil properties and crop yield. Whereas numerous attempts have been made to estimate SQI for major soils across the World, there is no standard method established and thus, a strong need exists for developing a user-friendly and credible SQI through comparison of various available methods. Therefore, the objective of this article is to compare three widely used methods to estimate SQI using the data collected from 72 soil samples from three on-farm study sites in Ohio. Additionally, challenge lies in establishing a correlation between crop yield versus SQI calculated either depth wise or in combination of soil layers as standard methodology is not yet available and was not given much attention to date. Predominant soils of the study included one organic (Mc), and two mineral (CrB, Ko) soils. Three methods used to estimate SQI were: (i) simple additive SQI (SQI-1), (ii) weighted additive SQI (SQI-2), and (iii) statistically modeled SQI (SQI-3) based on principal component analysis (PCA). The SQI varied between treatments and soil types and ranged between 0–0.9 (1 being the maximum SQI). In general, SQIs did not significantly differ at depths under any method suggesting that soil quality did not significantly differ for different depths at the studied sites. Additionally, data indicate that SQI-3 was most strongly correlated with crop yield, the correlation coefficient ranged between 0.74–0.78. All three SQIs were significantly correlated ($r = 0.92$–0.97) to each other and with crop yield ($r = 0.65$–0.79). Separate analyses by crop variety revealed that correlation was low indicating that some key aspects of soil quality related to crop response are important requirements for estimating SQI.

Editor: Upendra M. Sainju, Agricultural Research Service, United States of America

Funding: This research was part of a collaborative project supported by the USDA-NIFA, award number: 2011-68002-30190, "Cropping Systems Coordinated Agricultural Project (CAP): Climate Change, Mitigation, and Adaptation in Corn-based Cropping Systems" (sustainablecorn.org). The funders had no role in study design, data collection and analysis, decision to publish, or preparation of the manuscript.

Competing Interests: The authors have declared that no competing interests exist.

* Email: mukherjee.70@osu.edu

Introduction

A wide range of agricultural soils represents diversely managed arable lands while the main goal to improve soil quality, crop yield, and reduce the ecological foot print. Soil quality is defined as the soil's capacity to function within natural or managed ecosystem boundaries and to sustain plant productivity while reducing soil degradation [1–4]. As soil quality is a complex functional concept and cannot be measured directly in the field or laboratory [5] but can only be inferred from soil characteristics [6], a range of soil parameters or indicators has been identified to estimate soil quality. However, soil quality is often related to the management goal and practices as well to soil characteristics. Thus, a mathematical or statistical framework was put forward in early 1990s to estimate soil quality index (SQI) [1,3,4]. The SQI was assessed so that the management goals are not only focused on productivity per se, which may result in soil degradation [7], but also on environmental issues. Thus, an appropriate SQI may have three component goals: environmental quality, agronomic sustainability, and socio-economic viability [8].

Estimation of SQI is a complex process and difficult task [9], especially when linked with several functional goals. Yet a considerable progress has been made towards estimating SQI across a number of soil types and management practices [8–14]. Most studies indexed soil quality employing only one method, with

a few exceptions [8,15,16]. As computation of SQI is difficult, a strong need exists for developing a user-friendly and credible SQI through comparison of various available methods. Thus, the objective of this study was to compare SQIs computed by three methods which are conceptually different from each other. The study is based on the hypothesis that SQIs computed from three methods have similar relationship with crop yields. Data are scarce on validation of SQI against crop yield as most studies focused on the environmental aspects of the soil as end point variable and that SQI of various soil layers has not been computed [13,16–18]. Nevertheless, there is a challenge in validating SQI against crop yield as SQI computed only from surface soils (0–10 and/or 0–20 cm) may not evaluate realistic relationship between soil quality and crop yield because root system can extend to deeper layers [19]. Thus, the other objective of this study was to compute SQI from multiple depths of the soil to examine its relationship with crop yield.

Materials and Methods

Soil sampling and analyses

Soil samples from the field (on-farm) were collected from Logan county, Ohio that included an organic (Martisco Variant silt loam: Mc, organic parent material, >26% organic matter, 40°25′12.4″N, 83°40′55.9″W), and a mineral soil (Crosby silt loam: CrB: *Fine, mixed, active, mesic Aeric Epiaqualfs*, sand: 23%, silt: 35%, clay: 42%, 40°24′52.4″N, 83°39′21.9″W), and from Franklin county, Ohio that included a Kokomo silty clay loam (Ko: *Fine, mixed, superactive, mesic Typic Argiaquolls*, sand: 9%, silt: 62%, clay: 29%, 40°00′41.2″N, 83°12′23.1″W) in May, 2013. These locations were farmer-owned field sites and before collecting the soils from specific sites permissions of the farmers were obtained from the land-owners or farmers. Initial permission was granted through The Ohio State University's extension managers and thereafter communication was established directly by the primary author. Additionally, no endangered or protected species were involved in the current study sites, and thus no such permission was required from any other regulatory agencies. The properties of the soils and other management details were presented elsewhere [20,21] and briefly presented in the discussion section. Soil samples were obtained from 0–10, 10–20, 20–40, and 40–60 cm depths under no-tillage (NT) and conventional tillage (CT) practices for Mc and CrB soil series and under NT for one year cover crop (CC) and no-cover crop (NCC) for Ko series. The organic soil (Mc) has been under practice of CT corn (*Zea mays*) for 10 years which was compared with the adjacent grassland soil (termed as NT). The mineral soil (CrB) was under NT and recent introduction (one year) of CT corn practices. Both soils received fertilizer (N:P:K as 5.5:26:30) with the rates of 336 (30, 142 and 164 kg ha^{-1} of N, P, and K, respectively) and 280 (25, 118 and 137 kg ha^{-1} of N, P, and K, respectively) kg ha^{-1} for corn, for Mc and CrB soils, respectively. The other site (Ko) was under NT corn and soybean (*Glycin max L.*) annual rotation. A mixture of pea (*Pisum sativum L.*) and turnip (*Brassica rapa L.*) cover crops with the seeding rate of 140 kg ha^{-1} was seeded prior to growing soybean in 2013. Roundup-ready soybean was seeded on 20 April, 2013 and harvested on 30 September, 2013. A row spacing of 38 cm was used during planting for corn and soybean. Soil core (inside diameter: 5.3 cm, height: 5.9 cm) samples were collected manually using soil sampler before planting (7 May, 2013) from four depths (0–10, 10–20, 20–40, and 40–60) for determination of bulk density (BD) by dividing dry mass of soil in the core by core volume [22] and hydrologic properties. Undisturbed soil cores were used to determine the water retention at field capacity (0.033 MPa), while sieved samples (<2-mm size) were used to determine the permanent wilting point (1.5 MPa). The potential available water capacity (AWC) of the soil was calculated as the difference in volumetric water content at 0.033 and 1.5 MPa moisture potentials. The water content of a soil layer was calculated by multiplying thickness of the layer with the volumetric water capacity. A minimum of three field penetration resistance (PR) measurements for soil's mechanical strength were made for each depth using a CP40II cone penetrometer [23]. Bulk samples were obtained to measure the aggregate size distributions and fraction of water stable aggregates (WSA) using the wet sieving method [24]. Five sieves of diameter sizes 4.75, 2, 1, 0.5, and 0.25 mm openings were placed into a Yoder apparatus. Air-dried soils (5–8 mm, 51 g) were slowly wetted by capillarity action by adjusting the water level in the container so that the base of the top sieve just touched the water. The sieve combination was oscillated mechanically in the water at 60 oscillations per minute for 30 min. Aggregates retained in the sieves were transferred to glass beakers and the weight of each of five fractions was measured after drying at 60°C overnight. The data were used to compute WSA, mean weight diameter (MWD), and geometric mean diameter (GMD) [25]. Soil water retention at matrix potentials of 0.033 and 1.5 MPa were measured using a pressure plate apparatus [26]. The pH and electrical conductivity (EC) of soil were determined in 1:2 soil:water ratio slurry using a Thermo-scientific Orion Star

Series pH/Conductivity Meter. Concentrations of soil organic C and total N were determined using an Elemental analyzer (Vario Max, Elementar Americas, Inc., Germany) by the dry combustion method (900°C) after grinding subsamples to 0.25 mm. Total C and N stocks were calculated by multiplying the respective elemental concentrations by BD and the thickness of the soil layer [27].

Soil quality index (SQI) calculations

All the SQI methods involved a set of 72 soil samples and a number of soil quality indicators as parameters. The 13 parameters used for developing SQIs were pH, EC, BD, WSA, GMD, MWD, PR, SOC concentration, N concentration, C-Stock, N-Stock, AWC, and soil water content. As some of these parameters were synthesized and redundant, only nine parameters were chosen omitting MWD, C and N concentrations, and water content from the SQI-1 and SQI-2 calculations to avoid redundancy. However, SQI-3 was not a primarily additive approach and redundancy of the parameters was eliminated through the processes of elimination as prescribed before. Thus, all 13 parameters were initially included in SQI-3 model and only four non-redundant soil parameters with maximum variations in the dataset were finally retained in the model as described in the following section. The average and standard deviation values of these parameters grouped by soil types are presented in Table 1. Under the proposed framework an ideal soil would have SQI value of 1 for the highest quality soil and 0 for the severely degraded soil [1–4].

Simple additive SQI (SQI-1)

Simple additive SQI was estimated following the method outlined by Amacher et al. [28]. In this method, soil parameters were given threshold values based primarily on the literature review and expert opinion of the authors. The threshold levels, interpretations, and associated unitless soil index score values are listed in Table 2. The individual index values were then summed up to obtain a total SQI:

$$\sum SQI = \sum Individual\ soil\ parameter\ index\ values \quad (1)$$

The scaled SQI (SQI-1) of individual soil, was computed by Eq. 2:

$$SQI-1 = (\sum SQI - SQI_{Min})/(SQI_{Max} - SQI_{Min}), \quad (2)$$

whereas, SQI_{Min} = Minimum value of SQI, and SQI_{Max} = Maximum value of SQI from the total dataset.

Weighted additive SQI (SQI-2)

In this approach, each soil parameter was first assigned unitless score ranging from o to 1 by employing linear scoring functions [8]. Non-linear scoring functions were avoided because of their lower capacity of predicting the end point variable or crop yield [8]. Soil parameters were divided into groups based on three mathematical algorithm functions: (a) 'more is better' (e.g., WSA, GMD, C-Stock, N-Stock, and AWC) (b) 'less is better' (e.g., BD, PR), and (c) 'optimum' (e.g., pH and EC). 'Optimum' properties are those which have positive influence up to a certain level beyond which the influence could be considered detrimental [13]. For 'more is better' parameters, each observation was divided by the highest observed value of the entire dataset so that the highest observed value would have a score of 1; for 'less is better'

Table 1. Descriptive statistics of all the soil indicators collected from four depths under three soil types used to estimate SQI.

Ko

	0–10 cm		10–20 cm		20–40 cm		40–60 cm	
	Mean	SD	Mean	SD	Mean	SD	Mean	SD
pH	6.0	0.5	6.2	0.6	6.4	0.6	6.4	0.6
EC (μs cm^{-1})	190.6	151.1	165.6	84.2	151.7	71.8	175.9	92.2
BD (Mg m^{-3})	1.4	0.2	1.5	0.1	1.6	0.1	1.6	0.1
WSA %	73.1	14.8	79.5	12.0	73.3	22.5	84.7	5.6
GMD (mm)	1.4	0.4	1.6	0.4	1.4	0.3	1.5	0.3
MWD (mm)	2.6	1.7	3.1	1.8	2.3	1.7	3.4	1.1
PR (Mpa)	2.2	0.3	2.4	0.2	2.3	0.2	2.2	0.4
N (%)	0.2	0.0	0.2	0.0	0.2	0.0	0.2	0.0
SOC (%)	2.4	0.2	2.1	0.2	2.0	0.2	2.0	0.3
N-Stock (Mg/ha)	3.5	0.5	3.0	0.3	6.4	0.6	6.4	1.1
C-Stock (Mg/ha)	33.7	3.7	30.6	2.8	62.5	6.8	63.7	7.2
AWC (%)	38.5	4.1	36.7	2.1	35.8	6.4	39.9	5.8
Water content (cm)	3.8	0.4	3.7	0.2	7.2	1.3	7.9	1.0

CrB

	0–10 cm		10–20 cm		20–40 cm		40–60 cm	
	Mean	SD	Mean	SD	Mean	SD	Mean	SD
pH	7.4	0.2	7.4	0.2	7.4	0.2	7.4	0.1
EC (μs cm^{-1})	250.3	44.6	217.8	26.8	212.3	31.8	223.5	45.0
BD (Mg m^{-3})	1.5	0.2	1.5	0.1	1.5	0.3	1.5	0.3
WSA %	82.4	4.4	76.4	9.7	76.5	9.5	76.5	17.7
GMD (mm)	1.2	0.1	1.1	0.2	1.1	0.2	1.2	0.3
MWD (mm)	2.0	0.5	1.7	0.8	1.5	0.7	2.0	1.1
PR (Mpa)	2.4	0.2	2.5	0.5	2.6	0.2	2.9	0.7
N (%)	0.2	0.0	0.2	0.0	0.2	0.0	0.2	0.0
SOC (%)	2.4	0.3	2.1	0.4	1.8	0.4	1.8	0.5
N-Stock (Mg/ha)	3.2	0.2	3.1	0.3	5.2	1.6	5.1	2.2
C-Stock (Mg/ha)	34.3	3.2	32.2	4.1	53.8	16.8	54.6	24.4
AWC (%)	17.8	9.0	10.1	6.8	10.7	8.2	9.0	3.2
Water content (cm)	1.8	0.9	1.0	0.7	2.1	1.6	1.8	0.6

Mc

Table 1. Cont.

Ko

	0–10 cm		10–20 cm		20–40 cm		40–60 cm	
	Mean	SD	Mean	SD	Mean	SD	Mean	SD
pH	7.4	0.1	7.4	0.1	7.4	0.1	7.5	0.1
EC (μs cm^{-1})	616.3	40.1	570.0	92.8	633.2	126.4	599.2	129.2
BD (Mg m^{-3})	0.6	0.0	0.7	0.0	0.6	0.1	0.6	0.1
WSA %	91.2	0.5	91.0	2.5	89.7	2.5	82.5	8.7
GMD (mm)	2.1	0.0	2.2	0.0	2.2	0.0	2.2	0.1
MWD (mm)	5.5	0.1	5.5	0.2	5.7	0.4	5.0	0.5
PR (Mpa)	1.3	0.1	1.5	0.3	1.6	0.3	1.9	0.1
N (%)	1.0	0.1	0.9	0.1	0.9	0.1	1.0	0.1
SOC (%)	15.1	1.1	15.0	0.8	14.9	0.9	15.9	0.6
N-Stock (Mg/ha)	6.1	0.8	6.3	0.7	11.2	1.7	12.6	1.9
C-Stock (Mg/ha)	95.7	11.7	102.8	8.8	186.6	36.2	196.6	28.0
AWC (%)	248.3	151.3	84.8	42.4	48.2	13.8	36.1	22.1
Water content (cm)	24.8	15.1	8.5	4.2	9.6	2.8	7.2	4.4

Abbreviations: SD: standard deviation, EC: electrical conductivity, BD: bulk density, WSA: water stable aggregates, GMD: geometrical mean diameter, MWD: mean weight diameter, PR: penetration resistance, N: nitrogen, SOC: soil organic carbon, AWC: available water capacity.

Table 2. Soil indicators, threshold values, interpretations and scores.

Indicators	Range	Interpretation	Score	Reference
pH	5.5–7.2	Slightly acidic to neutral: Optimum for plant growth	2	[28]
	>7.2<8.0	Slightly to moderately alkaline: Preferred by some plants, possible P and some metal deficiencies	1	
EC (us/cm)	<200	Low salt level	0	[42]
	200–500	Optimum salt level for plants	1	
	>500	High salt level, adverse effect likely	0	
BD (Mg/m³)	<1.0	High organic soil, supports plant roots	2	[28,45–47]
	1.0–1.5	Adverse effects unlikely	1	
	>1.5	Adverse effects likely	0	
WSA (%)	<50	Infiltration and soil erosion problems likely	0	[30,42]
	50–70	Moderate constraints	1	
	70–90	Good soil	2	
	>90	Excellent soil	3	
GMD (mm)	<1.0	Infiltration and soil erosion problems likely	0	[30,42]
	1–2	Moderate limitations	1	
	>2	No limitation	2	
MWD (mm)	<1.0	Infiltration and soil erosion problems likely	0	[30,42]
	1–2	Moderate limitations	1	
	2–5	Slight limitations	2	
	>5.0	No limitation	3	
PR (Mpa)	1–2	Adverse effect on plant root unlikely	2	[42,48]
	2–3	Moderate adverse effect on plant root	1	
	>3.0	Severe adverse effect on plant roots	0	
N (%)	0.2–0.3	Moderate limitation	1	[41,42]
	>0.3	Slight to no limitation	2	
SOC (%)	2–3	Moderate limitation	1	[41,42]
	>3.0	Slight to no limitation	2	
N-Stock (Mg/ha)	<5.0	N deficient	1	Authors' opinion
	5–10	Moderate to optimum N level	2	
	>10.0	N-rich soil	3	
C-Stock (Mg/ha)	<50.0	C deficient	1	Authors' opinion
	50–100	Moderate to optimum C level	2	
	>100	C-rich soil	3	
AWC (%)	<20	Water-stress to plants	0	[31,42]
	20–50	Moderate water availability	1	
	>50	Good water capacity for plants	2	
Water content(cm)	<5.0	Water-stress to plants	0	Authors' opinion
	5–10	Moderate water availability	1	
	>10	Good water capacity for plants	2	

Abbreviations are same as Table 1.

parameters, the lowest observed value in the entire dataset was divided by each observation so that the lowest observed value received a score of 1; and 'optimum' parameters were scored up to a threshold value as 'more is better', and thereafter above the threshold values were scored as 'less is better' [8,29]. For example, pHs up to 5.5–7.2 were scored as 'more is better', and pHs >7.2 were scored as 'less is better'.

After normalizing soil parameters, the scores were integrated into a single index value for each soil using a weighted additive approach initially suggested by Karlen and Stott [4], but modified later by Fernandes et al. [13]. The following weighted additive function was used for development of SQI-2 (Eq. 3):

$$SQI - 2 = [(\text{Weight1}) * \text{RDC}] + [(\text{Weight2}) * \text{WSC}] + [(\text{Weight3}) * \text{NSC}], \tag{3}$$

where, RDC (root development capacity) is the rating for the soil's ability to allow plant root development, WSC (water storage capacity) is the rating for the soil's ability to store water, NSC

Table 3. Model of SQI-2.

Soil function	Weight	Soil Indicators	Sub-weight	Scaled score	B×C	∑B×C	D×A	%	SQI
	A		B	C		D			
RDC	0.4	BD	0.35	0.41	0.14	0.46	0.18	52.94	0.34
		PR	0.35	0.52	0.18				
		WSA	0.20	0.54	0.11				
		GMD	0.10	0.34	0.03				
WSC	0.2	AWC	1.00	0.06	0.06	0.06	0.01	2.94	
NSC	0.4	pH	0.30	0.79	0.24	0.38	0.15	44.12	
		EC	0.30	0.25	0.08				
		C-stock	0.20	0.11	0.02				
		N-stock	0.20	0.18	0.04				

Abbreviations: RDC: root development capacity, WSC: water storage capacity, NSC: nutrient storage capacity; all other abbreviations are same as Table 1.

(nutrient supply capacity) is the rating for the soil's ability to supply nutrients. Weight 1, 2 and 3 are the respective numerical weights for each soil function (RDC, WSC, and NSC). The numerical weights were assigned to each soil function according to their importance in fulfilling the management goal(s) of maintaining soil quality. While the summation of all these numerical weights (Weight 1, 2 and 3) are supposed to be 1 and distributed evenly as 0.33, 0.33, and 0.34 [4], however, a modified approach was taken for the current study following Fernandes et al. [13], where lower weightage was given to the functional attribute which had lower number of representative indicators in the model. Similar to Fernandes et al. [13] weight values were arbitrarily chosen for RDC, WSC, and NSC as 0.4, 0.2, and 0.4 in the current study. The WSC received lower weight value than others because of low number of representative indicators (only AWC) [13] among all included parameters.

The soil parameters or indicators selected for (i) RDC were BD, PR, WSA, and GMD, (ii) WSC was AWC and (iii) NSC were pH, EC, C-Stock, and N-Stock. Within this network, the sub-weight values were given to each indicator based on their importance under the particular soil functional property, field versus laboratory measurements and scope of redundancy. The sub-weight values of different soil indicators or parameters were summed up to 1 under each soil functional property [13]. The rank of the subweight values were as follows: field measurement indicators (BD, PR) > laboratory measurement indicators (WSA, AWC, pH, EC) >> synthesized parameters (GMD, C-Stock, and N-Stock). Field indicators were given the maximum weightage as these were the most representative of the soil's natural conditions and synthesized parameters received the lowest weightage to avoid data redundancy in the model. Thereafter, scaled SQI (SQI-2) of individual soil was computed by Eq. 2. An example of this model of development of SQI is shown in Table 3.

Statistically modeled SQI (SQI-3)

A statistics-based model was used to estimate SQI using principal component analysis (PCA) [8,12,14,18,30,31]. The PCA-model is used to create a minimum data set (MDS) to reduce the indicator load in the model and avoid data redundancy [8]. The main difference between the first two versus the PCA method is that the first two rely mainly on subjective expert opinion and literature review, while the PCA method is more objective of using a number of statistical tools (multiple correlation, factor and cluster analyses) which could avoid any biasness and data redundancy by choosing an MDS using mathematical formulae [8,32]. A relative advantage and disadvantage of each SQI method is explained in the later section. The preliminary function of PCA is to reduce the dimensionality of the entire data set consisting of a large number of interrelated variables, while retaining as much as possible of the variations present in the data set. This is achieved by transformation to a new set of variables, the principal components (PCs), which are uncorrelated, and ordered so that the first few retain most of the variation present in all of the original variables [33,34]. In other words, the PCA method was chosen as a data reduction tool to select the most appropriate indicator(s) to represent and estimate SQI [18].

All the original observations (untransformed data) of each soil were included in the PCA model using SPSS, version 21.0 [35]. The PCs with high eigenvalues represented the maximum variation in the dataset [8,12,14,18,30,31]. While most studies have assumed to examine PCs with eigenvalue >1.0 following Kaiser [36], the present study had the third and fourth components with eigenvalues of 0.98 and 0.89 and variances >5% (Table 4, Fig. 1). These were examined with first two PCs (eigenvalues 8.23, and 1.67) as

Table 4. Results of principal component analyses (PCA).

	PC-1	PC-2	PC-3	PC-4
Eigenvalues	8.23	1.67	0.98	0.89
% Variance	63.34	12.87	7.56	6.82
Cumulative variance	63.3	76.2	83.8	90.6
Eigen vectors or factor loading				
pH	0.423	0.351	−0.378	**_0.711_**
EC	**0.925**	0.113	−0.153	0.147
BD	**−0.881**	−0.034	0.212	0.002
WSA	0.590	−0.020	**_0.669_**	0.343
GMD	**0.927**	0.024	0.276	−0.024
MWD	**0.904**	0.001	0.342	0.046
PR	−0.770	0.268	−0.131	0.092
SOC	**_0.975_**	0.073	−0.109	−0.025
N	**0.972**	0.050	−0.103	−0.053
C-Stock	0.831	0.412	−0.107	−0.268
N-Stock	0.722	0.440	−0.067	−0.385
AWC	0.555	**−0.782**	−0.189	0.057
Water content	0.624	−0.693	−0.199	−0.066

Abbreviations are same as Table 1; PC: principal component; bold values under each component are highly weighted and underlined bold values are selected in minimum data set.

prescribed for the cases of fewer than three components with eigenvalue >1.0 [8,37]. Under a given PC, each variable had corresponding eigenvector weight value or factor loading (Table 2). Only the 'highly weighted' variables were retained to include in the MDS (Table 2). The 'highly weighted' variables were defined as the highest weighted variable under a certain PC and absolute factor loading value within 10% of the highest values under the same PC [37]. Thus, the bold-face values (Table 4: EC, BD, GMD, MWD, SOC, and N for PC-1, AWC for PC-2, WSA for PC-3, and pH for PC-4) were considered highly weighted eigenvectors and therefore were initially selected in the MDS. However, when more than one variable was retained under a particular PC, multivariate correlation matrix (Table 5) was used to determine the correlation coefficients between the parameters [8,12]. If the parameters were significantly correlated (r>0.60, $p<0.05$), then the one with the highest loading factor was retained in the MDS and all others were eliminated from the MDS to avoid redundancy. Following this procedure, except SOC, all other eigenvectors from PC-1 were eliminated from MDS due to high and significant correlation between each other (Table 5). The non-correlated parameters under a particular PC were considered important and retained in the MDS [10,12]. All the bold-faced and underlined soil parameters in Table 4 were selected in the final MDS.

After selection of parameters for the MDS, all selected observations were transformed using linear scoring functions (less is better, more is better and optimum) as described in SQI-2 method. Once the selected observations were transformed in numerical scores (ranged 0–1), a weighted additive approach was used to integrate them into indices for each soil [8,12]. Each PC explained a certain amount of variation in the dataset (Table 4), which was divided by the maximum total variation of the all PCs selected for the MDS to get a certain weightage value under a

particular PC [8]. For example, the % variance (63.3) was divided by total cumulative variance (90.6) to obtain the weight value of 0.7 for PC-1 (Table 4). Thereafter, the weighted additive SQI was computed using Eq. 4:

$$SQI - 3\,(PCA) = \sum Weight * Individual\ oil\ parameter\ score \quad (4)$$

Validation of SQI

The SQIs estimated from three different methods were validated against soybean (*Glycin max*) and corn (*Zea mays*) yield data of that particular year of collection of soil and crops by computing correlation coefficients [8]. The SQI values were also compared through pearson correlation to understand the effectiveness of each other.

Statistical analyses

All values are presented as means ± standard deviations of three field or laboratory measurements. Significant differences between treatments were analyzed using Tukey's test in PROC GLM in SAS version 9.2 [38]. Treatment differences were deemed significant at $p<0.05$. The PCA was performed in SPSS version 21 [35]. Descriptive statistics and linear regressions were computed in Microsoft Excel [39] and all the figures were obtained using Sigmaplot Version 12.0 [40].

Results and Discussions

Most data of three farm sites have been presented and the detailed discussions of soil type, crop and management effects on soil properties are discussed elsewhere [20,21]. The mean and

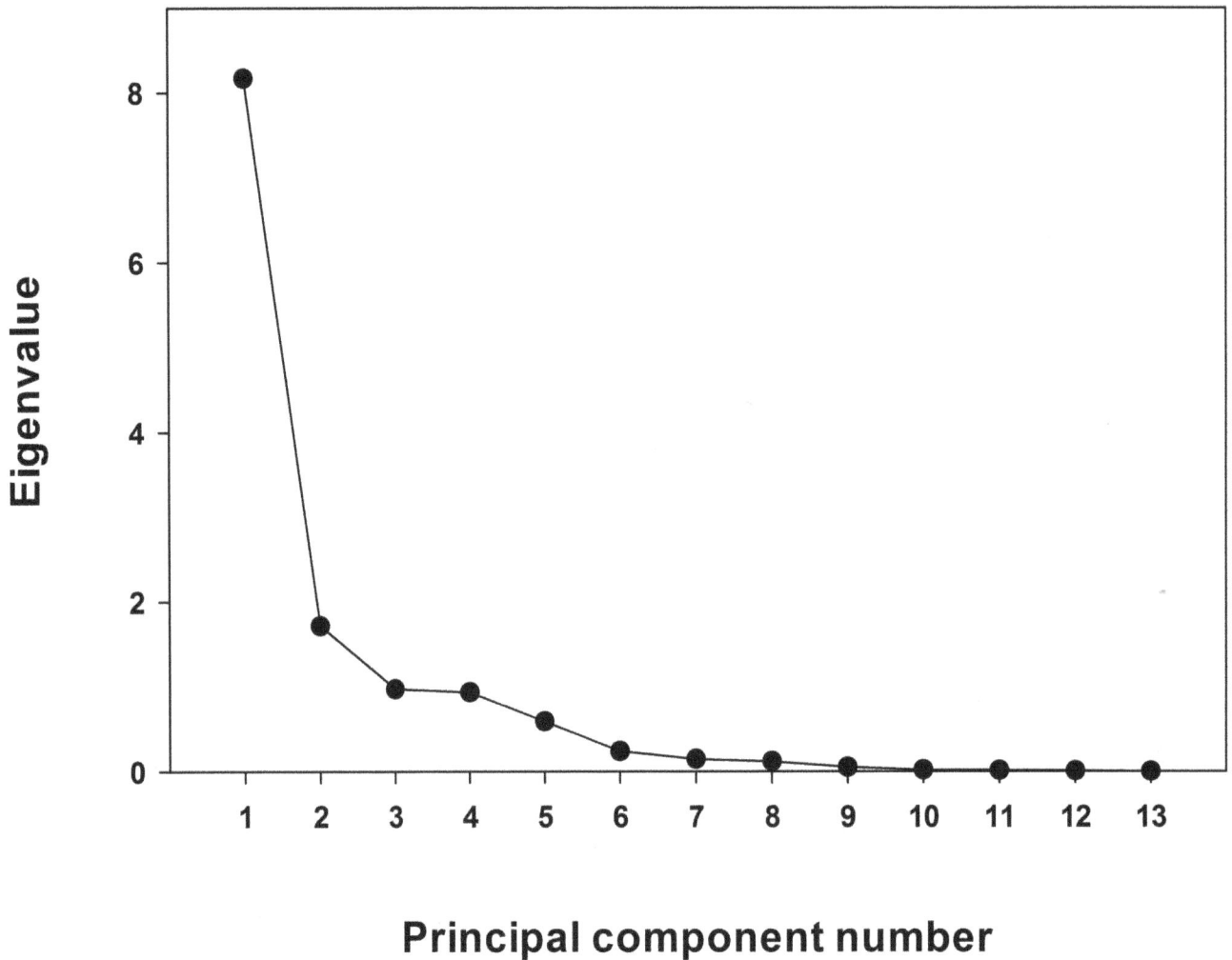

Figure 1. Scree plot of principal component analyses.

standard deviation values of the soil parameters by depth under three soil types are listed in Table 1. Most properties of mineral soils (Ko and CrB) were significantly different compared to those of the organic soil (Mc) and except for AWC and water content two mineral soils' characteristics were similar across soil profile up to 60 cm depth (Table 1). The water characteristics of Ko soil were significantly higher than those of CrB. On the other hand, organic soil (Mc) had much improved (lower BD and PR, and higher WSA, SOC, N concentrations and stock, AWC and water content) soil properties compared to those of the mineral counterparts in all four different depths (Ko and CrB). Impacts of all these parameters on overall soil quality in four soil depths calculated by three indexing methods are discussed in the following sections.

Under the framework of SQI-1, nine soil quality indicators were integrated numerically after scoring them primarily using the information from literature review. However, scoring data on some physicochemical indicators (e.g., C and N stocks) is scarce in the literature and thus, authors' opinion was used in those cases. Further, different studies used different scoring on the same indicator based on soil type, and management goals. For an example, while Amacher and Perry [28] had eight different range classes and scores for pH, Feiza et al. [41] had only four classes

and scores available for pH. Thus, scoring on indicators in the current study (Table 2) had importance on experts' opinion although the knowledge was primarily based on available literature and authors' experience [28,42]. The high variability in the observations (Table 1) was reflected in the SQI-1 values. There was no statistical difference ($p>0.05$) across treatments (NCC versus CC, and CT versus NT) and depth under specific soil type (Table 6). Overall performance of indexing methods was determined from averaging SQI values obtained from calculations for each depth and presented in Fig. 2. The, SQI-1 was significantly higher in the following order: Mc>Ko>CrB (Fig. 2), indicating that soil quality was influenced more by soil type than management and sampling depth under the SQI-1 approach.

Overall percentage of various soil functional influences in SQI-2 is presented in Fig. 3 and depthwise percentage of influence in SQI-2 is presented in Fig. S1. Generally, mineral soils (Ko and CrB) did not have any depthwise significant changes for WSC but 0–10 cm layer of organic (Mc) soil had significantly higher influence of WSC in SQI-2 compared to lower depths. On the other hand, while both Ko and Mc had significantly higher influence of RDC in SQI-2 in the upper layers than the lower depths, the opposite trend was found for the NSC (Fig. S1). In the

Table 5. Pearson correlation coefficients (r) for all soil indicators.

Correlation (r) matrix	pH	EC	BD	WSA	GMD	MWD	PR	N	SOC	N-Stock	C-Stock	AWC
EC	0.547											
BD	−0.394	−0.88										
WSA	0.248	0.483	−0.338									
GMD	0.279	0.8	−0.762	0.673								
MWD	0.286	0.782	−0.725	0.734	0.975							
PR	−0.129	−0.655	0.684	−0.456	−0.71	−0.687						
N	0.405	0.922	−0.908	0.47	0.877	0.835	−0.737					
SOC	0.438	0.93	−0.914	0.473	0.881	0.842	−0.731	0.995				
N-Stock	0.271	0.632	−0.551	0.32	0.642	0.595	−0.396	0.709	0.707			
C-Stock	0.382	0.764	−0.702	0.367	0.742	0.695	−0.489	0.822	0.833	0.965		
AWC	0.091	0.447	−0.449	0.275	0.439	0.441	−0.525	0.504	0.489	0.106	0.181	
Water content (cm)	0.092	0.487	−0.485	0.282	0.501	0.496	−0.557	0.558	0.546	0.28	0.333	0.967
Significance level												
EC	<0.001											
BD	<0.001	<0.001										
WSA	0.018	0.001	0.002									
GMD	0.009	<0.001	<0.001	<0.001								
MWD	0.007	<0.001	<0.001	<0.001	<0.001							
PR	0.141	<0.001	<0.001	<0.001	<0.001	<0.001						
N	<0.001	<0.001	<0.001	<0.001	<0.001	<0.001	<0.001					
SOC	<0.001	<0.001	<0.001	<0.001	<0.001	<0.001	<0.001	<0.001				
N-Stock	0.011	<0.001	<0.001	0.004	<0.001	<0.001	<0.001	0.001	0.001			
C-Stock	<0.001	<0.001	<0.001	0.001	<0.001	<0.001	<0.001	<0.001	<0.001	<0.001		
AWC	0.224	<0.001	<0.001	0.011	<0.001	<0.001	<0.001	<0.001	<0.001	0.195	0.07	
Water content (cm)	0.22	<0.001	<0.001	0.009	<0.001	<0.001	<0.001	<0.001	<0.001	0.01	0.003	<0.001

Abbreviations are same as Table 1.

Table 6. Treatment effects on soil quality index (SQI) under three soil types (Ko: Kokomo, CrB: Crosby silty loam, Mc: Muck) and four soil depths.

Treatments		0–10 cm		10–20 cm		20–40 cm		40–60 cm	
		NCC	CC	NCC	CC	NCC	CC	NCC	CC
Ko		CT	NT	CT	NT	CT	NT	CT	NT
	SQI-1	0.36a	0.36a	0.40a	0.40a	0.45a	0.40a	0.48a	0.50a
	SQI-2	0.40a	0.37a	0.39a	0.36a	0.40a	0.36a	0.43a	0.38a
	SQI-3	0.21a	0.23a	0.20a	0.22a	0.19a	0.22a	0.21a	0.21a
CrB		CT	NT	CT	NT	CT	NT	CT	NT
	SQI-1	0.36a	0.38a	0.36a	0.17a	0.45a	0.29a	0.38a	0.26a
	SQI-2	0.42a	0.40b	0.40a	0.37a	0.41a	0.40a	0.41a	0.40a
	SQI-3	0.26a	0.24a	0.24a	0.22a	0.22a	0.22a	0.22a	0.21a
Mc		CT	NT	CT	NT	CT	NT	CT	NT
	SQI-1	0.88a	0.88a	0.90a	0.90a	0.90a	0.90a	0.90a	0.81a
	SQI-2	0.70a	0.72a	0.65a	0.61a	0.65a	0.66a	0.67a	0.65a
	SQI-3	0.85a	0.85a	0.82a	0.78a	0.78a	0.77a	0.82a	0.81a

Abbreviations: NCC: no cover crop, CC: cover crop, CT: conventional tillage, NT: no tillage.
Different letters under specific row and depth indicate significant differences at $p < 0.05$ level, underline value is significant at $p < 0.1$ level.

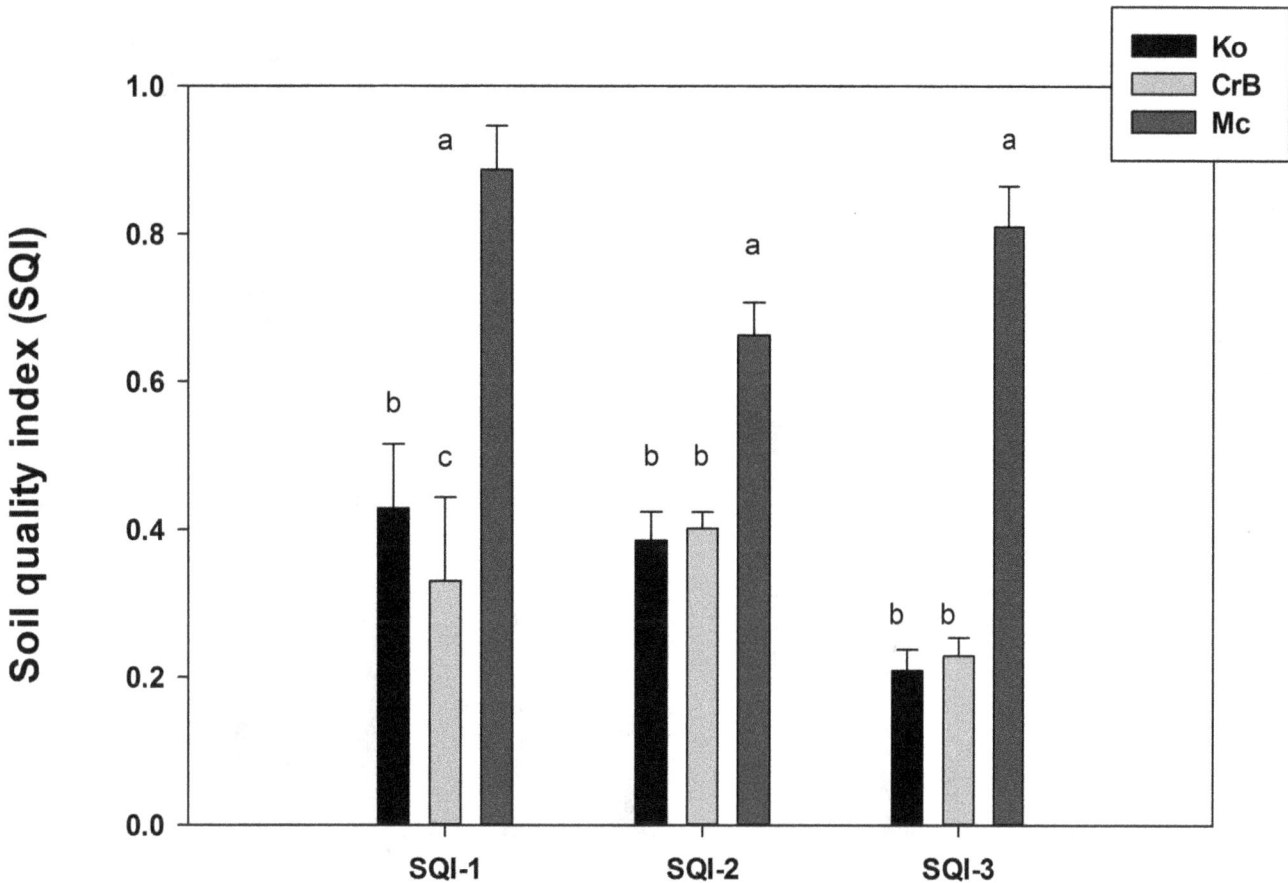

Figure 2. Overall soil quality index (SQI) values under different soil types. Different letters indicate significant differences at p<0.05 level for particular indexing method.

general composition of SQI-2, the main contributors were RDC, which ranged from 39–57% and NSC ranged from 35–59%. However, the WSC was relatively small and ranged merely from 0.1–12% (Fig. 3). The relative influence of root development functions in SQI-2 has been significantly higher in Ko than CrB, and Mc by 8 and 9%, respectively and the same for nutrient storage functions in SQI-2 has been significantly higher in CrB than Ko and Mc by 13 and 8%, respectively (Fig. 3). On the other hand, Mc soil had significantly higher water storage functional contribution to SQI-2 by 53 and 500%, respectively, compared to that of mineral (Ko and CrB) soils (Figs. 3). Thus, while the mineral soil quality was more sensitive to the functional attributes dedicated to root development and nutrient storage, muck soil quality was more influenced by soil attributes related to water storage under the scheme of SQI-2. Further, there was no significant effect on management practices on %RDC and %NSC to SQI-2 in different soil/management/crop combinations (Fig. 4). Note that CC/NCC management was used on field cultivated to soybean but CT/NT was used to soil under corn. However, contribution of WSC to SQI-2 for the case of CrB soil under NT management was significantly higher (178%) than that of CT practice (Fig. 4).

Correlation coefficients of SQIs to each other and depthwise SQIs and dry grain and straw are given in Table 7. Crop yield is obviously related to soil quality but soil quality in different layers may not be same and can influence crop yield accordingly. Data

on SQI variation for different soil layers are scarce and thus no comparison to the data presented in the current study (Table 6) was possible. However, while numerical values of SQI were almost unchangeable for different depths at Mc site, SQI generally variable with depth in Ko and CrB sites (Table 6). One goal of this study was to validate SQI against crop yield as only a few studies have done so with some exceptions [8,14]. Challenge lies in establishing a correlation between crop yield versus SQI calculated either depthwise or in combination of soil layers as standard methodology is not yet available and was not given much attention to date. For example, one can have only one crop yield value for multiple soil layers and thus to date crop yield has been correlated only against SQI calculated from surface layer. Low (ranged from 0.11–0.32) and high (ranged from 0.83–0.96) correlation coefficients of SQI and crop yield across a number of soil/crop combinations were reported [8,31]. The reasons attributed to low correlation between yield and SQI were: (i) the assumptions that the indicators considered good for a soil quality may not always lead to desired outcomes and flexibility in scoring functions needed [8], (ii) indicators which may not directly relate to crop performance were given more weightage by the method employed, or (iii) some key soil indicators were not included in the MDS due to the design of the study [9,43]. Nevertheless, in order to evaluate the index performance in the current study, SQIs were correlated against end point variable (i.e., yield) at variable depths (Table 7). In the present study, depth of soil sampling had

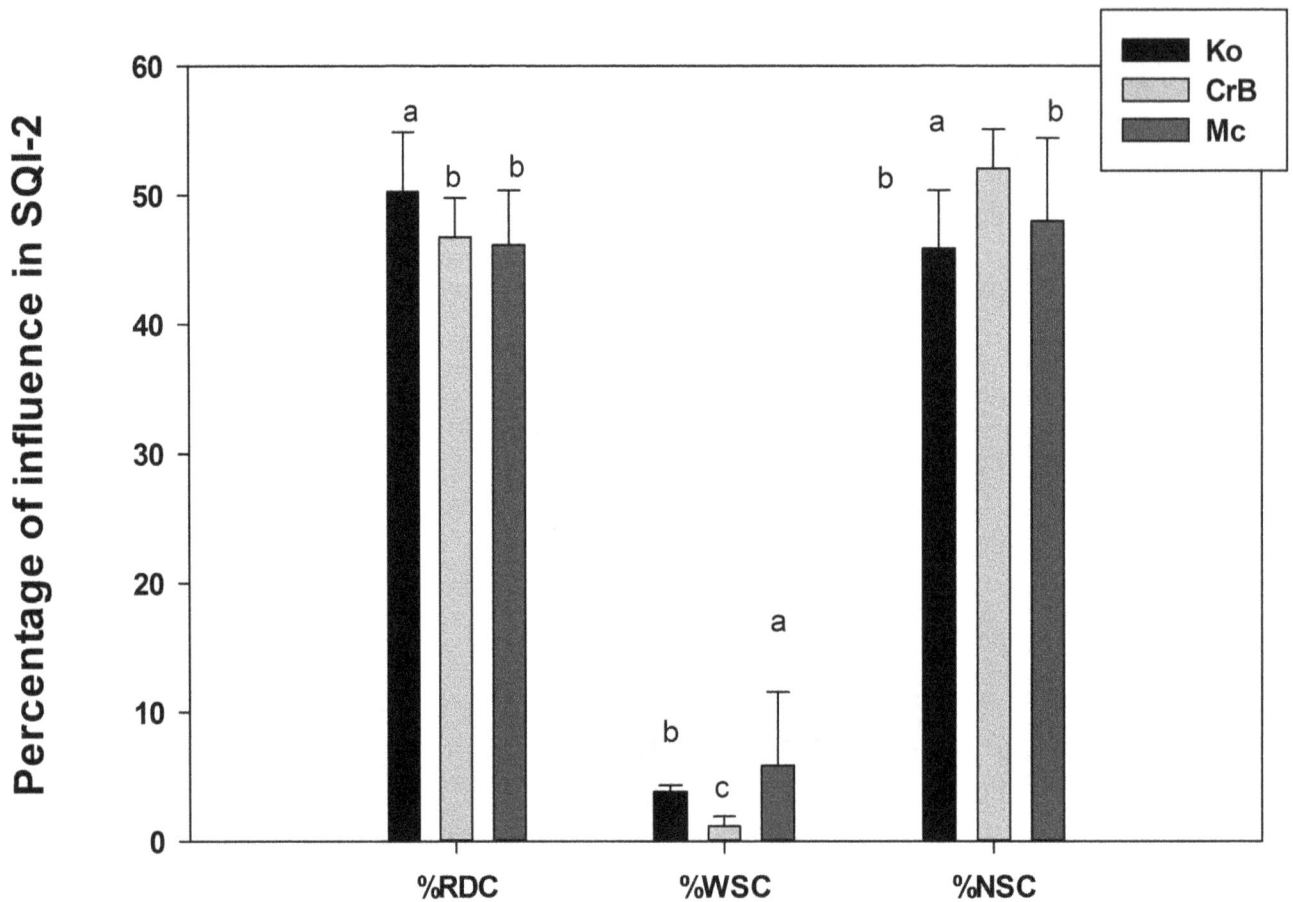

Figure 3. Percentage contribution of each soil function in SQI-2 under different soils. Different letters indicate significant differences at p<0.05 level for particular soil function. Abbreviations: RDC: root development capacity, WSC: water storage capacity, NSC: nutrient storage capacity.

no significant effect in SQI values (Table 6) and thus the individual SQI values were integrated by combining 0–20 cm and 0–60 cm profiles by averaging 0–10, and 10–20 cm for obtaining values for 0–20 cm depth and 0–10, 10–20, 20–40 and 40–60 cm for obtaining values for 0–60 cm depths and correlated the combined values with crop yield (Table 7). In the present study, grain and biomass yields were all significantly (p<0.05, r = 0.65–0.79) correlated under any SQI methods employed indicating that either of these methods was successful in predicting crop yield (Table 7). However, correlation coefficients in Table 7 are derived from the complete datasets including both corn and soybean yield but the correlation was considerably low (r<0.20 for soybean grain and r<0.60 for corn grain; data not shown), when dataset was further divided into corn and soybean. As discussed above, low correlation of SQI and crop yield was observed before and one of the listed reasons was some key soil indicators were not included in the dataset as per the experimental design and objective [8]. Crop yield is invariably related to soil fertility status and thus SQI may not always lead to high correlation with yield if some of the related characterizations are not observed as identified before [9,43] which is supported by the data of the current study (Table 7). Thus, the low correlation coefficient values may be because soil fertility (ion exchange capacities, macro and micro nutrients) and microbial (microbial biomass, microbial C, and N, and soil respiration) aspects were not monitored in the current studied sites, especially in the case of soybean grain yield.

Nevertheless, relatively higher correlation coefficient was observed in the 0–10 cm soil layer than the deeper soil profile (Table 7) probably due to higher plant root density in the top layer of the soil [44].

Additionally, SQI computed from different methods were also highly correlated with each other (Table 7) suggesting that (i) a relatively easy and user-friendly SQI (SQI-1) can be computed to evaluate and compare soil quality, which is similarly useful to other approaches, (ii) giving certain appropriate weightage on scores (SQI-2) could be similarly useful in predicting particular soil quality and (ii) a minimum group of carefully selected soil indicators (SQI-3) may be used to evaluate soil quality across various soil types and management practices. As all three SQI methods were significantly correlated to each other, it is difficult to conclude which one is the best approach, however, it highly depends on factors such as design of the study, choice of soil parameters included in the model to compute SQI and end point variable (environmental aspects versus crop yield). The advantage of using SQI-1 is that the soil quality could be assessed after measuring any number (low to high) of soil parameters and this procedure is relatively easier compared to other methods as the scoring requires literature review and expert opinions only. The disadvantage of SQI-1 is that it is subjective and relies mainly on researcher's point of view. On the other hand, advantage of SQI-2 is that it includes weightage based on the design of the study, system or the dataset to offset the subjectivity of the approach

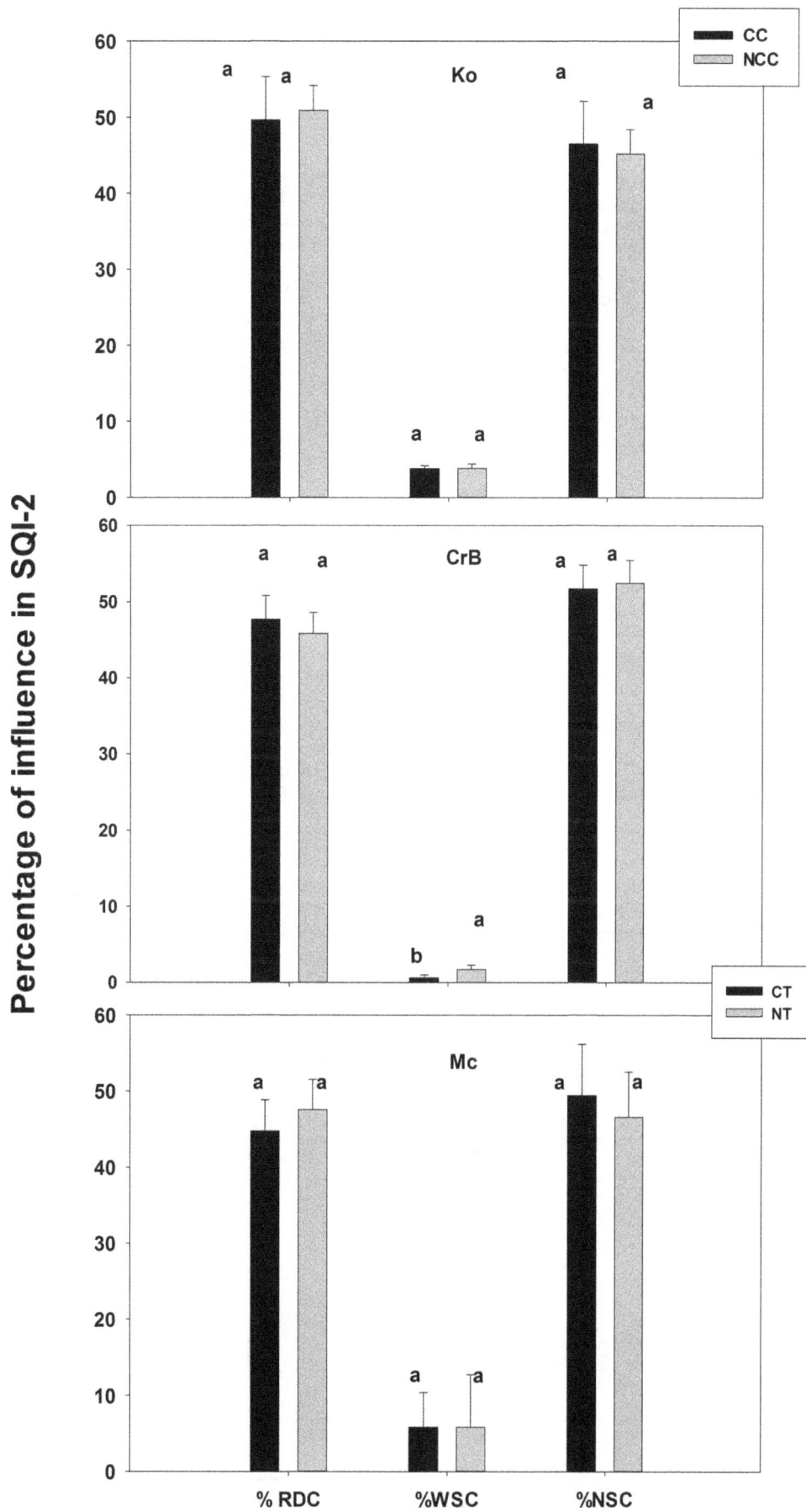

Figure 4. Percentage contribution of each soil function in SQI-2 under various management practices in different soil. Different letters indicate significant differences at p<0.05 level for particular soil function. Abbreviations: RDC: root development capacity, WSC: water storage capacity, NSC: nutrient storage capacity; NCC: no cover crop, CC: cover crop, CT: conventional tillage, NT: no tillage.

present in SQI-2. However, the disadvantage of SQI-2 is that it requires multiple numbers of soil parameters under different soil functional systems which may be expensive and time consuming in practical case. The SQI-3 is advantageous in the aspect of its ability to predict soil quality based on a reduced dataset with low number of soil parameters. Additionally, it is mostly objective approach as the statistical procedure would select a low number of soil parameters needed to calculate SQI based on the variances present in the whole dataset. So, in a long-term aspect within a particular soil/crop system, SQI-3 can be used effectively once it evaluated the most influential soil parameters required to assess soil quality of a particular soil/crop/management scenario. However, current data suggest that relative higher SQI values were obtained for Ko and Mc by SQI-1 than SQI-2 or SQI-3 and for CrB by SQI-2 than SQI-1 or SQI-3 (Fig. 2). Although SQI-1 and SQI-2 had higher values than SQI-3 in different scenarios, however, based on the current experimental design, and selected soil parameters SQI-3 successfully predicted crop yield relatively higher [Table 7 and also when grouped by crops (data not shown)] than SQI-1 and 2. Depth had little influence in any SQIs in the present study and additional chemical and biological parameters are also needed to strengthen the validity of SQI. Thus, SQI-3 appears to be the best method among the three under long-term scenario, especially due to its objective approach, relative higher correlations with crop yield and lower number of indicator selection which is more cost and time-effective over time than other approaches.

Conclusions

The data presented support the following conclusions: The SQI calculated by three different methods indicated that studied muck soil has significantly higher soil quality than that of mineral soils under on-farm conditions. The SQI was affected more by management and soil type than by depths in the studied on-farm sites. The SQI computed using three established methods were all significantly correlated to each other indicating that relatively easy and user-friendly SQI (SQI-1) can be similarly useful to evaluate soil quality, appropriate weightage on scores can predict soil's quality (SQI-2) with high performance but requires a number of soil parameters under different soil functional components and carefully chosen MDS with small numbers of soil variables (SQI-3) may adequately evaluate its quality. All three SQIs were highly and significantly correlated with crop yield. In general, however, analyses by crop type (corn and soybean) revealed that correlation was low ($r<0.20$ or 0.60 for soybean or corn grain yield, respectively; data not shown) suggesting choice of crop-specific and key soil parameters used in computing SQI. Under the current experimental on-farm conditions SQI-1 (for Ko and Mc) and SQI-2 (for CrB) values were higher than SQI-3, however, SQI-3 can be regarded as the best and easiest model given its relatively higher success to predict crop yield and objectivity approach with lower number of indicator selection ability which should be regarded as a relatively less expensive procedure over time compared to SQI-1 and 2. In addition, in order to effectively predict particular crop yield one must include soil fertility and microbial parameters in the model of SQI.

Table 7. Pearson correlation coefficients (r) of soil quality index (SQI) versus crop yield and correlations between different SQI values which were averaged up to a certain depth; all numbers are significant at $p<0.05$ level.

SQI-1	Dry Grain	Dry Straw
0–60 cm	0.65	0.67
0–20 cm	0.71	0.71
0–10 cm	0.73	0.71
SQI-2	**Dry Grain**	**Dry Straw**
0–60 cm	0.75	0.73
0–20 cm	0.75	0.72
0–10 cm	0.79	0.75
SQI-3	**Dry Grain**	**Dry Straw**
0–60 cm	0.76	0.74
0–20 cm	0.76	0.74
0–10 cm	0.78	0.76
	SQI-1	**SQI-2**
SQI-2	0.92	
SQI-3	0.93	0.97

Supporting Information

Figure S1 Percentage contribution of each soil function in SQI-2 under different soils in four soil layers; different letters indicate significant differences at p<0.05 level for particular soil function. Abbreviations: RDC: root development capacity, WSC: water storage capacity, NSC: nutrient storage capacity.
(TIF)

Acknowledgments

Authors would like to thank Dr. Shiguo Jiang for his help with PCA and Mr. Basant Rimal for his assistance in the laboratory. Authors would also like to thank three anonymous reviewers and an academic editor for their constructive suggestions to improve the manuscript.

Author Contributions

Conceived and designed the experiments: AM RL. Performed the experiments: AM. Analyzed the data: AM. Contributed reagents/materials/analysis tools: AM. Contributed to the writing of the manuscript: AM RL.

References

1. Doran J, Coleman D, Bezdicek D, Stewart B (1994) A framework for evaluating physical and chemical indicators of soil quality.
2. Doran JW, Parkin TB (1994) Defining and assessing soil quality. SSSA special publication 35: 3–3.
3. Karlen DL, Mausbach MJ, Doran JW, Cline RG, Harris RF, et al. (1997) Soil Quality: A Concept, Definition, and Framework for Evaluation (A Guest Editorial). Soil Sci Soc Am J 61: 4–10.
4. Karlen DL, Stott DE (1994) A framework for evaluating physical and chemical indicators of soil quality. In: Doran JW, Coleman DC, Bezdicek DF, Stewart BA, editors. Defining soil quality for a sustainable environment. Madison, WI: Soil Science Society of America. pp. 53–72.
5. Stocking MA (2003) Tropical Soils and Food Security: The Next 50 Years. Science 302: 1356–1359.
6. Diack M, Stott D (2001) Development of a soil quality index for the Chalmers Silty Clay Loam from the Midwest USA. Purdue University: USDA-ARS National Soil Erosion Research Laboratory. pp. 550–555.
7. Larson WE, Pierce FJ (1991) Conservation and enhancement of soil quality. Evaluation of Sustainable Land Management in the Developing World. International Board for Soil Research and Management, Bangkok, Thailand.
8. Andrews S, Karlen D, Mitchell J (2002) A comparison of soil quality indexing methods for vegetable production systems in Northern California. Agriculture, Ecosystems & Environment 90: 25–45.
9. Bhardwaj AK, Jasrotia P, Hamilton SK, Robertson GP (2011) Ecological management of intensively cropped agro-ecosystems improves soil quality with sustained productivity. Agriculture Ecosystems & Environment 140: 419–429.
10. Andrews SS, Carroll CR (2001) Designing a soil quality assessment tool for sustainable agroecosystem management. Ecological Applications 11: 1573–1585.
11. Andrews SS, Karlen DL, Cambardella CA (2004) The soil management assessment framework: A quantitative soil quality evaluation method. Soil Science Society of America Journal 68: 1945–1962.
12. Andrews SS, Mitchell JP, Mancinelli R, Karlen DL, Hartz TK, et al. (2002) On-Farm Assessment of Soil Quality in California's Central Valley. Agron J 94: 12–23.
13. Fernandes JC, Gamero CA, Rodrigues JGL, Mirás-Avalos JM (2011) Determination of the quality index of a Paleudult under sunflower culture and different management systems. Soil and Tillage Research 112: 167–174.
14. Mandal UK, Ramachandran K, Sharma K, Satyam B, Venkanna K, et al. (2011) Assessing Soil Quality in a Semiarid Tropical Watershed Using a Geographic Information System. Soil Science Society of America Journal 75: 1144–1160.
15. Zobeck TM, Halvorson AD, Wienhold B, Acosta-Martinez V, Karlen DL (2008) Comparison of two soil quality indexes to evaluate cropping systems in northern Colorado. Journal of Soil and Water Conservation 63: 329–338.
16. Zornoza R, Mataix-Solera J, Guerrero C, Arcenegui V, Mataix-Beneyto J, et al. (2008) Validating the effectiveness and sensitivity of two soil quality indices based on natural forest soils under Mediterranean conditions. Soil Biology and Biochemistry 40: 2079–2087.
17. Jokela W, Posner J, Hedtcke J, Balser T, Read H (2011) Midwest Cropping System Effects on Soil Properties and on a Soil Quality Index. Agronomy Journal 103: 1552–1562.
18. Navas M, Benito M, Rodriguez I, Masaguer A (2011) Effect of five forage legume covers on soil quality at the Eastern plains of Venezuela. Applied Soil Ecology 49: 242–249.
19. Brady NC, Weil RR (1984) The nature and properties of soils New York: Macmillan.
20. Mukherjee A, Lal R (2014) Long and short-term tillage effects on quality of organic and mineral soils under on-farm conditions in Ohio, Submitted. Soil and Tillage Research.
21. Mukherjee A, Lal R (2014) Short-term effects of cover cropping on quality of a Typic Argiaquolls in Central Ohio, Submitted.

22. Grossman RB, Reinsch TG (2002) Bulk density and linear extensibility. In: Dane JH, Topp GC, editors. Methods of soil analysis, Part 4: Physical methods. Madison, WI: SSSA. pp. 201–254.
23. Herrick JE, Jones TL (2002) A dynamic cone penetrometer for measuring soil penetration resistance. Soil Science Society of America Journal 66: 1320–1324.
24. Yoder RE (1936) A direct method of aggregate analysis of soils and a study of the physical nature of erosion losses. J Am Soc Agron 28: 337–351.
25. Youker RE, McGuinness JL (1957) A short method of obtaining mean weight-diameter values of aggregate analyses of soils. Soil Science 83: 291–294.
26. Dane JH, Hopmans JH (2002) Water retention and storage. In: Dane JH, Topp GC, editors. Methods of soil analysis Part 4, SSSA Book Ser 5. Madison, WI.: SSSA. pp. 671–717.
27. Chhabra A, Palria S, Dadhwal VK (2003) Soil organic carbon pool in Indian forests. Forest Ecology and Management 173: 187–199.
28. Amacher MC, Perry CH (2007) Soil vital signs: A new Soil Quality Index (SQI) for assessing forest soil health.
29. Liebig MA, Varvel G, Doran J (2001) A simple performance-based index for assessing multiple agroecosystem functions. Agronomy Journal 93: 313–318.
30. Li Q, Xu MX, Liu GB, Zhao YG, Tuo DF (2013) Cumulative effects of a 17-year chemical fertilization on the soil quality of cropping system in the Loess Hilly Region, China. Journal of Plant Nutrition and Soil Science 176: 249–259.
31. Masto RE, Chhonkar PK, Singh D, Patra AK (2008) Alternative soil quality indices for evaluating the effect of intensive cropping, fertilisation and manuring for 31 years in the semi-arid soils of India. Environmental monitoring and assessment 136: 419–435.
32. Doran JW, Parkin TB (1996) Quantitative indicators of soil quality: a minimum data set. SSSA special publication 49: 25–38.
33. Dunteman GH (1989) Principal components analysis: Sage.
34. Jolliffe I (2005) Principal component analysis: Wiley Online Library.
35. SPSS (2014) IBM SPSS, Version 21.0, Chicago, USA.
36. Kaiser HF (1960) The application of electronic computers to factor analysis. Educational and Psychological Measurement 20: 141–151.
37. Wander MM, Bollero GA (1999) Soil quality assessment of tillage impacts in Illinois. Soil Science Society of America Journal 63: 961–971.
38. SAS (2012) SAS Institute Inc.; SAS version 9.2.
39. Microsoft (2007) Microsoft Excel. Redmond, Washington.
40. Sigmaplot (2012) Scientific Software Solutions Internationals; Sigmaplot version 12.
41. Feiza V, Feiziene D, Kadziene G, Lazauskas S, Deveikyte I, et al. (2011) Soil state in the 11th year of three tillage systems application on a cambisol. Journal of Food, Agriculture & Environment 9: 1088–1095.
42. Lal R (1994) Methods and guidelines for assessing sustainable use of soil and water resources in the tropics; Washington D.C.: USDA/SMSS Technical Monograph 21.
43. Zornoza R, Mataix-Solera J, Guerrero C, Arcenegui V, García-Orenes F, et al. (2007) Evaluation of soil quality using multiple lineal regression based on physical, chemical and biochemical properties. Science of The Total Environment 378: 233–237.
44. Maurya PR, Lal R (1980) Effects of No-Tillage and Ploughing on Roots of Maize and Leguminous Crops. Experimental Agriculture 16: 185–193.
45. Brzezinska M, Sokolowska Z, Alekseeva T, Alekseev A, Hajnos M, et al. (2011) Some characteristics of organic soils irrigated with municipal wastewater. Land Degradation & Development 22: 586–595.
46. Sokoowska Z (2011) Effect of phosphates on dissolved organic matter release from peat-muck soils.
47. Walczak R, Rovdan E, Witkowska-Walczak B (2002) Water retention characteristics of peat and sand mixtures. International Agrophysics 16: 161–166.
48. Carter MR (2006) Quality critical limits and standardization. In: Lal R, editor. Encyclopedia of Soil Science. New York: Marcel Dekker. pp. 1412–1415.

21

Modeling Spatial Patterns of Soil Respiration in Maize Fields from Vegetation and Soil Property Factors with the Use of Remote Sensing and Geographical Information System

Ni Huang[1], Li Wang[1]*, Yiqiang Guo[2], Pengyu Hao[1], Zheng Niu[1]

1 The State Key Laboratory of Remote Sensing Science, Institute of Remote Sensing and Digital Earth, Chinese Academy of Sciences, Beijing, China, **2** Land Consolidation and Rehabilitation Center, Ministry of Land and Resources, Beijing, China

Abstract

To examine the method for estimating the spatial patterns of soil respiration (R_s) in agricultural ecosystems using remote sensing and geographical information system (GIS), R_s rates were measured at 53 sites during the peak growing season of maize in three counties in North China. Through Pearson's correlation analysis, leaf area index (LAI), canopy chlorophyll content, aboveground biomass, soil organic carbon (SOC) content, and soil total nitrogen content were selected as the factors that affected spatial variability in R_s during the peak growing season of maize. The use of a structural equation modeling approach revealed that only LAI and SOC content directly affected R_s. Meanwhile, other factors indirectly affected R_s through LAI and SOC content. When three greenness vegetation indices were extracted from an optical image of an environmental and disaster mitigation satellite in China, enhanced vegetation index (EVI) showed the best correlation with LAI and was thus used as a proxy for LAI to estimate R_s at the regional scale. The spatial distribution of SOC content was obtained by extrapolating the SOC content at the plot scale based on the kriging interpolation method in GIS. When data were pooled for 38 plots, a first-order exponential analysis indicated that approximately 73% of the spatial variability in R_s during the peak growing season of maize can be explained by EVI and SOC content. Further test analysis based on independent data from 15 plots showed that the simple exponential model had acceptable accuracy in estimating the spatial patterns of R_s in maize fields on the basis of remotely sensed EVI and GIS-interpolated SOC content, with R^2 of 0.69 and root-mean-square error of 0.51 μmol CO_2 m^{-2} s^{-1}. The conclusions from this study provide valuable information for estimates of R_s during the peak growing season of maize in three counties in North China.

Editor: Ben Bond-Lamberty, DOE Pacific Northwest National Laboratory, United States of America

Funding: This work was supported by the National Natural Science Foundation of China (41301498), the Public Service Sectors (Ministry of Land and Resources) Special Fund Research (201311127), the Special Foundation for Young Scientists of the State Laboratory of Remote Sensing Science (13RC-07), and the Major State Basic Research Development Program of China (2013CB733405). The funders had no role in study design, data collection and analysis, decision to publish, or preparation of the manuscript.

Competing Interests: The authors have declared that no competing interests exist.

* Email: wangli@radi.ac.cn

Introduction

Soil CO_2 efflux from terrestrial ecosystems to the atmosphere has been considered the second largest global carbon flux and is a vital component of ecosystem respiration [1]. In recent decades, significant progress has been made in identifying the biophysical factors that influence soil respiration (R_s) to predict soil CO_2 emission accurately in time and space [2–4].

The majority of R_s arises from root and microbial tissue. Therefore, understanding the spatial and temporal changes of these sources will facilitate the modeling of R_s. However, the large spatial and temporal heterogeneity of root and microbial activity within the landscape and the covariation of potentially important factors (i.e., temperature and water content) pose great challenges to the development of mechanistically based models that account for spatial and temporal variability in R_s [2]. Thus, many different

statistical models of R_s have been developed on the basis of data collected from different ecosystems [5]. Numerous studies have established R_s models based on soil temperature, soil moisture, or both [6,7]. Aside from soil temperature and moisture, plant productivity proxies [e.g., leaf area index (LAI), canopy chlorophyll content (Chl_{canopy}), and plant biomass] [8–10] and soil properties [e.g., soil organic carbon (SOC) content, soil total nitrogen (STN) content, and soil C and N ratio (soil C/N)] [11,12] also potentially influence R_s and are often included in models of R_s. However, most of the factors that affect variations in R_s tend to be derived through field measurements [13]. Furthermore, direct observation of these variables across long time spans or large spatial scales is expensive because of the required manpower and material resources. A simple method to derive data related to variations in R_s is necessary to facilitate the determination of the spatial and temporal distribution of R_s.

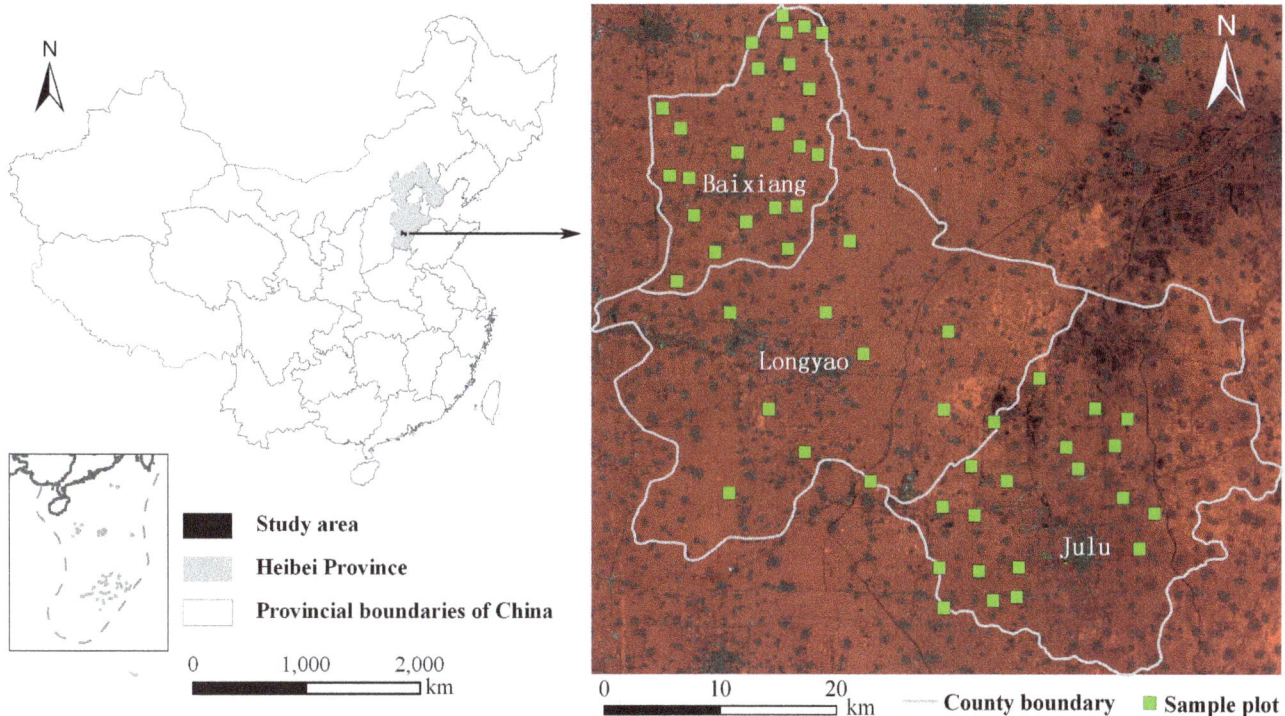

Figure 1. Spatial location of the sample plots for field experiments in three counties in North China. The box in the bottom left corner of Figure 1 shows the South China Sea islands.

Remote sensing and geographical information system (GIS) provide powerful tools for data acquisition, spatial analysis, and graphical display [14–16]. In the field of global change research, significant advances have been made in the development and application of remote sensing and GIS. These advances include land cover and land-use changes [17,18], environmental vulnerability and risk assessment [19,20], ecological restoration and management [21–23], and terrestrial ecosystem carbon cycle [24–26]. However, applying the data derived from remote sensing and GIS into R_s modeling remains controversial, especially for remote sensing data, because remotely sensed data in principle are independent measurements of site properties, not functionally important variables (e.g., soil temperature, soil moisture, and plant growth variables) that control R_s [3,27,28]. On the basis of statistical analysis of field experiments, previous studies found that remotely sensed vegetation indices (VIs) correlate with R_s in crop sites that lack drought stress [10] and can be used to model the spatial patterns of R_s during the peak growing season of alpine grasslands in the Tibetan Plateau [26]. However, few studies explore the potential of remote sensing and GIS data for estimating the spatial patterns of R_s in agricultural land, which may be affected by more complex factors than natural grasslands because of the influence of human activity. Although modern agriculture has successfully increased food production, the processes involved have profoundly affected the global carbon cycle through tillage, drainage and conversion of natural to agricultural ecosystems [29,30]. Therefore, a simple method should be identified to study the spatial characteristics of R_s in agricultural ecosystems.

This study aims to examine a potential new approach for estimating the spatial patterns of R_s during the peak growing season of maize by using remote sensing and GIS technology in Baixiang, Longyao and Julu Counties, which are typical agricul-

tural areas in the north plain of China. Studying the spatial characteristics of soil CO_2 efflux in maize fields will contribute to eco-agricultural development.

Materials and Methods

Ethics Statement

No specific permissions were required for the 53 sample plots in this study. We confirmed that the field studies did not involve endangered or protected species, and the specific location of the sample plots was provided in the manuscript (Fig. 1).

Study Site

The study site is situated within three counties (Baixiang, Longyao and Julu) in Southern Hebei Province of North China (Fig. 1). The total area of the study site is 1.64×10^3 km^2. This area is located in the North China Plain with a flat open terrain, single landform type, and a mean elevation of 30 m above sea level. Calcareous alluvial soil with high capacity to retain water and fertilizer is the main soil type in the study area. The study site is suitable for farming, and maize is the main crop. The climate is continental monsoon with four distinct seasons and adequate light and heat resources. Long records of meteorological data near the study site (http://cdc.cma.gov.cn) indicate that the mean annual temperature is 13.5°C with the coldest temperatures in January and the hottest in July. The mean annual precipitation is 502.8 mm, but precipitation is distributed unevenly in the four seasons with the greatest precipitation occurring in summer (362.5 mm). Therefore, drought influences agricultural development, and agriculture mainly involves irrigation in this study site.

Fifty-three sample plots located in the maize fields were identified within the study site (Fig. 1). The distance between any two sample plots was larger than 2 km. Each sample plot (greater than 100 m×100 m) has a large maize area, flat terrain, and maize under uniform growing conditions. All measurements were performed from August 11, 2013 to August 20, 2013, which corresponded to the tassel stage and peak growing period of maize. During the 10 days of field measurements, continuous measurements were performed, except on August 12 because of a minor precipitation event. Therefore, all field measurements required 9 days.

Field measurements

Soil respiration measurements. In each sample plot, R_s was measured by using a soil respiration chamber (LI-6400-09; LiCor, Lincoln, Nebraska, USA) connected to a portable photosynthesis system (LI-6400; LiCor, Lincoln, Nebraska, USA). The soil respiration chamber was mounted on a PVC soil collar that was sharpened at the bottom. Each PVC collar (5 cm long, 11 cm inside diameter) was inserted 2 cm to 3 cm into the ground and was installed at least 24 h prior to performing any measurements. To reduce the difference in root biomass, soil collars were placed in three locations on the basis of their distance to the maize plant: near a maize plant, inter-plant, and inter-row. Two collars were placed in each of the three positions for each R_s measurement. At least three to four consecutive measurements on each collar were performed to prevent any systematic error in the R_s estimates. An average R_s value was used for each collar, and the average value from six collars was used to represent the R_s value at plot level. Each R_s measurement was conducted between 09:00 h and 15:00 h (local time) because fluxes measured during this time interval are usually representative of the daily mean flux.

Soil temperature and soil moisture measurements. After the soil respiration measurement on a PVC soil collar in each plot, soil temperature and soil moisture were measured in this collar to minimize sample difference. Soil temperature was measured at a 10 cm depth (T_{s10}) by using a ground thermometer. Volumetric soil moisture at a depth of 0 cm to 20 cm (SM_{20}) was determined by using a portable time domain reflectometry probe (HydroSense, Campbell, USA). Thus, six soil temperature and moisture measurements were performed in each plot. The average value was used to represent soil temperature or soil moisture at the plot level.

Maize biophysical parameter measurements. LAI was measured by using an LAI-2000 (LI-COR Inc., Lincoln, Nebraska). In each plot, six representative positions were selected for LAI measurement, and in every position, two repeated measurements were performed. Leaf chlorophyll content (Chl$_{leaf}$) was determined by using a portable chlorophyll meter (SPAD-502, New Jersey, USA). Fully expanded leaves, which depended on the height of the maize plant, were randomly selected from three locations that corresponded to the upper, middle, and lower parts

of the maize plant. For each leaf location, 10 SPAD values were randomly collected. The vertical leaf area distribution in maize canopy was analyzed by measuring the area of each green leaf from the bottom to the top of eight randomly distributed maize plants with the use of an area meter (LI-3100, LI-COR, Lincoln, Nebraska). The area-weighted mean SPAD reading was used to derive Chl$_{leaf}$. However, the SPAD reading was in arbitrary units rather than in actual amounts of chlorophyll per unit area of the leaf tissue. A transform relationship exists between the SPAD readings and the actual chlorophyll content in maize [31]. To convert the SPAD readings to chlorophyll content per unit leaf area ($\mu g\ cm^{-2}$), this study used the transform relationship (Chl$_{leaf}$ = 0.95 × SPAD reading − 3.25) derived by Wu et al. [32] in maize plots, and the same SPAD meter was employed in this study. Chl$_{canopy}$ was then determined by using the following equation:

$$Chl_{canopy} = Chl_{leaf} \times GLAI \qquad (1)$$

where Chl$_{canopy}$ is the canopy chlorophyll content ($g\ m^{-2}$), Chl$_{leaf}$ is the leaf chlorophyll content of maize ($g\ m^{-2}$), and GLAI represents the green leaf area per unit ground area.

In each sample plot, three representative maize plants were harvested for aboveground biomass (AGB) measurement. These fresh maize plants were sealed in a plastic bag and immediately transported to a nearby laboratory for subsequent analysis. Thereafter, fresh samples were oven dried at 65°C until the mass of the sample reached a constant weight. The AGB in each plot can be derived by multiplying the average dry weight per plant (g plant^{-1}) and the average plant density of maize (plants m^{-2}).

Soil property measurements. Soil within the six PVC collars in each plot was destructively sampled after measuring R_s, soil temperature and soil moisture. Soil was sampled to a depth of approximately 20 cm by a cylindrical soil driller (4 cm diameter, 20 cm height), in which fine root biomass and microbial activity are the highest [33,34]. These collected soil samples were sealed in plastic bags and stored at room temperature while being transported to the laboratory. Six collected soil samples in each plot were uniformly mixed to form a composite sample for laboratory analysis. The composite sample was air-dried in the laboratory to a constant weight for soil chemical analyses. The air-dried soil samples were ground to pass through a 0.2 mm sieve after any visible plant tissues and debris were manually removed. The SOC content was estimated by using the standard Mebius method [35]. The STN content was analyzed by using the Kjeldahl digestion procedure [36]. In this study, soil C/N was calculated by the ratio of SOC and STN contents.

Table 1. Calculation for vegetation indices [a].

Vegetation index	Formula	Reference
Normalized difference vegetation index	$NDVI = \dfrac{R_{Nir}-R_{Red}}{R_{Nir}+R_{Red}}$	Rouse et al. [47], Gamon et al. [48]
Modified soil adjusted vegetation index	$MSAVI = \dfrac{2R_{Nir}+1-\sqrt{(2R_{Nir}+1)^2-8(R_{Nir}-R_{Red})}}{2}$	Qi et al. [49]
Enhanced vegetation index	$EVI = 2.5 \times \dfrac{R_{Nir}-R_{Red}}{1+R_{Nir}+6\times R_{Red}-7.5\times R_{Blue}}$	Huete et al. [50]

[a]R_{Blue}, R_{Red}, and R_{Nir} are reflectance of blue, red, and NIR band in the HJ-1A CCD optical image, respectively.

Table 2. Pearson's correlation among soil respiration and factors affecting soil respiration in maize fields during the peak growing season in three counties in North China.

	R_s	T_{s10}	SWC_{20}	Chl_{canopy}	LAI	AGB	SOC content	STN content	Soil C/N
R_s	1.00	-0.27	-0.18	0.54***	0.75***	0.59***	0.76***	0.59***	-0.23
T_{s10}		1.00	0.18	-0.15	-0.28	-0.27	-0.49**	-0.66***	0.51**
SWC_{20}			1.00	-0.17	-0.05	-0.07	0.16	-0.00	0.06
Chl_{canopy}				1.00	0.83***	0.81***	0.26	0.20	-0.18
LAI					1.00	0.76***	0.44**	0.38*	-0.28
AGB						1.00	0.45**	0.34*	-0.15
SOC content							1.00	0.78***	-0.29
STN content								1.00	-0.79***
Soil C/N									1.00

R_s is the daily mean soil respiration rate (μmol CO_2 m^{-2} s^{-1}), T_{s10} is the soil temperature at 10 cm depth (°C), SWC_{20} is the soil water content at 0 cm to 20 cm depth (m^3 m^{-3}), Chl_{canopy} is the canopy chlorophyll content (g m^{-2}), LAI is the leaf area index, AGB is the aboveground biomass (kg m^{-2}), SOC content is the soil organic carbon content (g kg^{-1}), STN content is the soil total nitrogen content (g kg^{-1}), and soil C/N is the soil C: N ratio. Significance levels:

*p<0.05,
**p<0.01,
***p<0.001.

Spatial data acquisition

Maize classification data. This study aimed to derive the spatial distribution of R_s in maize fields based on the field measurements at the plot scale. Maize classification data is necessary to spatially extrapolate R_s at the plot scale to the whole study area. Multi-temporal normalized difference vegetation index (NDVI) data collected over the growing season were used to classify maize at the study site [37–39]. Clouds are common occurrences in the study area during the growing season. Thus, obtaining a time sequence of cloud-free scenes is difficult. Two types of satellite data were used to establish the time-series NDVI data. One was the Operational Land Imager (OLI) image of Landsat 8, and the other was the small constellation for environmental and disaster mitigation (HJ-1A and B) charge coupled device (CCD) image [40–42]. Five scenes of OLI images acquired on May 3, 2013, May 19, 2013, July 6, 2013, October 10, 2013, and October 26, 2013 were downloaded from the U.S. Geological Survey (http://earthexplorer.usgs.gov/). Three HJ-1A and B CCD optical images acquired on June 6, 2013, August 17, 2013, and September 15, 2013 were downloaded from the China Center for Resource Satellite Data and Applications (http://www.cresda.com). The two types of remote sensing images exhibit same spatial resolution (30 m). The 30 m spatial resolution is appropriate for classifying maize patterns in the study area given the relatively large field in the region, which could spatially corresponded to five or more 30 m pixels. The strong relationship of the NDVI with biophysical vegetation characteristics, such as LAI and green biomass [43,44], enables the discrimination of land cover types on the basis of their unique phenological responses. Before land-use classification, pre-processing (i.e., radiometric calibration, atmospheric correction and geometric correction) of OLI images and HJ-1A and B CCD optical images was accomplished by using the Environment for Visualizing Images (ENVI) software (Version 4.7, Research Systems Inc., Boulder, Colorado, USA) [45,46]. This process ensured the consistency between the two types of remote sensing data and the seasonality of the NDVI time series. The maximum likelihood classification method, integrated in the ENVI software, was applied to the eight-date NDVI time series that spanned one maize growing season of the study site.

Spectral vegetation index for vegetation biophysical parameter estimation. Three greenness indices, namely, NDVI, enhanced vegetation index (EVI), and modified soil adjusted vegetation index (MSAVI), were derived from the HJ-1A CCD optical image acquired on August 17, 2013 (Table 1) for vegetation biophysical parameter estimation. Previous studies reported that greenness VIs offer important and convenient measures for vegetation biophysical parameters, such as LAI and Chl_{canopy} [51–54]. Meanwhile, LAI and Chl_{canopy} are also found to be good indicators of plant canopy photosynthesis [55–57] and are used in the modeling of R_s [58]. To obtain the spatial patterns of vegetation biophysical parameters in maize fields, the spatial distribution of vegetation biophysical parameters over the whole study area was overlapped with the maize classification data.

Quantifying the spatial pattern of SOC content. Statistics and geostatistics have been widely applied to quantify the spatial distribution patterns of SOC at a regional scale [59–61]. Based on the theory of regionalized variables, geostatistics provides advanced tools to quantify the spatial features of soil parameters and to conduct spatial interpolation [62,63]. In this study, geostatistical analyses were performed by using the geostatistical analyst module of ArcGIS software (Version 9.3, 2008) to quantify the spatial pattern of SOC content. To obtain the spatial pattern of the SOC content in the maize fields, the spatial distribution of

Table 3. Spatial characteristics of soil respiration (R_s, μmol CO_2 m^{-2} s^{-1}), soil temperature at 10 cm depth (T_{s10}, °C), soil water content at 0 cm to 20 cm depth (SWC_{20}, m^3 m^{-3}), canopy chlorophyll content (Chl_{canopy}, g m^{-2}), leaf area index (LAI), aboveground biomass (AGB, kg m^{-2}), soil organic carbon content (SOC content, g kg^{-1}), soil total nitrogen content (STN content, g kg^{-1}) and soil C: N ratio (soil C/N) in maize fields during the peak growing season in three counties in North China.

Variables	Mean	Maximum	Minimum	CV (%)
R_s	5.43	7.33	2.64	15.45
T_{s10}	28.32	30.93	25.78	4.73
SWC_{20}	27.54	33.27	19.54	12.48
Chl_{canopy}	0.18	0.21	0.16	6.54
LAI	3.75	4.53	2.81	8.64
AGB	0.94	1.89	0.44	31.93
SOC content	11.86	17.26	6.40	16.71
STN content	1.25	1.78	0.53	24.47
Soil C/N	9.82	14.38	7.07	18.53

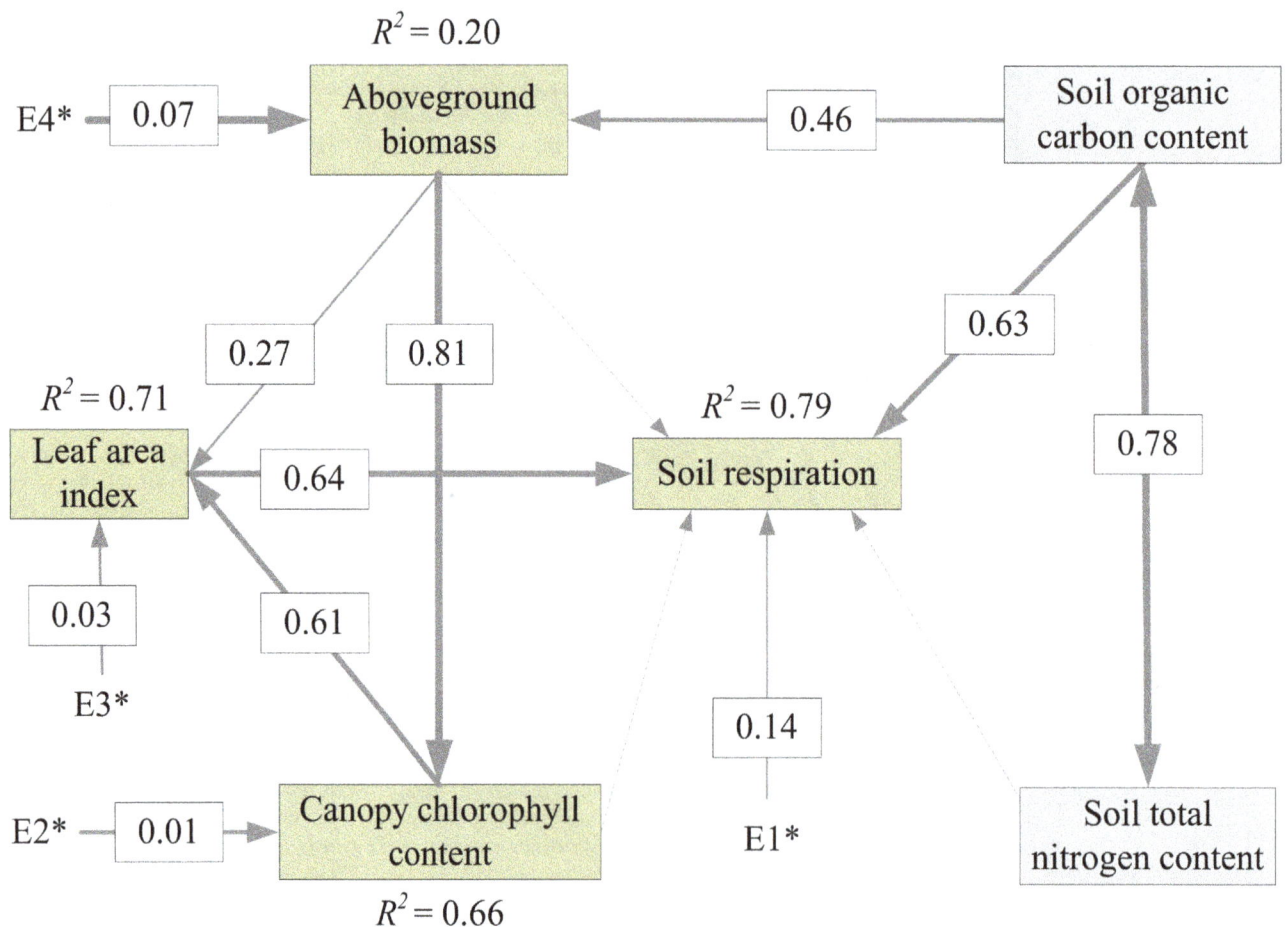

Figure 2. Final structural equation modeling (SEM) for soil respiration. Non-significant paths are shown in dashed line. The thickness of the solid arrows reflects the magnitude of the standardized SEM coefficients. Standardized coefficients are listed on each significant path. * represents error terms for observed variables, among them, E1, E2, E3, and E4 represent measurement errors for soil respiration, canopy chlorophyll content, leaf area index, and aboveground biomass, respectively.

Table 4. Total, direct, and indirect effects in the structural equation modeling.

Variable	Direct effect	Indirect effect	Total
Soil respiration			
Aboveground biomass	−0.10ns	0.46	0.36
Soil organic carbon content	0.63	0.16	0.79
Soil total nitrogen content	−0.09ns	0.30	0. 21
Leaf area index	0.64	-	0.64
Canopy chlorophyll content	−0.04ns	0.39	0.35
Aboveground biomass			
Soil organic carbon content	0.46	-	0.46
Soil total nitrogen content	−0.01ns	-	−0.01ns
Leaf area index			
Aboveground biomass	0.27	0.50	0.77
Soil organic carbon content	-	0.35	0.35
Soil total nitrogen content	-	−0.01ns	−0.01ns
Canopy chlorophyll content	0.61	-	0.61
Canopy chlorophyll content			
Aboveground biomass	0.81	-	0.81
Soil organic carbon content	-	0.37	0.37
Soil total nitrogen content	-	−0.01ns	−0.01ns

These effects were calculated using standardized path coefficients. Non-significant effects are indicated by "ns".

the SOC content over the whole study area was overlapped with the maize classification data.

Modeling spatial patterns of soil respiration

Identifying factors affecting spatial variability of soil respiration. The variables that explain the spatial variability of R_s are as follows: (1) soil properties, measured by SOC content, STN content and soil C/N; (2) environmental factors, encompassing T_{s10} and SM_{20}, and (3) plant photosynthesis proxy factors, including AGB, LAI and Chl_{canopy}. Pearson's correlation requires variables to be normally distributed and mutually independent. Each variable was tested for normal distribution by using the Shapiro–Wilk normality test and for randomness by the runs test of the Statistical Package for the Social Sciences (SPSS, Chicago, Illinois, USA). The results of the statistical analysis showed that each of these measured variables followed a normal distribution (Shapiro-Wilk, p>0.05) and showed randomness (runs test, p> 0.05). Thus, Pearson's correlation analysis, as implemented in the SPSS software, was used to screen important variables that influence R_s. Five variables with statistically significant correlation (p<0.05) with R_s, namely, SOC content, STN content, LAI, AGB, and Chl_{canopy}, were screened out (Table 2). However, these variables were cross-correlated [64–66] and included both direct and indirect effects. To solve this problem, structural equation modeling (SEM) was used to evaluate explicitly the causal relationships among these interacting variables [67–69] and to divide the total effects of variables on R_s into direct and indirect effects. On the basis of the theoretical knowledge on the major factors that influence spatial patterns of R_s at regional scales [8,13,26], we developed an SEM model to relate R_s to SOC content, STN content, LAI, AGB, and Chl_{canopy}. This SEM model was used to identify the direct effect factors for R_s estimation. The SEM model was fitted by using AMOS 18.0 for Windows [70]. After using the SEM, the fit indices, namely, comparative fit

index = 0.984 and goodness-of-fit index = 0.946. Thus, the theoretical model showed a good fit with the sample data.

Quantifying the spatial patterns of soil respiration in maize fields. In this study, the direct effect factors of R_s identified by SEM were used to estimate R_s. The spatial distribution data of these direct effect factors were first obtained on the basis of remote sensing or GIS to quantify the spatial patterns of R_s in maize fields. A simple exponential model that used the proxy data was then employed to estimate the spatial pattern of R_s during the peak growing season of maize. The accuracy of this method was examined by separating the observed data into two datasets through a random generator. One dataset consisted of 38 sample plots for analysis, whereas the other consisted of 15 for testing the accuracy of the R_s estimation.

Result

Spatial characteristics of soil respiration

Based on field-measured data at 38 plots, the daily mean R_s of maize during the peak growing season was 5.43 μmol CO_2 m^{-2} s^{-1} with a range of 2.64 μmol CO_2 m^{-2} s^{-1} to 7.33 μmol CO_2 m^{-2} s^{-1} and a coefficient of variation (CV) of 15.45% (Table 3). The spatial variability of soil temperature at 10 cm depth (T_{s10}) was relatively small at the study site with a CV of 4.73% and was far less than the spatial variation in soil water content at 0 cm to 20 cm depth (SWC_{20}). The AGB of maize showed greater spatial variability (CV = 31.93%) than LAI (CV = 8.64%) and Chl_{canopy} (CV = 6.54%).

Mean SOC content, STN content, and soil C/N at 0 cm to 20 cm depth in maize fields of the study site were 11.86 g kg^{-1} (ranged from 6.40 g kg^{-1} to 17.26 g kg^{-1}), 1.25 g kg^{-1} (ranged from 0.53 g kg^{-1} to 1.78 g kg^{-1}), and 9.82 (ranged from 7.07 to 14.38), respectively. Their CVs were not similar with the STN content which showed greater spatial variability than the SOC content and soil C/N.

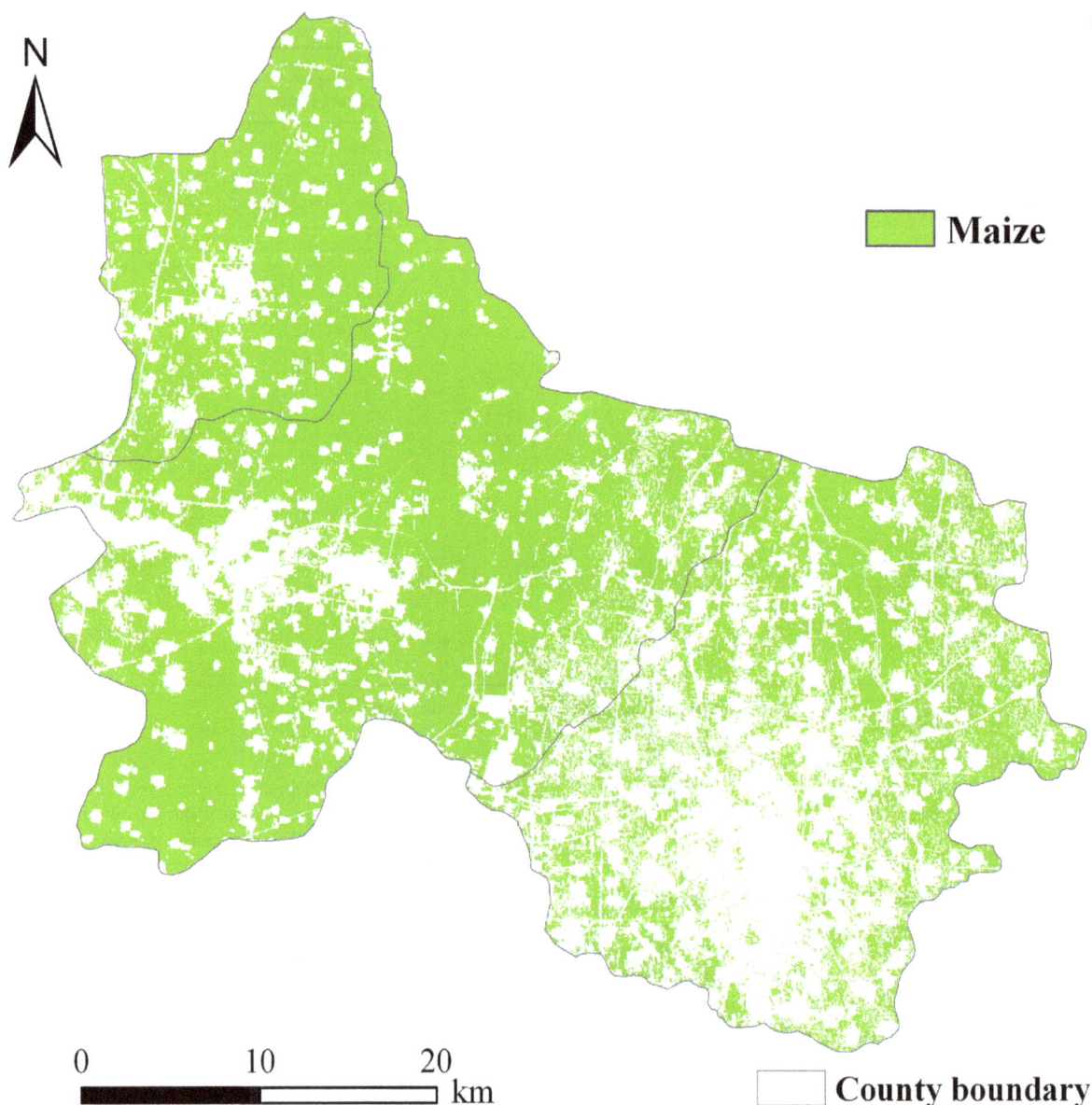

Figure 3. Maize classification map in three counties in North China.

Factors driving spatial variability of soil respiration

Based on Pearson's correlation analysis, five variables with significant correlation with R_s, namely, Chl_{canopy}, LAI, AGB, SOC content, and STN content, were selected (Table 2). However, the five selected variables were intercorrelated (Table 2), and their relationships with R_s combined both direct and indirect correlations. Thus, an SEM model was further used to evaluate the causal relationships among these interacting variables. The final SEM explained 79% of the variation in R_s (Fig. 2). The direct, indirect, and total effects of the variables are shown in Table 4. Among the five selected variables, LAI and SOC content directly affected R_s and can be used to predict R_s with relatively high accuracy ($R^2 = 0.79$). The other three variables (i.e., Chl_{canopy}, ABG, and STN content), despite having a significant correlation with R_s, only affected R_s indirectly through their direct relationship with SOC content and LAI. Thus, the two direct effect factors were used to estimate R_s, and the spatially distributed data proxies of

these two factors were used to quantify the spatial patterns of R_s in maize fields during the peak growing season.

Spatial data used for soil respiration estimation

Maize classification. The maize classification map of the study area is shown in Figure 3. The classification accuracy for maize at the study site could not be quantitatively assessed because of the limitation of the sample data. However, 53 sample plots were all located in the maize classification map, and the county-level maize patterns classified in the map were consistent with the general maize patterns across the three counties. In addition, the classified maize area was close to the maize area reported by the China County Statistical Yearbook [71]. Thus, the classification accuracy of maize was believed to be reasonable, and the maize classification map was then used to predict the spatial pattern of R_s during the peak growing season of maize.

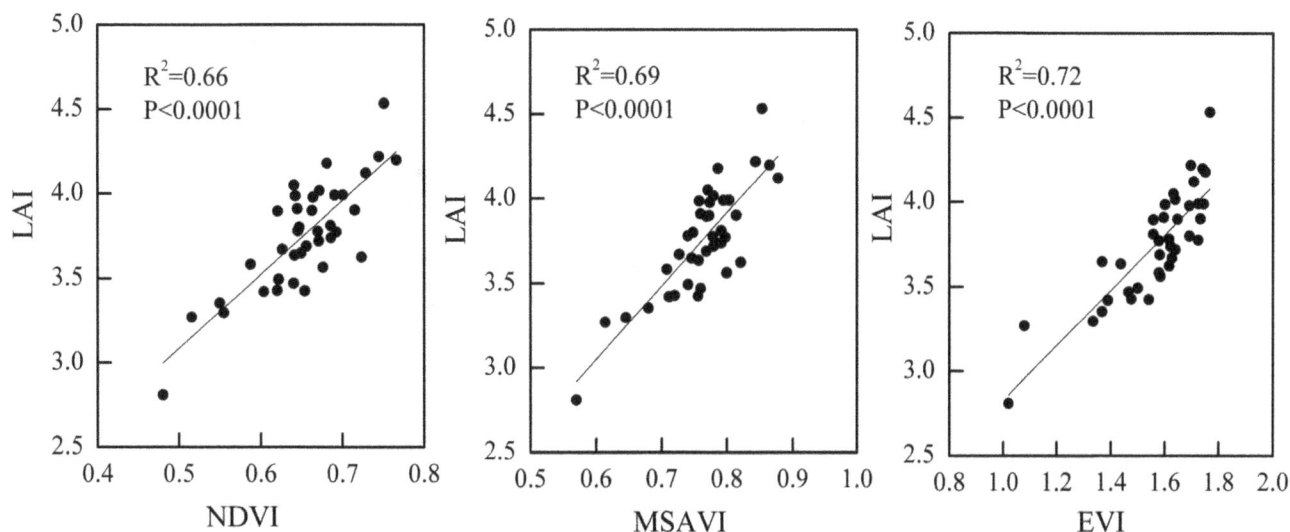

Figure 4. Linear relationships between three vegetation indices (VIs) and leaf area index (LAI) during the peak growing season of maize in three counties in North China (n = 38). The VIs are: normalized difference vegetation index (NDVI), enhanced vegetation index (EVI), and modified soil adjusted vegetation index (MSAVI).

LAI estimation from spectral vegetation index. Among the three greenness indices calculated from the optical image of HJ-1A satellite, EVI showed the best linear relationship with LAI, with a determination coefficient (R^2) of 0.72, followed by MSAVI and NDVI (Fig. 4). The explanation of LAI variance increased from 66% to 72% when EVI was used instead of NDVI for LAI estimation, and this increase was statistically significant (p<0.05). However, EVI and MSAVI did not exhibit a significant difference in explaining the variation in LAI, despite EVI having a slightly better relationship with LAI than MSAVI. Thus, EVI was used as a proxy for LAI to estimate R_s during the peak growing season of maize for simplicity. The spatially distributed EVI during the peak growing season of maize exhibited relatively small variability (Fig. 5). Overall, the EVI in the north and southwest parts of the study site (i.e., Baixiang and Longyao Counties) showed a high value. Relatively low EVI values mainly occurred in the southeast parts of the study site (i.e., Julu County), especially the northwest Julu County (Fig. 5).

Spatial distribution of SOC content. Kriging interpolation was performed by using ArcGIS 9.3 software to produce the spatial distribution map of the SOC content in maize fields of the study area. A cell size of 30 m×30 m was selected for the spatial interpolation to match the spatial resolution of images from OLI and HJ-1A/B. The final result of this spatial interpolation process is shown in Figure 6. Based on the spatial distribution map of the SOC content in maize fields, SOC content values were higher in the northwest and southwest parts of the study area than in the southeastern part.

Spatial distribution of soil respiration

The EVI and SOC content were used to estimate the spatial pattern of R_s during the peak growing season of maize on the basis of a simple exponential model. The geo-location information (latitude and longitude) of the 38 sample plots was used in the extraction of pixels. Pixels that contained these plots from the spatial distribution maps of EVI and SOC content data (Figs. 5 and 6) were extracted. These data were used to determine the model parameters by least-squares fitting. The resulting model was as follows:

$$R_s = 1.57 \times \exp(0.44 \times EVI + 0.05 \times SOC\ content) \quad (2)$$

$$(n = 38,\ R^2 = 0.73)$$

where R_s refers to the daily mean soil respiration rate in µmol CO_2 m^{-2} s^{-1}; EVI refers to enhanced vegetation index, as a proxy for LAI; and SOC content is the soil organic carbon content (g kg^{-1}) in maize fields of the study area. Eq. (2) was employed to predict the spatial pattern of R_s from spatially distributed EVI and SOC content data during the peak growing season of maize (Figs. 5 and 6). The spatial variation in R_s showed a pattern similar to that in SOC content (Figs. 6 and 7).

Figure 8 shows the accuracy assessment result of the R_s prediction model. The field measured R_s was comparable with the spatial data predicted R_s. Based on the independent test dataset, EVI and SOC content accounted for 69% of the spatial variation in ground-measured R_s, and the RMSE was 0.51 µmol CO_2 m^{-2} s^{-1}. The result of the accuracy assessment suggests that the prediction model, which used EVI and SOC content as the dependent variables, was effective in estimating R_s in maize fields during the peak growing season.

Discussion

Relationships between LAI and three VIs

In this study, in situ measured data were obtained during the peak growing period of maize (corresponding to the tassel stage of maize). The effect of soil background on the spectral reflectance of remote sensing images was negligible during this period because the maize cover was higher with LAI ranging from 2.81 to 4.53. The difference in the capability of spectral vegetation index (VI) responding to LAI variation mainly depended on the sensitivity of VI to the canopy structural variation of maize. Thus, the VI modified the effect of soil reflectance (i.e. MSAVI) did not exhibit a significantly greater advantage than NDVI, which is strongly affected by soil reflectance in sparsely vegetated areas [50]. EVI, which is more sensitive to variation in dense vegetation than

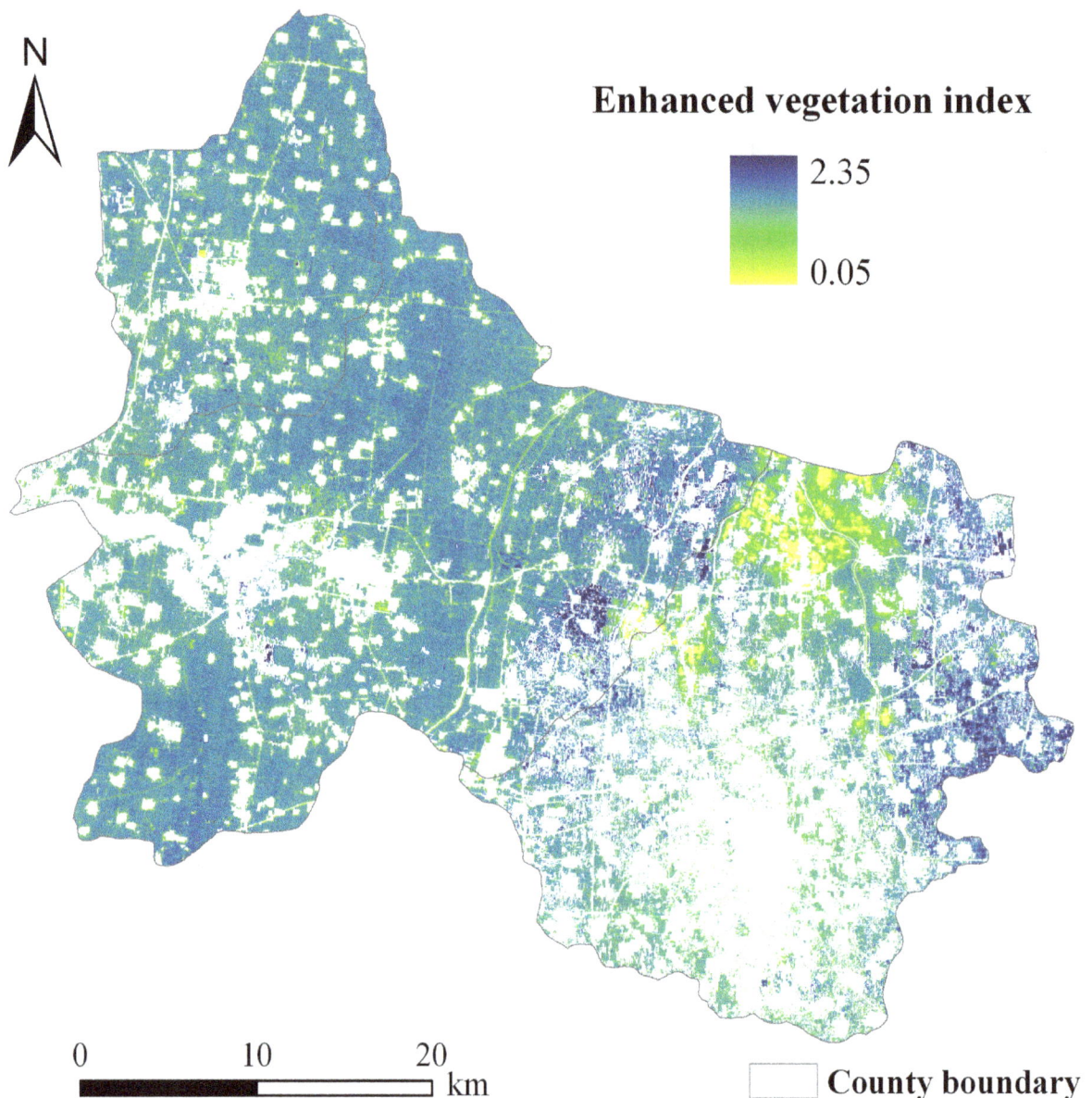

Figure 5. Spatial distribution map of enhanced vegetation index in maize fields in three counties in North China.

NDVI [50], showed the best relationship with the LAI of maize. This result was consistent with our previous study [58] that was conducted in irrigated and rainfed maize fields located at the University of Nebraska, Agricultural and Research Development Center, Mead, Eastern Nebraska, USA.

Measurement accuracy of SOC content

Field measurement data revealed that the SOC content at 0 cm to 20 cm depth in the maize fields ranged from 6.4 g kg^{-1} to 17.3 g kg^{-1}, and the mean value was 12.01 g kg^{-1}. For the mean dry land SOC content in North China, the value appeared to be higher than the previous estimate (0.83 from the average of 268 sample points) [72]. This difference was partly attributed to the fact that only the SOC content in maize fields, not in all dry land types, was considered. Most maize fields in the study site were on a winter wheat/maize rotation, and wheat straw was returned to the soil. The high productivity of maize crops contributed to the

development of a thick A horizon and high SOC content [73,74]. Additionally, only the SOC content in maize fields at 0 cm to 20 cm depth was analyzed, whereas previous studies estimated the SOC content on the basis of organic carbon content to a depth of 1 m [72,75,76]. In agricultural land, soil depth at 0 cm to 20 cm is located in the cultivation layer and has a higher SOC content than the SOC content at the deeper soil layers [34]. This condition contributed to the higher SOC content from the measured soil property data than the previous estimate.

Factors affecting spatial pattern of soil respiration

The spatial differences in R_s at the study site can be mainly attributed to the differences in vegetation productivity and soil property factors among the sample plots, whereas soil temperature and soil moisture served a minor function in regulating the spatial pattern of R_s. A previous study also demonstrated that site variables that reflect site productivity (e.g., LAI or aboveground

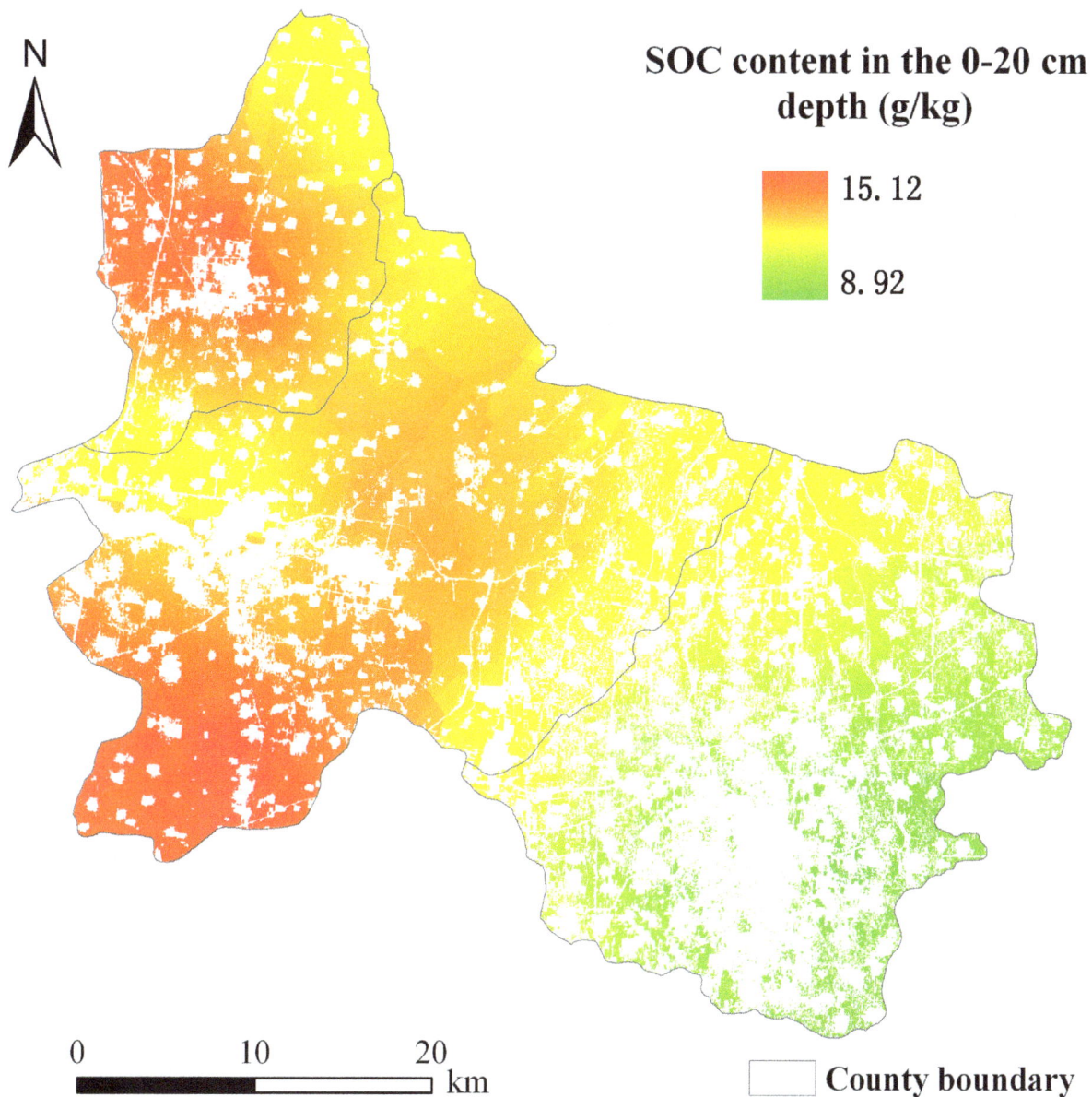

Figure 6. Spatial distribution map of soil organic carbon (SOC) content in the 0–20 cm depth in maize fields in three counties in North China.

net primary productivity) will provide a useful approach for large-scale estimates of regional R_s in terrestrial ecosystems [8]. Soil temperature evidently serves a predominant function in the spatial variations of R_s across sites of climatically contrasting environments [4]. However, at a local scale or under similar climatic conditions, other biological and biophysical factors, such as vegetation productivity and the size of organic carbon pools, may prevail as dominant drivers of R_s [4,77]. At a local scale, the spatial variation in T_{s10} in the study site was small (CV = 4.73%). Thus, soil temperature did not affect the spatial pattern of R_s. Although soil moisture in the maize fields showed a relatively large spatial variation (CV = 12.48%), this variation did not reach a degree that will affect the spatial dynamics of R_s. The soil C quantity and substrate quality factors (i.e., SOC and STN contents) were consistently and strongly correlated with one another and significantly affected the variation in R_s [5,12,13].

However, SEM results showed that the STN content only affected R_s indirectly through the direct effect on the SOC content at the study site.

During the peak growing season of maize, biophysical parameters, such as LAI, Chl_{canopy}, and AGB, were important variables that determined the size of the photosynthetic capacity [56,78]. However, these variables are not truly independent, and a correlation between one of them and R_s may lead to a correlation of the other with R_s. In this study, R_s was strongly correlated with LAI, Chl_{canopy} and AGB of maize fields, whereas LAI was the only variable directly related to R_s during the peak growing season of maize on the basis of SEM analysis.

The direct effect factors of R_s were used to estimate the spatial variability of R_s during the peak growing season of maize in three counties in North China. A simple exponential model, which included the corresponding spatial proxies from remote sensing

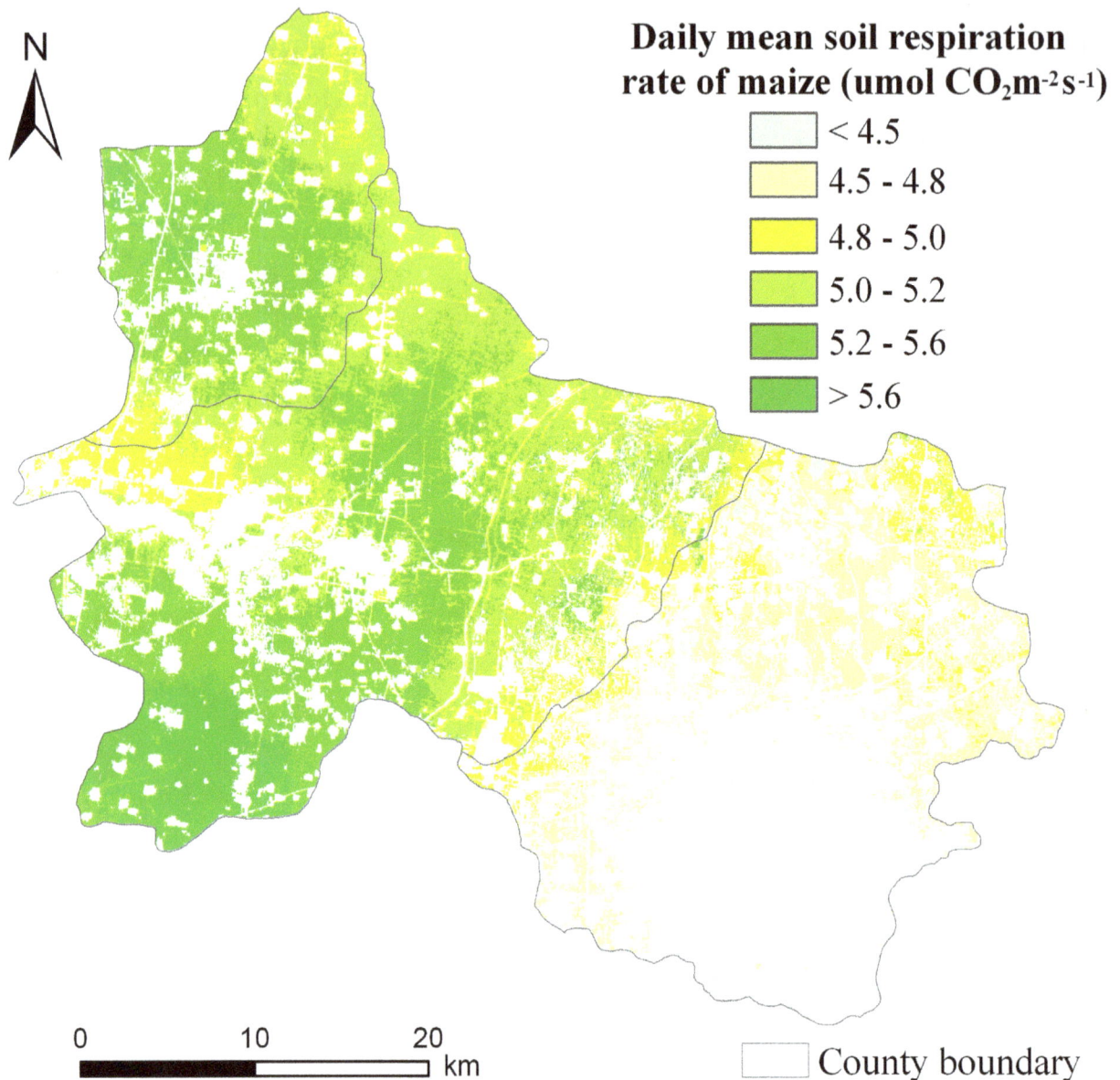

Figure 7. Spatial pattern of daily mean soil respiration rate during the peak growing season of maize in three counties in North China.

and GIS (i.e., EVI and spatially interpolated SOC content), was employed. A similar method was applied to a deciduous broadleaf forest site in the Midwest USA [79]. The independent test data also demonstrated the rationality of this method at the study site to a certain extent (Fig. 8). Regardless of the form of the R_s model, the relationship between LAI and EVI, as well as the kriging interpolation precision of the SOC content, affected the predictive accuracy of the R_s model. A moderate correlation between EVI and LAI (Fig. 4) affected the test accuracy of the exponential model with an R^2 value of 0.69 and an RMSE value of 0.51 μmol CO_2 m^{-2} s^{-1} (Fig. 8). The tendency of kriging to overestimate small values is supported by previous studies [80–82]. This tendency may help explain the bias toward overestimating R_s at low values (Fig. 8). Therefore, improving the accuracy of input parameters from remote sensing or GIS will increase the predictive capability of the R_s model.

Notably, the R_s model developed in this study was applicable to maize fields during the peak growth period in the three counties in North China. However, the model employed in this study does not consider temperature, a main driver of R_s that has high spatial variability. This model may be not used anywhere else or in other stages of the growing season. Furthermore, when spatially distributed data were used in the R_s model, a simple alternative method was employed to estimate the maize LAI by using the remotely sensed EVI, which may be problematic. Verstraeten et al. [83] highlighted that the assimilation of remotely sensed geophysical products into a carbon model is a complex process, and simply exchanging conventional input data for their remotely sensed counterparts is insufficient. Therefore, future research should focus on an integrating spatially distributed R_s datasets and geophysical products from remote sensing and GIS by using the data assimilation method, which has been extensively applied in

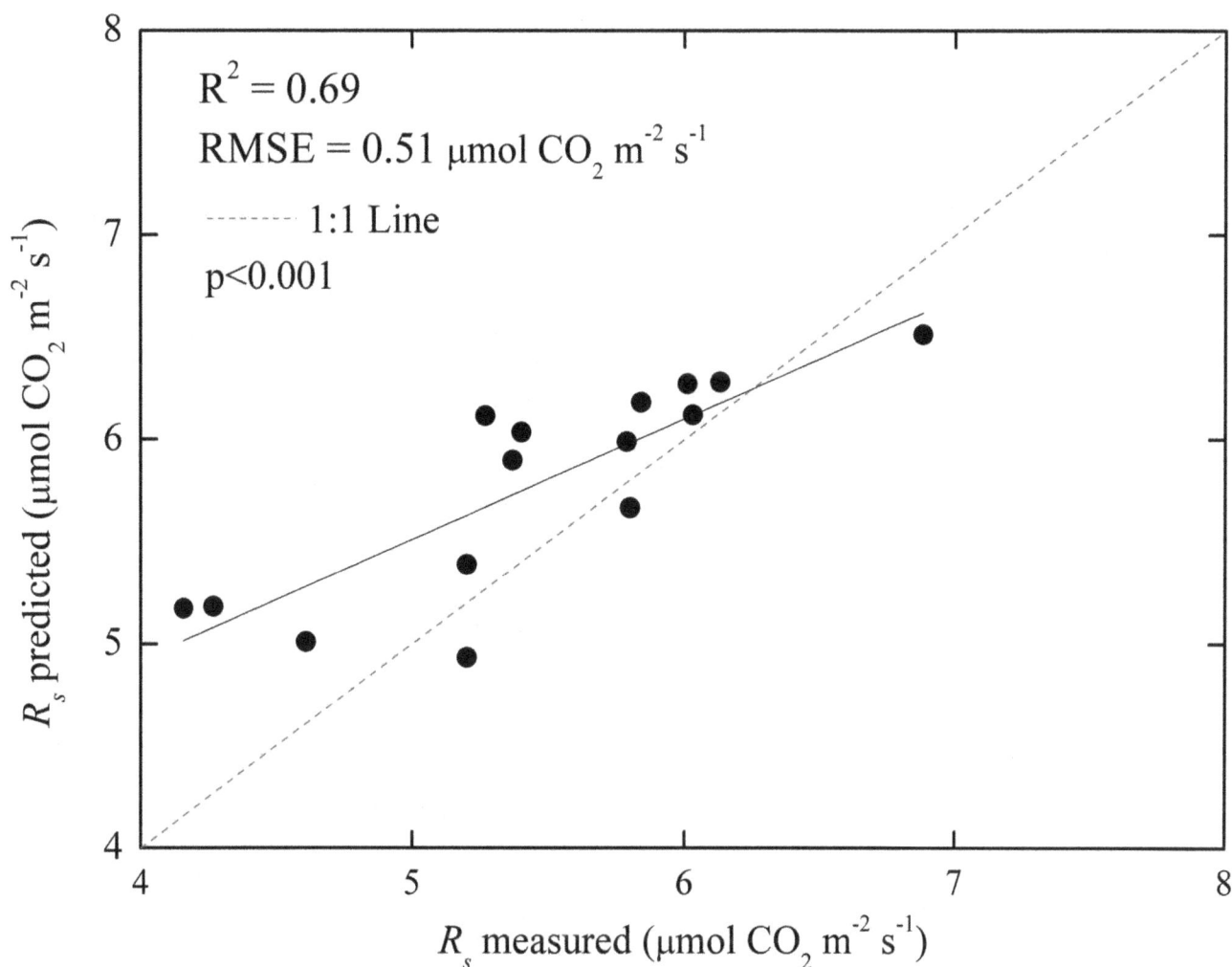

Figure 8. Spatial data predicted soil respiration (R_s) and corresponding ground-based measurements with R^2 and RMSE (μmol CO$_2$ m^{-2} s^{-1}) during the peak growing season of maize in three counties in North China ($n = 15$). The predicted soil respiration was attained with an exponential model that used EVI and SOC content as dependent variables.

terrestrial carbon cycle research [84–86]. However, this method lack the integration of R_s and spatially distributed data.

Conclusions

This study investigated the potential of spatial data from remote sensing and GIS for estimating the spatial patterns of R_s during the peak growing season of maize in three counties in North China. Based on in situ measurements, plant productivity (i.e., LAI) and soil property (i.e. SOC content) factors were identified as the most important determinants of spatial variability in R_s during the peak growing season of maize, and R_s was weakly related to soil temperature and soil moisture. Spectral VIs calculated from an HJ-1A CCD optical image were used to estimate LAI and EVI was found to be the best proxy for LAI. To derive the spatial pattern of R_s during the peak growing season of maize, a simple

exponential model, which included remotely sensed EVI and GIS spatially interpolated SOC content, was employed. This method was tested by using an independent sample dataset and was shown to be reasonable at the study site.

Acknowledgments

We sincerely thank the anonymous reviewers for their important and constructive revision advices on the manuscript.

Author Contributions

Conceived and designed the experiments: NH LW YQG. Performed the experiments: NH LW PYH. Analyzed the data: NH LW PYH. Contributed reagents/materials/analysis tools: NH LW YQG PYH ZN. Wrote the paper: NH.

References

1. Raich JW, Potter CS (1995) Global patterns of carbon dioxide emissions from soils. Global Biogeochemical Cycles 9: 23–36.
2. Davidson EA, Belk E, Boone RD (1998) Soil water content and temperature as independent or confound factors controlling soil respiration in a temperature mixed hardwood forest. Global Change Biology 4: 217–227.

3. Buchmann N (2000) Biotic and abiotic factors controlling soil respiration rates in Picea abies stands. Soil Biology and Biochemistry 32: 1625–1635.
4. Campbell JL, Sun OJ, Law BE (2004) Supply-side controls on soil respiration among Oregon forests. Global Change Biology 10: 1857–1869.

5. Webster KL, Creed IF, Skowronski MD, Kaheil YH (2009) Comparison of the performance of statistical models that predict soil respiration from forests. Soil Science Society of America Journal 73: 1157–1167.

6. Gaumont-Guay D, Black TA, Griffis TJ, Barr AG, Jassal RS, et al. (2006) Interpreting the dependence of soil respiration on soil temperature and water content in a boreal aspen stand. Agricultural and Forest Meteorology 140: 220–235.

7. Phillips SC, Varner RK, Frolking S, Munger JW, Bubier JL, et al. (2010). Interannual, seasonal, and diel variation in soil respiration relative to ecosystem respiration at a wetland to upland slope at Harvard Forest. Journal of Geophysical Research: Biogeosciences 115: G02019. doi: 10.1029/2008JG000858.

8. Reichstein M, Rey A, Freibauer A, Tenhunen J, Valentini R, et al. (2003) Modeling temporal and large-scale spatial variability of soil respiration from soil water availability, temperature and vegetation productivity indices. Global Biogeochemical Cycles 17: 1104. doi: 10.1029/2003GB002035.

9. Geng Y, Wang Y, Yang K, Wang S, Zeng H, et al. (2012) Soil respiration in Tibetan alpine grasslands: belowground biomass and soil moisture, but not soil temperature, best explain the large-scale patterns. PloS one 7: e34968. doi: 10.1371/journal.pone.0034968.

10. Huang N, Niu Z, Zhan YL, Tappertc MC, Wu CY, et al. (2012) Relationships between soil respiration and photosynthesis-related spectral vegetation indices in two cropland ecosystems. Agricultural and Forest Meteorology 160: 80–89.

11. Chen ST, Huang Y, Zou JW, Shen QR, Hu ZH, et al. (2010) Modeling interannual variability of global soil respiration from climate and soil properties. Agricultural and Forest Meteorology 150: 590–605.

12. Almagro M, Querejeta JI, Boix-Fayos C, Martínez-Mena M (2013) Links between vegetation patterns, soil C and N pools and respiration rate under three different land uses in a dry Mediterranean ecosystem. Journal of Soils and Sediments 13: 641–653.

13. Martin JG, Bolstad PV, Ryu SR, Chen J (2009) Modeling soil respiration based on carbon, nitrogen, and root mass across diverse Great Lake forests. Agricultural and Forest Meteorology 149: 1722–1729.

14. Longley P (Ed.) (2005) Geographic information systems and science. John Wiley & Sons.

15. Weng Q (2001) Modeling urban growth effects on surface runoff with the integration of remote sensing and GIS. Environmental Management 28: 737–748.

16. Lin ML, Chen CW (2011) Using GIS-based spatial geocomputation from remotely sensed data for drought risk-sensitive assessment. International Journal of Innovative Computing, Information and Control 7: 657–668.

17. Shalaby A, Tateishi R (2007) Remote sensing and GIS for mapping and monitoring land cover and land-use changes in the Northwestern coastal zone of Egypt. Applied Geography 27: 28–41.

18. Dewan AM, Yamaguchi Y (2009). Land use and land cover change in Greater Dhaka, Bangladesh: using remote sensing to promote sustainable urbanization. Applied Geography 29: 390–401.

19. Wang XD, Zhong XH, Liu SZ, Liu JG, Wang ZY, et al. (2008). Regional assessment of environmental vulnerability in the Tibetan Plateau: development and application of a new method. Journal of Arid environments 72: 1929–1939.

20. Ceccato P, Connor SJ, Jeanne I, Thomson MC (2005) Application of geographical information systems and remote sensing technologies for assessing and monitoring malaria risk. Parassitologia 47: 81–96.

21. Franklin J (1995) Predictive vegetation mapping: geographic modelling of biospatial patterns in relation to environmental gradients. Progress in Physical Geography 19: 474–499.

22. Raup B, Kääb A, Kargel JS, Bishop MP, Hamilton G, et al. (2007) Remote sensing and GIS technology in the Global Land Ice Measurements from Space (GLIMS) project. Computers & Geosciences 33: 104–125.

23. Keane RE, Burgan R, van Wagtendonk J (2001) Mapping wildland fuels for fire management across multiple scales: integrating remote sensing, GIS, and biophysical modeling. International Journal of Wildland Fire 10: 301–319.

24. He C, Wang S, Xu J, Zhou C (2002) Using remote sensing to estimate the change of carbon storage: a case study in the estuary of Yellow River delta. International Journal of Remote Sensing 23: 1565–1580.

25. Yuan W, Liu S, Yu G, Bonnefond JM, Chen J, et al. (2010) Global estimates of evapotranspiration and gross primary production based on MODIS and global meteorology data. Remote Sensing of Environment 114: 1416–1431.

26. Huang N, He JS, Niu Z (2013) Estimating the spatial pattern of soil respiration in Tibetan alpine grasslands using Landsat TM images and MODIS data. Ecological Indicators 26: 117–125.

27. Davidson EA, Richardson AD, Savage KE, Hollinger DY (2006) A distinct seasonal pattern of the ratio of soil respiration to total ecosystem respiration in a spruce-dominated forest. Global Change Biology 12, 230–239.

28. Vargas R, Baldocchi DD, Allen MF, Bahn M, Black TA, et al. (2010) Looking deeper into the soil: biophysical controls and seasonal lags of soil CO_2 production and efflux. Ecological Applications 20: 1569–1582.

29. Bondeau A, Smith PC, Zaehle S, Schaphoff S, Lucht W, et al. (2007) Modelling the role of agriculture for the 20th century global terrestrial carbon balance. Global Change Biology 13: 679–706.

30. Foley JA, DeFries RS, Asner GP, Barford C, Bonan G, et al. (2005) Global consequences of land use. Science 309: 570–574.

31. Krugh B, Bickham L, Miles D (1994) The solid-state chlorophyll meter: a novel instrument for rapidly and accurately determining the chlorophyll concentrations in seedling leaves. Maize Genetics Cooperation Newsletter 68: 25–27.

32. Wu CY, Wang L, Niu Z, Gao S, Wu MQ (2010) Nondestructive estimation of canopy chlorophyll content using Hyperion and Landsat/TM images. International Journal of Remote Sensing 31: 2159–2167.

33. Gao Y, Xie Y, Jiang H, Wu B, Niu J (2014) Soil water status and root distribution across the rooting zone in maize with plastic film mulching. Field Crops Research 156: 40–47.

34. Kou TJ, Zhu P, Huang S, Peng XX, Song ZW, et al. (2012) Effects of long-term cropping regimes on soil carbon sequestration and aggregate composition in rainfed farmland of Northeast China. Soil and Tillage Research 118: 132–138.

35. Nelson DW, Sommers LE (1982) Total carbon, organic carbon, and organic matter. In: Page AL, Miller RH, Keeney DR. (Eds.), Methods of Soil Analysis. American Society of Agronomy and Soil Science Society of American, Madison. 101–129.

36. Gallaher RN, Weldon CO, Boswell FC (1976) A semiautomated procedure for total nitrogen in plant and soil samples. Soil Science Society of America Journal 40: 887–889.

37. Wardlow BD, Egbert SL, Kastens JH (2007) Analysis of time-series MODIS 250 m vegetation index data for crop classification in the US Central Great Plains. Remote Sensing of Environment 108: 290–310.

38. Wilson EH, Sader SA (2002) Detection of forest harvest type using multiple dates of Landsat TM imagery. Remote Sensing of Environment 80: 385–396.

39. Zhong B, Ma P, Nie A, Yang A, Yao Y, et al. (2014) Land cover mapping using time series HJ-1/CCD data. Science China: Earth Sciences doi: 10.1007/s11430-014-4877-5.

40. Bian JH, Li AN, Jin HA, Lei GB, Huang CQ, et al. (2013) Auto-registration and orthorecification algorithm for the time series HJ-1A/B CCD images. Journal of Mountain Science 10: 754–767.

41. Liu Y, Li M, Mao L, Cheng L, Chen K (2013) Seasonal pattern of tidal-flat topography along the Jiangsu middle coast, China, using HJ-1 optical images. Wetlands 33: 871–886.

42. Wang SD, Miao LL, Peng GX (2012) An Improved Algorithm for Forest Fire Detection Using HJ Data. Procedia Environmental Sciences 13: 140–150.

43. Gamon JA, Field CB, Goulden ML, Griffin KL, Hartley AE, et al. (1995) Relationship between NDVI, canopy structure and photosynthesis in three Californian vegetation types. Ecological Applications 5: 28–41.

44. Hansen PM, Schjoerring JK (2003) Reflectance measurement of canopy biomass and nitrogen status in wheat crops using normalized difference vegetation indices and partial least squares regression. Remote Sensing of Environment 86: 542–553.

45. Yu X, Yan Q, Liu Z (2010) Atmospheric correction of HJ-1A multi-spectral and hyper-spectral images. Image and Signal Processing (CISP), 2010 3rd International Congress on. IEEE 5: 2125–2129.

46. Li P, Jiang L, Feng Z (2013) Cross-Comparison of Vegetation Indices Derived from Landsat-7 Enhanced Thematic Mapper Plus (ETM+) and Landsat-8 Operational Land Imager (OLI) Sensors. Remote Sensing 6: 310–329.

47. Rouse JW, Haas RH, Schell JA, Deering DW, Harlan JC (1974) Monitoring the vernal advancements and retrogradation of natural vegetation, In: NASA/GSFC, Final Report, Greenbelt, MD, USA, 1–137.

48. Gamon JA, Field CB, Goulden ML, Griffin KL, Hartley AE, et al. (1995) Relationship between NDVI, canopy structure and photosynthesis in three Californian vegetation types. Ecological Applications 5: 28–41.

49. Qi J, Chehbouni A, Huete AR, Kerr YH, Sorooshian S (1994) A modified soil adjusted vegetation index (MSAVI). Remote Sensing of Environment 48: 119–126.

50. Huete A, Didan K, Miura T, Rodriguez EP, Gao X, et al. (2002) Overview of the radiometric and biophysical performance of the MODIS vegetation indices. Remote Sensing of Environment 83: 195–213.

51. Broge NH, Leblanc E (2001) Comparing prediction power and stability of broadband and hyperspectral vegetation indices for estimation of green leaf area index and canopy chlorophyll density. Remote Sensing of Environment 76: 156–172.

52. Haboudane D, Miller JR, Tremblay N, Zarco-Tejada PJ, Dextraze L (2002) Integrated narrow-band vegetation indices for prediction of crop chlorophyll content for application to precision agriculture. Remote Sensing of Environment 81: 416–426.

53. Gitelson AA, Vina A, Ciganda V, Rundquist DC, Arkebauer TJ (2005) Remote estimation of canopy chlorophyll content in crops. Geophysical Research Letters 32: L08403. doi: 10.1029/2005GL022688.

54. Wu C, Niu Z, Tang Q, Huang W (2008) Estimating chlorophyll content from hyperspectral vegetation indices: Modeling and validation. Agricultural and Forest Meteorology 148: 1230–1241.

55. Hirose T, Ackerly DD, Traw MB, Ramseier D, Bazzaz FA (1997) CO_2 elevation, canopy photosynthesis, and optimal leaf area index. Ecology 78: 2339–2350.

56. Gitelson AA, Vina A, Verma SB, Rundquist DC, Arkebauer TJ, et al. (2006) Relationship between gross primary production and chlorophyll content in crops: Implications for the synoptic monitoring of vegetation productivity. Journal of Geophysical Research-Atmospheres 111: D08S11. doi: 10.1029/2005JD006017.

57. Glenn EP, Huete AR, Nagler PL, Nelson SG (2008) Relationship between remotely-sensed vegetation indices, canopy attributes and plant physiological

processes: what vegetation indices can and cannot tell us about the landscape. Sensors 8: 2136–2160.

58. Huang N, Niu Z (2013) Estimating soil respiration using spectral vegetation indices and abiotic factors in irrigated and rainfed agroecosystems. Plant and Soil 367: 535–550.

59. Chevallier T, Voltz M, Blanchart E, Chotte JL, Eschenbrenner V, et al. (2000) Spatial and temporal changes of soil C after establishment of a pasture on a long-term cultivated vertisol (Martinique). Geoderma 94: 43–58.

60. McGrath D, Zhang C (2003) Spatial distribution of soil organic carbon concentrations in grassland of Ireland. Applied Geochemistry 18: 1629–1639.

61. Liu D, Wang Z, Zhang B, Song K, Li X, et al. (2006) Spatial distribution of soil organic carbon and analysis of related factors in croplands of the black soil region, Northeast China. Agriculture, Ecosystems & Environment 113: 73–81.

62. Matheron G (1963) Principles of geostatistics. Economic geology 58: 1246–1266.

63. Webster R, Oliver MA (2007) Geostatistics for environmental scientists. John Wiley & Sons.

64. Raich JW, Tufekciogul A (2000) Vegetation and soil respiration: correlations and controls. Biogeochemistry 48: 71–90.

65. Schaefer DA, Feng W, Zou X (2009) Plant carbon inputs and environmental factors strongly affect soil respiration in a subtropical forest of southwestern China. Soil Biology and Biochemistry 41: 1000–1007.

66. Curiel Yuste J, Baldocchi DD, Gershenson A, Goldstein A, Misson L, et al. (2007) Microbial soil respiration and its dependency on carbon inputs, soil temperature and moisture. Global Change Biology 13: 2018–2035.

67. Pugesek BH, Tomer A, Von Eye A (Eds.) (2003) Structural equation modeling: applications in ecological and evolutionary biology. Cambridge University Press.

68. Iriondo JM, Albert MJ, Escudero A (2003) Structural equation modelling: an alternative for assessing causal relationships in threatened plant populations. Biological Conservation 113: 367–377.

69. Jonsson M, Wardle DA (2010) Structural equation modelling reveals plant-community drivers of carbon storage in boreal forest ecosystems. Biology Letters 6: 116–119.

70. Kim GS (2010) AMOS 18.0: Structural Equation Modeling. Seoul: Hannarae Publishing Co.

71. National Bureau of statistics of China (2006) China social-economic statistical yearbooks for China's counties and cities. China Statistics Press, Beijing.

72. Wang S, Tian H, Liu J, Pan S (2003) Pattern and change of soil organic carbon storage in China: 1960s–1980s. Tellus B 55: 416–427.

73. West TO, Post WM (2002) Soil organic carbon sequestration rates by tillage and crop rotation. Soil Science Society of America Journal 66: 1930–1946.

74. Wilhelm WW, Johnson JM, Karlen DL, Lightle DT (2007) Corn stover to sustain soil organic carbon further constrains biomass supply. Agronomy journal 99: 1665–1667.

75. Foley JA (1995) An equilibrium model of the terrestrial carbon budget. Tellus B 47: 310–319.

76. Lal R (1999) Soil management and restoration for C sequestration to mitigate the accelerated greenhouse effect. Progress in Environmental Science 1: 307–326.

77. Epron D, Bosc A, Bonal D, Freycon V (2006) Spatial variation of soil respiration across a topographic gradient in a tropical rain forest in French Guiana. Journal of Tropical Ecology 22: 565–574.

78. Suyker AE, Verma SB, Burba GG, Arkebauer TJ (2005) Gross primary production and ecosystem respiration of irrigated maize and irrigated soybean during a growing season. Agricultural and Forest Meteorology 131: 180–190.

79. Huang N, Gu L, Niu Z (2014) Estimating soil respiration using spatial data products: A case study in a deciduous broadleaf forest in the Midwest USA. Journal of Geophysical Research: Atmospheres 119. doi:10.1002/2013JD020515.

80. Hudak AT, Lefsky MA, Cohen WB, Berterretche M (2002) Integration of lidar and Landsat ETM+ data for estimating and mapping forest canopy height. Remote Sensing of Environment 82: 397–416.

81. Meng Q, Cieszewski C, Madden M (2009) Large area forest inventory using Landsat ETM+: a geostatistical approach. ISPRS Journal of Photogrammetry and Remote Sensing 64: 27–36.

82. Tsui OW, Coops NC, Wulder MA, Marshall PL (2013) Integrating airborne LiDAR and space-borne radar via multivariate kriging to estimate above-ground biomass. Remote Sensing of Environment 139: 340–352.

83. Verstraeten WW, Veroustraete F, Wagner W, Van Roey T, Heyns W, et al. (2010) Remotely sensed soil moisture integration in an ecosystem carbon flux model-The spatial implication. Climatic Change 103:117–136.

84. Rayner PJ, Scholze M, Knorr W, Kaminski T, Giering R, et al. (2005) Two decades of terrestrial carbon fluxes from a carbon cycle data assimilation system (CCDAS). Global Biogeochemical Cycles 19.

85. Chevallier F, Bréon FM, Rayner PJ (2007) Contribution of the Orbiting Carbon Observatory to the estimation of CO_2 sources and sinks: Theoretical study in a variational data assimilation framework. Journal of Geophysical Research: Atmospheres 112: D09307. doi: 10.1029/2006JD007375.

86. Knorr W, Kaminski T, Scholze M, Gobron N, Pinty B, et al. (2010) Carbon cycle data assimilation with a generic phenology model. Journal of Geophysical Research: Biogeosciences 115: G04017. doi: 10.1029/2009JG001119.

Permissions

All chapters in this book were first published in PLOS ONE, by The Public Library of Science; hereby published with permission under the Creative Commons Attribution License or equivalent. Every chapter published in this book has been scrutinized by our experts. Their significance has been extensively debated. The topics covered herein carry significant findings which will fuel the growth of the discipline. They may even be implemented as practical applications or may be referred to as a beginning point for another development.

The contributors of this book come from diverse backgrounds, making this book a truly international effort. This book will bring forth new frontiers with its revolutionizing research information and detailed analysis of the nascent developments around the world.

We would like to thank all the contributing authors for lending their expertise to make the book truly unique. They have played a crucial role in the development of this book. Without their invaluable contributions this book wouldn't have been possible. They have made vital efforts to compile up to date information on the varied aspects of this subject to make this book a valuable addition to the collection of many professionals and students.

This book was conceptualized with the vision of imparting up-to-date information and advanced data in this field. To ensure the same, a matchless editorial board was set up. Every individual on the board went through rigorous rounds of assessment to prove their worth. After which they invested a large part of their time researching and compiling the most relevant data for our readers.

The editorial board has been involved in producing this book since its inception. They have spent rigorous hours researching and exploring the diverse topics which have resulted in the successful publishing of this book. They have passed on their knowledge of decades through this book. To expedite this challenging task, the publisher supported the team at every step. A small team of assistant editors was also appointed to further simplify the editing procedure and attain best results for the readers.

Apart from the editorial board, the designing team has also invested a significant amount of their time in understanding the subject and creating the most relevant covers. They scrutinized every image to scout for the most suitable representation of the subject and create an appropriate cover for the book.

The publishing team has been an ardent support to the editorial, designing and production team. Their endless efforts to recruit the best for this project, has resulted in the accomplishment of this book. They are a veteran in the field of academics and their pool of knowledge is as vast as their experience in printing. Their expertise and guidance has proved useful at every step. Their uncompromising quality standards have made this book an exceptional effort. Their encouragement from time to time has been an inspiration for everyone.

The publisher and the editorial board hope that this book will prove to be a valuable piece of knowledge for researchers, students, practitioners and scholars across the globe.

List of Contributors

Stefan Wirtz and Johannes B. Ries
Department of Physical Geography, Trier University, Trier, Germany

Manuel Seeger
Department of Physical Geography, Trier University, Trier, Germany
Department of Land Degradation and Development, Wageningen University, Wageningen, The Netherlands

Andreas Zell and Christian Wagner
Department 7.3- Technical Physics, Saarland University, Saarbrücken, Germany

Jean-Frank Wagner
Department of Geology, Trier University, Trier, Germany

Jilili Abuduwaili
State Key Laboratory of Desert and Oasis Ecology, Xinjiang Institute of Ecology and Geography, Chinese Academy of Sciences, Urumqi, China

Zhang Zhaoyong
State Key Laboratory of Desert and Oasis Ecology, Xinjiang Institute of Ecology and Geography, Chinese Academy of Sciences, Urumqi, China
University of the Chinese Academy of Sciences, Beijing, China

Hamid Yimit
Key Laboratory of Xingjiang Arid Land Lake Environment and Resource, Xinjiang Normal University, Urumqi, China

Rajan Ghimire, Jay B. Norton and Peter D. Stahl
Department of Ecosystem Science and Management, University of Wyoming, Laramie, Wyoming, United States of America

Urszula Norton
Department of Plant Sciences, University of Wyoming, Laramie, Wyoming, United States of America

Annamaria Bevivino, Patrizia Paganin, Maite Sampedro Pellicer and Claudia Dalmastri
ENEA (Italian National Agency for New Technologies, Energy and Sustainable Economic Development) Casaccia Research Center, Technical Unit for Sustainable Development and Innovation of Agro-Industrial System, Rome, Italy

Alessandro Florio and Anna Benedetti
Consiglio per la Ricerca e la Sperimentazione in Agricoltura - Research Centre for the Soil-Plant System, Rome, Italy

Giovanni Bacci
Consiglio per la Ricerca e la Sperimentazione in Agricoltura - Research Centre for the Soil-Plant System, Rome, Italy
Laboratory of Microbial and Molecular Evolution, Department of Biology, University of Florence, Florence, Italy

Maria Cristiana Papaleo, Alessio Mengoni and Renato Fani
Laboratory of Microbial and Molecular Evolution, Department of Biology, University of Florence, Florence, Italy

Luigi Ledda
Dipartimento di Agraria, University of Sassari, Sassari, Italy

Jean-François Silvain
Institut de Recherche pour le Développement (IRD), UR 072, Laboratoire Evolution, Génomes et Spéciation, UPR 9034, Centre National de la Recherche Scientifique (CNRS), Gif sur Yvette, France et Université Paris-Sud 11, Orsay, France

Emile Faye
Institut de Recherche pour le Développement (IRD), UR 072, Laboratoire Evolution, Génomes et Spéciation, UPR 9034, Centre National de la Recherche Scientifique (CNRS), Gif sur Yvette, France et UniversitéParis-Sud 11, Orsay, France
UPMC Univ Paris06, Sorbonne Universités, Paris, France
Facultad de Ciencias Exactas y Naturales, Pontificia Universidad Católica del Ecuador, Quito, Ecuador

Olivier Dangles
Institut de Recherche pour le Développement (IRD), UR 072, Laboratoire Evolution, Génomes et Spéciation, UPR 9034, Centre National de la Recherche Scientifique (CNRS), Gif sur Yvette, France et Université Paris-Sud 11, Orsay, France
Facultad de Ciencias Exactas y Naturales, Pontificia Universidad Católica del Ecuador, Quito, Ecuador
Instituto de Ecología, Universidad Mayor San Andrés, Cotacota, La Paz, Bolivia

Mario Herrera
Facultad de Ciencias Exactas y Naturales, Pontificia Universidad Católica del Ecuador, Quito, Ecuador

Lucio Bellomo
Mediterranean Institute of Oceanography (MIO) CNRS/INSU, IRD, UM 110, Universitéde Toulon, La Garde, France

Fiona Walsh and Brion Duffy
Bacteriology Research Laboratory, Federal Department of Economic Affairs, Education and Research EAER, Research Station Agroscope Changins Wädenswil ACW, Wädenswil, Switzerland

Janet Macfall, Paul Robinette and David Welch
Center for Environmental Studies, Elon University, Elon, North Carolina, United States of America

Kabindra Adhikari and Alfred E. Hartemink
Department of Soil Science, University of Wisconsin2Madison, Madison, Wisconsin, United States of America

Budiman Minasny
Department of Environmental Sciences, The University of Sydney, Sydney, New South Wales, Australia

Rania Bou Kheir, Mette B. Greve and Mogens H. Greve
Department of Agro-ecology, Aarhus University, Tjele, Denmark

Baoru Sun, Yi Peng, Hongyu Yang, Zhijian Li, Yingzhi Gao, Chao Wang, Yuli Yan and Yanmei Liu
Key Laboratory of Vegetation Ecology, Northeast Normal University, Changchun, China

Syed Tahir Ata-Ul-Karim, Xia Yao, Xiaojun Liu, Weixing Cao, Yan Zhu
National Engineering and Technology Center for Information Agriculture, Jiangsu Key Laboratory for Information Agriculture, Nanjing Agricultural University, Nanjing, Jiangsu, P. R. China

Hai-Lin Zhang, Jian-Fu Xue, Zhong-Du Chen and Fu Chen
College of Agronomy and Biotechnology, China Agricultural University, Key Laboratory of Farming System, Ministry of Agriculture, Beijing, China

Xiao-Lin Bai
Patent Examination Cooperation Center of the Patent Office, SIPO, Beijing, China

Hai-Ming Tang
Soil and Fertilizer Institute of Hunan Province, Changsha, China

Jingang Liang, Jun Ji, Fang Meng, Xiaobo Zheng and, Zhengguang Zhang
Department of Plant Pathology, College of Plant Protection, Nanjing Agricultural University, and Key Laboratory of Integrated Management of Crop Diseases and Pests, Ministry of Education, Nanjing, China

Shi Sun and Cunxiang Wu
The National Key Facility for Crop Gene Resources and Genetic Improvement (NFCRI), MOA Key Laboratory of Soybean Biology (Beijing), Institute of Crop Science, The Chinese Academy of Agricultural Sciences, Beijing, China

Haiying Wu and Mingrong Zhang
Nanchong Academy of Agricultural Science, Nanchong, China

Yann Surget-Groba
Key Laboratory of Tropical Forest Ecology, Xishuangbanna Tropical Botanical Garden (XTBG), Chinese Academy of Sciences, Kunming, Yunnan, P. R. China

Shi Lingling
Key Laboratory of Tropical Forest Ecology, Xishuangbanna Tropical Botanical Garden (XTBG), Chinese Academy of Sciences, Kunming, Yunnan, P. R. China
University of the Chinese Academy of Sciences, Beijing, P. R. China

Charles H. Cannon
Key Laboratory of Tropical Forest Ecology, Xishuangbanna Tropical Botanical Garden (XTBG), Chinese Academy of Sciences, Kunming, Yunnan, P. R. China
Texas Tech University, Lubbock, Texas, United States of America

Peter O. Alele
Key Laboratory of Tropical Forest Ecology, Xishuangbanna Tropical Botanical Garden (XTBG), Chinese Academy of Sciences, Kunming, Yunnan, P. R. China
University of the Chinese Academy of Sciences, Beijing, P. R. China Great Nile Conservation Centre (GNCC), Lira, Uganda
Institute of Tropical Forest Conservation (ITFC), Mbarara University of Science and Technology (MUST), Kabale, Uganda

Douglas Sheil
Department of Ecology and Natural Resource Management, Norwegian University of Life Sciences, Ås, Norway

Center for International Forestry Research (CIFOR), Bogor, Indonesia
Department of Ecology and Natural Resource Management, School of Environment, Science and Engineering, Southern Cross University, Lismore, New South Wales, Australia
Institute of Tropical Forest Conservation (ITFC), Mbarara University of Science and Technology (MUST), Kabale, Uganda

G. W. Wieger Wamelink and Joep Y. Frissel
Alterra, Wageningen UR, Wageningen, the Netherlands

Wilfred H. J. Krijnen and M. Rinie Verwoert
Unifarm, Wageningen UR, Wageningen, the Netherlands

Paul W. Goedhart
Biometris, Wageningen UR, Wageningen, the Netherlands

Ji Wenjun and Li Shuo
Institute of Agricultural Remote Sensing and Information Technology, College of Environmental and Resource Sciences, Zhejiang University, Hangzhou, China

Shi Zhou
Institute of Agricultural Remote Sensing and Information Technology, College of Environmental and Resource Sciences, Zhejiang University, Hangzhou, China
Zhejiang Provincial Key Laboratory of Subtropical Soil and Plant Nutrition, Zhejiang University, Hangzhou, China

Huang Jingyi
School of Biological, Earth and Environmental Science, The University of New South Wales, Kensington, Australia

Kornelia Smalla
Julius Kühn-Institut – Federal Research Centre for Cultivated Plants (JKI), Institute for Epidemiology and Pathogen Diagnostics, Braunschweig, Germany

Susanne Schreiter
Julius Kühn-Institut – Federal Research Centre for Cultivated Plants (JKI), Institute for Epidemiology and Pathogen Diagnostics, Braunschweig, Germany
Leibniz Institute of Vegetable and Ornamental Crops Gro beeren/Erfurt e.V., Department Plant Health, Gro beeren, German

Martin Sandmann and Rita Grosch
Leibniz Institute of Vegetable and Ornamental Crops Gro beeren/Erfurt e.V., Department Plant Health, Gro beeren, German

Jun Wang
College of Urban and Environmental Sciences, Northwest University, Xian, Shaanxi Province, China

Upendra M. Sainju
U.S. Department of Agriculture, Agricultural Research Service, Northern Plains Agricultural Research Laboratory, Sidney, Montana, United States of America

Xianni Chen and Xudong Wang
College of Resources and Environment, Northwest A&F University, Yangling, Shaanxi, PR China

Matt Liebman
Department of Agronomy, Agronomy Hall, Iowa State University, Ames, Iowa, United States of America

Michel Cavigelli
Sustainable Agricultural Systems Lab, Agricultural Research Center, Beltsville, Maryland, United States of America

Michelle Wander
Department of NRES, University of Illinois, Urbana-Champaign, Illinois, United States of America

Xiaodong Nie, Zhongwu Li,Bin Huang, Yan Zhang, Wenming Ma, Yanbiao Hu and Guangming Zeng
College of Environmental Science and Engineering, Hunan University, Changsha, PR China
Key Laboratory of Environmental Biology and Pollution Control (Hunan University), Ministry of Education, Changsha, PR China

Jinquan Huang
Department of Soil and Water Conservation, Yangtze River Scientific Research Institute, Wuhan, PR China

Atanu Mukherjee and Rattan Lal
Carbon Management and Sequestration Center, School of Environment and Natural Resources, The Ohio State University, Columbus, Ohio, United States of America

Ni Huang, Li Wang,Pengyu Hao and Zheng Niu
The State Key Laboratory of Remote Sensing Science, Institute of Remote Sensing and Digital Earth, Chinese Academy of Sciences, Beijing, China

Yiqiang Guo
Land Consolidation and Rehabilitation Center, Ministry of Land and Resources, Beijing, China

Index

A
Agricultural Ecosystems, 38, 234-235
Agricultural Management, 24, 36, 38-39, 41, 43, 45, 47, 49, 51, 124, 142
Alfalfa (medicago Sativa L.), 25, 99, 101, 103, 105, 107, 109
Alfalfa Light Transmission, 102
Animal Husbandry, 99, 108-109

B
Bank Geomorphology, 74-75, 77-79, 81, 83, 85
Biocontrol Of Two Bacterial Inoculant Strains, 177, 179, 181, 183, 185, 187

C
Carbon Mineralization In Soils, 199, 201, 203, 205, 207, 209
Carbon Sequestration, 36-37, 86, 197-198, 208, 218, 246-247
Coarse-scale Climate Models, 52
Comparison Of Soil Quality Index, 219, 221, 223, 225, 227, 229, 231, 233
Critical Nitrogen Dilution Curve, 111, 113, 115, 117, 119, 121-122
Crop Types, 61, 188-189, 191, 193, 195-197
Crop Yields, 20, 110, 188-189, 191, 193, 195, 197, 219
Culturable Soil Antibiotic Resistome, 63, 65, 67, 69, 71, 73

D
Different Hydraulic Parameters, 1, 3, 5, 7, 9, 11
Different Texture And Cropping Histories, 199
Different Tillage Systems, 123-125, 127, 129, 131, 133, 188, 218
Digital Mapping, 86-87, 89, 91, 93, 95, 97
Double-cropped Paddy Fields, 123, 125, 127, 129, 131, 133
Dry Grain And Straw, 229

E
Ectothermic Crop Pest, 52
Emissions Of Ch4 And N2o, 123-125, 127, 129, 131, 133
Enhanced Vegetation Index (evi), 234, 237, 241
Erosion Of The Haw River, 74-75, 77, 79, 81, 83, 85
European Settlement, 74, 85

F
Farming And Pastoral Areas, 99
Field Simulated Rainfall Events, 210-211, 213, 215, 217
Field-grown Lettuce, 177, 186
Forage Production Systems, 24-25, 33, 35

Fungal Communities In The Nile River Watershed, 144

G
Geographical Information System, 234-235
Gis Statistical Toolbox, 52
Greenhouse Gas Emissions, 86, 132, 188, 198
Greenhouse Gases (ghg) Emissions, 123
Growth Experiment, 157, 159, 161, 163, 165

H
High Order River, 74-75, 77, 79, 81, 83-85
High-methionine Transgenic Line, 134

I
In Situ Measurement Of Some Soil Properties, 166-167, 169, 171, 173, 175
Intercropping, 99-110
Interpolated Climatic Grids, 52-53, 55, 57, 59, 61
Irrigated Organic, 24-25, 27, 29, 31, 33, 35, 37

L
Local Temperature Mapping, 52-53, 55, 57, 59, 61

M
Maize (zea Mays L.), 99, 101, 103, 105, 107, 109
Mars And Moon Soil Simulants, 157, 159, 161, 163, 165
Mediterranean Region, 38-39, 50
Microbial Biomass, 24-29, 31, 33, 35-36, 43, 46-47, 50-51, 186, 188, 190, 194-198, 208-209, 230
Microbial Community Responses, 24-25, 27, 29, 31, 33, 35, 37
Multi-drug Resistant Bacteria, 63, 65, 67, 69, 71, 73

N
Natural Tropical Rainforest Ecosystems, 144-145, 147, 149, 151, 153, 155
No-till (nt), 123
Northeast China, 13, 99, 109, 217, 246-247

P
Paddy Soil, 127, 132, 166-167, 169, 171, 173, 175
Phenotypic Enzyme Assay, 63
Phospholipid Fatty Acids (plfas), 24
Plant Analysis, 111
Plant Productivity, 155, 219, 245

R
Reduced-tillage Crop, 24-25, 28

Remote Sensing, 22, 111, 121, 166, 176, 234-235, 237, 239, 241, 243-244, 246-247

Residue And Nitrogen Fertilizer, 198-199, 201, 203, 205, 207, 209

Residue Placement, 188-189, 191-197

Rhizosphere Bacterial Communities, 134-135, 137, 139, 141-143, 177

Rhizosphere Microbial Community, 134, 177-178, 186

Rotary Tillage (rt), 123-124

S
Seasonal Changes, 38-39, 41, 43, 45, 47, 49, 51

Selective Transportation, 210-211, 213, 215, 217

Soc Accrual, 24, 32-33, 35

Soil Bacterial, 27, 38-39, 41, 43-45, 47-51, 63-65, 72, 137, 144-145, 147, 149, 151, 153-156, 178, 186-187

Soil Bacterial Community Response, 38-39, 41, 43, 45, 47, 49, 51

Soil Carbon And Nitrogen Fractions, 188-191, 193, 195, 197

Soil Detachment In Eroding Rills, 1, 3, 5, 7, 9, 11

Soil Microbial Substrate Properties, 24-25, 27, 29, 31, 33, 35, 37

Soil Microbiotic Properties, 24, 28-29, 33

Soil Organic Carbon Contents, 86-87, 89, 91, 93, 95, 97, 133

Soil Organic Carbon Loss, 210-211, 213, 215, 217

Soil Property Factors, 234-235, 237, 239, 241, 243, 245, 247

Soil Respiration, 25-27, 31, 35, 201, 208-209, 230, 234-247

Soil Salinization Risk, 12-13, 15, 17, 19-21, 23

Soil Salt And Assessment, 12-13, 15, 17, 19, 21, 23

Soil Sustainability, 199

Soil Type Dependent Rhizosphere, 177, 179, 181, 183, 185, 187

Stem Dry Matter In Rice, 111, 113, 115, 117, 119, 121

Stocks In Denmark, 86-87, 89, 91, 93, 95, 97

T
Tropical Mountainous Agricultural Landscapes, 52

U
Upstream Urbanization, 74-75, 82

V
Visible And Near-infrared Spectroscopy, 166-167, 169, 171, 173, 175

Y
Yanqi Basin, 12-14, 16-23

Yield And Economic Incomes, 99, 106

Z
Zigongdongdou Soybean, 134-135, 137, 139, 141, 143